Multisensory Object Perception in the Primate Brain

Marcus J. Naumer · Jochen Kaiser
Editors

Multisensory Object Perception in the Primate Brain

Foreword by Barry E. Stein

Springer

Editors
Marcus J. Naumer
Goethe University
Institute of Medical Psychology
Heinrich-Hoffmann-Str. 10
Frankfurt 60528
Germany
m.j.naumer@med.uni-frankfurt.de

Jochen Kaiser
Goethe University
Institute of Medical Psychology
Heinrich-Hoffmann-Str. 10
Frankfurt 60528
Germany
j.kaiser@med.uni-frankfurt.de

ISBN 978-1-4899-8206-3 ISBN 978-1-4419-5615-6 (eBook)
DOI 10.1007/978-1-4419-5615-6
Springer New York Dordrecht Heidelberg London

Cover illustration: By courtesy of Dr. James Lewis

Printed on acid-free paper

Springer is part of Springer Science+Business Media (www.springer.com)

Foreword

It should come as no surprise to those interested in sensory processes that its research history is among the longest and richest of the many systematic efforts to understand how our bodies function. The continuing obsession with sensory systems is as much a reflection of the fundamental need to understand how we experience the physical world as it is to understand how we become who we are based on those very experiences. The senses function as both portal and teacher, and their individual and collective properties have fascinated scientists and philosophers for millennia.

In this context, the attention directed toward specifying their properties on a sense-by-sense basis that dominated sensory research in the 20th century seems a prelude to our current preoccupation with how they function in concert. Nevertheless, it was the concentrated effort on the operational principles of individual senses that provided the depth of understanding necessary to inform current efforts to reveal how they act cooperatively.

We know that the information provided by any individual sensory modality is not always veridical, but is subject to a myriad of modality-specific distortions. Thus, the brain's ability to compare across the senses and to integrate the information they provide is not only a way to examine the accuracy of any individual sensory channel but also a way to enhance the collective information they make available to the brain. For each sense provides different information about the same event and can inform its counterparts. As a result, their aggregate neural product is more salient and more accurate than that provided by any individual sense. It is this interaction among multiple senses and the fusion of their information content that is captured by the term "multisensory integration."

Viewed from this perspective, multisensory integration is a process by which the brain makes maximal use of whatever sensory information is available at the moment and is able to compare that information to the body of knowledge that it has already acquired. Thus, the ubiquitous presence of multiple senses in extant species seems as likely a reflection of the survival value of multisensory integration itself as it is of having sensors that can substitute for one another in different circumstances, as when hearing and touch substitute for vision in the dark. Coupling these two survival factors appears to be an ancient strategy that is believed to date from our earliest single-cell progenitor; an organism whose different sensory receptors, being tuned to different environmental stimuli, could function independently. However, because they were all embedded in the same cell membrane, they used ion fluxes that accessed the same intracellular milieu. As a result, when they were active collectively the organism was rendered an obligate multisensory integrator. The same basic theme of coupling these two survival strategies has been elaborated repeatedly during sensory evolution, so that while species benefited from selective pressures that helped them craft each of their individual senses to meet specific ecological challenges (sometimes doing so in seemingly extravagant ways), they retained the ability to use them synergistically.

That the senses function together was recognized by our scientific forbears. They understood the importance of this cooperative arrangement but didn't know how it worked and could never have anticipated the technological revolution that underlies modern efforts to reveal its mysteries. Although the concept of cooperation among the senses had never been forgotten, the emphasis on the processes underlying their synergy has never been as great as it is today. Fortunately, we now have a wide variety of physiological and psychophysical approaches with which we can examine these processes. Studies at the level of the individual neuron and among complexes of neurons, as well as a host of psychophysical approaches in adult and developing animals, make accessible information that could not have been imagined in earlier eras.

Based on these new technologies we now understand that multisensory integration at the neural level not only speeds responses, and increases their salience and reliability, but also provides the basis for unique experiences that arise from the binding of different sensory components. We also now appreciate that these multisensory experiences are exceedingly common and add a depth to perception that might not otherwise be possible.

The current volume includes the results derived from many of the most active and productive laboratories using the latest anatomical, physiological, and psychophysical techniques to explore some of these issues as they relate to object perception. Of necessity the scope is limited to interactions among several senses (visual, auditory, somatosensory), but deals with general principles that are likely to be applicable to interactions among all the senses. Given that the study of multisensory processes is among the most active in the neurosciences, it is rapidly changing. Thus, the present volume should be viewed as an introduction to the many compelling issues in this field, as a foundation for appreciating the importance of the research currently being conducted, and as an encouragement for even greater research efforts in the future.

Winston-Salem, NC Barry E. Stein

Contents

Contributors

Amir Amedi Department of Medical Neurobiology, Institute for Medical Research Israel-Canada (IMRIC), Hebrew University-Hadassah Medical School, Jerusalem 91220, Israel, amir.amedi@ekmd.huji.ac.il

Julien Besle INSERM – UNITE 821 "Brain Dynamics and Cognition", Centre Hospitalier le Vinatier, University Lyon, 95, Bd Pinel, 69500 Bron, France, julien.besle@inserm.fr

Andreja Bubic Department of Medical Neurobiology, Institute for Medical Research Israel-Canada (IMRIC), Hebrew University-Hadassah Medical School, Jerusalem 91220, Israel, bubic@cbs.mpg.de

H. Ruth Clemo Department of Anatomy and Neurobiology, Virginia Commonwealth University School of Medicine, Richmond, VA 23298, USA, rclemo@vcu.edu

Asif A. Ghazanfar Departments of Psychology and Ecology & Evolutionary Biology, Neuroscience Institute, Princeton University, Princeton, NJ 08540, USA, asifg@princeton.edu

Marie-Hélène Giard INSERM – UNITE 821 "Brain Dynamics and Cognition", Centre Hospitalier le Vinatier, University Lyon, 95, Bd Pinel, 69500 Bron, France, mh.giard@inserm.fr

Thomas W. James Department of Psychological and Brain Sciences, Cognitive Science Program, Indiana University, 1101 E Tenth St., Bloomington, IN 47405, USA, thwjames@indiana.edu

Jochen Kaiser Institute of Medical Psychology, Faculty of Medicine, Goethe University, Heinrich-Hoffmann-Str. 10, 60528 Frankfurt am Main, Germany, j.kaiser@med.uni-frankfurt.de

Christoph Kayser Max Planck Institute for Biological Cybernetics, Spemannstrasse 38, 72076 Tübingen, Germany, christoph.kayser@tuebingen.mpg.de

Sunah Kim Department of Psychological and Brain Sciences, Cognitive Science Program, Indiana University, 1101 E Tenth St., Bloomington, IN 47405, USA, suakim@indiana.edu

Ryo Kitada Division of Cerebral Integration, National Institute for Physiological Sciences (NIPS), Myodaiji, Okazaki 444-8585, Japan, kitada@nips.ac.jp

Roberta L. Klatzky Department of Psychology, Carnegie Mellon University, Pittsburgh, PA 15213-3890, USA, klatzky@cmu.edu

Juliane Krueger Neuroscience Graduate Program, Vanderbilt University, Nashville, TN 37232, USA; Vanderbilt Kennedy Center for Research on Human Development, Vanderbilt University, Nashville, TN 37232, USA, juliane.krueger@vanderbilt.edu

Susan J. Lederman Department of Psychology, Queen's University, Kingston, ON K7L 3N6, Canada, susan.lederman@queensu.ca

James W. Lewis Department of Physiology and Pharmacology, Sensory Neuroscience Research Center, and Center for Advanced Imaging, West Virginia University, PO Box 9229, Morgantown, WV 26506, USA, jlewis@hsc.wvu.edu

David J. Lewkowicz Department of Psychology, Florida Atlantic University, 777 Glades Rd., Boca Raton, FL, USA, lewkowic@fau.edu

Nikos K. Logothetis Max Planck Institute for Biological Cybernetics, Spemannstrasse 38, 72076 Tübingen, Germany; Division of Imaging Science and Biomedical Engineering, University of Manchester, Manchester M13 9PT, UK, nikos.logothetis@tuebingen.mpg.de

M. Alex Meredith Department of Anatomy and Neurobiology, Virginia Commonwealth University School of Medicine, Richmond, VA 23298, USA, mameredi@vcu.edu

Micah M. Murray Centre for Biomedical Imaging, Department of Clinical Neurosciences, Department of Radiology, Vaudois University Hospital Centre and University of Lausanne, BH08.078, Rue du Bugnon 46, 1011, Lausanne, Switzerland; Department of Hearing and Speech Sciences, Vanderbilt University Medical Center, Nashville, Tennessee, USA, micah.murray@chuv.ch

Marcus J. Naumer Institute of Medical Psychology, Faculty of Medicine, Goethe University, Heinrich-Hoffmann-Str. 10, 60528 Frankfurt am Main, Germany, m.j.naumer@med.uni-frankfurt.de

Fiona N. Newell School of Psychology and Institute of Neuroscience, Lloyd Building, Trinity College, Dublin 2, Ireland, fiona.newell@tcd.ie

Christopher I. Petkov Max Planck Institute for Biological Cybernetics, Spemannstrasse 38, 72076 Tübingen, Germany, chris.petkov@ncl.ac.uk

Andrea Polony Institute of Medical Psychology, Faculty of Medicine, Goethe University, Heinrich-Hoffmann-Str. 10, 60528 Frankfurt am Main, Germany, apolony@hotmail.com

Constantin Rothkopf Frankfurt Institute for Advanced Studies (FIAS), Goethe University Frankfurt, Ruth-Moufang-Str. 1, 60438 Frankfurt am Main, Germany, rothkopf@fias.uni-frankfurt.de

David W. Royal Vanderbilt Kennedy Center for Research on Human Development, Vanderbilt University, Nashville, TN 37232, USA, david.royal@vanderbilt.edu

Charles Spence Crossmodal Research Laboratory, Department of Experimental Psychology, University of Oxford, South Parks Road, Oxford OX1 3UD, UK, charles.spence@psy.ox.ac.uk

Holger F. Sperdin Neuropsychology and Neurorehabilitation Service, Centre Hospitalier Universitaire Vaudois and University of Lausanne, Hôpital Nestlé, 5 av. Pierre Decker, 1011 Lausanne, Switzerland, holger-franz.sperdin@chuv.ch

Ella Striem-Amit Department of Medical Neurobiology, Institute for Medical Research Israel-Canada (IMRIC), Hebrew University-Hadassah Medical School, Jerusalem 91220, Israel, ella.striem@mail.huji.ac.il

Jochen Triesch Frankfurt Institute for Advanced Studies (FIAS), Goethe University Frankfurt, Ruth-Moufang-Str. 1, 60438 Frankfurt am Main, Germany, triesch@fias.uni-frankfurt.de

Jasper J. F. van den Bosch Institute of Medical Psychology, Faculty of Medicine, Goethe University, Heinrich-Hoffmann-Str. 10, 60528 Frankfurt am Main, Germany, vandenbosch@med.uni-frankfurt.de

Argiro Vatakis Institute for Language and Speech Processing, Research Centers "Athena", Artemidos 6 & Epidavrou, Athens 151 25, Greece, argiro.vatakis@gmail.com

Mark T. Wallace Vanderbilt Brain Institute, Vanderbilt University, Nashville, TN 37232, USA, Vanderbilt Kennedy Center for Research on Human Development, Vanderbilt University, Nashville, TN 37232, USA; Department of Hearing & Speech Sciences, Vanderbilt University, Nashville, TN 37232, USA; Department of Psychology, Vanderbilt University, Nashville, TN 37232, USA, mark.wallace@vanderbilt.edu

Thomas Weisswange Frankfurt Institute for Advanced Studies (FIAS), Goethe University Frankfurt, Ruth-Moufang-Str. 1, 60438 Frankfurt am Main, Germany, weisswange@fias.uni-frankfurt.de

Chapter 1
General Introduction

Marcus J. Naumer and Jochen Kaiser

Traditionally a large proportion of perceptual research has assumed a specialization of cortical regions for the processing of stimuli in a single-sensory modality. However, perception in everyday life usually consists of inputs from multiple sensory channels. Recently the question of how the brain integrates multisensory information has become the focus of a growing number of neuroscientific investigations. This work has identified both multisensory integration in heteromodal brain regions and crossmodal influences in regions traditionally thought to be specific to one sensory modality. Furthermore, several factors have been identified that enhance integration such as spatiotemporal stimulus coincidence and semantic congruency.

The present volume aims at elucidating the mechanisms of multisensory integration of object-related information with a focus on the visual, auditory, and tactile sensory modalities. Levels of analysis range from intracranial electrophysiological recordings to noninvasive electro- or magnetoencephalography, functional magnetic resonance imaging, behavioral approaches, and computational modeling. Worldwide leading experts in primate multisensory object perception review the current state of the field in 18 chapters, which have been aggregated in four sections: mechanisms, audio-visual processing, visuo-tactile processing, and plasticity.

The first section on mechanisms starts with the chapter by Meredith and Clemo who summarize neuroanatomical investigations of multisensory convergence. While historically multisensory neurons were thought to be confined to multisensory areas like the superior temporal sulcus of monkeys or the feline anterior ectosylvian sulcus, recent research has identified subthreshold multisensory cells in regions traditionally considered 'unimodal.' Such subthreshold multisensory neurons are activated by a single modality but show modulations by crossmodal inputs. For example, crossmodal suppression effects could be found in somatosensory areas by

M.J. Naumer (✉)
Institute of Medical Psychology, Faculty of Medicine, Goethe University, Heinrich-Hoffmann-Str. 10, 60528 Frankfurt am Main, Germany
e-mail: m.j.naumer@med.uni-frankfurt.de

M.J. Naumer, J. Kaiser (eds.), *Multisensory Object Perception in the Primate Brain*,
DOI 10.1007/978-1-4419-5615-6_1, © Springer Science+Business Media, LLC 2010

auditory input and vice versa. Meredith and Clemo propose a continuum of multisensory convergence from 'bimodal' via 'subthreshold multisensory' to 'unimodal' neurons.

The third chapter by Rothkopf, Weisswange, and Triesch deals with the computational modeling of multisensory object perception. Bayesian approaches are favored that take existing knowledge into account by defining 'prior distributions' and that allow testing how well a model fits the observed data. Importantly, neuronal response variability is not treated as 'noise' but as reflecting the probabilistic nature of the stimuli. The chapter lists a number of open questions that need to be addressed by future research.

In Chapter 4, Giard and Besle focus on methodological issues pertaining to the identification of multisensory interactions in human electro- or magnetoencephalography (EEG or MEG) data. In the first part of their chapter, they discuss possible biases and artifacts such as activities common to all stimuli, unimodal deactivations of sensory cortices and attentional effects, and suggest ways to overcome these problems. They argue in favor of applying the additive model to human electrophysiological data, e.g., because it serves to correct for volume conduction effects. In the second part, Giard and Besle discuss the interpretation of additive effects in data from other sources like hemodynamic imaging or single-neuron recordings. Finally, they propose a statistical procedure to assess crossmodal interactions.

In the following chapter, we review studies assessing the role of gamma-band activity in EEG or MEG for multisensory integration. High-frequency oscillatory signals have been proposed as a correlate of cortical network formation and might thus be involved in the integration of information processed in distant parts of the brain. Several investigations have found increases of gamma-band activity during the processing of temporally, spatially, or semantically congruent bimodal stimuli. However, findings of enhanced oscillatory signals to nonmatching stimuli suggest that the relationship between gamma-band activity and multisensory integration is strongly modulated by task context.

In Chapter 6, we discuss the use of functional magnetic resonance imaging (fMRI) in the field of multisensory research. Traditionally, discussions in the multisensory fMRI community have mainly focused on principles of and statistical criteria for multisensory integration. More recently, however, the use of advanced experimental designs and increasingly sensitive (multivariate) analysis tools even allows to noninvasively differentiate between regional and neuronal convergence and to reveal the connectional basis of multisensory integration in human subjects.

The section on audio-visual processing starts with a chapter on temporal integration of auditory and visual stimuli. Vatakis and Spence introduce theories and (psychophysical) methods to assess audio-visual temporal perception. They describe the various factors that influence the shape of the temporal window of integration, like spatial distance, stimulus complexity, or familiarity. In the second part of their chapter Vatakis and Spence summarize the main results from their own work. For example, they describe influences of the type of materials (objects, speech, music) on temporal asynchrony judgments and the dominance of the modality providing the more reliable information.

In Chapter 8, Kayser, Petkov, and Logothetis review the available macaque monkey fMRI literature on multisensory processing. The authors favor a high-resolution localization technique, which provides a functional map of individual fields of auditory cortex. They report crossmodal influences occurring already in secondary auditory cortices and increasing along the auditory processing hierarchy. Finally, Kayser, Petkov, and Logothetis also discuss the interpretability of their fMRI findings with regard to the underlying neuronal activity.

In the next chapter, Ghazanfar argues that the default mode of both human and nonhuman vocal communication is multisensory and that this mode of communication is reflected by the organization of the neocortex. Importantly, associations are mediated not solely through association regions but also through large-scale networks including unisensory areas. Behavioral evidence for an integration of audio-visual communication in monkeys is presented. Gamma-band correlations between the lateral belt of the auditory cortex and the upper bank of the superior temporal sulcus suggest that the auditory cortex may act as an integration region for face/voice information. In the final part of his chapter, Ghazanfar compares developmental patterns of multisensory integration between humans and monkeys.

In Chapter 10, Lewis reviews fMRI studies which have explicitly examined audio-visual interactions in the human brain. In a series of meta-analyses he compares diverse networks of (mainly) cortical regions that have been preferentially activated under different types of audio-visual interactions. Finally, Lewis discusses how these multiple parallel pathways which appear to mediate audio-visual perception might be related to representations of object knowledge.

In the final chapter of this section, Murray and Sperdin review the literature on the impact of prior multisensory processing on subsequent unisensory representations. Studies by the authors have demonstrated, e.g., enhanced recognition of visual images previously paired with semantically congruent sounds (but not incongruent sounds). Such behavioral effects were observed after single-trial exposures. They were associated with differences in the topographies of event-related potentials and fMRI activations between previously paired versus unpaired stimuli suggesting a differential involvement of lateral occipital cortex.

The third section of this volume deals with visuo-tactile integration. It starts with Chapter 12, in which Klatzky and Lederman describe differences and commonalities in surface texture perception by touch, vision, and audition. When visual and tactile texture judgments are systematically compared, vision is biased toward encoding geometric patterns and touch toward intensive cues. When both modalities are available, long-term biases and the immediate context are reflected in the relative weights. Klatzky and Lederman close with a brief discussion of the relevant neuroimaging literature.

In Chapter 13, James and Kim apply the perspective of two sensory stream theories to visuo-haptic shape processing. They focus on multisensory integration in ventral perception and dorsal action pathways. To assess potential sites of visuo-haptic integration, James and Kim propose an additive factors approach, which assesses how a systematic variation of stimulus salience affects the observed multisensory gain both at the perceptual and at the neural levels.

In the next chapter, Newell reviews evidence on how multisensory object information can resolve perceptual core problems such as maintaining shape constancy or updating spatial representations. She also reports neuroimaging evidence suggesting shared neural resources for object and spatial recognition. Newell concludes that the visual and haptic sensory systems can modulate and affect high-level perceptual outcomes, as they are highly interactive all along the processing hierarchy.

In Chapter 15, Lederman, Klatzky, and Kitada address the nature of haptic face perception. They concentrate on how the haptic system processes and represents facial identity and emotional expressions both functionally and neurally. With respect to face perception, the authors consider issues that pertain to configural versus feature-based processing, visual mediation versus multisensory processing, and intersensory transfer. With respect to face representations, they consider the influence of orientation and the relative importance of different facial regions. Finally, the authors also address the underlying neural mechanisms that subserve haptic face perception.

The final section of this volume focuses on core aspects of object-related multisensory plasticity. It starts with the chapter by Lewkowicz who argues that a multisensory object concept develops gradually during infancy. While an infant's ability to perceive low-level audio-visual relations such as synchrony is present at an early age, the ability to perceive higher level and arbitrary relations emerges considerably later.

In Chapter 17, Wallace, Krueger, and Royal review recent work in both nonhuman primates and cats on the neural bases of multisensory development. They report evidence for a gradual growth in the size of multisensory neuronal pools during early postnatal life as well as these neurons' transition from an early non-integrative mode to a later integrative mode. In addition, the authors illustrate diverse approaches such as dark-rearing to assess the role of sensory experience for the development of mature multisensory perception.

In the final chapter, Bubic, Striem-Amit, and Amedi explore the topic of human brain plasticity. The authors concentrate on vision and blindness and describe the different types of neuroplastic changes that are reflected in altered cognitive functions observed in the blind. Furthermore, they report on current sensory substitution devices which supply visual information to the blind through other (auditory or tactile) modalities and more invasive sensory restoration techniques which attempt to convey visual information directly to the visual pathways.

At the end of this introduction, we would like to express our deep gratitude to all colleagues for contributing excellent chapters to this volume. Working jointly with you on this book has always been intellectually enriching as well as joyful. We would also like to thank all chapter reviewers for their generous support of this project. And finally, we would like to express our gratitude for all the encouragement, patience, and assistance that we have received from Ann Avouris and Melissa Higgs (Springer, NYC), without whom this volume would not have been realized.

Part I
Mechanisms

Chapter 2
Corticocortical Connectivity Subserving Different Forms of Multisensory Convergence

M. Alex Meredith and H. Ruth Clemo

2.1 Introduction

The body possesses an impressive array of receptors with which to detect external events. Perhaps because these effects occupy such a large portion of our consciousness (and contribute significantly to adaptation and survival), a great deal of investigative effort has been directed at understanding how the signals generated by exteroreceptors produce behaviors and perception. It has long been understood that the activity generated by the eyes, ears, and/or skin must eventually be shared between the different systems, as it would be extremely inefficient to require a separate system to direct one's eyes toward the source of a sound and another for eye movements toward an itch on the skin, etc. However, our notion of how the nervous system shares information between the different sensory systems has undergone a dramatic shift in recent years. Not long ago, most evidence suggested that such information sharing between the sensory systems took place at the higher levels of processing (e.g., Jones and Powell, 1970). In support of this perspective was the apparent lack of multiple sensory effects in lower level regions of the brain (but see below), as well as the clear presence of excitatory effects from different modalities in the higher, association regions of cortex such as the monkey superior temporal sulcus (STS).

2.2 Bimodal Properties: The Superior Temporal Sulcus

Multisensory processing is defined as the influence of one sensory modality on activity generated by another modality. Hence, areas in which neurons respond overtly to more than one sensory stimulus presented alone (e.g., were bimodal or

M.A. Meredith (✉)
Department of Anatomy and Neurobiology, Virginia Commonwealth University School of Medicine, Richmond, VA 23298, USA
e-mail: mameredi@vcu.edu

M.J. Naumer, J. Kaiser (eds.), *Multisensory Object Perception in the Primate Brain*,
DOI 10.1007/978-1-4419-5615-6_2, © Springer Science+Business Media, LLC 2010

trimodal) clearly meet this criterion. One of the first cortical areas identified as 'polysensory' was the upper bank of the monkey STS (e.g., Benevento et al., 1977; Bruce et al., 1981; Hikosaka et al., 1988). The presence of neurons responsive to visual–auditory or visual–somatosensory cues naturally led to questioning the sources of these inputs. With tracer injections into the upper 'polysensory' STS bank, retrogradely labeled neurons were identified in adjoining auditory areas of the STS, superior temporal gyrus, and supratemporal plane, and in visual areas of the inferior parietal lobule, the lateral intraparietal sulcus, the parahippocampal gyrus, and the inferotemporal visual area, as illustrated in Fig. 2.1 (top; Saleem et al., 2000; Seltzer and Pandya, 1994). Although inconclusive about potential somatosensory inputs to the STS, this study did mention the presence of retrogradely labeled neurons in the inferior parietal lobule, an area that processes both visual and somatosensory information (e.g., Seltzer and Pandya, 1980). The fact that many of these inputs were derived from areas that had been identified as higher level processing areas, especially in the visual modality, appeared to confirm the correlation between higher cortical processing and multisensory properties.

2.2.1 Bimodal Properties: The Anterior Ectosylvian Sulcus

Like the STS, the feline anterior ectosylvian sulcus (AES) was found to contain multisensory neurons (e.g., Jiang et al., 1994a, b; Rauschecker and Korte, 1993; Wallace et al., 1992). However, the AES is distinguished from the STS by its modality-specific regions (somatosensory=area SIV, visual= ectosylvian visual area or AEV, auditory= field of the anterior ectosylvian sulcus, or FAES), with multisensory neurons generally found at the intersections between these different representations (Carriere et al., 2007; Meredith, 2004; Meredith and Allman, 2009; Wallace et al., 2004). Further distinctions between STS and the AES reside in the cortical connectivity of the latter, as depicted in Fig. 2.1 (bottom). Robust somatosensory inputs reach the AES from somatosensory areas SI–SIII (Burton and Kopf, 1984; Reinoso-Suarez and Roda, 1985) and SV (Clemo and Meredith, 2004; Mori et al., 1996); inputs to AEV arrive from extrastriate visual area PLLS (posterolateral lateral suprasylvian visual area), with smaller contributions from ALLS and PMLS (anterolateral and posteromedial lateral suprasylvian visual areas; Olson and Graybiel, 1987); auditory inputs to FAES project from RSS (rostral suprasylvian sulcus), AII (secondary auditory area), and PAF (posterior auditory field; Clemo et al., 2007; Lee and Winer, 2008). Each of these major sources of input to the AES has been regarded as unisensory (i.e., responsive to only one sensory modality), lending support to the notion that the multisensory processing observed in the AES was the result of convergence occurring at that site. In addition, some of the afferents to the AES arose from lower levels of processing that included primary (SI; Reinoso-Suarez and Roda, 1985) or secondary (SII: Reinoso-Suarez and Roda, 1985; AII: Lee and Winer, 2008) somatosensory areas. In cat cortex, the AES is not alone as a cortical site of convergence of inputs from representations of different sensory modalities: the posterior ectosylvian gyrus (an auditory–visual area; Bowman and Olson, 1988)

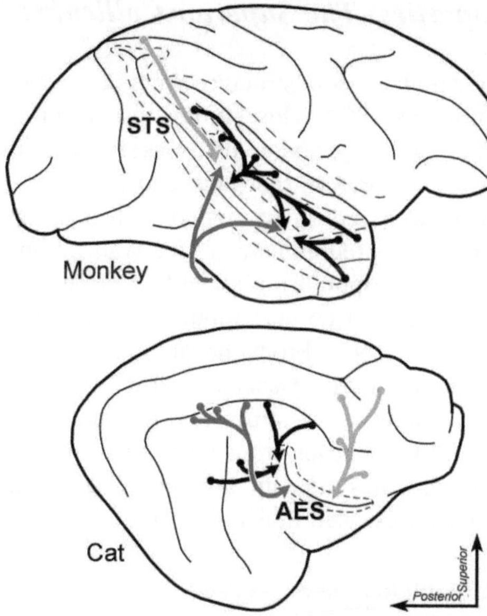

Fig. 2.1 *Top*: Cortical afferents to the monkey superior temporal sulcus (STS). On the lateral view of the monkey brain, the STS is opened (*dashed lines*). The multisensory regions TP0-4 are located on the upper bank (not depicted). Auditory inputs (*black arrows*) from the neighboring superior temporal gyrus, the planum temporale, preferentially target the anterior portions of the upper bank. Visual inputs, primarily from the parahippocampal gyrus (*medium gray arrow*) and also from the inferior parietal lobule (*light gray arrow*), also target the upper STS bank. Somatosensory inputs were comparatively sparse, limited to the posterior aspects of the STS, and may arise from part of the inferior parietal lobule (Pandya and Seltzer, 1994). Note that the different inputs intermingle within their areas of termination, thus potentially providing a substrate for multisensory convergence on individual neurons there. *Bottom*: Cortical afferents to the cat anterior ectosylvian sulcus (AES). On the lateral view of the cat cortex, the AES is opened (*dashed lines*). On the anterior dorsal bank is the somatosensory representation SIV, which receives inputs (*light gray arrow*) from somatosensory areas SI, SII, SIII, and SV (Clemo and Meredith, 2004; Reinoso-Suarez and Roda, 1985). The posterior end of the sulcus is occupied by the auditory field of the AES (FAES), with inputs primarily from the rostral suprasylvian auditory field, the sulcal portion of the anterior auditory field (Montiero et al., 2003), as well as portions of the dorsal zone of auditory cortex, AII, and the posterior auditory field (*black arrows*; Lee and Winer, 2008). The ventral bank is visual, and the ectosylvian visual (AEV) area receives visual inputs (*medium gray arrows*) primarily from the posterolateral lateral suprasylvian (PLLS) area, and to a lesser extent from adjacent anterolateral lateral suprasylvian (ALLS), and posteromedial lateral suprasylvian (PMLS) visual areas (Olson and Graybiel, 1987). Note that the SIV, FAES and AEV domains, as well as their inputs, are largely segregated from one another with the potential for multisensory convergence occurring primarily at their points of intersection (Carierre et al., 2007; Meredith, 2004). Brains are not drawn to scale

and the RSS (an auditory–somatosensory area; Clemo et al., 2007) also exhibit similar connectional features. However, it should be pointed out that, despite the volume of anatomical studies over the last decades, identification of the source of multiple sensory inputs to an area reveals relatively little insight into how the recipient area processes that multisensory information.

2.2.2 Bimodal Properties: The Superior Colliculus

Perhaps the most studied multisensory structure is not in the cortex, but the midbrain structure: superior colliculus (SC). This six-layered region contains spatiotopic representations of visual, auditory, and somatosensory modalities within its intermediate and deep layers (for review, see Stein and Meredith, 1993). Unisensory, bimodal, and trimodal neurons are intermingled with one another in this region, but the multisensory neurons predominate (63%; Wallace and Stein, 1997). The SC receives multiple sensory inputs from ascending projections (e.g., see Harting et al., 1992) but is dependent on cortical inputs for multisensory function (Jiang et al., 2001; Wallace et al., 1993). From the myriads of connectional studies of the SC, it seems clear that the multisensory properties of this structure are derived from inputs largely from unisensory structures and that not only their different inputs elicit suprathreshold responses from more than one sensory modality, but their combination often generates dramatic levels of response integration (Meredith and Stein, 1983, 1986), as illustrated in Fig. 2.2 (bimodal SC neuron). Despite this

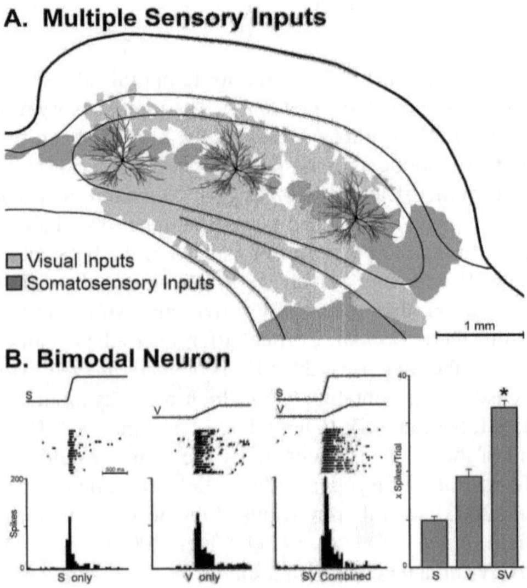

Fig. 2.2 Multisensory convergence in the superior colliculus. (**a**) The coronal section shows alternating cellular and fibrous layers of the cat superior colliculus. Terminal boutons form a discontinuous distribution across the deeper layers with somatosensory (*dark gray*, from SIV) and visual (*light gray*, from AEV) that largely occupy distinct, non-overlapping domains (redrawn from Harting et al., 1992). A tecto-reticulo-spinal neuron (redrawn from Behan et al., 1988) is shown, to scale, repeated at different locations across the intermediate layer where the dendritic arbor virtually cannot avoid contacting multiple input domains from the different modalities. Accordingly, tecto-reticulo-spinal neurons are known for their multisensory properties (Meredith and Stein, 1986). For a bimodal SC neuron (**b**) shows its responses (raster 1 dot= 1 spike, each row = 1 trial; histogram = 10 ms time bins) to a somatosensory stimulus, a visual stimulus, and to the combination of the same somatosensory and visual stimuli. Vigorous suprathreshold responses were evoked under each sensory condition and, as indicated in the bar graph (mean and SEM), the combined stimulation elicited response enhancement (* = statistically significant, $p < 0.05$; paired *t* test)

investigative attention, structure–function relationships have been identified for only a few multisensory neurons. Often those SC neurons most readily identifiable, histologically or by recording, are the tecto-spinal and tecto-reticulo-spinal neurons, with large somata averaging 35–40 μm in diameter and extensive dendritic arbors extending up to 1.4 mm (Behan et al., 1988; Moschovakis and Karabelas, 1985). As a population, these large multipolar neurons have a high incidence of exhibiting multisensory properties, usually as visual–auditory or visual–somatosensory bimodal neurons (Meredith and Stein, 1986). Indeed, from the overlapping relationship of their dendritic arbor with the distribution of input terminals from different sensory modalities, like that illustrated in Fig. 2.2, it is easy to imagine how these neurons receive their multisensory inputs. Another form of morphologically distinct SC neuron also shows multisensory properties: the NOS-positive interneuron. These excitatory local circuit neurons have been shown to receive bimodal inputs largely from the visual and auditory modalities (Fuentes-Santamaria et al., 2008). Thus, the SC contains morphological classes of neurons that highly correlate with multisensory activity. In addition, given the correlation between the sensory specificity of these different afferent sources and the clear presence of neurons in the SC that were overtly responsive to combinations of those inputs, these observations serve to reinforce the notion that the bimodal (and trimodal) neuron represented the fundamental unit of multisensory processing.

2.3 Non-bimodal Forms of Multisensory Processing

The terms 'bimodal' and 'multisensory' have sometimes been used synonymously. However, the widely accepted definition of multisensory processing stipulates that the effects of one sensory modality *influence* those of another modality, and 'influence' can have a variety of neurophysiological manifestations that include, but are not limited to, suprathreshold activation by two or more modalities. Accordingly, alternate forms of multisensory processing have been identified, mostly in recent studies of cat and ferret cortices.

The first, systematic examination of non-bimodal forms of multisensory convergence occurred in a region known not for its multisensory, but unisensory tactile properties. Somatosensory area SIV was found to receive crossmodal inputs from the neighboring auditory field of the anterior ectosylvian sulcus (FAES; Dehner et al., 2004). However, the bimodal auditory–somatosensory neurons that would be expected from such convergence rarely occur in SIV (see also Clemo and Stein, 1983; Jiang et al., 1994a; Rauschecker and Korte, 1993), and orthodromic stimulation of this crossmodal pathway curiously failed to elicit suprathreshold spiking from SIV neurons. Instead, when somatosensory stimulation was paired with electrical activation of the FAES, approximately 66% of the SIV neurons had their somatosensory responses significantly suppressed. Since these multisensory effects were only observed as a modulation of responses to the effective somatosensory modality, this 'new' form of multisensory neuron was designated as 'subthreshold' (Dehner et al., 2004; Meredith, 2002, 2004). It must be emphasized that these subthreshold multisensory neurons did not represent bimodal neurons that were inadequately stimulated by a second modality. Indeed, there are numerous examples of bimodal neuronal responses that, due to spatial or temporal relations, failed to demonstrate suprathreshold activation by that configuration (e.g., see Meredith and Stein, 1986; Meredith et al., 1987). However, in these cases responses in both modalities could be elicited when the stimulation parameters were changed. In contrast, subthreshold multisensory neurons consistently fail to show suprathreshold activation via a second modality despite the wide variety of stimulus parameters offered. Presumably, these subthreshold effects are mediated through inhibitory mechanisms or the simple lack of sufficient inputs from the second modality to trigger spiking activity.

Because subthreshold multisensory effects essentially modulate the activity of a dominant modality, these neurons appear to be unimodal unless tested with multisensory stimuli. Such subthreshold multisensory neurons have now been identified in apparently 'unimodal' areas outside SIV. For example, over 25% of apparently 'unimodal' auditory responses in the FAES were suppressed by somatosensory stimulation (Meredith et al., 2006). Similarly, subthreshold multisensory responses, like those illustrated in Fig. 2.3, have also been identified in the visual posterolateral lateral suprasylvian cortex (PLLS; Allman and Meredith, 2007). Importantly, these subthreshold effects in PLLS were shown to be sensitive to parametric changes in stimulus quality, indicating that they are sensory, not non-specific (e.g., alerting, arousal) responses (Allman et al., 2008b). Subthreshold multisensory responses have been observed in the cat rostral suprasylvian sulcal cortex (Clemo et al., 2007),

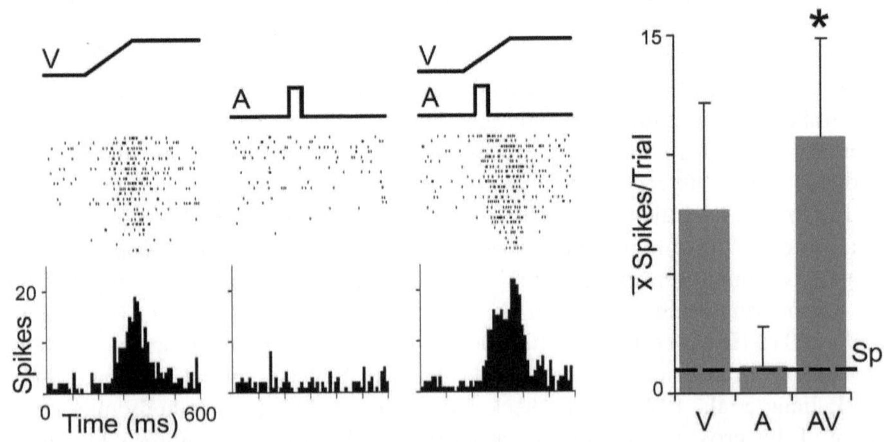

Fig. 2.3 Subthreshold multisensory processing. This neuron was highly responsive to a visual stimulus (ramp labeled 'V'), as indicated by the raster (1 dot = 1 spike; each row = 1 trial) and histogram (10 ms time bin). It was not activated by an auditory stimulus (*square* wave labeled 'A'). However, when the visual and auditory stimuli were combined, the response was significantly facilitated (*, $p < 0.05$ paired t test) as summarized in the bar graph. Sp = spontaneous levels, error bars = standard deviation

anterior ectosylvian visual area (AEV; Meredith and Allman, 2009), and anterior ectosylvian cortex (Carriere et al., 2007). In addition, recent studies of ferret cortex have identified a consistent population of subthreshold multisensory neurons in areas of the rostral suprasylvian sulcus (Keniston et al., 2008, 2009). Furthermore, examples of subthreshold multisensory processing appear in figures from studies of primate STS (Barraclough et al., 2005) and prefrontal cortex (Sugihara et al., 2006) as well as rattlesnake optic tectum (Newman and Hartline, 1981). Thus subthreshold forms (also called 'modulatory'; Carriere et al., 2007; Driver and Noesselt, 2008) of multisensory processing occur in different neural regions and in different species, lending support to the notion that this is not an unusual, but a general form of multisensory convergence.

2.3.1 Subthreshold Multisensory Processing and Crossmodal Cortical Connectivity

The cortical projections that presumably generate subthreshold multisensory convergence were examined in several areas where subthreshold multisensory neurons have been identified. Tracer injections placed in somatosensory SIV revealed retrogradely labeled neurons in auditory FAES as the source of its crossmodal inputs (Dehner et al., 2004). Similarly, tracers injected into visual PLLS identified retrogradely labeled neurons in auditory FAES and, to a lesser extent, AI and the posterior auditory field. These connections were confirmed by placing tracers in the regions of origin for the projection and observing the presence of labeled axon terminals within SIV (Dehner et al., 2004) and the PLLS (Clemo et al., 2008), respectively. These terminal projection patterns are illustrated in Fig. 2.4, along with crossmodal terminations in other cortical regions such as the FAES and the rostral suprasylvian sulcus (Clemo et al., 2007; Meredith et al., 2006). When these terminal projections are compared, it becomes obvious that the overwhelming majority of crossmodal projections are comparatively sparse (e.g., contain tens to few hundreds of labeled boutons/section rather than thousands) and preferentially terminate in the supragranular layers of their target regions. Therefore, it is tempting to speculate that such supragranular crossmodal terminations might underlie the subthreshold forms of multisensory convergence.

2.3.2 Crossmodal Cortical Connections and Multisensory Integration

Many of the crossmodal projections described above were modest to sparse in the density of their terminations in a given target region. Accordingly, it has been suggested that this relative reduction in projection strength may be an identifying property of convergence that generates subthreshold multisensory effects (Allman and Meredith, 2009). Other reports of cortical crossmodal projections between

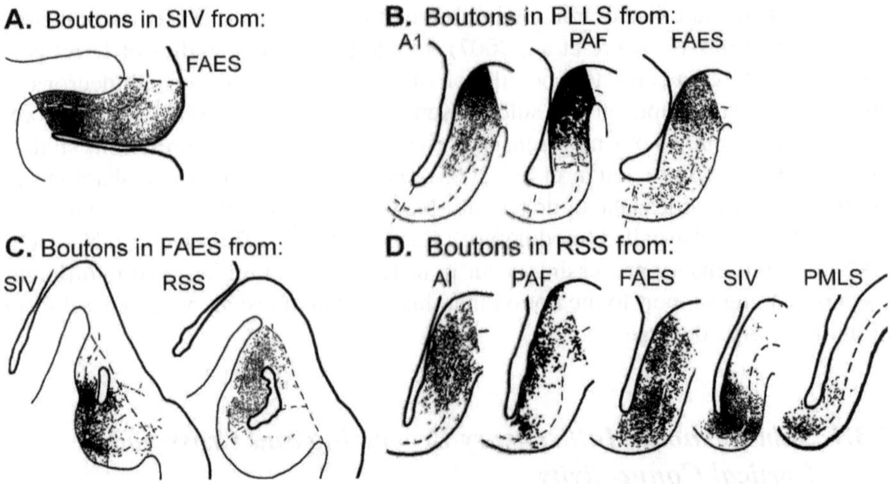

Fig. 2.4 Crossmodal corticocortical projections preferentially terminate in supragranular layers. (**a**) A tracer injection into auditory FAES produced labeled axon terminals (1 dot = 1 bouton) in somatosensory SIV. *Curved dashed line* indicates layer IV. (**b**) The PLLS visual area with labeled axon terminals from injections into auditory cortex (A1, primary), posterior auditory field (PAF), and the FAES. (**c**) The auditory FAES receives terminal projections, primarily in the supragranular layers, from somatosensory areas SIV and in the rostral suprasylvian sulcus (RSS). (**d**) The lateral bank of the rostral suprasylvian sulcus and the largely supragranular location of terminal label from auditory AI, PAF, and FAES, from somatosensory SIV, and from the postermedial lateral suprasylvian visual area. Redrawn from Clemo et al. (2007, 2008) and Meredith et al. (2006)

auditory and visual cortex in monkeys (Falchier et al., 2002; Rockland and Ojima, 2003) have also been characterized by the same sparseness of projection. Thus, although the presence of crossmodal corticocortical projections appears to be consistent, the functional effects of such sparse crossmodal projections had not been evaluated. To this end, a recent study (Allman et al., 2008a) examined the functional effects of a modest crossmodal projection from auditory to visual cortices in ferrets. Tracer injections centered on ferret primary auditory cortex (A1) labeled terminal projections in the supragranular layers of visual Area 21. However, single-unit recordings in Area 21 were unable to identify the result of that crossmodal convergence: neither bimodal nor subthreshold multisensory neurons were observed. Only when local inhibition was pharmacologically blocked (via iontophoresis of the GABA-A antagonist bicuculline methiodide) was there an indication of crossmodal influence on visual processing. Under these conditions, concurrent auditory stimulation subtly facilitated responses to visual stimuli. These observations support the notion that multisensory convergence does lead to multisensory processing effects, but those effects may be subtle and may manifest themselves in non-traditional forms. This interpretation is supported by recent observations of the effects of auditory stimulation on visual processing in V1 of awake, behaving monkeys (Wang et al., 2008) where no bimodal neurons were observed, but responses to visual–auditory stimuli were significantly shorter in latency when compared with those elicited by visual stimuli alone.

Given the relative sparseness of the crossmodal projections that underlie the subtle, subthreshold multisensory processing effects, it would be incorrect to assume that these forms of multisensory processing are, on the whole, negligible or insignificant. A recent comparison of subthreshold and bimodal processing in the PLLS demonstrated an altogether different conclusion (Allman et al., 2009). In this study, every neuron at a predetermined interval along full-thickness penetrations across the PLLS was examined for its responses to separate- and combined-modality stimuli. The bimodal neurons that were encountered showed the highest levels of response enhancement in response to the combined stimuli, and subthreshold neurons generated lower, more subtle levels of facilitation to the same stimuli. However, when the population responses were compared, so many more subthreshold neurons participated in signaling the sensory event that they generated over 60% of the multisensory signal in the PLLS. Thus, a neuronal population that traditionally would not be detected as multisensory (i.e., are activated by only on sensory modality) in reality contributes substantially to multisensory processing.

2.4 Multisensory Convergence: A Continuum?

These studies described above demonstrate that multisensory convergence is not restricted to bimodal neurons. Accordingly, the well-known pattern of convergence underlying bimodal neurons, as depicted in Fig. 2.5 (left), can be modified to include subthreshold multisensory neurons whose functional behavior might be defined by an imbalance of inputs from the two different modalities, which is also illustrated in the same figure. Furthermore, if this pattern of connectivity were to be modified to represent that observed in visual Area 21, the further reduction of the inputs subserving the subthreshold modality could result in subthreshold multisensory influences that occur only under specific contexts or conditions. Ultimately, by reducing the second set of inputs further toward zero essentially converts a multisensory circuit (albeit weak one) to unisensory, as shown in Fig. 2.5 (right). Given that these different patterns of convergence (and multisensory processing effects) could hypothetically result from only subtle but systematic changes in connectivity, it seems apparent that multisensory convergence patterns of connectivity actually represent a continuum. As depicted in Fig. 2.5, this continuum is represented by profuse levels of inputs from different modalities that generate bimodal neurons, on one end, and the complete lack of inputs from a second modality that defines unisensory neurons on the other.

2.4.1 Cortical Diversity: Differential Distributions of Multisensory Neuron Types?

The fact that some neural areas contain a preponderance of bimodal multisensory neurons (SC=63%, Wallace and Stein, 1997) while others include a majority of

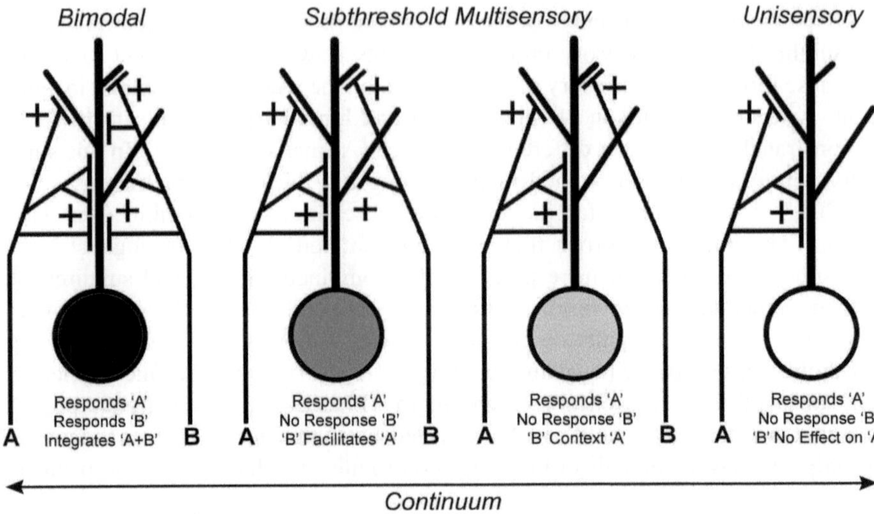

Fig. 2.5 Patterns of sensory convergence (from modality 'A' or 'B') onto individual neurons result in different forms of processing. On the left, robust and nearly balanced inputs from modalities 'A' and 'B' render this neuron responsive to both, the definitive property of a bimodal neuron (*black*). In addition, some bimodal neurons are capable of integrated responses when both modalities are presented together. If the pattern of inputs is slightly adjusted for the *dark gray* neuron, such that those from modality 'B' are not capable of generating suprathreshold activation, this neuron will be unimodal. In addition, if the different stimuli are combined, modality 'B' significantly influences the activity induced by 'A' revealing the subthreshold multisensory nature of the neuron. For the *light gray* neuron, if the inputs from 'B' are further minimized, their effect might become apparent when combined with inputs from modality 'A' only under specific contexts or conditions. Ultimately, if inputs from modality 'B' are completely eliminated, the neuron (*white*) is activated and affected only by inputs from modality 'A,' indicative of a unisensory neuron. Because each of the illustrated effects results from simple yet systematic changes in synaptic arrangement, these patterns suggest that multisensory convergence occurs over a continuum synaptic patterns that, on the one end, produce bimodal multisensory properties while, on the other, underlie only unisensory processing

subthreshold multisensory neurons (SIV=66%, with few bimodal types; Dehner et al., 2004) indicates that multisensory properties are differentially distributed in different regions. For example, as illustrated in Fig. 2.6 (left), some areas exhibit a high proportion of bimodal neurons. As a consequence, the output signal from these regions could be strongly influenced by multisensory integrative processes. Other regions that contain fewer bimodal neurons, or primarily contain subthreshold forms of multisensory neurons, would reveal a substantially lower multisensory effect (Fig. 2.6, center panels). In fact, bimodal neurons constitute only 25–30% of most cortical areas examined (Allman and Meredith, 2007; Carriere et al., 2007; Clemo et al., 2007; Jiang et al., 1994a, b; Meredith et al., 2006; Rauschecker and Korte, 1993). Thus, regardless of their bimodal or subthreshold nature, the multisensory signal from a population of multisensory neurons is likely to be further diluted by the fact that only a small proportion of neurons contribute to that

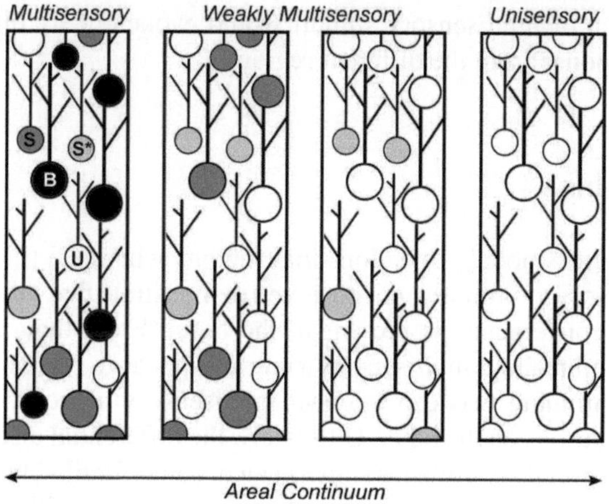

Fig. 2.6 Neurons with different multisensory ranges, when distributed in different proportions in different areas, produce regions with different multisensory properties. Each panel shows the same array of neurons, except that the proportions of bimodal (B, *black*), subthreshold (S, *dark gray*), minimal or contextual-subthreshold (S*, *light gray*), and unisensory (U, *white*) neurons are different. Areas in which bimodal proportions are largest have the potential to show the highest overall levels of multisensory integration, while those with lower proportions of bimodal or subthreshold multisensory neurons have lower potential levels of multisensory processing. Ultimately, these arrangements may underlie a range of multisensory distributions that occur along a continuum from one extreme (highly integrative) to the other (no integration) and indicate that different multisensory regions are likely to respond differently to a given multisensory stimulus

signal. Furthermore, not all bimodal neurons are capable of the same level of multisensory integration: some bimodal neurons are highly integrative (e.g., stimulus combinations evoke large response changes when compared with their separate presentations), while others apparently lack an integrative capacity altogether (e.g., show no response change when separate stimuli are combined). Such bimodal neurons have been observed in the SC (Perrault et al., 2005) as well as in the cortex (Clemo et al., 2007; Keniston et al., 2009). Ultimately, the combination of several factors (proportion of multisensory neurons, types of multisensory neurons present, and their integrative capacity) makes it extremely unlikely that the multisensory signal generated by a given area is as strong as that produced by its constituent bimodal neurons.

In the context of the behavioral/perceptual role of cortex, modest levels of multisensory integration levels may be appropriate. For example, when combining visual and auditory inputs to facilitate speech perception (e.g., like the cocktail party effect), it is difficult to imagine how a stable perception of the event could be maintained if every neuron showed a response change in excess of 1200%. On the other hand, for behaviors in which survival is involved (e.g., detection, orientation, or escape), multisensory areas reporting interactions >1200% would clearly provide an adaptive advantage. Therefore, because the range and proportion of multisensory integration is different in different neural areas, it should be expected that the same combination of multisensory stimuli would evoke widely different levels of multisensory response from the different regions.

2.5 Conclusions

Observations of crossmodal projections into regions where the traditional bimodal form of multisensory neuron was not observed demonstrate that more than one form of multisensory convergence is present in the brain. From their connections and activity, these subthreshold multisensory neurons appear to represent a transitional form along a continuum between bimodal multisensory neurons, on the one end, and unisensory neurons on the other. Ultimately, the differential distribution of these neuronal types across different neural areas can account for likelihood that different levels of multisensory response can be evoked by the same multisensory event.

Acknowledgments We appreciate the comments of Dr. B. Allman on the chapter. Supported by NIH grant NS039460.

References

Allman BL, Bittencourt-Navarrete RE, Keniston LP, Medina A, Wang MY, Meredith MA (2008a) Do cross-modal projections always result in multisensory integration? Cereb Cortex 18: 2066–2076

Allman BL, Keniston LP, Meredith MA (2008b) Subthreshold auditory inputs to extrastriate visual neurons are responsive to parametric changes in stimulus quality: Sensory-specific versus non-specific coding. Brain Res 1232:95–102

Allman BL, Keniston LP, Meredith MA (2009) Not just for bimodal neurons anymore: the contribution of unimodal neurons to cortical multisensory processing. Brain Topogr. 21:157–167, doi: 10.1007/s10548-009-0088-3

Allman BL, Meredith MA (2007) Multisensory processing in 'unimodal' neurons: cross-modal subthreshold auditory effects in cat extrastriate visual cortex. J Neurophysiol 98:545–549

Barraclough NE, Xliao D, Baker Cl, Oram MW, Perret DI (2005) Integration of visual and auditory information by superior temporal sulcus neurons responsive to the sight of actions. J Cognit Neurosci 17:377–391

Behan M, Appell PP, Graper MJ (1988) Ultrastructural study of large efferent neurons in the superior colliculus of the cat after retrograde labeling with horseradish peroxidase. J Comp Neurol 270:171–184

Benevento LA, Fallon JH, Davis B, Rezak M (1977) Auditory-visual interaction in single cells in the cortex of the superior temporal sulcus and the orbital frontal cortex of the macaque monkey. Exp Neurol 57:849–872

Bowman EM, Olson CR (1988) Visual and auditory association areas of the cat's posterior ectosylvian gyrus: cortical afferents. J Comp Neurol 272:30–42

Bruce C, Desimone R, Gross CG (1981) Visual properties of neurons in a polysensory area in superior temporal sulcus of the macaque. J Neurophysiol 46:369–384

Burton H, Kopf EM (1984) Ipsilateral cortical connections from the 2nd and 4th somatic sensory areas in the cat. J Comp Neurol 225:527–553

Carriere BN, Royal DW, Perrault TJ, Morrison SP, Vaughan JW, Stein BE, Wallace MT (2007) Visual deprivation alters the development of cortical multisensory integration. J Neurophysiol 98:2858–2867

Clemo HR, Allman BL, Donlan MA, Meredith MA (2007) Sensory and multisensory representations within the cat rostral suprasylvian cortices. J Comp Neurol 503:110–127

Clemo HR, Meredith MA (2004) Cortico-ortical relations of cat somatosensory areas SIV and SV. Somat Mot Res 21:199–209

Clemo HR, Sharma GK, Allman BL, Meredith MA (2008) Auditory projections to extrastriate visual cortex: connectional basis for multisensory processing in 'unimodal' visual neurons. Exp Brain Res 191:37–47

Clemo HR, Stein BE (1983) Organization of a fourth somatosensory area of cortex in cat. J Neurophysiol 50:910–925

Dehner LR, Keniston LP, Clemo HR, Meredith MA (2004) Cross-modal circuitry between auditory and somatosensory areas of the cat anterior ectosylvian sulcal cortex: a 'new' inhibitory form of multisensory convergence. Cereb Cortex 14:387–403

Driver J, Noesselt T (2008) Multisensory interplay reveals crossmodal influences on 'sensory-specific' brain regions, neural responses, and judgments. Neuron 57:11–23

Falchier A, Clavagnier C, Barone P, Kennedy H (2002) Anatomical evidence of multimodal integration in primate striate cortex. J Neurosci 22:5749–5759

Fuentes-Santamaria V, Alvarado JC, Stein BE, McHaffie JG (2008) Cortex contacts both output neurons and nitrergic interneurons in the superior colliculus: direct and indirect routes for multisensory integration. Cereb Cortex 18:1640–1652

Harting JK, Updyke BV, Van Lieshout DP (1992) Corticotectal projections in the cat: anterograde transport studies of twenty-five cortical areas. J Comp Neurol 324:379–414

Hikosaka K, Iwai E, Saito H, Tanaka K (1988) Polysensory properties of neurons in the anterior bank of the caudal superior temporal sulcus of the macaque monkey. J Neurophysiol 60: 1615–1637

Jiang H, Lepore F, Ptito M, Guillemot JP (1994a) Sensory interactions in the anterior ectosylvian cortex of cats. Exp Brain Res 101:385–396

Jiang H, Lepore F, Ptito M, Guillermot JP (1994b) Sensory modality distribution in the anterior ectosylvian cortex (AEC) of cats. Exp Brain Res 97:404–414

Jiang W, Wallace MT, Jiang H, Vaughan W, Stein BE (2001) Two cortical areas mediate multisensory integration in superior colliculus neurons. J Neurophysiol 85:506–522

Jones EG, Powell TPS (1970) An anatomical study of converging sensory pathways within the cerebral cortex of the monkey. Brain 93:793–820

Keniston LP, Allman BL, Meredith MA (2008) Multisensory character of the lateral bank of the rostral suprasylvian sulcus in ferret. Soc Neurosci Abstr 38:457.10

Keniston LP, Allman BA, Meredith MA, Clemo HR (2009) Somatosensory and multisensory properties of the medial bank of the ferret rostral suprasylvian sulcus. Exp Brain Res 196:239–251

Lee CC, Winer JA (2008) Connections of cat auditory cortex: III. Corticocortical system. J Comp Neurol 507:1920–1943

Meredith MA (2002) On the Neuronal Basis for multisensory convergence: a brief overview. Cognit Brain Res 14:31–40

Meredith MA (2004) Cortico-cortical connectivity and the architecture of cross-modal circuits. In: Spence C, Calvert G, Stein B (eds) Handbook of multisensory processes. MIT Press, Cambridge, pp 343–355

Meredith MA, Allman BL (2009) Subthreshold multisensory processing in cat auditory cortex. Neuroreport 20:126–131

Meredith MA, Keniston LR, Dehner LR, Clemo HR (2006) Cross-modal projections from somatosensory area SIV to the auditory field of the anterior ectosylvian sulcus (FAES) in cat: further evidence for subthreshold forms of multisensory processing. Exp Brain Res 172:472–484

Meredith MA, Nemitz JW, Stein BE (1987) Determinants of multisensory integration in superior colliculus neurons. I. Temporal factors. J Neurosci 7:3215–3229

Meredith MA, Stein BE (1983) Interactions among converging sensory inputs in the superior colliculus. Science 221:389–391

Meredith MA, Stein BE (1986) Visual, auditory, and somatosensory convergence on cells in the superior colliculus results in multisensory integration. J Neurophysiol 56:640–662
Monteiro G, Clemo HR, Meredith MA (2003) Auditory cortical projections to the rostral suprasylvian sulcal cortex in the cat: implications for its sensory and multisensory organization. Neuroreport 14:2139–2145
Mori A, Fuwa T, Kawai A, Yoshimoto T, Hiraba Y, Uchiyama Y, Minejima T (1996) The ipsilateral and contralateral connections of the fifth somatosensory area (SV) in the cat cerebral cortex. Neuroreport 7:2385–2387
Moschovakis AK, Karabelas AB (1985) Observations on the somatodendritic morphology and axonal trajectory of intracellularly HRP-labeled efferent neurons located in the deeper layers of the superior colliculus of the cat. J Comp Neurol 239:276–308
Newman EA, Hartline PH (1981) Integration of visual and infrared information in bimodal neurons of the rattlesnake optic tectum. Science 213:789–791
Olson CR, Graybiel AM (1987) Ectosylvian visual area of the cat: location, retinotopic organization, and connections. J Comp Neurol 261:277–294
Perrault TJ Jr, Vaughan JW, Stein BE, Wallace MT (2005) Superior colliculus neurons use distinct operational modes in the integration of multisensory stimuli. J Neurophysiol 93:2575–2586
Rauschecker JP, Korte M (1993) Auditory compensation for early blindness in cat cerebral cortex. J Neurosci 13:4538–4548
Reinoso-Suarez F, Roda JM (1985) Topographical organization of the cortical afferent connections to the cortex of the anterior ectosylvian sulcus in the cat. Exp Brain Res 59:313–324
Rockland KS, Ojima H (2003) Multisensory convergence in calcarine visual areas in macaque monkey. Int J Psychophysiol 50:19–26
Saleem KS, Suzuki W, Tanaka K, Hashikawa T (2000) Connections between anterior inferotemporal cortex and superior temporal sulcus regions in the macaque monkey. J Neurosci 20:5083–5101
Seltzer B, Pandya DN (1980) Converging visual and somatic sensory input to the intraparietal sulcus of the rhesus monkey. Brain Res 192:339–351
Selzer B, Pandya DN (1994) Parietal, temporal, and occipital projections to cortex of the superior temporal sulcus in the rhesus monkey: a retrograde tracer study. J Comp Neurol 343:445–463
Stein BE, Meredith MA (1993) Merging of the senses. MIT Press, Cambridge, MA
Sugihara T, Diltz MD, Averbeck BB, Romanski LM (2006) Integration of auditory and visual communication information in the primate ventrolateral prefrontal cortex. J Neurosci 26: 11138–11147
Wallace MT, Meredith MA, Stein BE (1992) Integration of multiple sensory modalities in cat cortex. Exp Brain Res 91:484–488
Wallace MT, Meredith MA, Stein BE (1993) Converging influences from visual, auditory, and somatosensory cortices onto output neurons of the superior colliculus. J Neurophysiol 69: 1797–1809
Wallace MT, Ramachandran R, Stein BE (2004) A revised view of sensory cortical parcellation. Proc Nat Acad Sci U S A 101:2167–2172
Wallace MT, Stein BE (1997) Development of multisensory neurons and multisensory integration in cat superior colliculus. J Neurosci 17:2429–2444
Wang Y, Celebrini S, Trotter Y, Barone P (2008) Visuo-auditory interactions in the primary visual cortex of the behaving monkey: electrophysiological evidence. BMC Neurosci 9:79

Chapter 3
Computational Modeling of Multisensory Object Perception

Constantin Rothkopf, Thomas Weisswange, and Jochen Triesch

3.1 Introduction

The brain receives a vast number of sensory signals that relate to a multitude of external and internal states. From these signals it has to somehow compute meaningful internal representations, reach useful decisions, and carry out actions. Because of the inherently probabilistic nature of all sensing processes one of their fundamental properties is their associated uncertainty. Sources of uncertainty include the neural noise due to physical processes of transduction in early stages of neural encoding, noise due to physical constraints such as unavoidable aberrations of every imaging device, and uncertainties because many environmental states can give rise to the same sensory measurement as well as many different sensory measurements can be evoked by the same state of the world. All these uncertainties render the inverse computation from sensory signals to states of the world difficult, but the brain gives humans the perception of a stable and mostly unambiguous world. Vision, audition, touch, proprioception, and all other senses suggest to us a world of individual objects and well-defined states. How can the brain do this?

Over the last decades ample data has been assembled in order to shed light on how the human and primate brains are able to accomplish this feat. The literature on empirical investigations of cue combination, cue integration, perception as an inference process, and many related aspects is vast (see, e.g., Ernst and Bülthoff, 2004; Kersten et al., 2004; Yuille and Kersten, 2006 for reviews). Psychophysical, neurophysiological, and imaging studies have quantified human and primate performance and knowledge on the neural implementation has accumulated. Nevertheless, the question on how the brain merges sensory inputs into complete percepts offers many unsolved problems. Sensory processing has been traditionally thought to be separate in the respective modalities. The segregation has been applied even within modalities, as, e.g., in the separation of ventral and dorsal streams in vision,

C. Rothkopf (✉)
Frankfurt Institute for Advanced Studies (FIAS), Goethe University Frankfurt,
Ruth-Moufang-Str. 1, 60438 Frankfurt am Main, Germany
e-mail: rothkopf@fias.uni-frankfurt.de

M.J. Naumer, J. Kaiser (eds.), *Multisensory Object Perception in the Primate Brain*,
DOI 10.1007/978-1-4419-5615-6_3, © Springer Science+Business Media, LLC 2010

promoting a hierarchical and modular view of sensory processing. But recent experimental results have emphasized the multimodal processing of sensory stimuli even in areas previously regarded as unimodal.

Theoretical work largely based on advances in artificial intelligence and machine learning has not only furthered the understanding of some of the principles and mechanisms of this process but also led to important questions and new experimental paradigms. The last 20 years have seen an increasing emphasis on Bayesian techniques in multimodal perception, mostly because such models explicitly represent uncertainties, a crucial aspect of the relation between sensory signals and states of the world. Bayesian models allow for the formulation of such relationships and also of explicit optimality criteria against which human performance can be compared. They therefore allow answering the question, how close human performance comes to a specific formulation of best performance. Maybe even more importantly, Bayesian methods allow comparing quantitatively different models by how well they account for observed data.

The success of Bayesian techniques in explaining a large variety of perceptual phenomena has also led to a large number of additional open questions, especially about how the brain is able to perform computations that are consistent with the functional models and also about the origin of these models. For this reason, and because comprehensive review articles of perceptual processes and their modeling have been published in the past years (see, e.g., Ernst and Bülthoff, 2004; Kersten et al., 2004; Knill and Pouget, 2004; Shams and Seitz, 2008; Yuille and Kersten, 2006), this chapter specifically emphasizes open questions in theoretical models of multisensory object perception.

3.2 Empirical Evidence for Crossmodal Object Perception

Considerable evidence has been collected over the last century supporting the fact that primates use input from multiple sensory modalities in sophisticated ways when perceiving their environment. While other chapters in this book focus explicitly on neurophysiological, psychophysical, and imaging studies demonstrating this ability, a few key studies are listed here with an emphasis on results that have guided theoretical work on modeling the processes manifest in crossmodal object perception.

Traditionally, it was assumed that sensory inputs mostly are processed separately and that separate cortical, subcortical regions are specialized in the integration of the distinct modalities, and that such computations were late in the processing hierarchy (Fellemann and Van Essen, 1991; Jones and Powell, 1970). It was assumed that multisensory processing was confined to specific areas in so-called later stages of the processing hierarchy, a hypothesis that is reflected by the fact that such cortical areas were termed 'associative cortices.'

Multimodal processes had been reported early on (von Schiller, 1932) and behavioral evidence for multisensory integration was shown all along, e.g., it was shown that it can speed up reaction times (Gielen et al., 1983; Hershenson, 1962), it can improve detection of faint stimuli (Frens and Van Opstal, 1995), and it can change percepts in a strong way as in the ventriloquist illusion (Pick et al., 1969), the McGurk effect (McGurk and Mc Donald, 1976), and the parchment skin illusion (Jousmaki and Hari, 1998). These studies were paralleled by neurophysiological investigations demonstrating multisensory processing in primates (Bruce et al., 1981; Lomo and Mollica, 1959; Murata and Bach-y-Rita, 1965). While most of these studies were able do describe these multimodal effects qualitatively, during the last 20 years, psychophysics together with quantitative models of computations involving uncertainties had a prominent role in the understanding of multisensory perception. This was achieved by quantifying how behavior reflects quantities describing the stimuli in different modalities across a wide range of tasks. One of the main stories emerging from these investigations was the central role of uncertainty in explaining observed psychophysical performance.

Neurophysiological investigations of multisensory processing also shifted from an early emphasis on the segregation of sensory processing in separate processing streams to the rediscovery of multimodal integration in primary sensory areas. Evidence now has been found that responses of neurons in so-called early sensory areas can indeed be modulated by stimuli from other modalities.

Now, multisensory responses in early areas have been found repeatedly using a wide variety of analysis techniques including blood oxygenation level-dependent (BOLD) signals in functional magnetic resonance imaging (fMRI) (Calvert et al., 1997), event-related potential (ERP) (Foxe et al., 2000), single-cell recordings in macaque (Ghazanfar et al., 2005; Schroeder and Foxe, 2002), in anesthetized ferrets (Bizley et al., 2006), and in awake behaving monkeys (Ghazanfar et al., 2005). Such interactions are not restricted to so-called early sensory areas, but have instead been shown to be common throughout the cortex including audio-visual responses in inferotemporal cortex (IT) (Gibson and Maunsell, 1997; Poremba et al., 2003), auditory responses in cat visual cortex (Morrell, 1972), lateral occipital cortex (LOC) responses during tactile perception (James et al., 2002), mediotemporal (MT) activation from tactile motion (Hagen et al., 2002), and visual and auditory responses in somatosensory areas (Zhou and Fuster, 2000). Furthermore, plasticity of multisensory processing can lead to remapping between modalities, as in congenitally blind and deaf individuals (Finney et al., 2001; Kujala et al., 1995; Sadato et al., 1996).

3.3 Computational Principles Evident from the Experimental Data

Theoretical modeling inherently needs to abstract from the full richness of the studied system. One of the central tools in developing models of multisensory object perception has been the formalization of the environment in terms of cues. Although a definition of a cue is highly difficult and ambiguous, the idea is to take a physical entity in the world as given and attribute it certain properties such as size, reflectance, weight, and surface structure. These measurable physical states of the world are unknown to the brain, which instead obtains signals from its sensory apparatus. Individual variables that can be recovered from this sensory input and which somehow reflect the physical state of the world are termed cues. Thus, contrast, spatial frequency, pressure sensation, sensation of acceleration, and loudness have all been considered as cues. A multitude of cues have been considered in the modality of vision, where more than a dozen cues are known for depth alone including disparity, occlusion, texture gradients, and linear perspective. Using this formalism, a variety of different tasks and modes of sensory processing have been distinguished.

3.3.1 Sensory Integration

When there are two cues available for inferring a quantity of interest, it is advantageous to use both measurements in order to infer the unknown cause, because the common cause will usually influence both measurements in a structured way. The hope is that by somehow combining the two measurements in a sensible way, the uncertainty in the unknown quantity can be reduced by taking advantage of this structure. This operation is called cue integration and is certainly the best-studied multimodal inference task.

A wide variety of experiments have been devised to test how individual cues are combined to a single estimate of the unknown quantity. While some experiments have used different cues within the same modality, such as different depth cues in vision (Jacobs, 1999; Knill and Saunders, 2003), there are a variety of investigations of how such integration is carried out across different modalities, including audio-visual (Battaglia et al., 2003), visual–haptic (Ernst and Banks, 2002), and visual–proprioceptive (van Beers, 1999) tasks. Trimodal cue integration has also been studied as, e.g., in Wozny et al. (2008). The results of these investigations have demonstrated that subjects combine the individual measurements in order to increase the reliability of their estimate of the unobserved quantity.

3.3.2 Sensory Combination

The literature on multisensory processing has made the distinction between sensory integration and sensory combination (e.g., Ernst and Bülthoff, 2004). This distinguishes between cases in which different sensory signals are commensurate in the sense that they are somehow represented in the same coordinate frame, have the same units, and are of the same type, i.e., discrete or continuous, and cases in which they are not. Ernst and Bülthoff (2004) provide an example of the latter case of sensory combination involving visual and haptic information to complement each other. When touching and viewing an object, visual and haptic information are often obtained from different sides of the object, where vision codes for the front of an object while haptic codes for the back of an object (Newell et al., 2001). These different cues that are not commensurate can nevertheless be combined to aid in object perception.

3.3.3 Integration of Measurements with Prior Knowledge

As mentioned previously, one of the central problems of the sensory apparatus is that the signals that it obtains are ambiguous with respect to what caused them. Because of this ambiguity it is advantageous to make use of the regularities encountered in the world, as these can be used to bias the interpretation away from very unlikely causes. Direct evidence that such biases are at work in everyday perception comes from visual illusions involving shading. When viewing spherical objects whose upper part is brighter these are interpreted as emerging from the plane. This percept is very stable, although the image itself is highly ambiguous. The sphere could be interpreted as being convex and illuminated from above but also as being concave and illuminated from below. The explanation for this perceptual phenomenon is that the visual system biases its interpretation of such ambiguous input toward a scene configuration in which the light source is located above the object instead of below the object, because this is the more common configuration under natural circumstances. Furthermore, Mamassian and Landy (2001) showed how two priors can interact, in their case priors for the direction of illumination and for the viewpoint of the observer.

3.3.4 Explaining Away

There are perceptual tasks involving several independent causes and different cues. In such situations, explaining away describes the fact that one of the causes for observed data becomes less probable when new data is observed, which makes an alternative cause more probable. Another way of viewing this phenomenon is that causes compete for an explanation of the data. As an example, consider a stimulus with two rectangular image regions both containing the same luminance gradient, which are placed next to each other. These two regions give rise to an illusion in which the lightness at the boundary between the two regions is perceived as being different across the same positions on the identical brightness distributions. Knill and Kersten (1991) reported that placing two ellipses at the bottom of the two rectangular regions strongly reduced this illusion as the ensemble image is now perceived as two equally shaped cylinders illuminated from the side. This additional scene parameter explains away the previous interpretation.

A more elaborate example involving visual and haptic measurements comes from a study by Battaglia et al. (2005). Subjects were intercepting moving targets based on monocular visual information in a virtual reality environment. When observing the target object along its trajectory, the uncertainty about the true object size transfers to uncertainty about the distance of the object and thereby renders interception uncertain, too. To test for the possibility of additional measurements to disambiguate the causes of the target's apparent size, subjects had to catch the target in some trials only based on the visual information or in different trials using visual and haptic information. The data showed that interception performance was indeed better if subjects had previously touched the target, suggesting that the additional haptic measurement enabled subjects to explain away the influence of ball size on image size leading to a better estimate of target distance.

While there is evidence for perceptual processes in which a cue is able to bias the percept away from an alternative percept rendered more likely by another cue, the literature also reports cases in which a disambiguating cue does not explain away an alternative cause. Mamassian et al. (1998) report such cases involving cast shadows. In one such example, a folded paper card is imaged from above in such a way that its shape can either be perceived as a W or as an M. The addition of a pencil lying on top of the folded card together with its shadow falling on the card should in principle exclude one of the possible shapes but fails to do so. Instead, the card is still ambiguous and can be seen to flip between both interpretations.

3.3.5 Using the Appropriate Model

Most of the classical cue combination studies were designed in such a way that the model describing the inference process in a particular experiment is known or controlled by the experimenter and the correct relationship between the observed quantities and the unobserved causes are fully specified within that model. Crucially, in most of these studies, the sensory input is always interpreted using this single model. Some recent experiments have investigated cases in which observations may have different interpretations under different models.

Wallace and Stein (2007) and Hairston et al. (2004) considered the case of a paired auditory and visual stimulus. Here, a briefly played tone and a light flash may be interpreted as having a common cause, if the positions of the two signals are close enough in space, but may also be perceived as originating from two different sources, if their spatial separation is sufficiently large. The perceptual system may select somehow whether the appropriate model upon which the inference is to be based is that of a single common source for the two signals or whether it is more likely that the signals came from two separate sources, which is the approach taken by Koerding et al. (2007). Alternatively, it may compute the a posteriori distribution of position according to the two models separately and then integrate these by somehow taking into account the full uncertainties of each models' appropriateness in explaining the observations.

3.3.6 Decision Making

While subjects have an implicit reward structure when they are instructed to carry out a perceptual task in psychophysical experiments in laboratory settings, it is also important to consider different explicit reward structures. Trommershäuser et al. (2003) varied the reward structure in a series of fast pointing tasks and demonstrated that human subjects not only take their respective motor uncertainties into account but also adjust their pointing movements so as to maximize the average reward. In their case, reward was made explicit through the amount of monetary compensation for the subject's participation based on their performance. These results show that the human brain is able to not only integrate sensory cues by taking their respective uncertainties into account, but that it can also apply different cost functions to the performed inferences when taking decisions.

3.3.7 Learning of Cue Reliabilities and Priors

The above-mentioned perceptual experiments varied the reliabilities of individual cues within the stimulus, i.e., a visual cue can be made more uncertain by reducing its contrast, or by adding noise to its spatial distribution. A different form of uncertainty can be manipulated when controlling the reliabilities across trials.

Such experiments often use virtual reality environments, as these allow for the full control of different stimulus modalities and allow controlling cue conflicts in a precise manner. Such experiments have shown that observers can learn the reliabilities of individual cues through extensive training (Jacobs and Fine, 1999) and that cues in the haptic modality can be used as standard against which the reliabilities of two visual cues are judged (Atkins et al., 2001). Furthermore, humans can quickly adjust to changing reliabilities of individual cues (Triesch et al., 2002) across trials. So, if the system is able to reweigh individual cues by their reliabilities, it must somehow compute with the individual estimates over time.

3.3.8 Task Dependence

While a visual scene contains a vast amount of information in terms of all possible variables that a perceptual system could infer from it, in a specific task only few variables really matter to obtain a judgment or plan an action. When guiding the reach of a hand in grasping a cup, it may not be necessary to infer the source of illumination or the slant of the surface on which the cup rests, but it may well be necessary to infer the shape and orientation of the cup's handle. But visual cues in such scenes may depend on both the light source and the slant of the surface. By contrast, for placing a cup on an inclined surface the slant of the surface is an important scene variable. Thus, dependent on the task at hand, some of the quantities determining the visual scene may be of direct interest and need to be inferred whereas in other tasks they do not provide important information. Explicit investigations of such task dependencies have been done by Schrater and Kersten (2000) and Greenwald and Knill (2009).

3.3.9 Developmental Learning

It may be important for the choice of the modeling framework to consider the question of whether multisensory object perception is learnt over developmental timescales or whether it is innate. Developmental studies have shown a wide range of phenomena in multisensory processing ranging from cue integration being present at an early age (Lewkowicz, 2000) to cases in which young children up to the age of 8 years did not integrate cues (Gori, 2008; Nardini et al., 2006). Gori et al. (2008) have reported that children instead relied on only a single modality's cue when integrating visual and haptic information would have been advantageous. Which modality they relied on depended on the specific task. The range of these different results has suggested that at least some aspects of multisensory object perception are learned over developmental timescales.

3.4 Models of Multimodal Object Perception

Computational models of cognitive processes such as multimodal object perception can be built at different levels of description. It may be helpful to adopt the notion of Marr's (1982) three levels of description in order to distinguish current models of multimodal perception. At the level of *computational theory*, it can be asked what the goal of the perceptual system's computations is. While it is difficult in principle to separate the perceptual system from the goals of the entire living system, the currently dominating view is that the perceptual system aims at inferring the causes of the sensory input. This idea is often attributed to von Helmholtz (1867) but really can be traced back to Al Hazen around the year 1000 (Smith, 2001). With uncertainty about these causes being the predominant factor these computations are therefore considered to perform inference and the computational framework for modeling multimodal object perception becomes that of an ideal observer. This theoretical construct answers the question, how well could any computing device possibly do on the inference problem at hand? The problem is solved in an optimal way without any restrictions on how such computations could actually be implemented in the primate brain. In a Bayesian framework this entails computing a probability distribution over the causes given the sensory signals. Finally, if a decision has to be taken in a normative way, such a distribution has to be combined with a cost function, which specifies the cost for all possible errors between the true value of a cause and its estimate.

At the *algorithmic level*, one may consider different approaches of how to solve an inferential task and how the sensory uncertainties might be represented. The area of machine learning has made significant advances in recent years (Bishop, 2006; MacKay, 2003) and has brought about a multitude of different algorithmic approaches on how to solve inference problems. Nevertheless, it is still an active research area with many open problems. An important lesson from this field is that even with fairly simple looking problems that involve only a few quantities that need to be represented probabilistically the inference problem can quickly become analytically intractable. This requires using approximations in order to obtain results. Indeed, many questions still remain regarding how uncertainties are represented in the brain, how computations involving uncertainties could be carried out, and which representations and approximations are most suitable for implementation in biological system.

Finally, at the *implementation level* the question is how the neural substrate may be able to implement multisensory perception. This remains a daunting problem, given that even the nature of the neural code is still debated and many questions regarding the biophysical and neurochemical processes in the nervous system are still unanswered. More importantly, while more and more tools with high sophistication become available such as transcranial direct current stimulation, in vivo fast-scan cyclic voltammetry, the use of transgenic species, photolysis by two-photon excitation, use of neurotropic viruses, the data that these methods produce rarely clarify their function in terms of the computations the brain is implementing. Nevertheless, all the approaches to understanding how the primate brain accomplishes multisensory object perception need to take into account what type of computations the brain must be able to accomplish based on the observed psychophysical evidence. In the following we describe the most prominent directions of such computational models.

3.4.1 Ideal Observer Models of Multimodal Object Perception

The ideal observer framework formulates an inference task by specifying the involved quantities as random variables and expressing their probabilistic relationship in a generative model. Such generative models describe the process of how a particular sensory variable is caused by attributes of the world. When considering a visual scene, the illumination source and the reflectance properties of an object are causes for the appearance of the object in the image. The generative model also needs to specify the exact probabilistic relationship between the individual variables. When the generative model has been specified, the configuration of a scene determines which of the variables in the generative model are observed and which are hidden or latent. This will determine the statistical dependencies between the involved quantities and accordingly how the uncertainties between individual variables are related. The task at hand will furthermore determine which variable needs to be computed. While in some circumstances a particular variable may be the goal of the computation, it may be discounted in other cases. A convenient way of depicting the statistical relationships between individual variables is shown in Fig. 3.1, in which observed variables are depicted as shaded circles, non-observed or latent variables are shown in a clear circle, and arrows represent causal relationships (see Bishop, 2006; Pearl, 1988).

Finally, when considering the full decision process based on the inference, it is necessary to apply a cost function that assigns a cost to errors in each variable dimension that is estimated. In the following, all these computations are described in more detail. The Bayesian framework can accommodate many of the above computational principles evident from the psychophysical literature. Cue combination, cue integration, integration with prior knowledge, explaining away, and certain forms of task dependence all fit in this framework (see Kersten et al., 2004 for a review).

The essence of the Bayesian approach is to formulate a model that explicitly represents the uncertainties as probability distributions for scene variables and

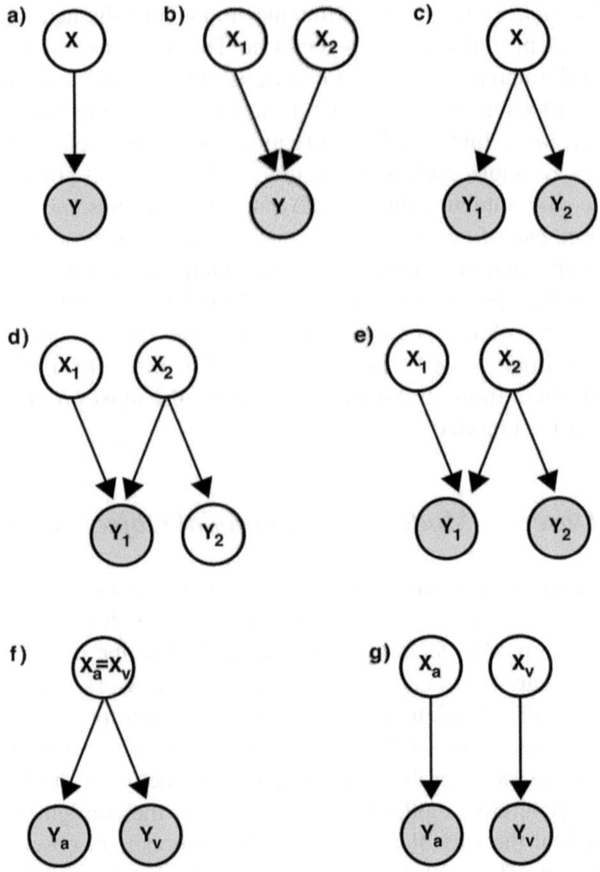

Fig. 3.1 Representation of different statistical dependencies between non-observed (latent) scene parameters, shown as *clear circles*, and measurement variables, shown as *shaded circles*, as Bayesian nets. See text for more detail

sensory variables that are statistically dependent on the scene variables that need to be estimated in the specific task. In the most general form, a scene variable such as the shape of an object is related probabilistically to a particular appearance image of that shape, because many different shapes can give rise to the same image or a single shape can result in many different images and because of the additional noise introduced by the imaging and transduction processes. Such a situation is depicted in Fig. 3.1a, where the latent variable X corresponds to the shape and the observed variable Y corresponds to the sensed image. Figure 3.1b corresponds to the case in which two latent variables, such as shape and illumination direction, are made explicit in their influence on the observed image.

As an example for multimodal object recognition we will consider the laboratory version of an audio-visual orienting task (Thomas, 1941), i.e., the task of inferring the position of an object, which is only made apparent through a brief light flash and a brief tone (Alais and Burr, 2004). In such a setting the observer has the task of inferring the position X from two measurements, the auditory cue Y_a and the visual cue Y_v, as depicted in Fig. 3.1c. The Bayesian view is that there is not enough certainty in order to know the position of the object exactly, but that instead one needs to report the full probability distribution representing how likely it is that the object is at a particular position $X=x$ given the two observations. Accordingly, the ideal observer obtains two measurements Y_a and Y_v and now infers the full probability distribution over all possible positions $X=x$ so that we assign a probability to each position of the source being at that location. This probability distribution can be computed using Bayes theorem:

$$P(X\,|\,Y_a, Y_v) = \frac{P(Y_a, Y_v\,|X)P(X)}{P(Y_a, Y_v)} \tag{3.1}$$

The term on the left-hand side of the equation is the posterior probability that the object is at some position $X=x$ given the numerical values of the actually observed auditory and visual measurements. Note that this posterior distribution may have a complex structure and that there are a variety of choices of how to obtain a decision from such a distribution, if only a single percept or a single target for an action is required.

The right-hand side contains first the term $P(Y_a, Y_v\,|X)$, which in the generative model view fully specifies how a single cause, or in the case of a vector-valued X multiple causes, give rise to the observed signals. In this view, a cause takes on a particular value $X=x$ and the term $P(Y_a, Y_v\,|X)$ specifies the generative process by which observable quantities are generated. This may be by linear superposition of individual causes or by some geometric dependency of visual features on the causing scene parameters. The probabilistic nature of this term may reflect the fact that multiple different causes can lead to the same scene parameters. It can also reflect the sensory noise that renders the observation stochastic and therefore the true value of the scene variable in question uncertain. In the chosen example of audio-visual localization, this term fully specifies the uncertainty in the measurements of Y_a and Y_v when the true object location is $X=x$, i.e., it assigns a probability to each measurement pair of Y_a and Y_b when the true value of X is x. Or, in the frequentist view, it describes the variability in the measurements of Y_a and Y_v on repeated trials, when the true value of X equals x.

On the other hand, when Y_a and Y_v have been observed then the expression $P(Y_a, Y_v\,|X)$ is viewed as a function of X and is often called the likelihood function. Note that the posterior distribution is a proper probability distribution given by conditioning on the observed values Y_a and Y_v. The likelihood function instead is conditioned on the unknown variable X. It expresses the relationship between observing a particular value for the two cues when a certain true position of the object is given. This term therefore fully describes the uncertainty associated with a particular measurement, i.e., it specifies the likelihood of sensing the actually observed values Y_a and Y_v for all possible values of X. Therefore, it may not be a proper probability distribution summing to one.

The prior probability $p(X)$ has the special role of representing additional knowledge. In the specific case of audio-visual localization this may be knowledge about common and unusual positions of the object emitting the sensory signals. While in some experimental conditions this distribution can be uniform so that it has no additional effect on the likelihood of the position of the object, the prior is often important for more naturalistic stimuli, where the unknown property of the object causing the observations follows strong regularities encountered in the natural environment. Therefore, it may be very advantageous to use this additional knowledge in inferring the causes of the sensed stimuli, especially for large uncertainties in the sensory stimuli. In the provided example, the laboratory setup is usually such that the source of the auditory and visual stimuli are equiprobable within a predefined range, but in a natural environment it may very well be that additional priors can be taken into account. One example for such regularities is that the sounds of birds may well have an associated prior with higher probability in the upper hemisphere of the visual field. Finally, the term $P(Y_a, Y_v)$ is not dependent on the value of the variable X but is a normalization factor that is required in order for the posterior to be a proper probability distribution summing to 1.

The above equation can be simplified under the assumption that the uncertainties in the two estimates are conditionally independent given the position of the object. One way of thinking about this is that the noise in the auditory cue is not dependent on the noise in the visual cue. More generally, this means that if the true position of the object is known, the uncertainties in the visual and auditory cue do not influence each other. Under this independence assumption the above equation can be rewritten as

$$P(X \mid Y_a, Y_v) = \frac{P(Y_a \mid X)P(Y_v \mid X)P(X)}{P(Y_a, Y_v)} \tag{3.2}$$

The advantage of this expression compared to Eq. (3.1) is that due to this factorization, instead of having to characterize the distribution $P(Y_a, Y_v \mid X)$, it is only necessary to know $P(Y_a \mid X)$ and $P(Y_v \mid X)$. The advantage lies in the fact that the joint distribution is three dimensional and would require to specify a probability value for each combination of X, Y_a, and Y_b whereas the two factored distributions $P(Y_a \mid X)$ and $P(Y_v \mid X)$ are only two dimensional. These types of independences are of central importance in Bayesian inference involving Bayes nets (Bishop, 2006; Pearl, 1988) and become particularly important when many random variables are involved and when variables change over time. Accordingly, in Eq. (3.2) the ideal Bayesian observer will base its decision on a quantity that is proportional to the product of the three terms in the numerator on the right-hand side. Again, note that this quantity expresses the probability given to each of all possible values of the variable X, the position of the object. For the particular case of audio-visual object localization, the above simplification means that only the uncertainties in object localization when either a tone alone or a flash alone is presented need to be measured in order to calculate the posterior.

Depending on the particular scene parameters and measurements considered in an inference process, it is necessary to specify the mathematical form of the involved probability distributions. A computationally particularly straightforward case is that of cue combination when the individual cues have independent additive Gaussian noise. It can be easily shown that under such circumstances the optimal inference about the causes in a cue combination experiment combines the measurements linearly. Although this is only a special case, this makes computations much easier. While Gaussian noise may not be the appropriate model for many real-world processes, many laboratory experiments in which the researcher can control the perturbations of the stimuli use this paradigm for simplicity. The optimal linear cue weights by which the individual estimates have to be weighted to obtain the best combined estimate can be shown to be proportional to the inverse variances of their respective cues:

$$w_{\mathrm{a}} = \frac{\frac{1}{\sigma_{\mathrm{a}}^2}}{\frac{1}{\sigma_{\mathrm{a}}^2} + \frac{1}{\sigma_{\mathrm{v}}^2}} \qquad w_{\mathrm{v}} = \frac{\frac{1}{\sigma_{\mathrm{v}}^2}}{\frac{1}{\sigma_{\mathrm{a}}^2} + \frac{1}{\sigma_{\mathrm{v}}^2}} \tag{3.3}$$

Intuitively, this is a satisfying solution in that it means that cues that have a higher associated uncertainty are relied upon less, while cues that have a lower uncertainty influence the final result more. Importantly, the variance of the combined estimate is smaller than the individual variances and can be shown to be

$$\sigma_{\mathrm{av}}^2 = \frac{\sigma_{\mathrm{a}}^2 \sigma_{\mathrm{v}}^2}{\sigma_{\mathrm{a}}^2 + \sigma_{\mathrm{v}}^2} \tag{3.4}$$

Nonlinear cue combination is in principle nothing special, as the Bayesian formulation in Eqs. (3.1) and (3.2) makes it clear, that the full likelihood functions have to be considered in general to compute the posterior at each value of *X*. Nevertheless, because of the computational simplicity, the vast majority of cue combination experiments have used stimuli such that linear cue combination can be used. Nonlinear cue combination can arise in different ways. The simplest case is that one of the component likelihood functions is non-Gaussian. It may also be that the individual likelihood functions are Gaussian, but that the task at hand does not require estimating an additional scene parameter, which nevertheless influences the measurements. This leads to an estimate that discounts the uninteresting variable and the associated marginalization procedure, which collapses the probability in the uninteresting variable onto the distributions of the cues, leads to the computation of an integral for the combined likelihood function that usually results in a highly non-Gaussian shape (e.g., Saunders and Knill 2001). A similar situation arises when the likelihood function is modeled as a mixture distribution in order to accommodate different scene configurations (Knill, 2003).

The prior distribution *P*(*X*) has a particularly interesting role in many modeling situations. The physical world is characterized by a large number of structural regularities, which manifest themselves in certain states of scene parameters having much higher probability than other states. A good example for the impact of the prior distribution comes from a study by Weiss, Simoncelli, and Adelson (2002) in which human motion perception was considered. It is well known from the area of computer vision that the direction and the speed of motion is ambiguous when inferred from local image motion, that is, a single measurement of a single motion cue gives rise to an infinite number of possible true motions in the scene. This relationship is expressed in a constraint equation, which linearly relates the velocity in horizontal direction with the velocity in vertical direction given a local velocity measurement. The likelihood function expresses this linear relationship and additionally reflects the fact, that the uncertainty about the image motion increases when the contrast in the image display is reduced. Importantly, Weiss, Simoncelli, and Adelson (2002) noticed, that a wide variety of perceptual phenomena in motion perception and visual illusions with moving stimuli could be explained by assuming that humans use a prior distribution over motion velocities that makes zero velocity and low velocities more likely. There is considerable literature on characterizing prior distributions both by finding those distributions that best explain psychophysical performance (e.g., Stocker and Simoncelli, 2006; Weiss et al., 2002;) as well as directly measuring distributions of physical states of the world (e.g., Geisler et al., 2001). These studies have shown that the prior distribution is often non-Gaussian in real-world settings and that picking a particular prior distribution is the key in explaining observed behavior.

After the above computations have been executed, the ideal observer has obtained a posterior probability distribution over the unknown scene parameters that are of interest, as the position x of the audio-visual source in the localization example. Given that perceptually we experience a single state of the world instead of a posterior distribution over possible scene parameters, the posterior distribution has to be utilized to come to a single estimate corresponding to the perceived scene. Similarly, when an action is required such as reaching for a point in the scene, a single direction of motion of the hand has to be obtained from the entire probability distribution. The system therefore has to collapse the posterior probability distribution over all possible states of the relevant scene parameter to a single value, which can be done in different ways, depending on the used criterion. The literature on optimal decision making and utility functions is vast not only in the areas of cognitive science and statistics but especially in economics (Neumann and Morgenstern, 1944) and can be tracked back to Bernoulli (1738). Bayesian decision theory deals with how to come to such a single decision given a distribution over X and a loss function $L(\hat{x})$, which measures the penalty of deciding that the value of X is $\hat{x}$ when the true value is x. Here, the relevant distribution is the posterior distribution over the scene parameters. The expected loss, which quantifies how much we expect to lose with a decision given the uncertainty in the variable X, can be expressed as follows in the case of the audio-visual localization task:

$$L(\hat{X} \mid Y_{\text{a}}, Y_{\text{v}}) = \int L(\hat{X}, X) P(X \mid Y_{\text{a}}, Y_{\text{v}}) dX \tag{3.5}$$

By calculating this point-by-point product of the loss function with the posterior and then integrating over all possible positions x, one obtains the expected loss as a function of $\hat{x}$. The optimal decision is then to choose that value $\hat{x}$ that minimizes the expected loss. While a wide variety of cost functions are possible, only a few are commonly used in practice, such as a cost function that assigns a one to the decision corresponding to the true value and a zero otherwise, or linear costs, or quadratic costs. In the special case of a Gaussian distribution, these cost functions give the same result, corresponding to choosing the maximum a posteriori (MAP) estimate, which is the value of highest posterior probability. This sequence of computations is depicted in the Fig. 3.2a for the audio-visual localization task.

Bayesian ideal observers have been designed for a large variety of tasks involving many different environmental properties and cues and they can accommodate the

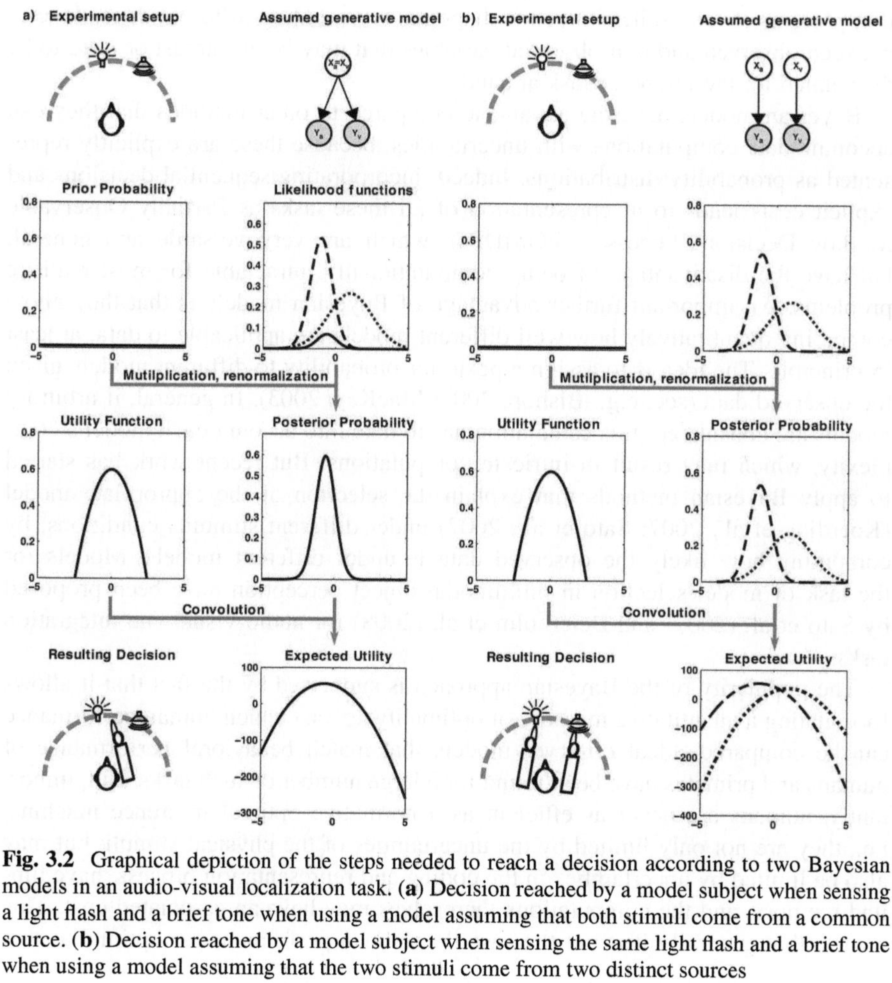

Fig. 3.2 Graphical depiction of the steps needed to reach a decision according to two Bayesian models in an audio-visual localization task. (**a**) Decision reached by a model subject when sensing a light flash and a brief tone when using a model assuming that both stimuli come from a common source. (**b**) Decision reached by a model subject when sensing the same light flash and a brief tone when using a model assuming that the two stimuli come from two distinct sources

computational principles mentioned in the above section such as cue combination (Fig. 3.1c), cue integration (Fig. 3.1c), explaining away (Fig. 3.1d, e). In principle, there is no limit to the number of different variables that make up the generative model, although exact computations may have to be replaced by approximations due to complexity. As an example, consider the case in which the observer needs to infer the identity of an object but obtains image measurements that are dependent on the position of the illumination source, the orientation of the object toward the observer, the specularity of the surface of the object, and many additional variables describing the scene. In such a case, the inferential process needs to compute the posterior probability of the potential target objects by discounting the non relevant variables such as illumination direction in a probabilistically optimal way. Such marginalization calculations can be very involved, but it has been shown, that there are situations in which human subjects take such uncertainties into account (Kersten, 1999). For further examples see the review by Kersten and Yuille (2003) that presents several other influence diagrams representing different dependencies between observed and non-observed variables that may be of interest or need to be discounted for the inference task at hand.

Bayesian models have the advantage compared to other methods that they can accommodate computations with uncertainties, because these are explicitly represented as probability distributions. Indeed, incorporating sequential decisions and explicit costs leads to a representation of all these tasks as Partially Observable Markov Decision Processes (POMDPs), which are very versatile and general, but have the disadvantage of being computationally intractable for most realistic problems. An important further advantage of Bayesian models is that they allow computing quantitatively how well different models are applicable to data, at least in principle. The idea is to assign a posterior probability to different models given the observed data (see, e.g., Bishop, 2006; MacKay, 2003). In general, if arbitrary models are considered, this calculation has to take into account each model's complexity, which may result in intricate computations. But recent work has started to apply Bayesian methods that explain the selection of the appropriate model (Koerding et al., 2007; Sato et al., 2007) under different stimulus conditions, by computing how likely the observed data is under different models. Models for the task of model selection in multimodal object perception have been proposed by Sato et al. (2007) and Beierholm et al. (2008) for audio-visual cue integration tasks.

The popularity of the Bayesian approach is supported by the fact that it allows formulating a quantitative measure of optimality against which human performance can be compared. Ideal observer models that match behavioral performance of humans and primates have been found for a large number of such tasks. But, importantly, humans are never as efficient as a normative optimal inference machine, i.e., they are not only limited by the uncertainties of the physical stimuli, but may also be limited by uncertainties in the coding and representation process, have limited memory, and the computations themselves may have an associated cost. This means, that generally it is not the case that human performance is optimal as that of an ideal observer, but it simply means that it is tested whether the brain executes computations by taking uncertainties into account. Nevertheless, it is worth stressing the importance of this basic fact and its historic relevance, as previous research, e.g., by Kahneman and Tversky (2000) showed, that in many cognitive tasks, humans do not correctly take the respective uncertainties into account. This means that subjects in cognitive decision tasks fail to minimize expected loss but in mathematically equivalent movement tasks often choose nearly optimal (Trommershäuser et al., 2008).

Unfortunately, the normative formulation also has a number of limitations. First, while it is important to formulate a model of optimal behavior, such models are always bound by their respective assumptions. As an example, most cue integration studies assume that the underlying probability distributions are implicitly known by the subjects, but it is not clear, how these distributions are learned. Accordingly, many of such studies do not report and are not interested in the actual learning process during the cue integration experiments. An interesting aspect of this problem is, that the often referred to concept of internal noise can actually veil a number of possible sources of uncertainty. As one example, if a human subject used a likelihood function that is not the correct one describing the generative process of the data, performance on an inference task may be suboptimal and the experimenter might interpret this as evidence for internal noise.

An optimality formulation may seem to be straightforward to come about for a range of experiments, in which the experimenter can fully control impoverished and constrained stimuli. But in Bayesian terms it is often possible to model the respective data in several ways. Consider a linear cue integration paradigm. The assumption in such experiments is, that the system knows the respective uncertainties in the underlying Gaussian distributions. But presumably, in experiments where these uncertainties have to be learned, these uncertainties need to be estimated over trials. In a Bayesian model, it would be necessary to model the task with explicitly representing the uncertainties over the parameters, i.e., one could model each cue in terms of a Gaussian distribution with unknown mean and unknown variance and introduce a distribution over the two unknown parameters, such as an Normal-scaled inverse gamma distribution. But many alternatives exist, and it is currently not clear, how to decide which model best describes human and animal performance.

Furthermore, it may be straightforward to write down a generative model for a specific task, but the inversion of the model, which is required for the inference process, may be a rather daunting task that requires extensive approximations. For example, in model selection it is often difficult to calculate the posterior distribution of the data given different models and usually approximation techniques have to be used. While it has been suggested that the brain could carry out computations such as Markov chain Monte Carlo (MCMC) (Hinton and Sejnowski, 1986; Hoyer and Hyvärinen, 2003) and recent work has used such algorithms to model subjects' behavior in cognitive tasks (Sanborn et al., 2006), this remains an open question that will require further research.

3.4.2 Intermediate Models of Multimodal Object Perception

There is considerable literature that has been interested in modeling multisensory object perception at an intermediate level of description. This discussion has been partly motivated by attempting to map algorithms from the domain of machine perception and especially computer vision to human visual perception. It has also been motivated by the hypothesis that individual areas along the cortical hierarchy compute separate perceptual estimates. A central topic in this literature has been that of computational modules. If one assumes that distinct cortical regions are responsible for specific computations involved in the extraction of distinct features, it is natural to ask how the individual results of computations can be combined to obtain a globally optimal estimate of a scene parameter. While such models have also employed Bayesian techniques, the emphasis here is more on how an optimal estimate of a scene parameter may be obtained if specific separate computations by individual modules have to be combined to a globally optimal estimate.

Yuille and Bülthoff (1996) consider the case where the posterior distribution of a scene variable given two cues is proportional to the product of the two posterior distributions given the individual cues, i.e., $P(X\,|Y_1, Y_2) = c\,P(X\,|Y_1)\;P(X\,|Y_2)$. In terms of the ideal observer analysis, note that in cue combination paradigms with independent Gaussian noise an estimate for the unknown scene property can be computed by obtaining the mean and variance of the posterior for each individual cue separately. These separate estimates can be weighted linearly to obtain the mean and variance of the maximum a posteriori combined estimate. The question therefore arises whether these types of calculations are carried out separately first before combination or whether these computations are always carried out jointly.

Consider an example by Ernst and Bülthoff (2004) in which a human subject estimates the location of its hand knocking on a surface. There are visual, auditory, and proprioceptive cues to this location, but the signals are represented in different coordinate frames and units. The auditory and visual signals have to be combined with the postural signals of the head into a coordinate frame that allows combination with the proprioceptive signal from the knocking hand. Again, the question arises whether the auditory and visual signals are first separately transformed into different coordinate systems followed by a combination with the proprioceptive signal. The alternative is that these computations are done by a single calculation in which all separate signals are directly combined to an optimal estimate. In this context, weak fusion refers to independent processing of individual cues followed by a linear combination of these estimates, while strong fusion refers to computations in which the assumed cue processing modules interact nonlinearly. Extensions to these types of models have been proposed in the literature (Landy et al., 1995; Yuille and Bülthoff, 1996).

While it is important to ask how the brain actually carries out such computations, it should be noted that the ideal observer framework can accommodate these views in a straightforward way by using the Bayesian network formalism. Considering the above example, if the sensory noise in the two cues is conditionally independent, the ideal observer model expresses this as a factorized distribution, i.e., $P(Y_1 | X)P(Y_2 | X)$. If instead the noise is not independent, the full joint probability distribution $P(Y_1, Y_2 | X)$ needs to be specified. Thus, weak and strong fusion can be matched to factorizations underlying the generative model of the data. Similarly, nonlinearities can be introduced by marginalization computations when discounting scene variables that are not of interest in the task but determine the values of the observed cues. Finally, internal computations such as sensorimotor transformations can be incorporated in Bayesian ideal observers by choosing appropriate probability distributions in the corresponding likelihoods of cues or explicitly as an additional intermediate random variable.

3.4.3 Neural Models Implementing Multimodal Perception

There is a line of research that has pursued the modeling of multimodal perception at the level of individual neurons and networks of neurons. This research has often approached this goal by looking at response properties of neurons in specific brain regions and trying to obtain similar behavior starting from neuronal dynamics and learning rules. Work in this category has, e.g., considered the multisensory responses of single neurons in the superior colliculus, which only fire when signals from different modalities are close in time and space (Meredith and Stein, 1986; Wallace et al., 1996). Some neurons also show an enhanced response to a multisensory stimulation that is larger than the linear sum of the responses to the individual signals, often called Multisensory Enhancement (MSE) (e.g., Alvarado et al. 2007). Models by Anastasio and Patton (Anastasio and Patton, 2003; Patton and Anastasio, 2003) try to explain the development of MSE through a two-stage unsupervised learning mechanism, which includes cortical modulation of the sensory inputs. Rowland, Standord, and Stein (2007) developed a model of the processes in a single multisensory neuron that also shows MSE. It is motivated by the characteristics of NMDA and AMPAR receptors in neurons as well as inspired by the fact that superior colliculus neurons obtain their inputs from 'higher' cortical areas as well as from sensory areas. These two models used artificial inputs and looked at a limited set of processes, but they nevertheless show that some computations for a possible implementation of Bayesian principles in the brain could be accomplished with basic neuronal mechanisms (Anastasio et al., 2000).

The success of Bayesian models in accounting for a large number of experimental data has lead to the more general idea that neural computations must be inherently probabilistic and maybe even explicitly represent probability distributions. Accordingly, a different line of theoretical research has started investigating models in which neuronal activity represents and computes with probability distributions. In principle, the reviewed studies have demonstrated that human and primate behavior is taking the relative uncertainties of the cues into account. This does not imply that the computations carried out in the brain have to be Bayes optimal, but at least this implies that the relative uncertainties have to be represented somehow. Nevertheless, major efforts are currently directed at proposing neural representations that explicitly encode uncertainties in sensory representations together with suggestions how computations such as cue integration, maximum-likelihood estimation, and Bayes optimal model selection could be carried out.

Coding schemes have been proposed in which the activity of populations of neurons directly represents probability distributions. We briefly review the key ideas of such coding models and refer the interested reader to the review by Knill and Pouget (2004) and the book by Doya et al. (2007) for more detailed descriptions. The starting point of the proposed coding schemes is a new interpretation of the probabilistic nature of neuronal responses. Responses of neurons to the repeated presentations of a stimulus to which they are tuned show considerable variability from trial to trial, i.e., their activity has been characterized as being noisy. This response variability has now been thought instead to be inherently related to the probabilistic nature of the stimuli itself. Several models have proposed different ways of how to map this idea to different random variables, which can be continuous or discrete.

A straightforward model is to have a neuron's firing rate be some function of the probability of an event. Gold and Shadlen (2001) have proposed such a model based on neurophysiological data obtained from LIP neurons during a motion discrimination tasks, in which monkeys had to decide which of two possible motion directions a random dot kinematogram had. Decisions taken by the monkey could be predicted by a decision variable, which was calculated as the logarithm of the difference between firing rates of neurons. By relating this decision variable to the probability of motion direction, this result was taken as evidence for a direct encoding of a function of the posterior probability of motion direction given sensory stimuli. Deneve (2005) extended this model to spiking neurons, in which each neuron represents the ratio of the logarithms of the probability of two preferred states and the neuron only fires a spike, if the difference between a prediction and a current observation is exceeded.

A different direction in modeling encodings of probability distributions is motivated by the fact that large populations of neurons jointly represent sensory variables. Individual neurons in such populations are distinct from each other by their respective tuning curve, i.e., by the mean firing rate as a function of the considered stimulus value. As an example, V1 simple cells' activities vary with the orientation of a stimulus and individual cells have different preferred orientations. There exist two closely related representations of probability distributions over scene parameters referred to as 'convolution encoding' and 'convolution decoding.' Both proposals are based on the mathematical idea of a change of bases. In essence, one can approximately synthesize a function or probability distribution by a linear sum of individual prototype functions, so-called basis functions. In order to find the linear coefficients for the sum of basis functions to return the original function, this has to be projected down onto the new basis set.

In both convolution encoding (Zemel et al., 1998) and convolution decoding (Anderson et al., 1994) a neuron's activity is related to an approximation of the true posterior $P(X\,|\,Y)$ using basis functions, which are often chosen to be Gaussians. While in convolution decoding neuronal firing rates represent the coefficients that are multiplied with their associated basis functions to give their respective contributions to the approximate posterior, in convolution encoding the posterior is projected down onto individual neuron's tuning curves. Because neuronal activity is modeled as being stochastic, the additional Poisson noise ends up having different influences on the resulting probabilistic encoding and decoding schemes. One of the well-known difficulties with these coding schemes is that highly peaked distributions corresponding to small amounts of uncertainty get much broader. Nevertheless, considerable work has investigated how computations necessary for multimodal inference could be carried out on the basis of these coding schemes.

A recent approach to the representation of probability distributions and importantly also to inference computations with these representations in neurons comes from the work of Ma et al. (2006). The fundamental idea of this probabilistic population code is that the activity of a population of neurons will at some point be decoded to obtain a posterior distribution over the stimulus dimension X given this neuronal population activity r. Given that the neuronal response is r, this can be formulated as inference of the stimulus value X based on Bayes theorem:

$$P(X\,|r) = \frac{P(r\,|X)P(X)}{P(r)} \tag{3.6}$$

The common way of quantifying neuronal population activity is by tuning curves describing the activity of individual neurons to the stimulus value. Given the near-Poisson variability across repeated trials, this mean response value given by the tuning curves corresponds to the variance of the activity, given that mean and variance have the same value for the Poisson distribution. Under certain conditions, such as independent Poisson variability across neurons, the posterior $P(X|r)$ converges to a Gaussian distribution, whose mean is close to the stimulus at which the population activity has its peak, while the variance in the posterior is inversely proportional to the amplitude of the activity hill. So, a stimulus with high uncertainty corresponds to a small height of the hill of activity, whereas low uncertainty corresponds to large amplitude.

In a situation requiring cue integration, the brain needs to compute an estimate from two such hills of activity and reach a result that corresponds to Eq. (3.2). Ma et al. (2006) showed that this point-by-point multiplication of probability distributions could be carried out simply by adding the two neuronal populations in the probabilistic population code framework. Furthermore they showed that only Poisson-like neuronal variability allows for such a computationally straightforward scheme, which can easily be implemented at the neuronal level. Thus, the interesting result from this model is that it provides a new interpretation of the near-Poisson variability of neuronal activities as a basis for computing with uncertainties. It is important to note, however, that these proposals encode those types of uncertainties that are readily given in the stimulus such as the uncertainty of a visual stimulus by virtue of its contrast. This is different from situations in which cues have an associated uncertainty that is in turn inferred from the reliability of a cue's statistic on a very fast timescale (Triesch et al., 2002) or when the reliabilities are learned over longer timescales (Fine and Jacobs, 1999).

3.5 Open Questions

3.5.1 How Are Model Parameters Learned?

If the brain is capable of computing with uncertainties in multisensory signals according to Bayesian inference, then it is of interest to understand where the parameters of the probability distributions in the generative model describing the relationship between the hidden causes and the observations come from. A variety of techniques in machine learning allow the computation of the parameters in a given generative model, if the model is somehow specified a priori (e.g., Bishop, 2006; MacKay, 2003). Maximum-likelihood techniques allow setting the parameters such that the probability of observed data is maximized under the given model. This can also be achieved, if some variables in the fully specified model are unobserved or latent by using the expectation maximization algorithm (EM) or related techniques.

The EM algorithm iteratively alternates between an expectation (E) step, which computes an expectation of the likelihood given the current guesses for the latent variables and a maximization (M) step, which maximizes the expected likelihood found in the E step by adjusting the parameters. Repeating this procedure can be shown to increase the likelihood of the data given the parameters, but may only find a local minimum in the likelihood. But how the brain can accomplish such computations is still unknown.

In case of the audio-visual localization task, the brain must somehow find the parameters of the Gaussian distributions that describe the uncertainty in the auditory and the visual location measurements. In a laboratory version of this task the uncertainties can be varied, e.g., by changing the contrast of the visual stimulus or the loudness of the sound. This means that the system needs to compute with the uncertainties associated with the respective contrast and loudness. Similarly, given that acuity is dependent on the eccentricity, these uncertainties are also dependent on the relative position of the stimuli with respect to the auditory and visual foveae. Furthermore, for ecologically valid contexts, these distributions are often dependent on many more parameters that the system needs to take into account.

A further complication under natural conditions is that the relative reliabilities of cues may change over time and could also be context dependent. Similarly, individual cues could acquire biases over time. Studies by Jacobs and Fine (1999) and Ernst et al. (2000) have shown that humans are able to adjust the relative cue reliabilities and biases when multiple cues are available and one of the individual cues is perturbed artificially. While these studies required subjects to learn to adjust the relative cue weights over timescales of hours and days, a study by Triesch et al. (2002) showed that such adaptation processes could also be observed on the timescale of seconds. In their experiment subjects had to track individual objects in a display containing several distractors and detect the target object after an occlusion. Objects were distinguishable by the three cues of color, shape, and size. While being occluded, some of the features defining the object identity were changed. By adapting to the relative reliabilities of individual features, subjects reweighed the cues depending on their respective reliabilities on the timescale of 1 s.

3.5.2 How Are the Likelihood Functions Learned?

Similarly, but even more difficult to answer with current knowledge is the question of how the brain is able to learn the likelihood functions describing the relationship between individual scene parameters. One of the main tasks of the modeler is often to come up with the correct likelihood function but it is not clear how the primate brain accomplishes this feat. As an example, how does the brain know that texture information about the slant of a planar surface is less reliable at low slants than at high slants (Knill and Saunders, 2003)? And how does the brain know the uncertainty in orientation of a visual stimulus as a function of its contrast?

3.5.3 Where Does the Prior Come From?

If the system is able to use prior distributions over scene parameters it must have stored such knowledge. Although it is conceivable that some such prior distributions are encoded genetically, there is also evidence that they can change due to learning. Adams et al. (2004) used the well-known human prior in the judgment of shape from shading which prefers interpretations of ambiguous two-dimensional shaded image regions as being lit from above. Subjects' shape judgments changed over time with visuo-haptic training, in which the haptic stimuli were consistent with a light source shifted by 30°. Moreover, a different subsequent lightness judgment task revealed that subjects transferred the shifted prior on illumination direction to new tasks. This suggests that subjects can learn to adjust their prior on illumination direction from visual–haptic stimuli. Similarly, Knill (2007) showed that humans learn to adapt their prior, which favors the interpretation of ellipses as slanted circles. The data was modeled qualitatively with a two-step Bayesian model in which a first process uses the current prior over elliptical shapes to estimate the aspect ratio of the currently presented ellipse. A second process then updates the current prior using the estimated shape of the viewed ellipse.

Additional complexities in learning priors can be found in considering the case of motion perception from local estimates of image motion. While the work by Weiss et al. (1999) shows that human experimental data can be well explained by assuming a prior distribution that favors velocities near zero, it is not clear how such a prior is learned, if at all. One could hypothesize that such a prior is learned only on the basis of the statistics of image motion but it is also conceivable that it is learned after image segmentation and by compensating for the image motion induced by self-motion of the observer, i.e., that the prior is learned on the basis of the estimated motion distributions of objects in the world.

While the vast majority of work on the origin of prior distributions has focused on characterizing the statistics of properties of the natural environment such as luminance, contrast, disparity, and sound wave amplitude, these investigations have almost exclusively characterized these distributions without taking into account the active exploration of the environment. But the actual input to the sensory system is without doubt dependent on the ongoing behavior, which often samples the environment in structured ways. Rothkopf and Ballard (2009) showed that the usage of the sensory system crucially influences the statistics of the sensory input to the visual system in a task-dependent manner and that therefore the input to the visual system significantly depends on behavior. Thus, when characterizing the prior distributions of scene variables, which are considered to be the input to the sensory system, it is important to take the organism's active behavior into account.

Indeed, a further study by Rothkopf et al. (2009) explored how the sensory apparatus itself and its active use during behavior determine the statistics of the input to the visual system. A virtual human agent was simulated navigating through a wooded environment under full control of its gaze allocation during walking. Independent causes for the images obtained during navigation were learned across the visual field with algorithms that have been shown to extract computationally useful representations similar to those encountered in the primary visual cortex of the mammalian brain. The distributions of properties of the learned simple cell-like units were in good agreement with a wealth of data on the visual system including the oblique effect, the meridional effect, properties of neurons in the macaque visual cortex, and functional magnetic resonance imaging (fMRI) data on orientation selectivity in humans and monkeys. Crucially, this was only the case if gaze was allocated overwhelmingly in the direction of locomotion, as is the case in natural human walking. But when gaze was directed mostly to a point on the ground plane, the distributions of properties of the learned simple cells differed significantly from those consistent with the empirical findings in humans and primates.

3.5.4 How Does the Brain Learn the Appropriate Generative Model, If at All?

This question extends the above points in that the full generative model that is required for inferring unobserved scene parameters obviously includes the structure of the variables describing the scene and all available cues. For example, one needs to know that shading is dependent on the illumination direction but independent from self-motion. With the very large number of multisensory variables that primates are capable of extracting from their environment this is a daunting task. How such independencies are acquired is still unknown. Again, the field of machine learning offers algorithms for learning the structure of a generative model given observed data. In its most general form this problem is NP hard and no analytical solution is available. Under certain restrictions on the types of possible models, i.e., on the way individual variables are independent of others, there are algorithms that allow for a more efficient calculation of the probability of observing the given data under a particular model, thus allowing selecting the best fitting model. Approximate structure learning in Bayesian Network models is an area of active research.

It is also important to note that some experimental data does not conclusively answer the question of whether subjects use the correct likelihood functions in the first place. As an example, a study by Oruç et al. (2003) showed that there are significant differences between individual subjects that can be explained by the assumption that they used different cue integration strategies. The study looked at human estimation of slant from linear perspective and texture gradient cues. The idea was that the noise in the two cues may be correlated, as the same visual neurons were involved in representing both cues. The behavioral data was best explained by the assumption that some subjects used a cue combination rule consistent with independent noise in the cues while others assumed correlated noise. This study also points toward the problem that it is difficult to assess whether a subject uses exactly a specific generative model in an inference task, as errors due to an approximate or simply wrong likelihood model could look like noise.

A further question that arises from these studies is whether human subjects can in principle learn arbitrary models. Two studies by Ernst (2007) and by Michel and Jacobs (2007) investigated this question. The idea was to involve human participants in a learning study in which they were exposed to artificial cue congruencies that are not encountered in natural environments and to test whether these cues were combined in a subsequent cue combination task. The former study used visual and haptic stimuli while the latter used several visual and auditory cues. Interestingly, while the former study found significant effects of learning and concluded that humans can learn arbitrary signals across modalities, the second study came to the conclusion that humans can learn the parameters of the implicit generative models, but not novel structures such as causal dependencies between usually independent variables.

A different approach to this problem was recently proposed by Weisswange et al. (2009). The task in this case is again an audio-visual orienting task, in which the auditory and visual sources may coincide or may be originating from different sources. A learner is rewarded for orienting toward the true location of the object on repeated trials, in which the two sensory stimuli can appear at different positions. This study used a basic reinforcement learner, i.e., a system that learns to do optimal control on the basis of exploring the environmental dynamics and obtaining feedback from the environment about the success of its actions. The optimal control in this case is to orient toward the true position of the light source. Interestingly, after learning, the agent was able to orient toward the most likely position of the target by taking the relative reliabilities of the auditory and visual cues into account for the integration. Furthermore, when the two sources are further apart, the agent learned to not integrate the two cues, but to rely only on the less uncertain visual measurement. Thus, although this study does not conclusively show that reinforcement learning alone is the basis for learning cue integration and causal inference, it provides a mechanism by which such abilities could be learned from interacting with the environment.

3.5.5 How Are Different Models Combined?

Recent work on causal inference in multisensory perception (Koerding et al., 2007) has suggested that humans have multiple generative models that describe data for different scene configurations. In their experiments, subjects are modeled as computing the likelihood of an observed flash and a tone under the assumption that these came from the same source (Fig. 3.2a) and under the assumption that they came from two different sources (Fig. 3.2b). While under such laboratory situations it is possible to construct very controlled stimuli that are compatible only with few potential scene configurations, in ecologically valid situations the number of potential models that have to be compared grows quickly. How the brain may choose the appropriate models for comparison or how the brain may learn the appropriate models is an open question.

Furthermore, model selection may not be the optimal way of combining different models. Indeed, the Bayesian method of combining the estimates of different models is again to weigh the different estimates according to the uncertainty associated with the data given each model. To make things even more complicated, it is in principle also possible to assume that the parameters of individual models change over time. This requires the system to obtain additional data over trials in order to estimate the changes in the parameters. Under such circumstances it is possible to develop models that use a strategy of probability matching. Currently, there is conflicting evidence as to what primates do under such circumstances.

3.5.6 Does the Brain Represent Full Probability Distributions or Implicit Measures of Uncertainties?

The jury on how the brain computes with uncertainties is still out. Some empirical studies have tried shedding more light on this issue. Work by Körding and Wolpert (2004) on sensorimotor learning used a task in which human subjects had to reach to a visual target with their index finger in a virtual reality setup. Visual feedback about the finger's position was given only through a set of light dots. This allowed introducing a conflict that displaced the center of the light cloud relative to the true position of the index finger. The lateral shifts on single trials were obtained as samples from different complex probability distributions in different learning experiments. Some subjects were exposed to shifts drawn from a Gaussian distribution, while others were exposed to shifts coming from a mixture of two or three Gaussian distributions. The overall result of the experiments was that subjects took the uncertainties in the visual feedback and of the distribution of applied shifts into account in a way that was compatible with a Bayesian strategy. But subjects did not behave optimally when a complex distribution such as the mixture of three Gaussians was used, at least within the allotted number of approximately 2000 trials given by the experimenters. This suggested to the authors that there are limits to the representation or computation of arbitrary probability distributions describing the uncertainties in a task.

At the neuronal level, the models of probabilistic population codes described above propose different ways how populations of neurons could encode full probability distributions or functions of them. Further research will need to develop experiments that can disambiguate different proposals. It should also be noted that there is evidence that the brain may have developed a number of different codes depending on the required task or stimulus dimension. Work by Ahissar and colleagues (see Knutsen and Ahissar, 2008 for a review) has demonstrated that tactile encoding of object location in rodents uses timing and intensity of firing separately for horizontal and radial coordinates. Accordingly, it may very well be that the brain uses different solutions to the problem of representing uncertainties.

3.5.7 How Ideal Is Human Learning?

The vast majority of cue combination experiments report averaged behavioral data and experiments that include a learning component rarely report individual learning curves. Indeed, the performance of humans is far from the Bayesian ideal observer in terms of utilizing the available information acquired during learning experiments, if one applies the true generative model from which the stimuli were generated. The problem here is that it is not exactly known what model is actually used by the brain or whether the brain maintains several models in parallel. These and similar observations from the animal learning literature have led to Bayesian learning models in the areas of cognitive tasks (Sanborn et al., 2006) and animal conditioning (Daw and Courville, 2008) that explicitly model how individual decisions on a trial may be modeled by assuming that the brain does inference using sequential Monte Carlo sampling. The idea is that, instead of maintaining a full belief distribution, at each decision one sample from one of the current hypotheses guides the decision. Further research will be needed to clarify what methods the primate brain actually implements.

3.5.8 How Do Laboratory Experiments Transfer to Natural Tasks?

In ecologically valid situations there is an abundance of 'cues,' e.g., there are at least a dozen depth cues in natural, rich scenes. While cue integration experiments have demonstrated that two or three such depth cues are integrated into a percept that takes the inverse variances into account, it is not known whether this strategy also transfers to the cases of rich scenes, in which a few cues are much more reliable than others. Are really all cues always integrated by default? If computations do not have an intrinsic associated cost or, equivalently, there are no restrictions on the durations for the computations to be terminated, it is of course of considerable advantage to have a multitude of different cues. Under such conditions, the respective biases and uncertainties can be corrected for by recalibrating those cues that disagree most with results that have been successful in the recent past. This is exactly the approach of Triesch and von der Malsburg (2001). In their democratic cue integration model, a multitude of individual sensory cues are brought into agreement and each cue adapts toward the agreed upon result.

3.5.9 The Role of Approximations

While ideal observer models often use techniques for approximate Bayesian inference, because there are no analytic solutions available, the role of approximations with regard to the brain has been underestimated. It is currently unknown how the brain accomplishes computations that are in accordance with Bayesian inference. But it may very well be that the brain is confronted with similar difficulties as researchers in machine learning in carrying out such computations and that the approximations that are used to solve the inference problems are of central importance. Similarly, the involved computations themselves may have different effects on the associated uncertainties. Thus, what are the approximations that the brain uses? Can these approximations be expressed in terms of cost functions, i.e., are there intrinsic behavioral costs that influence what type of approximations are adequate?

One particular uncertainty introduced by computations carried out by the brain itself stems from sensorimotor transformations. Consider again the earlier example of a task requiring the system to integrate two position estimates that are represented in different coordinate frames, such as an eye-centered and a head-centered reference frame. In order to plan and execute a reach, such information has to be remapped into a common coordinate frame, which has to take into account the noisy proprioceptive estimates of the involved joint angles. These computations introduce additional coordinate transformation uncertainty. A study by Schlicht and Schrater (2007) investigated whether human reaching movements take these uncertainties into account and found evidence that subjects adjusted their grip aperture reflecting the uncertainties in the visual stimuli and the uncertainties due to coordinate transformations. The behavioral human data was well accounted for by a Bayesian model built on the assumption that internal spatial representations used an eye-centered reference frame for this task. From a theoretical point of view, there are many more instances of uncertainties that are the result of computations and further research may reveal additional insight into whether the nervous system takes these into account. Considerable further research will have to determine the experimental and analytical tools required to address these questions.

3.6 Conclusion

Models of multisensory object perception have been developed over the last decades that have had great success at explaining a multitude of behavioral data. The main reason for this success has been to recur to models that not only explicitly represent a sensory estimate but also the uncertainties associated with them. Bayesian techniques have been especially successful, as they allow for the explicit representation of uncertainties of model parameters. Nevertheless, a wealth of open questions remains to be answered, ranging from how humans and primates learn parameters and models connecting sensory stimuli and world properties to how the neuronal substrate may compute with uncertainties.

References

Adams WJ, Graf EW, Ernst MO (2004) Experience can change the 'light-from-above' prior. Nat Neurosci 7(10):1057–1058

Alais D, Burr D (2004) The ventriloquist effect results from near-optimal bimodal integration. Curr Biol 14(3):257–262

Alvarado JC, Vaughan JW, Stanford TR, Stein BE (2007) Multisensory versus unisensory integration: contrasting modes in the superior colliculus. J Neurophysiol 97(5): 3193–3205

Anastasio TJ, Patton PE (2003) A two-stage unsupervised learning algorithm reproduces multisensory enhancement in a neural network model of the corticotectal system. J Neurosci 23(17):6713–6727

Anastasio TJ, Patton PE, Belkacem-Boussaid K (2000) Using Bayes' rule to model multisensory enhancement in the superior colliculus. Neural Comput 12(5):1165–1187

Anderson CH, Van Essen DC (1994) Neurobiological computational systems. In: Zureda JM, Marks RJ, Robinson CJ (eds) Computational intelligence imitating life. IEEE Press, New York, pp 213–222

Atkins JE, Fiser J, Jacobs RA (2001) Experience-dependent visual cue integration based on consistencies between visual and haptic percepts. Vision Res 41(4):449–461

Battaglia PW, Jacobs RA, Aslin RN (2003) Bayesian integration of visual and auditory signals for spatial localization. J Opt Soc Am A Opt Image Sci Vis 20(7):1391–1397

Battaglia PW, Schrater P, Kersten D (2005) Auxiliary object knowledge influences visually-guided interception behavior. In: Proceedings of the 2nd symposium on applied perception in graphics and visualization, ACM International Conference Proceeding Series. ACM, New York, NY, pp 145–152

Bernoulli D.; Originally published in 1738; (January 1954). "Exposition of a New Theory on the Measurement of Risk". Econometrica 22(1): 22–36 (trans: Lousie Sommer)

Beierholm U, Kording K, Shams L, Ma WJ (2008) Comparing Bayesian models for multisensory cue combination without mandatory integration. Advances in neural information processing systems 20. MIT Press, Cambridge, MA, vol. 1, pp 81–88

Bishop CM (2006) Pattern recognition and machine learning. Springer, Heidelberg

Bizley JK, Nodal FR, Bajo VM, Nelken I, King AJ (2007) Physiological and anatomical evidence for multisensory interactions in auditory cortex. Cereb Cortex 17(9):2172–2189

Bruce C, Desimone R, Gross CG (1981) Visual properties of neurons in a polysensory area in superior temporal sulcus of the macaque. J Neurophysiol 46(2):369–384

Calvert GA, Bullmore ET, Brammer MJ, Campbell R, Williams SC, McGuire PK, Woodruff PW, Iversen SD, David AS (1997) Activation of auditory cortex during silent lipreading. Science 276(5312):593–596

Daw N, Courville A (2008) The pigeon as particle filter. In: Advances in neural information processing systems 20 (NIPS 2007). MIT Press, Cambridge, MA, pp 369–376

Deneve S (2005) Bayesian inferences in spiking neurons. In: Advances in neural information processing systems 17 (NIPS 2004). MIT Press, Cambridge, MA, pp 353–360

Doya K, Ishii S, Pouget A, Rao RPN (2007) The Bayesian brain: probabilistic approaches to neural coding. MIT Press, Cambridge, MA

Ernst MO (2007) Learning to integrate arbitrary signals from vision and touch. J Vis 7(5):7.1–7.14

Ernst MO, Banks MS (2002) Humans integrate visual and haptic information in a statistically optimal fashion. Nature 415(6870):429–433

Ernst MO, Banks MS, Bülthoff HH (2000) Touch can change visual slant perception. Nat Neurosci 3:69–73

Ernst MO, Bülthoff HH (2004) Merging the senses into a robust percept'. Trends Cogn Sci 8(4):162–169

Fellemann DJ, Van Essen DC (1991) Distributed hierarchical processing in the primate cerebral cortex. Cereb Cortex 1(1):1–47

Fine I, Jacobs RA (1999) Modeling the combination of motion, stereo, and vergence angle cues to visual depth. Neural Comput 11(6):1297–1330

Finney EM, Fine I, Dobkins KR (2001) Visual stimuli activate auditory cortex in the deaf. Nat Neurosci 4(12):1171–1173

Foxe JJ, Morocz IA, Murray MM, Higgins BA, Javitt DC, Schroeder CE (2000) Multisensory auditory-somatosensory interactions in early cortical processing revealed by high-density electrical mapping. Brain Res Cogn Brain Res 10(1–2):77–83

Frens MA, Van Opstal AJ, Van der Willigen RF (1995) Spatial and temporal factors determine auditory-visual interactions in human saccadic eye movements. Percept Psychophys 57(6): 802–816

Geisler WS, Perry JS, Super BJ, Gallogly DP (2001) Edge co-occurrence in natural images predicts contour grouping performance. Vision Res 41(6):711–724

Ghazanfar AA, Maier JX, Hoffman KL, Logothetis NK (2005) Multisensory integration of dynamic faces and voices in rhesus monkey auditory cortex. J Neurosci 25(20):5004–5012

Gibson JR, Maunsell JH (1997) Sensory modality specificity of neural activity related to memory in visual cortex. J Neurophysiol 78(3):1263–1275

Gielen SC, Schmidt RA, Van den Heuvel PJ (1983) On the nature of intersensory facilitation of reaction time. Percept Psychophys 34(2):161–168

Gold JI, Shadlen MN (2001) Neural computations that underlie decisions about sensory stimuli. Trends Cog Sci 5:10–16

Gori M, Del Viva M, Sandini G, Burr DC (2008) Young children do not integrate visual and haptic form information. Curr Biol 18(9):694–698

Greenwald HS, Knill DC (2009) A comparison of visuomotor cue integration strategies for object placement and prehension. Vis Neurosci 26(1):63–72

Hagen MC, Franzén O, McGlone F, Essick G, Dancer C, Pardo JV (2002) Tactile motion activates the human middle temporal/V5 (MT/V5) complex. Eur J Neurosci 16(5):957–964

Hairston WD, Wallace MT, Vaughan JW, Stein BE, Norris JL, Schirillo JA (2003) Visual localization ability influences cross-modal bias. J Cogn Neurosci 15(1):20–29

Helmholtz H von (1867) Handbuch der physiologischen Optik. Brockhaus, Leipzig

Hershenson M (1962) Reaction time as a measure of intersensory facilitation. J Exp Psychol 63:289–293

Hinton GE, Sejnowski TJ (1986) Learning and relearning in Boltzmann machines, In: Rumelhart DE, McClelland JL (eds) Parallel distributed processing explorations in the microstructure of cognition volume foundations. MIT Press, Cambridge, MA

Hoyer PO, Hyvärinen A (2003) Interpreting neural response variability as Monte Carlo sampling of the posterior. In: Advances in neural information processing systems 15 (NIPS*2002). MIT Press, Cambridge, MA, pp 277–284

Jacobs RA (1999) Optimal integration of texture and motion cues to depth. Vision Res 39(21):3621–3629

Jacobs RA, Fine I (1999) Experience-dependent integration of texture and motion cues to depth. Vision Res 39(24):4062–4075

James TW, Humphrey GK, Gati JS, Servos P, Menon RS, Goodale MA (2002) Haptic study of three-dimensional objects activates extrastriate visual areas. Neuropsychologia 40(10): 1706–1714

Jones EG, Powell TP (1970) An anatomical study of converging sensory pathways within the cerebral cortex of the monkey. Brain 93(4):793–820

Jousmaki V, Hari R (1998) Parchment-skin illusion: sound-biased touch. Curr Biol 8(6): R190–R191

Kersten D (1999) High-level vision as statistical inference. In: Gazzaniga MS (ed) The new cognitive neurosciences, 2nd edn. MIT Press, Cambridge, MA, pp 352–364

Kahneman D, Tversky A (2000) Choices, values, and frames. Cambridge University Press, New York, NY

Kersten D, Mamassian P, Yuille A (2004) Object perception as Bayesian inference. Annu Rev Psychol 55:271–304

Kersten D, Yuille A (2003) Bayesian models of object perception. Curr Opin Neurobiol 13(2): 150–158

Knill DC (2003) Mixture models and the probabilistic structure of depth cues. Vision Res 43(7):831–854

Knill DC (2007) Learning Bayesian priors for depth perception. J Vis 7(8):13

Knill DC, Kersten D (1991) Apparent surface curvature affects lightness perception. Nature 351(6323):228–230

Knill DC, Saunders JA (2003) Do humans optimally integrate stereo and texture information for judgments of surface slant? Vision Res 43(24):2539–2558

Knill DC, Pouget A (2004) The Bayesian brain: the role of uncertainty in neural coding and computation. Trends Neurosci 27(12):712–719

Knill DC, Saunders JA (2003) Do humans optimally integrate stereo and texture information for judgments of surface slant? Vision Res 43:2539–2558

Knutsen PM, Ahissar E (2008) Orthogonal coding of object location. Trends Neurosci 32(2): 101–109

Koerding KP, Beierholm U, Ma WJ, Quartz S, Tenenbaum JB, Shams L (2007) Causal inference in multisensory perception. PLoS One 2(9):e943

Körding KP, Wolpert D (2004) Bayesian integration in sensorimotor learning. Nature 427:244–247

Kujala T, Huotilainen M, Sinkkonen J, Ahonen AI, Alho K, Hämäläinen MS, Ilmoniemi RJ, Kajola M, Knuutila JE, Lavikainen J, Salonend O, Simolab J, Standertskjöld-Nordenstamd, C-G, Tiitinena H, Tissarie SO, Näätänen R (1995) Visual cortex activation in blind humans during sound discrimination. Neurosci Lett 183(1–2):143–146

Landy MS, Maloney LT, Johnston EB, Young M (1995) Measurement and modeling of depth cue combination: in defense of weak fusion. Vision Res 35(3):389–412

Lewkowicz DJ (2000) Perceptual development in human infants. Am J Psychol 113(3):488–499

Lomo T, Mollica A (1959) Activity of single units of the primary optic cortex during stimulation by light, sound, smell and pain, in unanesthetized rabbits. Boll Soc Ital Biol Sper 35:1879–1882

Ma WJ, Beck JM, Latham PE, Pouget A (2006) Bayesian inference with probabilistic population codes. Nat Neurosci 9(11):1432–1438

MacKay D (2003) Information theory, inference, and learning algorithms. Cambridge University Press, New York, NY

Mamassian P, Knill DC, Kersten D (1998) The perception of cast shadows. Trends Cogn Sci 2(8):288–295

Mamassian P, Landy MS (2001) Interaction of visual prior constraints. Vision Res 41(20): 2653–2668

Marr D (1982) Vision: a computational investigation into the human representation and processing of visual information. W.H. Freeman & Co., San Francisco

McGurk H, MacDonald J (1976) Hearing lips and seeing voices. Nature 264(5588):746–748

Meredith MA, Stein BE (1986) Spatial factors determine the activity of multisensory neurons in cat superior colliculus. Brain Res 365(2):350–354

Michel MM, Jacobs RA (2007) Parameter learning but not structure learning: a Bayesian network model of constraints on early perceptual learning. J Vis 7(1):4

Morrell F (1972) Visual system's view of acoustic space. Nature 238:44–46

Murata K, Cramer H, Bach-y-Rita P (1965) Neuronal convergence of noxious, acoustic, and visual stimuli in the visual cortex of the cat. J Neurophysiol 28(6):1223–1239

Nardini M, Jones P, Bedford R, Braddick O (2006) Development of cue integration in human navigation. Curr Biol 18(9):689–693

Neumann Jv, Morgenstern O (1944) Theory of games and economic behavior. Princeton University Press, Princeton, pp 648

Newell FN, Ernst MO, Tjan BS, Bülthoff HH (2001) Viewpoint dependence in visual and haptic object recognition. Psychol Sci 12(1):37–42

OruçI, Maloney LT, Landy MS (2003) Weighted linear cue combination with possibly correlated error. Vision Res 43(23):2451–2468

Patton PE, Anastasio TJ (2003) Modeling cross-modal enhancement and modality-specific suppression in multisensory neurons. Neural Comput 15(4):783–810

Pearl J (1988) Probabilistic reasoning in intelligent systems: networks of plausible inference, 2nd edn. Morgan Kaufmann Publishers, San Mateo

Pick HL, Warren DH, Hay JC (1969): Sensory conflict in judgements of spatial direction. Percept Psychophys 6:203–205

Poremba A, Saunders RC, Crane AM, Cook M, Sokoloff L, Mishkin M (2003) Functional mapping of the primate auditory system. Science 299(5606):568–572

Rothkopf CA, Ballard DH (2009) Image statistics at the point of gaze during human navigation. Vis Neurosci 26(1):81–92

Rothkopf CA, Weisswange TH, Triesch J (2009) Learning independent causes in natural images explains the space variant oblique effect. In: Proceedings of the 8th International Conference on Development and Learning (ICDL 2009). Shanghai, China

Rowland BA, Stanford TR, Stein BE (2007) A model of the neural mechanisms underlying multisensory integration in the superior colliculus. Perception 36(10):1431–1443

Sadato N, Pascual-Leone A, Grafman J, Ibañez V, Deiber MP, Dold G, Hallett M (1996) Activation of the primary visual cortex by Braille reading in blind subjects. Nature 380(6574):526–528

Sanborn A, Griffiths T, Navarro DA (2006) A more rational model of categorization. Proc Cog Sci 2006:726–731

Sato Y, Toyoizumi T, Aihara K (2007) Bayesian inference explains perception of unity and ventriloquism aftereffect: identification of common sources of audiovisual stimuli. Neural Comput 19(12):3335–3355

Saunders JA, Knill DC (2001) Perception of 3d surface orientation from skew symmetry. Vision Res 41(24):3163–3183

Schlicht EJ, Schrater PR (2007) Effects of visual uncertainty on grasping movements. Exp Brain Res 182(1):47–57

Schrater PR, Kersten D (2000) How optimal depth cue integration depends on the task. Int J Comp Vis 40(1):71–89

Schroeder CE, Foxe JJ (2002) The timing and laminar profile of converging inputs to multisensory areas of the macaque neocortex. Brain Res Cogn Brain Res 14(1):187–198

Shams L, Seitz AR (2008) Benefits of multisensory learning. Trends Cogn Sci 12(11):411–417

Smith AM (ed and trans) (2001) Alhacen's theory of visual perception: a critical edition, Transactions of the American Philosophical Society, Philadelphia, 91(4–5)

Stocker AA, Simoncelli EP (2006) Noise characteristics and prior expectations in human visual speed perception. Nat Neurosci 9(4):578–585

Thomas G (1941) Experimental study of the influence of vision on sound localisation. J Exp Psychol 28:167177

Triesch J, Ballard DH, Jacobs RA (2002) Fast temporal dynamics of visual cue integration. Perception 31(4):421–434

Triesch J, von der Malsburg C (2001) Democratic integration: self-organized integration of adaptive cues. Neural Comput 13(9):2049–2074

Trommershäuser J, Maloney LT, Landy MS (2003) Statistical decision theory and trade-offs in the control of motor response. Spat Vis 16(3–4):255–275

Trommershäuser J, Maloney LT, Landy MS (2008) Decision making, movement planning and statistical decision theory. Trends Cogn Sci 12(8):291–297

van Beers RJ, Sittig AC, Gon JJ (1999) Integration of proprioceptive and visual position-information: an experimentally supported model. J Neurophysiol 81(3):1355–1364

von Schiller P (1932) Die Rauhigkeit als intermodale Erscheinung. Z Psychol Bd 127:265–289

Wallace MT, Stein BE (2007) Early experience determines how the senses will interact. J Neurophysiol 97(1):921–926

Wallace MT, Wilkinson LK, Stein BE (1996) Representation and integration of multiple sensory inputs in primate superior colliculus. J Neurophysiol 76(2):1246–1266

Weiss Y, Fleet DJ (2002) Velocity likelihoods in biological and machine vision. In: Rao RPN, Olshausen BA, Lewicki MS (eds) Probabilistic models of the brain. MIT Press, Cambridge, MA

Weiss Y, Simoncelli EP, Adelson EH (2002) Motion illusions as optimal percepts. Nat Neurosci 5(6):598–604

Weisswange TH, Rothkopf CA, Rodemann T, Triesch J (2009) Can reinforcement learning explain the development of casual inference in multisensory integration? In: Proceedings of the 8th International Conference on Development and Learning (ICDL 2009). Shanghai, China

Wozny DR, Beierholm UR, Shams L (2008) Human trimodal perception follows optimal statistical inference. J Vis 8(3):24, 1–11

Yuille AL, Bülthoff HH (1996) Bayesian theory and psychophysics. In: Knill D, Richards W (eds) Perception as Bayesian inference. Cambridge University Press, New York, NY, pp 123–161

Yuille A, Kersten D (2006). Vision as Bayesian inference: analysis by synthesis? Trends Cogn Sci 10(7):301–308

Zemel RS, Dayan P, Pouget A (1998) Probabilistic interpretation of population code. Neural Comput 10(2):403–430

Zhou YD, Fuster JM (2000) Visuo-tactile cross-modal associations in cortical somatosensory cells. Proc Natl Acad Sci U S A 97(17):9777–9782

Chapter 4
Methodological Considerations: Electrophysiology of Multisensory Interactions in Humans

Marie-Hélène Giard and Julien Besle

4.1 Introduction

The last 10 years have seen an explosion of research on the neurophysiological bases of multisensory processing, with a large variability in methodological approaches and issues addressed. Among those, one major research axis has concerned the genuine question of multisensory integration in perception, that is, how the sensory systems integrate separate features of an 'object' to form a unitary percept when this object is characterized by features stimulating several (say, two) sensory modalities simultaneously. Answering this question necessitates to understand how brain processes involved in the perception of a bimodal event differ from those underlying perception of the same event presented in either modality alone, that is, what are the neuronal operations specifically involved in bimodal processing as compared to each unisensory processing. Numerous studies have focused on the neural mechanisms of multisensory feature integration in humans using functional neuroimaging techniques, such as functional magnetic resonance imaging (fMRI) or positron emission tomography (PET), and electrophysiological recordings such as magnetoencephalography (MEG) or electroencephalography (EEG), and event-related potentials (ERPs) recorded from surface or intracerebral electrodes.

4.1.1 Specificities of the Electrophysiological Techniques

In contrast to PET and fMRI signals that measure task-induced changes in brain metabolism, electrophysiological techniques provide a direct measure of the electrical or electromagnetic fields generated by neuronal activity. More specifically, synchronous activation of a large set of neurons (post-synaptic transmembrane currents in pyramidal cells) generates intracerebral currents that spread to the head

M.-H. Giard (✉)
INSERM – UNITE 821 "Brain Dynamics and Cognition", Centre Hospitalier le Vinatier, University Lyon, 95, Bd Pinel, 69500 Bron, France
e-mail: mh.giard@inserm.fr

M.J. Naumer, J. Kaiser (eds.), *Multisensory Object Perception in the Primate Brain*,
DOI 10.1007/978-1-4419-5615-6_4,

surface through variable conductive media (volume conduction effects). These currents create, at any instant, a distribution of potentials that can be recorded between two (one reference, one active) electrodes and magnetic fields that can be recorded by SQUID (superconducting quantum interference devices) sensors. Recording from a large number of electrodes (or sensors) at the head surface allows to draw the topographic map of the resulting neuronal activities at each time sample. Thus, electrophysiological techniques provide fine-grained time information enabling to follow the underlying neuronal activities with a precision of the order of the millisecond. This allows for the dissociation of successive neuronal operations that are involved in the brain function studied. On the other hand, unlike PET and fMRI techniques which provide direct information on the brain structures that are activated, scalp EEG/ERP and MEG recordings pick up activities at the head surface. One thus needs inferences from scalp topographies or mathematical models to identify the brain structures that generated the surface signals. Electrophysiological neuroimaging therefore differs from metabolic neuroimaging both in terms of spatiotemporal resolution and biological origin and hence requires specific methods of signal analysis, with different consequences for the interpretation of the data.

4.2 The Additive Model in Human Electrophysiology

A great deal of human studies interested in identifying the neuronal operations specifically associated with multisensory integration have used paradigms in which the brain responses to events (objects, speech stimuli) presented in two sensory modalities (say, auditory and visual) are compared with the responses to the same events presented in each modality in isolation. The rationale is that if the auditory (A) and visual (V) dimensions of the bimodal stimulus are independently processed, the neural activities induced by the audiovisual (AV) stimulus should be equal to the algebraic sum of the responses generated separately by the two unisensory stimuli. Conversely, any neural activity departing from the mere summation of unimodal activities should be attributed to the bimodal nature of the stimulation, that is, to interactions between the processing of the auditory and visual cues. Using this model, the cross-modal interactions can thus be estimated in the differences between the brain responses to bimodal stimuli and the sum of the unimodal responses.

$$\text{AV interactions} = \text{response to (AV)} - [\text{response to (A)} + \text{response to (V)}]$$

This procedure was first used by Berman (1961) in event-related corticogram in cats, and later by Barth et al. (1995) in an evoked potential study in rat cortex to identify the brain regions that responded uniquely to bimodal AV stimuli as compared to unimodal responses. Later, the additive model has been widely used in human studies of cross-modal interactions (e.g., scalp ERP and MEG: Fort et al., 2002a, b; Foxe et al., 2000; Giard and Peronnet, 1999; Klucharev et al., 2003; Molholm et al., 2002; Möttönen et al., 2004; Raij et al., 2000; Senkowski et al., 2007; Stekelenburg and Vroomen, 2007; Talsma and Woldorff, 2005). It has been also criticized both by electrophysiologists (Gondan and Röder, 2006; Teder-Sälejärvi et al., 2002) and by authors using hemodynamic techniques (Calvert and Thesen, 2004) because of the multiple biases it can generate in the estimation of the cross-modal interactions if several important conditions are not fulfilled.

We will discuss below the following questions: (1) What these biases are and how to avoid or minimize them, with particular emphasis on electromagnetic (EEG/MEG) techniques; (2) Why the use of the additive model is not only suitable but also *necessary* to identify cross-modal interactions from EEG/MEG signals; and (3) How violation of the additive model implies different conclusions when used in human EEG/MEG data, fRMI data, or at the single-cell level in animal studies. Finally, we will discuss statistical methods that can be applied to test this model on experimental data.

4.2.1 Potential Biases and Artifacts Generated by the Additive Model

Non-biased estimation of multisensory interactions in the human cortex using the additive model [AV − (A + V)] requires taking important precautions both in the experimental design and in the data analysis.

4.2.1.1 Activities Common to All (Bimodal and Unimodal) Stimuli

The additive model makes the assumption that the analyzed brain responses do not include activities that would be common to all three types of stimuli (A, V, and AV). Indeed these activities will be added once but subtracted twice in the [AV − (A+V)] procedure, which can lead to spurious confound effects in the estimation of multisensory interactions. These common activities may be of several types.

First, they include general task-related neural activities associated with late cognitive processes, such as target processing (e.g., N2b/P3 waves in ERP/MEG recordings), response selection, or motor processes. The ERP literature has shown that these activities usually arise after about 200 ms post-stimulus whereas earlier latencies are characterized by sensory-specific responses (review in Hillyard et al., 1998). One way to avoid inclusion of these common activities is to restrict the analysis of brain responses to the early time period (<200 ms) of stimulus processing. While this procedure is very simple in ERP/MEG recordings since their time resolution is of the order of the millisecond, sorting the response components according to their latency is impossible at this timescale in hemodynamic imaging techniques. These common activities actually correspond to a processing stage at which auditory and visual processing are known to *converge* (e.g., response selection or motor processes), but the corresponding convergence structures or processes do not contribute anything to the question of genuine multisensory interactions or integration.

Second, in paradigms requiring speeded responses with rapidly presented stimuli, 'anticipatory' slow activities may arise before each (unimodal and bimodal) stimulus and continue for a time after stimulus onset. They appear, when present in ERP/MEG recordings, as slow ramp-like deflections in the prestimulus and early post-stimulus periods (Brunia and van Boxtel, 2001). These deflections can thus give rise to spurious residual effects in the [AV–(A+V)] signals that may be confused with early cross-modal interactions (Teder-Sälejärvi et al., 2002). Such anticipatory processes are independent of the technique used and could also be included in fMRI/PET responses (Voisin et al., 2006). Several procedures have been proposed to avoid or strongly reduce anticipatory processes.

At the level of the experimental design, a procedure valid whatever the neuroimaging technique used is to present the stimuli at random interstimulus time intervals during data acquisition. If one wishes to keep regular interstimulus times, another procedure is to estimate these anticipatory activities by adding trials in which no stimulus is presented ('no-stimulation' trials), while it is still expected. One can then subtract out this response from the interaction estimation (Talsma and Woldorff, 2005). It is an open question whether this method also removes some activities common to all conditions. Gondan and Röder (2006) have proposed to introduce a third, tactile (T) modality and to compare (T+TAV) vs (TA+TV), instead of comparing only AV vs (A+V). This procedure is formally correct to subtract out activities common to all modalities, if one makes the assumption that there is no interaction specific to presenting three modalities simultaneously. However, nothing guarantees that the brain integrates three modalities in the same way as two modalities. Indeed, several studies in our lab have shown that the integration mechanisms are extremely flexible and adapt to the best strategy according to the nature of the stimuli to be combined, the task at hand, attention, and preferred sensory modality of the subject.

At the level of the ERP/MEG signal analysis two other methods have been proposed to correct for anticipatory effects: modify the latency of the prestimulus period that will be used as the reference baseline and/or high-pass filter the data (e.g., 2-Hz cutoff frequency) to remove the slow wave effects (Teder-Sälejärvi et al., 2002). If the early latency effects in [AV−(A+V)] stand up to these control procedures, they are likely to reflect genuine audiovisual interactions.

4.2.1.2 Unimodal Deactivation of Sensory Cortices: Block vs Random Designs

Applying the additive model also requires the assumption that in the conditions of unimodal stimulus presentation (e.g., visual stimulation), the ongoing activity in the non-stimulated sensory cortex (e.g., auditory cortex) is not affected by stimulation in the other sensory (visual) modality. However, several functional imaging studies using block-designed paradigms have reported deactivations in sensory-specific cortex (e.g., the auditory cortex) when subjects were presented with continuous stimulation in another (e.g., visual) modality (Haxby et al., 1994; Kawashima et al., 1995; Laurienti et al., 2002). Indeed, when a particular sensory cortex is continuously and exclusively stimulated in one sensory modality during a whole block while the other modalities are not stimulated, sustained attentional resources are directed to the relevant modality (even in passive or little attention-demanding tasks), whereas the other modalities are more or less voluntarily ignored (deactivated) to optimize the processing in the relevant sensory cortex (see also Ghatan et al., 1998 for similar attentional effects; Kawashima et al., 1999). In studies of multisensory integration, the [AV − (A + V)] model should therefore not be used in experiments based on block-designed paradigms, since these unimodal deactivations would be subtracted from the bimodal activations, resulting in artificial increases of the 'cross-modal' effects. One way to eliminate or considerably reduce such sustained attention-related deactivations in unisensory cortices is to use paradigms in which the stimuli are randomly and equiprobably delivered across all modality conditions (e.g., Calvert et al., 2000, 2001; Fort et al., 2002a, b; Foxe et al., 2000; Giard and Peronnet, 1999; Laurienti et al., 2002; Molholm et al., 2002; Raij et al., 2000; Wright et al., 2003). More generally, random designs reduce the contribution of state-related confounding factors such as sustained attention, alertness, and fatigue, and unless one has specific reasons to use block designs (for example, to study sustained attention), random designs should be preferred.

4.2.1.3 More About Attentional Effects in the Additive Model

Randomly mixing stimuli from different conditions may however not be sufficient to avoid contaminating the cross-modal interactions with attentional effects. Although a widely used procedure to avoid attentional biases in [AV − (A+V)] is to require from the subject the same task in the three modalities, in some paradigms, the task may be easier and demand less attentional effort in one modality than in the other. In this case, as attention can quickly shift from one modality to the other, subjects can focus on visual-only trials, for example, more than on auditory or audiovisual trials.

This is typically the case in audiovisual integration studies of speech perception. Consider lip movements associated with syllable sounds randomly presented in the three A, V, and AV conditions: if a discrimination task (e.g., respond to target syllables) is required under the three modality conditions, the processing of syllables in the lip reading condition alone will include a tremendous visual attention effect that will not be present when the subject discriminates syllables in the auditory only or audiovisual condition, since in speech perception (unlike what is likely to occur for bimodal non-speech objects), normal subjects will *naturally* engage much less visual attention to process AV than V stimuli. In these conditions, the shift of attention, present only in the V condition, will not be eliminated in the [AV − (A+V)] computation, generating spurious effects in the computation of interactions. Alternatively, if the subjects are required to respond only when they hear a target syllable (A and AV conditions), then the experimental situation mimics the natural situation in which we focus on the auditory message, and accordingly, the shift in visual attention can be expected to be similar in visual-only trials and in audiovisual trials (similarly, shifts in auditory attention will be expressed rather similarly for A and AV stimuli and eliminated in the [AV − (A+V)] model). A general principle in dealing with attentional problems is therefore, in addition to systematically mixing conditions, to equate the attentional load between each unimodal condition and the bimodal condition, but *not necessarily* between the two unimodal conditions.

Other manifestations of attention shifts on the other hand can be legitimately considered parts of the audiovisual interactions, as is the case for cross-modal attentional shifts (McDonald et al., 2001; Schwartz et al., 2004). Cross-modal attentional shifts happen when a cue in one modality (say visual) draws attention to a subsequent stimulus in another modality. This effect might explain part of the interactions found in our own multisensory speech studies (Besle et al., 2004b, 2008) and has been recently tested by Stekelenburg and Vroomen (2007) using non-speech AV objects. This kind of attentional effect may be considered a genuine audiovisual interaction since it implies a crosstalk between sensory-specific processes.

In conclusion, the additive model can estimate non-biased multisensory interactions only if a number of conditions related to both experimental design and data analysis are fulfilled. We will see that if these conditions are respected, using the additive model presents strong advantages for the analysis of EEG/MEG signals.

4.2.2 Suitability and Necessity of the Additive Model in Human Electrophysiology

As we wrote above, a fundamental difference between fMRI/PET and EEG/MEG techniques is that in hemodynamic data, a voxel presenting significant activity provides direct information on the brain structure activated, whereas in ERP/MEG recordings a significant value at a particular electrode (sensor) potentially reflects activity from *distant* 'generators' located anywhere in the brain (cortical and subcortical structures). Indeed, EEG/MEG does not measure only the activity of the structure just beneath the electrode/sensor. Thus, interpretation of the surface signals needs to take into account volume conduction factors (using topographic analysis, generator modeling, etc.) and using the additive model to characterize the patterns of cross-modal interactions in EEG/MEG data is particularly suitable for this purpose.

A direct consequence of the volume conduction effects for the study of multisensory interactions is that, in ERP/MEG signals, one *cannot* directly compare a bimodal response with either unimodal response at a particular scalp site, since the bimodal response may include volume conduction effects from the response to the complementary modality. A striking example is given in Fig. 4.1, which shows the ERPs to speech syllables presented in the auditory modality alone, the visual modality alone (lip reading), or in both modalities at two electrode sites (Cz and P24). Looking at the 'N1peak' at Cz, an immediate interpretation would be that there is no difference between the bimodal and the unimodal auditory response since their amplitudes are essentially identical at Cz, and thus the [AV − (A + V)] effect at Cz would be an artifact of subtraction due to the visual response at that electrode site. However, for a correct interpretation (Besle et al., 2004a, b), one needs to take into account the diffusion of the visual response at Cz in the bimodal stimulus, that is, to remove the electrical fields due to the visual response in the bimodal response before comparing the bimodal response to the auditory response, which in turn means applying the additive model.

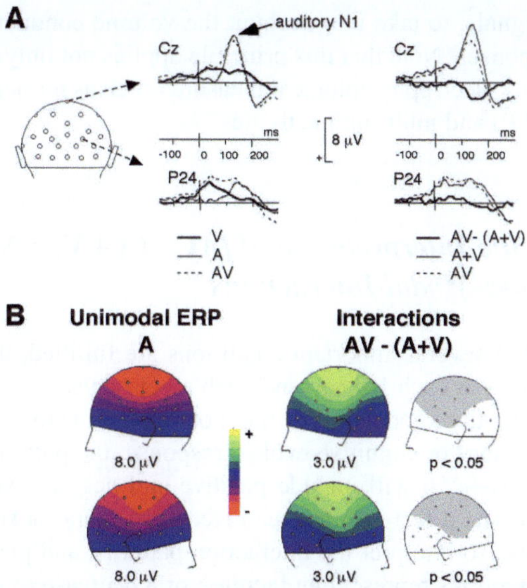

Fig. 4.1 Interpretation of the additive model in human EEG/MEG signals: an example of the necessity to reason in terms of topography, and not at single-electrode level. (**a**). *Left*: ERPs elicited by speech syllables presented in the auditory modality alone (A), the visual modality alone (V), and in bimodal condition (AV) at two electrodes (central Cz and posterior P24). At the latency of the unimodal auditory N1 peak (135 ms), the AV response at Cz appears rather similar to the auditory response, whereas a visual signal can be observed at that electrode. *Right*: superposition of the audiovisual (AV) ERP, the sum of the unimodal (A+V) ERPs, and the interaction term [AV − (A + V)] An immediate interpretation would be that the [AV − (A + V)] effect at Cz results from an artifact of subtraction due to the visual signal at that electrode. (**b**) Once the electrical fields (volume conduction effects) due to the visual response have been removed from the bimodal response at all the electrode sites (AV − V), the electrical fields specifically due to the interactions can be isolated by comparing (AV − V) to A, which in turns means computing [AV − (A + V)]. Whatever the volume conduction effects of the visual response, the topographic distribution of the interactions was found, around the N1 peak latency, similar to that of the unimodal auditory response with opposite polarity. (The gray areas indicate the scalp sites where the interactions are statistically significant.) The correct interpretation is thus that the interactions would reflect a decrease (depression) of the auditory unimodal response (Besle et al. 2004). Color scale: the value in µV below each map indicates half range of the color scale

What makes the additive model particularly suitable to situations where different subcomponents of the brain responses overlap is that it is valid whatever the number, nature, or configuration of the intracerebral generators. It is indeed based on the law of superposition of electrical fields, according to which the potentials from separate current sources in a conductive medium sum linearly. This enables to remove, in the bimodal response, the conduction effects present in the unimodal responses whatever the polarity (positive or negative) of the response and the site of recording. What is left in the interaction term is an estimation of the volume conduction effects specifically related to interactions, but excluding volume conduction specific to unimodal activities. In this sense, the use of the model is not only suitable, but *necessary* with ERP/MEG signals, to take into account the volume conduction effects of the different brain responses. Note that this principle applies not only to scalp EEG and MEG but also to any electrophysiological measures such as intracranial ERPs, local field potentials (LFP) and multi-unit activities.

4.2.3 Topographic Interpretation of [AV − (A + V)]: Nature of the Cross-Modal Interactions

We have seen that, if several important conditions are fulfilled, the additive model allows for estimating, at each latency and each electrode/sensor site, a measure of brain activity specific to bimodal processing, potentially due to multiple generators. Similarly to any sensory or cognitive evoked response, the pattern of interaction as revealed in [AV − (A + V)] will include positive and negative voltage values, the polarity of which has no direct significance in terms of supra- or subadditivity of the unimodal responses. To interpret the interaction patterns, and particularly to know whether these interactions represent an increase or a decrease of one or both of the unisensory responses, or a new activity, it is necessary to compare the topography of the interaction pattern with those of both unimodal responses around the same latency. If the interaction pattern has the same topography with the same polarity as that of one unisensory component, it is likely to reflect an *increased* amplitude of that sensory component in the bimodal condition (as compared to the response in the unimodal condition), that is, an enhanced activity of the underlying neural generator(s) in the corresponding sensory cortex. If the topographies of the interaction pattern and a sensory component are similar with opposite polarities, the interaction probably reflects a *decrease* of the unimodal response around the considered latency. If the interaction pattern presents a spatial distribution not seen in either unimodal response, then this interaction might reflect the activation of a *new* brain structure specifically activated by the bimodal nature of the stimulus. Note however that this is not a guarantee of the existence of activities specific to multisensory processing since these activities could still exist in the unimodal response but be hidden by other unimodal components of the response. It is interesting to note that all three types of interactions (enhancement or decrease of unisensory responses, new activity) can be found at different latencies along the processing chain. This is what we observed, for example, in our first study describing the cross-modal interactions generated during a bimodal object recognition task: the interaction patterns were expressed as enhanced activities in the visual cortex before 150 ms of processing, enhanced activities in the auditory cortex around 90–110 ms (i.e., larger amplitude of the unimodal auditory N1 component), decreased activities in the visual cortex around 150–200 ms (i.e., decrease of the unimodal visual N185 component), and finally new activities over the non-specific right temporo-frontal scalp sites around 140–165 ms (Giard and Peronnet, 1999). This method can be extended to other types of electrophysiological data where spatial information is available. For example, we used the spatiotemporal profiles of activity along stereotaxic depth multi-electrodes to show that audiovisual interactions took place in the human auditory cortex in the planum temporale (Besle et al., 2008).

4.3 The Additive Model in EEG/MEG, Hemodynamic, and Single-Neuron Data: One Model, Three Interpretations

So far, we focused our discussion on the additive model in EEG/MEG/intracranial EEG. However, the same or equivalent additivity computation can be and has been applied to other types of neurophysiological/functional data. It is important to understand to which extent these computations and their interpretations are comparable.

4.3.1 Comparison with Hemodynamic Data

In hemodynamic studies, the additive model consists in comparing the response to bimodal stimuli to the sum of unimodal responses, generally through regression of a theoretical hemodynamic response in the General Linear Model (e.g., Beauchamp, 2005). In essence, it consists in finding areas of the brain where the response is stronger for audiovisual stimuli than the sum of unimodal conditions (super-additivity) or vice versa (subadditivity). In spite of the apparent similarity with the additive model in human electrophysiology, the differences in assumptions and interpretation are numerous.

First, and most importantly, the mere validity of the model in hemodynamic studies is questionable: in human electrophysiological data, the violation of the additive model at the signal level logically implies some convergence or interaction at the biological level. Indeed, the law of superposition of electric fields states that if two current sources are separate, then the currents (and the differences of potential) they produce sum linearly at any point in the brain. It logically follows that if the bimodal response is not the linear sum of the unimodal responses, then the sources are not separate and hence interact in some way. For metabolic data, however, there is no reason to assume that metabolic activity is linearly related to the electrical activity produced by the neurons. Hence the rejection of additivity *does not* ensure interaction at the neuronal level.

Second, subadditivity and superaddivity have different interpretations in fMRI compared to EEG/MEG. In EEG/MEG, the sign of the [AV − (A + V)] term does not have any meaning a priori. In contrast, in fMRI, a positive difference (super-additivity) suggests that the activity in the bimodal condition is stronger than in unimodal conditions, whereas a negative difference (subadditivity) suggests that activity is suppressed in the bimodal condition. The equivalent of such a sub-additivity or super-additivity in EEG/MEG would be increases or decreases in sensory-specific responses that can be found by comparing the bimodal response to the unimodal components and thus necessitates the additional step of comparing topographies (as discussed above).

Note however that in fMRI, the interpretation of the sign of the interaction is not completely straightforward: super-additivity can also be found if a region is deactivated in one of the unimodal conditions. An additional criterion that has been proposed to exclude this possibility is to require that all conditions (A, V, and AV) show an increase in blood flow compared to baseline (Beauchamp, 2005). Subadditivity effects can be due either to processes common to bimodal and unimodal conditions or to genuine interactions, but these two possibilities cannot be distinguished on the basis of their latency as in EEG/MEG.

Some of those points have been discussed in more detail elsewhere (Beauchamp, 2005; Calvert and Thesen, 2004; Laurienti et al., 2005).

4.3.2 Comparison with Single-Unit Data

Similarly, in spite of the same terminology (additive model), the same algebraic operation [AV−(A+V)], and the same raw signal being recorded (intracranial EEG), the procedure refers to different phenomena when applied at the single-cell level in animal studies and in EEG/MEG signals.

In single-cell studies, the electrophysiological signal originally recorded is high-pass filtered and thresholded to select only action potentials. This process of selection allows to localize the activity in a unique source (the neural cell) but, at the same time, it introduces a non-linear transformation that removes it from the domain of application of the law of superposition of electrical fields.

Yet the additive model has been applied, at the cell level, to study interactions *within* a given cell (or at least the consequence of upstream interactions at this cell). Several authors have repeatedly shown that, 'multisensory' neurons (in the superior colliculus and several cortical areas of mammals) that fire in response to stimuli from two sensory modalities also increase their firing rate to a level exceeding that predicted by summing the responses to each unimodal cue when presented with a bimodal stimulus (King and Palmer, 1985; Meredith and Stein, 1983, 1986; Stein and Meredith, 1993; Wallace et al., 2004). In the same multisensory cells, spatially or temporally disparate multisensory cues can induce profound response depression. The additive model thus enables to reveal the *integrative properties* of these multisensory neurons.

On the other hand, as we have seen, the additive model in EEG/MEG/LFP or fMRI data applies at the *population* level and may reveal integration or interaction effects without implying the existence or activation of genuine 'multisensory neurons.' This fundamental distinction between the single-cell level and the neuronal population level also implies that the subadditive and super-additive criteria of multisensory integration defined at the neuron level do not apply in EEG/MEG or fMRI (contrary to the main assumption of Calvert et al., 2001). Particularly, as we have seen, it is possible to observe integration effects associated to behavioral enhancements that are expressed as response depression at the population level in human data (Besle et al., 2004b; van Wassenhove et al., 2005).

4.4 Statistical Assessment of Cross-Modal Interactions

To test for statistical significance of the interaction effects revealed using the additive model, the general method is to compare the estimate [$ERP_{AV} - (ERP_A + ERP_V)$] to 0 at any given time sample and any given electrode/sensor. The statistical tests that can be used to achieve this comparison depend on the experimental design, and on the assumptions one is ready to make about the distribution of electrophysiological data (or verification thereof).

4.4.1 At a Group Level

Most authors (including us) have used the Student's *t* test when testing data from a conventional EEG/MEG study, that is, a repeated measure design in which all subjects are submitted to all three conditions (A, V, and AV) and therefore one ERP is obtained per subject and condition. For each subject, one can estimate the interactions [$ERP_{AV} - (ERP_A + ERP_V)$] and test the difference between this term and 0. If one is not willing to make the assumption of normality, a permutation test can replace the *t* test. However, normality is generally a reasonable assumption for ERP group analyses and we have noted that permutation tests and *t* tests give similar results.

An alternative test, if 'no-stimulation' trials are available, is to use a 2-factor repeated measure ANOVA with a 'visual stimulus' factor (presence or absence) and an 'auditory stimulus' factor (presence or absence). The audiovisual interaction can then be assessed statistically as the interaction term between the auditory and the visual stimulus factors in the ANOVA. Indeed, the interaction corrected for common activity, as estimated by the ERP to 'no-stimulation' trials, is

$$(ERP_{AV} - ERP_0) - [(ERP_A - ERP_0) + (ERP_V - ERP_0)]$$

This is formally equivalent to the interaction term, which is compared to 0 in the ANOVA model:

$$(ERP_{AV} - ERP_A) - (ERP_V - ERP_0)$$

The advantage of the ANOVA is that it can take into account the variability of the common activity estimated by the 'no-stimulation trials' and solves the problem of comparing one condition to the sum of two conditions. The flipside is that it assumes that the 'no-stimulation' condition is a good model for the activity common to all conditions.

4.4.2 At an Individual Subject Level

Sometimes, the experimental design does not provide repeated measures that are comparable between subjects. This is the case for intracranial data acquired in different patients (e.g., Besle et al., 2008): since implantation of electrodes is specific for each patient and guided by clinical considerations, statistics have to be computed at the individual level. In this case, the units of the statistical test are not subjects, but single trials. Since single trials are independent, there cannot be any repeated measures and one has to use independent sample tests. If 'No-stimulation' trials are available, one could use a two-factor-independent sample ANOVA and assess the statistical significance of the interaction between the auditory and the visual stimulus factors, as described earlier. If no such trials have been recorded, parametric tests cannot be readily applied. Moreover, it is an open question whether normality is a reasonable assumption for single trial samples (except if the number of trials is large enough for the distribution to converge to a normal distribution).

For such a case, we have developed an ad hoc randomization test specifically aimed at testing interactions in the additive model (Besle et al., 2008). This test is graphically described in Fig. 4.2. It consists, for each time sample and at each electrode, in comparing the actual value of the violation of the additive model to the distribution of the value under the null hypothesis, estimated by resampling from the actual data.

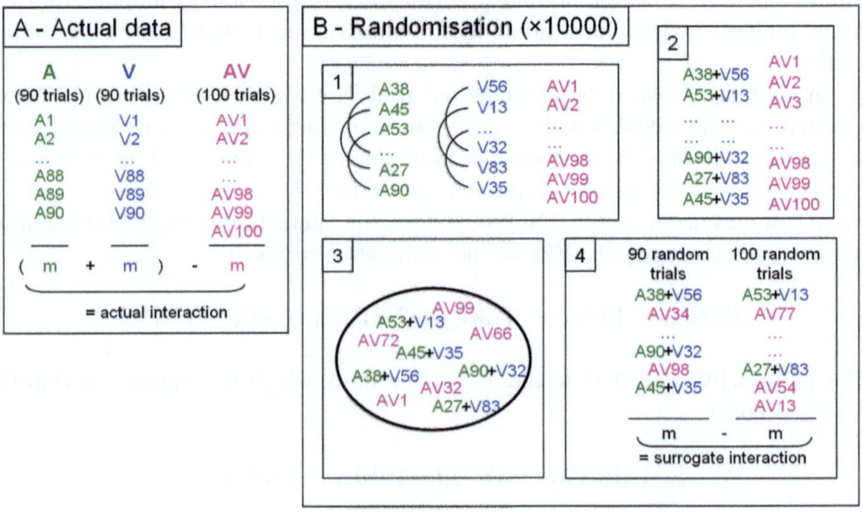

Fig. 4.2 Steps of the randomization test for assessing the statistical significance of audiovisual interactions as estimated in the additive model [AV − (A + V)] at an individual subject level (see text)

The null hypothesis is that the ERP (at any sample/electrode) in the AV condition is equal to the sum of the ERPs in the A and the V conditions, that is, AV trials and sums of A and V trials are drawn from the same distribution. This is modeled by pooling AV trials and sums of A and V trials together and drawing them at random (without replacement) into two re-samples. (Note that there is no need to have the same number of trials for the different modalities).

The test involves the following steps (depicted in Fig. 4.2):

Step A: compute the actual interaction as $[ERP_{AV} - (ERP_A + ERP_V)]$
Step B: randomization

1. Randomly re-order the N_A auditory trials and the N_V visual trials separately
2. Compute the N_{A+V} corresponding sums of unimodal trials (for simplicity, we assume that $N_{A+V} = N_A = N_V$
3. Pool the N_{A+V} sums with the N_{AV} audiovisual trials
4. From this pool, draw at random two samples of N_{A+V} and N_{AV} trials, respectively, and compute the difference between the means of these two samples (surrogate interaction)

Perform step B a large number of times (say 10,000) so as to obtain a distribution of surrogate interactions. Now this distribution is an estimate of the distribution of interactions under the null hypothesis. One can compute the associated probability value by counting the number of surrogate interaction values that are larger (in absolute value for a bilateral test) than the actual interaction value and dividing by the number of randomizations.

This method allowed us to find multiple interactions in intracranial data at the level of individual patients. To generalize the results to the population level, one has to resort to other (non-statistical) methods, like anatomical normalization.

Note that this method can also be advantageous compared to a group analysis for MEG in which there is much more variability between the measures in different subjects than in ERP recordings (more variability of the head position relative to the sensors, less volume conduction smearing).

4.4.3 Correction for Multiple Testing

The multiple testing problem is not specific to the additive model and has been discussed elsewhere (Blair and Karniski, 1993; Guthrie and Buchwald, 1991; Maris et al., 2007). However, the application of the additive model has naturally led to analyses of the exploratory type and has not been used, at least to our knowledge, as a tool for specific hypothesis testing (except in the form of the restriction of the time-window analysis to the first 200 ms). For exploratory analyses consisting in testing all time-samples at all electrodes/sensors with independent tests, it is very important to take into account the inflation of family-wise errors.

For the temporal dimension, several solutions have been proposed, some based on modeling the EEG signal (Guthrie and Buchwald, 1991) and others based on randomizations (Blair and Karniski, 1993). They both amount to estimating the length (number) of consecutive significant tests necessary to correct for the increase in false positives with the number of samples tested. Correcting for multiple testing in the spatial dimension follows the same principle, although it is trickier to implement because electrodes/sensors are defined in a bidimensional space (the head or the helmet). No published studies, to our knowledge, have used both corrections, which might bring into question the validity of part of the interactions found.

4.5 Conclusion

In summary, correct estimation of multisensory interactions in human electrophysiology requires (1) designing the experiment so that the conditions of application of the additive model are fulfilled; (2) determining the points in time and space where the bimodal response differs from the sum of the unimodal responses, and correcting for multiple comparisons in both spatial and temporal dimensions; and (3) comparing the patterns of multisensory interactions with the unimodal responses around the same latencies to know whether these interactions represent enhancement or depression of either unimodal response. This procedure will allow characterizing the nature and the time course of the cross-modal interactions along the sensory processing chain, for a better understanding of the full neural processes underlying multisensory integration.

References

Barth DS, Goldberg N, Brett B, Di S (1995) The spatiotemporal organization of auditory, visual and auditory-visual evoked potentials in rat cortex. Brain Res 678:177–190

Beauchamp MS (2005) Statistical criteria in FMRI studies of multisensory integration. Neuroinformatics 3:93–114

Berman AL (1961) Interaction of cortical responses to somatic and auditory stimuli in anterior ectosylvian gyrus of cat. J Neurophysiol 24:608–620

Besle J, Fort A, Giard MH (2004a) Interest and validity of the additive model in electrophysiological studies of multisensory interactions. Cogn Process 5:189–192

Besle J, Fort A, Delpuech C, Giard MH (2004b) Bimodal speech: early suppressive visual effects in the human auditory cortex. Eur J Neurosci 20:2225–2234

Besle J, Fischer C, Bidet-Caulet A, Lecaignard F, Bertrand O, Giard MH (2008) Visual activation and audiovisual interactions in the auditory cortex during speech perception: intracranial recordings in humans. J Neuroscience 28:14301–14310

Blair RC, Karniski W (1993) An alternative method for significance testing of waveform difference potentials. Psychophysiology 30:518–524

Brunia CH, van Boxtel GJ (2001) Wait and see. Int J Psychophysiol 43:59–75

Calvert GA, Thesen T (2004) Multisensory integration: methodological approaches and emerging principles in the human brain. J. Physiol Paris 98:191–205

Calvert GA, Campbell R, Brammer MJ (2000) Evidence from functional magnetic resonance imaging of crossmodal binding in the human heteromodal cortex. Curr Biol 10:649–657

Calvert GA, Hansen PC, Iversen SD, Brammer MJ (2001) Detection of audio-visual integration sites in humans by application of electrophysiological criteria to the BOLD effect. Neuroimage 14:427–438

Fort A, Delpuech C, Pernier J, Giard MH (2002a) Early auditory-visual interactions in human cortex during nonredundant target identification. Cogn Brain Res 14:20–30

Fort A, Delpuech C, Pernier J, Giard MH (2002b) Dynamics of cortico-subcorical cross-modal operations involved in audio-visual object detection in humans. Cereb Cortex 12: 1031–1039

Foxe JJ, Morocz IA, Murray MM, Higgins BA, Javitt DC, Schroeder CE (2000) Multisensory auditory-somatosensory interactions in early cortical processing revealed by high-density electrical mapping. Brain Res Cogn Brain Res 10:77–83

Ghatan PH, Hsieh JC, Petersson KM, Stone-Elander S, Ingvar M (1998) Coexistence of attention-based facilitation and inhibition in the human cortex. Neuroimage 7:23–29

Giard MH, Peronnet F (1999) Auditory-visual integration during multimodal object recognition in humans : a behavioral and electrophysiological study. J Cogn Neurosci 11: 473–490

Gondan M, Röder B (2006) A new method for detecting interactions between the senses in event-related potentials. Brain Res 1073–1074:389–397

Guthrie D, Buchwald JS (1991) Significance testing of difference potentials. Psychophysiology 28:240–244

Haxby JV, Horwitz B, Ungerleider LG, Maisog JM, Pietrini P, Grady CL (1994) The functional organization of human extrastriate cortex: a PET-rCBF study of selective attention to faces and locations. J Neurosci 14:6336–6353

Hillyard SA, Teder-Sälejärvi WA, Munte TF 1998. Temporal dynamics of early perceptual processing. Curr Opin Neurobiol 8:202–210

Kawashima R, O′Sullivan BT, Roland PE (1995) Positron-emission tomography studies of cross-modality inhibition in selective attentional tasks: closing the "mind's eye". Proc Natl Acad Sci U S A 92:5969–5972

Kawashima R, Imaizumi S, Mori K, Okada K, Goto R, Kiritani S, Ogawa A, Fukuda H (1999) Selective visual and auditory attention toward utterances - a PET study. Neuroimage 10: 209–215

King AJ, Palmer AR (1985) Integration of visual and auditory information in bimodal neurons in the guinea-pig superior colliculus. Exp Brain Res 60:492–500

Klucharev V, Möttönen R, Sams M (2003) Electrophysiological indicators of phonetic and non-phonetic multisensory interactions during audiovisual speech perception. Brain Res Cogn Brain Res 18:65–75

Laurienti PJ, Perrault TJ, Stanford TR, Wallace MT, Stein BE (2005) On the use of superadditivity as a metric for characterizing multisensory integration in functional neuroimaging studies. Exp Brain Res 166:289–297

Laurienti PJ, Burdette JH, Wallace MT, Yen YF, Field AS, Stein BE (2002) Deactivation of sensory-specific cortex by cross-modal stimuli. J Cogn Neurosci 14:420–429

Maris E, Oostenveld R (2007) Nonparametric statistical testing of EEG- and MEG-data. J Neurosci Meth 164:177–190

McDonald JJ, Teder-Sälejärvi WA, Heraldez D, Hillyard SA (2001) Electrophysiological evidence for the "missing link" in crossmodal attention. Can J Exp Psychol 55:141–149

Meredith MA, Stein BE (1983) Interactions among converging sensory inputs in the superior colliculus. Science 221:389–391

Meredith MA, Stein BE (1986) Visual, auditory, and somatosensory convergence on cells in superior colliculus results in multisensory integration. J Neurophysiol 56:640–662

Molholm S, Ritter W, Murray MM, Javitt DC, Schroeder CE, Foxe JJ (2002) Multisensory auditory-visual interactions during early sensory processing in humans: a high-density electrical mapping study. Cogn Brain Res 14:115–128

Möttönen R, Schurmann M, Sams M (2004) Time course of multisensory interactions during audiovisual speech perception in humans: a magnetoencephalographic study. Neurosci Lett 363:112–115

Raij T, Uutela K, Hari R (2000) Audiovisual integration of letters in the human brain. Neuron 28:617–625

Schwartz JL, Berthommier F, Savariaux C (2004) Seeing to hear better: evidence for early audio-visual interactions in speech identification. Cognition 93:B69–B78

Senkowski D, Saint-Amour D, Kelly SP, Foxe JJ (2007) Multisensory processing of naturalistic objects in motion: a high-density electrical mapping and source estimation study. Neuroimage 36:877–888

Stein BE, Meredith MA (1993) The merging of the senses, 1st edn. The MIT Press, Cambridge, MA

Stekelenburg JJ, Vroomen J (2007) Neural correlates of multisensory integration of ecologically valid audiovisual events. J Cogn Neurosci 19:1964–1973

Talsma D, Woldorff MG (2005) Selective attention and multisensory integration: multiple phases of effects on the evoked brain activity. J Cogn Neurosci 17:1098–1114

Teder-Sälejärvi WA, McDonald JJ, Di Russo F, Hillyard SA (2002) An analysis of audio-visual crossmodal integration by means of event-related potential (ERP) recordings. Cogn Brain Res 14:106–114

van Wassenhove V, Grant KW, Poeppel D (2005) Visual speech speeds up the neural processing of auditory speech. Proc Natl Acad Sci U S A 102:1181–1186

Voisin J, Bidet-Caulet A, Bertrand O, Fonlupt P (2006) Listening in silence activates auditory areas: a functional magnetic resonance imaging study. J Neurosci 26:273–278

Wallace MT, Ramachandran R, Stein BE (2004) A revised view of sensory cortical parcellation. Proc Natl Acad Sci USA 101:2167–72

Wright TM, Pelphrey KA, Allison T, McKeown MJ, McCarthy G (2003) Polysensory interactions along lateral temporal regions evoked by audiovisual speech. Cereb Cortex 13:1034–1043

Chapter 5
Cortical Oscillations and Multisensory Interactions in Humans

Jochen Kaiser and Marcus J. Naumer

5.1 Introduction

Oscillatory synchronization of neuronal assemblies has been proposed as a mechanism underlying the formation of coherent percepts (Engel and Singer, 2001; Singer et al., 1997). Electrophysiological research in animals has indicated a special role of synchronization at higher frequencies in the gamma range (~30–120 Hz) for the integration of object features processed by spatially distant cortical units (Eckhorn et al., 1988; Fries et al., 1997; Gray et al., 1989). Higher frequency oscillatory synchronization is thought to enable the fast and flexible information transfer between cortical assemblies (Fries, 2005; Fries et al., 2007) and to result in an efficient driving of downstream target neurons (Jensen et al., 2007; Womelsdorf et al., 2007). Consistent with the hypothesized role of gamma-band activity (GBA) for perception, electro- and magnetoencephalogram (EEG and MEG) studies in humans have demonstrated increased GBA during the processing of coherent/familiar compared with incoherent/unfamiliar stimuli (Busch et al., 2006; Lutzenberger et al., 1995; Tallon-Baudry et al., 1996). More recent research has shown gamma synchronization to be involved in cognitive functions beyond perception (Herrmann et al., 2004; Jensen et al., 2007). For example, human studies have found increased GBA related to attentional selection (Bauer et al., 2006; Gruber et al., 1999; Müller and Keil, 2004; Wyart and Tallon-Baudry, 2008) and both short- and long-term memory (Gruber et al., 2004; Jokisch and Jensen, 2007; Kaiser et al., 2003, 2008; Lutzenberger et al., 2002; Osipova et al., 2006; Tallon-Baudry et al., 1998).

In our natural environment we constantly receive inputs from different sensory channels. Fast and efficient processing of object characteristics like identity, spatial location, or motion requires the integration of information obtained from multiple sensory systems. Inspired by demonstrations of crossmodal interactions in subcortical structures like the superior colliculi (Meredith and Stein, 1983; Wallace et al.,

J. Kaiser (✉)
Institute of Medical Psychology, Faculty of Medicine, Goethe University, Heinrich-Hoffmann-Str. 10, 60528 Frankfurt am Main, Germany
e-mail: j.kaiser@med.uni-frankfurt.de

M.J. Naumer, J. Kaiser (eds.), *Multisensory Object Perception in the Primate Brain*,
DOI 10.1007/978-1-4419-5615-6_5, © Springer Science+Business Media, LLC 2010

1998), previous research on multisensory integration at the cortical level has focused on the identification of heteromodal convergence regions in higher order association cortex (Mesulam, 1998). Regions like the superior temporal sulcus/middle temporal gyrus (Beauchamp et al., 2004), posterior parietal cortex (Noppeney et al., 2008), and inferior frontal cortex (Hein et al., 2007) have been shown to be crucially involved in crossmodal integration. However, recently evidence has accumulated suggesting multisensory interactions already in early sensory regions traditionally considered 'unisensory' (Driver and Noesselt, 2008; Foxe et al., 2002; Ghazanfar and Schroeder, 2006; Kayser et al., 2005; Merabet et al., 2007; Molholm et al., 2002; Schroeder et al., 2003). While the vast majority of these findings has been obtained with event-related potentials and functional magnetic resonance imaging, a growing number of studies have investigated cortical oscillations as potential correlates of multisensory processing. Regardless of whether crossmodal interactions take place directly between early sensory areas or between early sensory and association cortices, one may assume that if oscillatory synchronization subserves feature integration *within* one modality, it should also be involved in interactions *between* sensory systems.

This chapter assesses the existing empirical evidence for the involvement of cortical oscillatory activity in multisensory processing. An authoritative review on this topic has been provided by Senkowski et al. (2008). We will focus on human studies using electro- or magnetoencephalography (EEG or MEG) and on the relevance of activity in the higher frequencies. As the conclusions concerning cortical generators or possible functional interactions that can be drawn from studies on oscillatory activity depend on the methodological approach, the present overview will be organized by the types of findings reported. These include (i) topographically unspecific or widespread spectral activity, (ii) topographically local oscillatory activity at the electrode/sensor level or at the level of cortical sources, and (iii) cortico-cortical interactions as revealed by coherence or phase synchrony.

5.2 Topographically Unspecific or Widespread Activity

This section describes studies using recordings with a limited spatial resolution or reporting topographically unspecific or widespread effects. Sakowitz et al. (2001, 2005) investigated spectral EEG responses during passive stimulation with basic audiovisual stimuli. Sinusoidal tones and rectangular light stimuli were presented in separate blocks either unimodally or simultaneously. Bisensory stimuli gave rise to super-additive evoked responses in the gamma range (Sakowitz et al., 2001), but also in lower frequency bands from theta to beta: amplitudes to bimodal stimuli exceeded the sum of the unimodal responses. Multisensory gamma components were observed as early as 50 ms after the onset of stimulation and were topographically widespread (Sakowitz et al., 2005). Other studies have assessed crossmodal influences on active stimulus processing. Senkowski et al. (2006) also used simple stimuli (a 1000-Hz tone and a red circular disk) that were presented either uni- or bimodally in a task requiring speeded motor responses whenever a stimulus in either sensory modality was detected. Early evoked beta activity showed super-additive amplitude enhancements to bimodal stimuli over frontal, occipital, and sensory–motor cortex. Negative correlations between beta activity and reaction time suggested a direct link between multisensory oscillatory responses and behavioral facilitation effects. The detection and localization of somatosensory stimuli can be facilitated or inhibited by spatially congruent or incongruent visual stimuli, respectively. Spatially congruent visuotactile stimuli were accompanied by GBA (40–60 Hz) increases over occipito-parietal and frontal regions, whereas an enhanced theta-band response was observed in the incongruent condition, suggesting separate systems of multisensory interaction (Kanayama and Ohira, 2009).

In addition to spatiotemporal coincidence (Stein and Meredith, 1993), semantic congruency is an important principle of multisensory integration (Amedi et al., 2005; Doehrmann and Naumer, 2008). Crossmodal interactions in familiar object recognition were investigated in a study involving simultaneous presentations of animal pictures and vocalizations (Yuval-Greenberg and Deouell, 2007). Participants were instructed to attend to either the visual or auditory modality and recognize the presented animal. GBA was analyzed for visually defined regions of interest. Whereas the evoked gamma response was unaffected by semantic congruence, induced broadband GBA showing a wide scalp distribution with a peak at the posterior–midline region was increased for congruent compared with incongruent audiovisual stimulus pairs. This was interpreted as reflecting a role of oscillatory synchronization for the formation of object representations at conceptual multisensory levels.

Crossmodal illusions are powerful examples of multisensory integrative processes. A striking example for the crossmodal biasing of perception is the audiovisual sound-induced flash illusion where an eccentrically presented single flash of light accompanied by two auditory beeps is frequently perceived as two flashes (Shams et al., 2000). Bhattacharya et al. (2002) recorded EEG responses from three occipital electrodes for unimodal and bimodal stimulation. They compared trials where the illusion was perceived with trials where it was not experienced, i.e., where subjects perceived a single flash only. Illusion trials were accompanied by generally higher gamma-band responses around 40 Hz that showed super-additivity at latencies of 150–225 and 425–500 ms. The authors concluded that GBA represents a neurophysiologic correlate of the illusion which is perceived only when this activity exceeds a critical threshold.

5.3 Localized Activity

In the following we will give an overview of studies providing more specific information about the topography of oscillatory activity either at the surface or at the level of estimated cortical sources. Attentional influences on multisensory processing have been investigated in two studies by Senkowski et al. (2005, 2007). In a spatial attention experiment (Senkowski et al., 2005), unisensory auditory (sinusoidal tones), visual (horizontal square wave gratings), or multisensory audiovisual stimuli were presented in the left and right hemifields. Participants were instructed to attend to one or the other hemispace to detect intensity deviants in either modality. Higher early evoked gamma amplitudes at latencies of 40–50 ms post-stimulus onset were observed over medial frontal cortex both for attended compared with unattended bimodal stimuli and for attended audiovisual compared with attended unimodal stimuli. The attentional enhancement was not observed for unisensory stimuli. This was interpreted as an early attention effect affecting the integrative processing of audiovisual stimuli. In a second study subjects had to attend to either auditory or visual stimuli presented at variable stimulus-onset asynchronies (Senkowski et al., 2007). Here the same stimuli as in the earlier, spatial attention study by the same group were presented centrally. While no attentional modulation of oscillatory activity was found, multisensory interactions were observed in the early evoked gamma responses when the auditory and visual stimuli were presented with the closest synchrony. These effects were found over midline frontal and occipital cortex at latencies of 30–80 and 60–120 ms, respectively. These findings were in line with the known importance of temporal synchrony for crossmodal interactions (Meredith, 2002). A study assessing crossmodal influences of tactile stimulation on behavioral and magnetoencephalographic responses to visual stimuli found both shorter response times and enhanced occipital gamma synchronization for bimodal compared with unimodal visual stimulation (Bauer et al., 2009). Moreover, the amplitude of induced occipital GBA served as a predictor of reaction times, supporting the relevance of oscillatory activity for crossmodal integration.

A topographically more fine-grained analysis of the sound-induced flash illusion was performed by Mishra et al. (2007). Single 5-ms light flashes presented at 20° of visual angle to the left of fixation could be accompanied by single 10-ms noise bursts or pairs presented with a stimulus-onset asynchrony of 70 ms. In addition to event-related potentials, oscillatory activity recorded from 64 EEG electrodes was investigated in those half of the subjects who experienced the illusion more frequently (on average in 57% of trials). Two components of increased GBA at occipital sites were observed for trials during which the illusion was perceived compared with when it was not experienced. Activity at frequencies of 25–35 Hz peaked at 130 ms, and a later GBA enhancement at 32–40 Hz was observed at latencies around 220 ms. These oscillatory components were interpreted as forming part of a sequence of rapid interactions between unisensory and heteromodal cortical areas underlying multisensory integration.

We investigated the processing of illusory acoustic percepts in an audiovisual paradigm based on a modified version of the McGurk effect (McGurk and MacDonald, 1976). A sequence of audiovisual stimuli was presented with frequent presentations of congruent combinations of the acoustic syllable /ta/ and a short video of the corresponding mouth movement (standard stimuli) and infrequent presentations of combinations including either an incongruent syllable or an incongruent mouth movement (acoustic or visual deviants, respectively) (Kaiser et al., 2005). While the former represented real acoustic deviance, the latter frequently induced illusory acoustic changes. The participants were instructed to detect any kind of perceived acoustic change. We employed a statistical probability mapping approach (Kaiser et al., 2000) to compare spectral responses in MEG to either type of deviant with the standard stimuli. Acoustic deviants gave rise to GBA increases at around 75 Hz in two sensors over left inferior frontal cortex. GBA (72–80 Hz) enhancements to visual deviants were localized over posterior parietal cortex peaking at 160 ms after the onset of the acoustic stimulus, over midline occipital cortex at 270 ms, and over left inferior frontal cortex at 320 ms (Fig. 5.1). The posterior GBA increases may have reflected networks involved in the processing of visual speech-related movements and the top-down modulation of earlier visual cortex by higher posterior parietal motion processing regions. The left inferior frontal activation common to both types of deviants was consistent with the topography

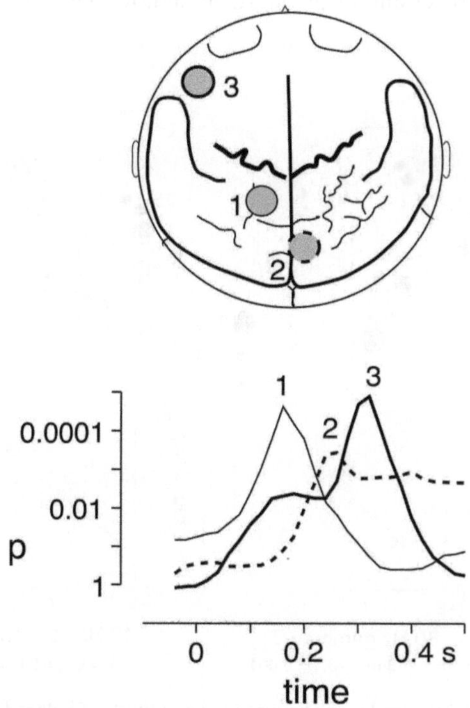

Fig. 5.1 Topographies and time courses of GBA enhancements in response to visual deviants in the audiovisual change detection study by Kaiser et al. (2005). *Top*: Locations of sensors showing spectral amplitude increases at ~72–80 Hz for visual deviants compared with congruent audiovisual standards. Sensors are projected onto two-dimensional cortical surface maps including some major anatomical landmarks (seen from above, nose up). GBA increases were localized in sensors over left posterior parietal cortex (1), midline occipital cortex (2) and left inferior frontal cortex (3). *Bottom*: statistical (p value) time courses of spectral amplitude differences for the three sensors depicted above between 50 ms before and 500 ms after stimulus onset. Reprinted from Kaiser et al. (2005) with permission from Oxford University Press

of GBA increases to purely acoustic phonetic mismatch found in an earlier study (Kaiser et al., 2002). This suggested crossmodal influences on networks signaling acoustic pattern changes even in the absence of physical sound deviance.

During this audiovisual oddball task, subjects showed a considerable variability in their ability to perceive acoustic deviance caused by changes in either the visual or the acoustic part of the audiovisual stimuli. We assessed the neural correlates of detection accuracy by correlating stimulus-related changes in oscillatory activity and behavioral performance (Kaiser et al., 2006). The detection of real acoustic deviants was positively associated with induced GBA at 42 Hz between 200 and 300 ms over left superior temporal cortex, and it was correlated negatively with evoked gamma responses at 41 Hz between 220 and 240 ms over occipital areas. In contrast, the detection of illusory acoustic changes induced by visual deviants was associated with increased oscillatory activity centered at frequencies around 80 Hz at latencies of 250–350 ms over midline occipital cortex (Fig. 5.2). These findings suggested that the processing of incongruent audiovisual stimuli relies on network

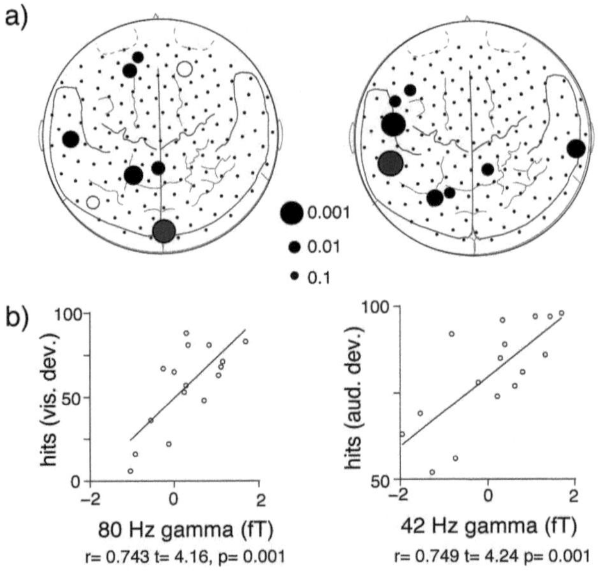

Fig. 5.2 (**a**) Topography of correlations between the amplitude of induced oscillatory activity and detection rate for visual deviants at 80 Hz (*left map*) and acoustic deviants at 42 Hz (*right map*) (Kaiser et al., 2006). *Circle size* symbolizes the statistical probability of correlation coefficients at each sensor. *Filled circles* show positive correlations; open circles depict negative correlations. The statistically most robust correlations were observed for a midline occipital sensor for visual deviants and for a left temporal sensor for acoustic deviants. (**b**) Correlation plots for the sensors with the most pronounced correlations for visual and acoustic deviants are shown in the left and right graphs, respectively. Percentages of correctly detected deviants (y-axis) and spectral amplitude difference values for deviant minus standard stimuli (x-axis) are plotted across participants. Reprinted from Kaiser et al. (2006), copyright (2006), with permission from Elsevier

synchronization in early sensory areas involved in the analysis of information in the modality where a change compared to the congruent standard stimulus took place.

Oscillatory responses to congruent versus incongruent audiovisual information were investigated in a symbolic matching task (Widmann et al., 2007). While in standard trials, simple musical score-like visual patterns predicted corresponding sound patterns, occasionally a single sound was incongruent with the corresponding visual symbol. Participants had to judge the congruence of the visual and auditory patterns. Here matching sounds were accompanied by an increase of evoked spectral EEG activity at 40 Hz peaking as early as 42 ms post-stimulus onset. Both the latency and the frontal topography of this activation were compatible with sources in the auditory cortices. This was followed by a power increase of induced GBA at a latency of around 150 ms. No gamma responses were observed for incongruent sounds. Apparently crossmodal matching between an expectation and a sensory input was performed at very early levels of auditory processing. The later GBA increase was interpreted as reflecting the formation of a coherent bimodal event representation. Semantic matching was investigated also in a visual-to-auditory object priming task (Schneider et al., 2008). Participants performed a semantic categorization task on objects designated by auditory stimuli that were preceded either by matching or by nonmatching visual primes. Sounds and photographs of natural objects were employed. Accuracy and speed of auditory recognition were enhanced by congruent visual primes. Increased GBA (40–50 Hz) was observed at fronto-central and bilateral temporal electrodes between 120 and 180 ms following auditory stimulus onset for semantically congruent stimulus pairs. Source analysis yielded a maximal increase in left middle temporal gyrus, an area known to be involved both in the processing of complex auditory stimuli (Doehrmann et al., 2008; Lewis et al., 2004) and in multisensory processing (Beauchamp, 2005). These findings were interpreted as support for the notion that GBA reflects crossmodal semantic matching in multisensory convergence regions. The same stimuli and temporal trial structure, but a different task, was used in a further study from the same laboratory (Senkowski et al., 2009). Here the subjects actively compared whether or not visual and auditory stimuli belonged to the same object. While there was no difference in reaction times, error rates were higher for congruent than incongruent trials. The topographic analysis of oscillatory signals in EEG revealed increased GBA (30–40 Hz) for matching compared with nonmatching stimulus pairs. This activity was found at latencies of 120–180 ms after the onset of the auditory stimulus over central and left frontal cortex. The involvement of GBA in multisensory matching in working memory was in line with the hypothesized role of oscillatory synchronization for the integrative processing of sensory inputs across modalities.

5.4 Cortico-cortical Interactions

This section deals with the evidence for synchronization between cortical areas underlying multisensory integration. Recent neurophysiologic studies in animals have demonstrated inter-areal synchronization. For example, Lakatos et al. (2007)

have shown that somatosensory inputs reset the phase of ongoing neuronal oscillations in primary auditory cortex, resulting in amplified neuronal responses for auditory inputs arriving during a high-excitability phase. Comparable findings have been obtained also for visual influences on oscillations in auditory cortex (Kayser et al., 2008). Recording concurrently from auditory cortex and superior temporal sulcus during presentations of dynamic faces and voices compared with unimodal stimulation, Ghazanfar et al. (2008) have found strengthened functional interactions between both areas as reflected by gamma-band correlations and tighter phase relationships, see Chapter 9. The same mechanism was found also to underlie the multisensory integration of auditory and visual looming signals (Maier et al., 2008).

In humans to our knowledge only three studies have demonstrated increased cortico-cortical interactions during multisensory processing. Hummel and Gerloff (2005) assessed interhemispheric synchronization in the form of EEG alpha coherence during a visuotactile integration paradigm involving the comparison of Braille stimuli and visual dot patterns. The comparison of trials with good versus poor performance revealed differences in task-related coherences for connections involving occipital and left lateral central electrodes. Apparently the ability to generate topographically specific long-range interregional synchrony was functionally relevant for behavioral success in this crossmodal integration task. The attentive processing of synchronous or asynchronous audiovisual speech streams leading to perceptually coherent or incoherent audiovisual speech, respectively, was investigated by Doesburg et al. (2008). Here asynchronous compared with synchronous audiovisual speech was associated with increased gamma-phase synchrony between frontal and left posterior EEG electrodes. This unexpected finding was interpreted in terms of increased processing resources needed to evaluate whether the streams could be integrated that determined the level of network activation. Oscillatory synchronization in EEG was assessed also during the rubber-hand illusion where synchronous visuotactile stimulation produces the illusory perception of a rubber hand belonging to one's own body (Kanayama et al., 2007). Supporting a role of high-frequency oscillations in the integrative processing of crossmodal stimuli, parietal gamma-band (40–50 Hz) power was found to increase 200–250 ms after stimulation in the congruent condition only. In addition, increased gamma-band phase synchrony across distant electrodes was observed in the congruent condition compared with the incongruent or unimodal conditions. Importantly, gamma-phase synchrony was correlated with behavioral indices of the intermodal illusion.

5.5 Conclusions

In summary, an increasing number of human EEG and MEG studies has reported modulations of oscillatory activity during multisensory processing. Effects were frequently observed in the higher frequencies (>20 Hz: high beta and gamma ranges) but slower frequencies, e.g., in the theta or alpha bands, also seem to be involved in crossmodal integration. The reported effects range from topographically widespread increases in spectral activity (Sakowitz et al., 2001; Yuval-Greenberg and Deouell,

2007) via more local enhancements involving sensory regions (Kaiser et al., 2005; Mishra et al., 2007; Widmann et al., 2007) or association cortices (Schneider et al., 2008; Senkowski et al., 2005) to evidence for interactions between cortical areas (Doesburg et al., 2008; Hummel and Gerloff, 2005; Kanayama et al., 2007). While studies in monkeys have provided direct evidence for oscillatory coupling between regions involving primary sensory cortex during multisensory processing (Ghazanfar et al., 2008; Maier et al., 2008), at scalp level, the role of oscillatory signals for multisensory integration is still far from clear. On the basis of the existing literature it is hard to decide whether activations reflect local processing in classical multisensory integration regions or in sensory areas or whether they indicate true interactions between specific task-relevant regions. In addition, conclusions about the localization of cortical generators of the observed signals should be treated with caution given the limited spatial resolution of EEG or MEG. More advanced source modeling and nonlinear approaches for the analysis of cortico-cortical interactions may help to elucidate these issues in the future.

While enhancements of oscillatory activity were reported for temporally (Senkowski et al., 2007), spatially (Kanayama and Ohira, 2009), or semantically matching bimodal stimulation (Schneider et al., 2008; Senkowski et al., 2009; Yuval-Greenberg and Deouell, 2007), interestingly some studies have found increased GBA to incongruent stimulus combinations (Doesburg et al., 2008; Kaiser et al., 2005). Apparently task requirements and attentional factors seem to modulate the relationship between crossmodal stimulus congruence and oscillatory responses. While there is little direct evidence for oscillatory activity underlying cortico-cortical synchronization in humans, the reported findings may be seen as compatible with a broader view of GBA reflecting a basic feature of cortical processing associated with a variety of cognitive functions (Jensen et al., 2007).

In general the search for correlates of crossmodal interactions might be most promising in situations where integration is both highly demanding and functionally relevant, as under conditions of poor signal-to-noise ratios and when the correct matching of stimuli between modalities is of high importance. Furthermore, the functional relevance of oscillatory activity for multisensory processing will be demonstrated most convincingly by showing its value as a predictor for behavioral performance (Hummel and Gerloff, 2005; Kaiser et al., 2006; Kanayama et al., 2007; Senkowski et al., 2006).

References

Amedi A, von Kriegstein K, van Atteveldt NM, Beauchamp MS, Naumer MJ (2005) Functional imaging of human crossmodal identification and object recognition. Exp Brain Res 166:559–571

Bauer M, Oostenveld R, Fries P (2009) Tactile stimulation accelerates behavioral responses to visual stimuli through enhancement of occipital gamma-band activity. Vision Res 49:931–942

Bauer M, Oostenveld R, Peeters M, Fries P (2006) Tactile spatial attention enhances gamma-band activity in somatosensory cortex and reduces low-frequency activity in parieto-occipital areas. J Neurosci 26:490–501

Beauchamp MS (2005) See me, hear me, touch me: multisensory integration in lateral occipital-temporal cortex. Curr Opin Neurobiol 15:145–153

Beauchamp MS, Lee KE, Argall BD, Martin A (2004) Integration of auditory and visual information about objects in superior temporal sulcus. Neuron 41:809–823

Bhattacharya J, Shams L, Shimojo S (2002) Sound-induced illusory flash perception: role of gamma band responses. Neuroreport 13:1727–1730

Busch NA, Herrmann CS, Müller MM, Lenz D, Gruber T (2006) A cross-laboratory study of event-related gamma activity in a standard object recognition paradigm. Neuroimage 33:1169–1177

Doehrmann O, Naumer MJ (2008) Semantics and the multisensory brain: how meaning modulates processes of audio-visual integration. Brain Res 1242:136–150

Doehrmann O, Naumer MJ, Volz S, Kaiser J, Altmann CF (2008) Probing category selectivity for environmental sounds in the human auditory brain. Neuropsychologia 46:2776–2786

Doesburg SM, Emberson LL, Rahi A, Cameron D, Ward LM (2008) Asynchrony from synchrony: long-range gamma-band neural synchrony accompanies perception of audiovisual speech asynchrony. Exp Brain Res 185:11–20

Driver J, Noesselt T (2008) Multisensory interplay reveals crossmodal influences on 'sensory-specific' brain regions, neural responses, and judgments. Neuron 57:11–23

Eckhorn R, Bauer R, Jordan W, Brosch M, Kruse W, Munk M, Reitboeck HJ (1988) Coherent oscillations: a mechanism of feature linking in the visual cortex? Multiple electrode and correlation analyses in the cat. Biol Cybern 60:121–130

Engel AK, Singer W (2001) Temporal binding and the neural correlates of sensory awareness. Trends Cogn Sci 5:16–25

Foxe JJ, Wylie GR, Martinez A, Schroeder CE, Javitt DC, Guilfoyle D, Ritter W, Murray MM (2002) Auditory-somatosensory multisensory processing in auditory association cortex: an fMRI study. J Neurophysiol 88:540–543

Fries P (2005) A mechanism for cognitive dynamics: neuronal communication through neuronal coherence. Trends Cogn Sci 9:474–480

Fries P, Nikolic D, Singer W (2007) The gamma cycle. Trends Neurosci 30:309–316

Fries P, Roelfsema PR, Engel AK, König P, Singer W (1997) Synchronization of oscillatory responses in visual cortex correlates with perception in interocular rivalry. Proc Natl Acad Sci U S A 94:12699–12704

Ghazanfar AA, Schroeder CE (2006) Is neocortex essentially multisensory? Trends Cogn Sci 10:278–285

Ghazanfar AA, Chandrasekaran C, Logothetis NK (2008) Interactions between the superior temporal sulcus and auditory cortex mediate dynamic face/voice integration in rhesus monkeys. J Neurosci 28:4457–4469

Gray CM, König P, Engel AK, Singer W (1989) Oscillatory responses in cat visual cortex exhibit inter-columnar synchronization which reflects global stimulus properties. Nature 338:334–337

Gruber T, Müller MM, Keil A, Elbert T (1999) Selective visual-spatial attention alters induced gamma band responses in the human EEG. Clin Neurophysiol 110:2074–2085

Gruber T, Tsivilis D, Montaldi D, Müller MM (2004) Induced gamma band responses: an early marker of memory encoding and retrieval. Neuroreport 15:1837–1841

Hein G, Doehrmann O, Muller NG, Kaiser J, Muckli L, Naumer MJ (2007) Object familiarity and semantic congruency modulate responses in cortical audiovisual integration areas. J Neurosci 27:7881–7887

Herrmann CS, Munk MH, Engel AK (2004) Cognitive functions of gamma-band activity: memory match and utilization. Trends Cogn Sci 8:347–355

Hummel F, Gerloff C (2005) Larger interregional synchrony is associated with greater behavioral success in a complex sensory integration task in humans. Cereb Cortex 15:670–678

Jensen O, Kaiser J, Lachaux JP (2007) Human gamma-frequency oscillations associated with attention and memory. Trends Neurosci 30:317–324

Jokisch D, Jensen O (2007) Modulation of gamma and alpha activity during a working memory task engaging the dorsal or ventral stream. J Neurosci 27:3244–3251

Kaiser J, Lutzenberger W, Ackermann H, Birbaumer N (2002) Dynamics of gamma-band activity induced by auditory pattern changes in humans. Cereb Cortex 12:212–221

Kaiser J, Ripper B, Birbaumer N, Lutzenberger W (2003) Dynamics of gamma-band activity in human magnetoencephalogram during auditory pattern working memory. Neuroimage 20: 816–827

Kaiser J, Hertrich I, Ackermann H, Lutzenberger W (2006) Gamma-band activity over early sensory areas predicts detection of changes in audiovisual speech stimuli. Neuroimage 30:1376–1382

Kaiser J, Lutzenberger W, Preissl H, Ackermann H, Birbaumer N (2000) Right-hemisphere dominance for the processing of sound-source lateralization. J Neurosci 20:6631–6639

Kaiser J, Hertrich I, Ackermann H, Mathiak K, Lutzenberger W (2005) Hearing lips: gamma-band activity during audiovisual speech perception. Cereb Cortex 15:646–653

Kaiser J, Heidegger T, Wibral M, Altmann CF, Lutzenberger W (2008) Distinct gamma-band components reflect the short-term memory maintenance of different sound lateralization angles. Cereb Cortex 18:2286–2295

Kanayama N, Ohira H (2009) Multisensory processing and neural oscillatory responses: separation of visuotactile congruency effect and corresponding electroencephalogram activities. Neuroreport 20:289–293

Kanayama N, Sato A, Ohira H (2007) Crossmodal effect with rubber hand illusion and gamma-band activity. Psychophysiology 44:392–402

Kayser C, Petkov CI, Logothetis NK (2008) Visual modulation of neurons in auditory cortex. Cereb Cortex 18:1560–1574

Kayser C, Petkov CI, Augath M, Logothetis NK (2005) Integration of touch and sound in auditory cortex. Neuron 48:373–384

Lakatos P, Chen CM, O'Connell MN, Mills A, Schroeder CE (2007) Neuronal oscillations and multisensory interaction in primary auditory cortex. Neuron 53:279–292

Lewis JW, Wightman FL, Brefczynski JA, Phinney RE, Binder JR, DeYoe EA (2004) Human brain regions involved in recognizing environmental sounds. Cereb Cortex 14:1008–1021

Lutzenberger W, Pulvermüller F, Elbert T, Birbaumer N (1995) Visual stimulation alters local 40-Hz responses in humans: an EEG-study. Neurosci Lett 183:39–42

Lutzenberger W, Ripper B, Busse L, Birbaumer N, Kaiser J (2002) Dynamics of gamma-band activity during an audiospatial working memory task in humans. J Neurosci 22: 5630–5638

Maier JX, Chandrasekaran C, Ghazanfar AA (2008) Integration of bimodal looming signals through neuronal coherence in the temporal lobe. Curr Biol 18:963–968

McGurk H, MacDonald J (1976) Hearing lips and seeing voices. Nature 264:746–748

Merabet LB, Swisher JD, McMains SA, Halko MA, Amedi A, Pascual-Leone A, Somers DC (2007) Combined activation and deactivation of visual cortex during tactile sensory processing. J Neurophysiol 97:1633–1641

Meredith MA (2002) On the neuronal basis for multisensory convergence: a brief overview. Cogn Brain Res 14:31–40

Meredith MA, Stein BE (1983) Interactions among converging sensory inputs in the superior colliculus. Science 221:389–391

Mesulam MM (1998) From sensation to cognition. Brain 121:1013–1052

Mishra J, Martinez A, Sejnowski TJ, Hillyard SA (2007) Early cross-modal interactions in auditory and visual cortex underlie a sound-induced visual illusion. J Neurosci 27:4120–4131

Molholm S, Ritter W, Murray MM, Javitt DC, Schroeder CE, Foxe JJ (2002) Multisensory auditory-visual interactions during early sensory processing in humans: a high-density electrical mapping study. Cogn Brain Res 14:115–128

Müller MM, Keil A (2004) Neuronal synchronization and selective color processing in the human brain. J Cogn Neurosci 16:503–522

Noppeney U, Josephs O, Hocking J, Price CJ, Friston KJ (2008) The effect of prior visual information on recognition of speech and sounds. Cereb Cortex 18:598–609

Osipova D, Takashima A, Oostenveld R, Fernandez G, Maris E, Jensen O (2006) Theta and gamma oscillations predict encoding and retrieval of declarative memory. J Neurosci 26:7523–7531
Sakowitz OW, Quiroga RQ, Schurmann M, Basar E (2001) Bisensory stimulation increases gamma-responses over multiple cortical regions. Cogn Brain Res 11:267–279
Sakowitz OW, Quian Quiroga R, Schurmann M, Basar E (2005) Spatio-temporal frequency characteristics of intersensory components in audiovisually evoked potentials. Cogn Brain Res 23:316–326
Schneider TR, Debener S, Oostenveld R, Engel AK (2008) Enhanced EEG gamma-band activity reflects multisensory semantic matching in visual-to-auditory object priming. Neuroimage 42:1244–1254
Schroeder CE, Smiley J, Fu KG, McGinnis T, O'Connell MN, Hackett TA (2003) Anatomical mechanisms and functional implications of multisensory convergence in early cortical processing. Int J Psychophysiol 50:5–17
Senkowski D, Talsma D, Herrmann CS, Woldorff MG (2005) Multisensory processing and oscillatory gamma responses: effects of spatial selective attention. Exp Brain Res 166:411–426
Senkowski D, Molholm S, Gomez-Ramirez M, Foxe JJ (2006) Oscillatory beta activity predicts response speed during a multisensory audiovisual reaction time task: a high-density electrical mapping study. Cereb Cortex 16:1556–1565
Senkowski D, Schneider TR, Foxe JJ, Engel AK (2008) Crossmodal binding through neural coherence: implications for multisensory processing. Trends Neurosci 31:401–409
Senkowski D, Schneider TR, Tandler F, Engel AK (2009) Gamma-band activity reflects multisensory matching in working memory. Exp Brain Res 198:363–372
Senkowski D, Talsma D, Grigutsch M, Herrmann CS, Woldorff MG (2007) Good times for multisensory integration: effects of the precision of temporal synchrony as revealed by gamma-band oscillations. Neuropsychologia 45:561–571
Shams L, Kamitani Y, Shimojo S (2000) What you see is what you hear. Nature 408:788
Singer W, Engel AK, Kreiter A, Munk MHJ, Neuenschwander S, Roelfsema PR (1997) Neuronal assemblies: necessity, signature and detectability. Trends Cogn Sci 1:252–261
Stein BE, Meredith MA (1993) The merging of the senses. MIT Press, Cambridge, MA
Tallon-Baudry C, Bertrand O, Delpuech C, Pernier J (1996) Stimulus specificity of phase-locked and non-phase-locked 40 Hz visual responses in human. J Neurosci 16:4240–4249
Tallon-Baudry C, Bertrand O, Peronnet F, Pernier J (1998) Induced gamma-band activity during the delay of a visual short-term memory task in humans. J Neurosci 18:4244–4254
Wallace MT, Meredith MA, Stein BE (1998) Multisensory integration in the superior colliculus of the alert cat. J Neurophysiol 80:1006–1010
Widmann A, Gruber T, Kujala T, Tervaniemi M, Schroger E (2007) Binding symbols and sounds: evidence from event-related oscillatory gamma-band activity. Cereb Cortex 17:2696–2702
Womelsdorf T, Schoffelen JM, Oostenveld R, Singer W, Desimone R, Engel AK, Fries P (2007) Modulation of neuronal interactions through neuronal synchronization. Science 316: 1609–1612
Wyart V, Tallon-Baudry C (2008) Neural dissociation between visual awareness and spatial attention. J Neurosci 28:2667–2679
Yuval-Greenberg S, Deouell LY (2007) What you see is not (always) what you hear: induced gamma band responses reflect cross-modal interactions in familiar object recognition. J Neurosci 27:1090–1096

Chapter 6
Multisensory Functional Magnetic Resonance Imaging

Marcus J. Naumer, Jasper J. F. van den Bosch, Andrea Polony, and Jochen Kaiser

6.1 Introduction

Research on multisensory integration in general and on multisensory object perception in particular is a relatively young scientific endeavor which started about 30 years ago. During the last decade this research field received rapidly growing interest which resulted in an exponential growth of publication numbers (Fig. 6.1a). This interest in multisensory research was especially boosted by two events: the publication of a seminal book *The Merging of the Senses* (Stein and Meredith, 1993) and the birth of the *International Multisensory Research Forum* (1999).

At the same time, functional magnetic resonance imaging (fMRI) became the prime research methodology in human neuroscience. Based on these parallel developments, it is not surprising that the relative importance of fMRI for multisensory research has been constantly growing at a high rate (Fig. 6.1b, c). Over recent years, fMRI has become the prime imaging modality for multisensory research (Fig. 6.1d). Based on combined improvements of scanner hardware, experimental designs, and data analysis tools, the general capabilities of fMRI continue to evolve. Discussions within the multisensory fMRI community traditionally focused on principles of and appropriate statistical criteria for multisensory integration. The recent availability of more sophisticated experimental designs and increasingly sensitive (multivariate) statistical tools has enabled multisensory researchers to (noninvasively) differentiate between regional and neuronal convergence and to reveal the connectional basis of multisensory processing. In this chapter we will start with a brief description of the particular strengths and weaknesses of fMRI in general. In its core sections, we will then review and discuss current approaches to differentiate between regional and neuronal convergence and to reveal the connectional basis of multisensory processing.

M.J. Naumer (✉)
Institute of Medical Psychology, Faculty of Medicine, Goethe University,
Heinrich-Hoffmann-Str. 10, 60528 Frankfurt am Main, Germany
e-mail: m.j.naumer@med.uni-frankfurt.de

M.J. Naumer, J. Kaiser (eds.), *Multisensory Object Perception in the Primate Brain*,
DOI 10.1007/978-1-4419-5615-6_6,

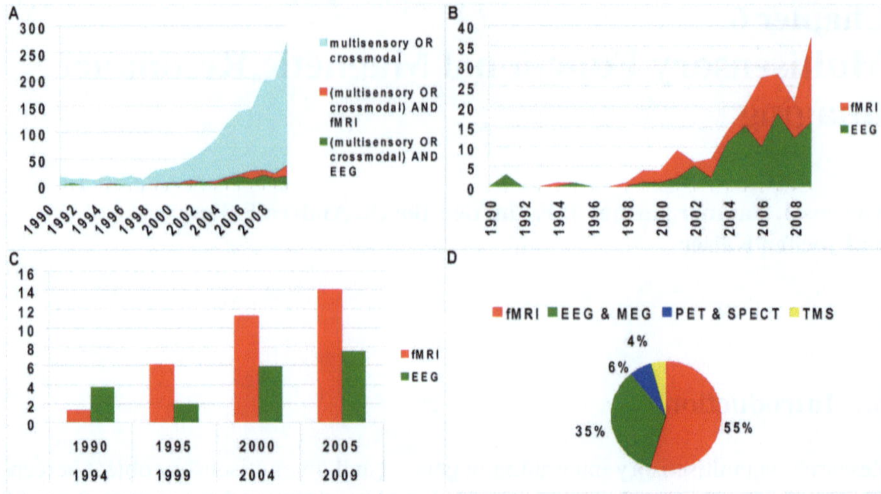

Fig. 6.1 Descriptive statistics on the relevance of fMRI for the field of multisensory research. (**a**) The publications per year (for the period of 1990–2009) according to PubMed searches for 'multisensory OR crossmodal' (in *light blue*), '(multisensory OR crossmodal) AND fMRI' (in *red*), and '(multisensory OR crossmodal) AND EEG' (in *green*). (**b**) The publications per year (for the period of 1990–2009) according to PubMed searches for '(multisensory OR crossmodal) AND fMRI' (in *red*) and '(multisensory OR crossmodal) AND EEG' (in *green*). (**c**) The percentage of 'multisensory OR crossmodal' publications which used fMRI (in *red*) or EEG (in *green*). (**d**) The relative frequency with which diverse noninvasive techniques were used in multisensory neuroimaging studies (in the period of 1990–2009)

6.2 Limitations and Strengths of fMRI

The blood-oxygen-level-dependent (BOLD) fMRI signal is a vascular measure which indirectly reflects neuronal activation with both a substantial delay (of several seconds) and a low temporal resolution (in the range of seconds). The exact neuronal basis of the fMRI BOLD response is still unclear (Logothetis, 2008; see also Chapter 8 of this volume by Kayser and colleagues). Due to inter-individual variations in brain size, analyses on the group-level require a substantial amount of spatial normalization, and different neuroanatomical reference frames (such as the Montreal Neurological Institute (MNI) and Talairach spaces) are used in parallel. Substantial progress has been made by adding sophisticated cortex-based morphing approaches (e.g., van Essen et al., 1998; Fischl et al., 1999; Goebel et al., 2006) which allow correcting the functional data for the substantial inter-individual variation in cortical gyrification.

The main strength of BOLD fMRI is its non-invasiveness allowing investigations in living human subjects without any problematic side effects. While fMRI enables the acquisition of functional data with a high spatial resolution (in the range of millimeters), it also allows to measure the entire brain including subcortical and cerebellar brain regions. This is especially important for the investigation of widely distributed brain networks and the connectivity between the involved brain regions. Based on the continuing development of scanners with higher magnetic field strengths ('ultra-high field fMRI') spatial resolution is expected to increase in the future.

6.3 Detection of Multisensory Integration Regions

Inspired by the intriguing results of multisensory single-cell studies in cats and monkeys (see Stein and Meredith, 1993; Stein and Stanford, 2008 for reviews), multisensory fMRI researchers aimed to test the spatial, temporal, and inverse effectiveness rules originally described by invasive electrophysiological studies. Calvert and colleagues (Calvert, 2001; Calvert et al., 2001) proposed that in order to detect multisensory integration sites, i.e., regions containing multisensory integrative (MSI) cells with a superadditive activation profile, researchers should apply the so-called *superadditivity* criterion, i.e., the multisensory (e.g., audio-visual, AV) response has to significantly exceed the linear sum of the two unisensory responses (i.e., AV > A + V). This proposal initiated a still ongoing debate regarding the appropriate statistical criteria for multisensory fMRI studies. As has been argued by Laurienti and colleagues (2005), the BOLD fMRI signal measures synaptic processes rather than the spiking output of neurons, and the relatively low proportion of superadditive neurons might not be resolvable at the level of a particular brain region using fMRI. Moreover, the limited dynamic range of the BOLD fMRI signal might preclude the detection of superadditivity due to ceiling effects (Haller et al., 2006).

Ceiling effects might be prevented by using less effective (i.e., degraded) experimental stimuli, which elicit smaller BOLD signal amplitudes in response to unimodal stimulation. Stevenson and colleagues (Stevenson and James, 2009; Stevenson et al., 2007) applied *stimulus degradation* in a series of audio-visual experiments including speech and nonspeech object stimuli and revealed superadditive effects. More recently, they applied this strategy also in the visuo-haptic domain (Kim and James, 2009; Stevenson et al., 2009) and generalized their approach by proposing an *additive factors design* (as described in detail by James and Kim in Chapter 13 of this volume) to differentiate between regional and neuronal convergence.

However, based on the arguments discussed above the majority of researchers in the field has favored the so-called *max criterion*, i.e., the bimodal response has to significantly exceed the maximal unisensory response in a given brain region in order to render this brain region multisensory (Beauchamp, 2005; Doehrmann and Naumer, 2008; Driver and Noesselt, 2008; Goebel and van Atteveldt, 2009; Laurienti et al., 2005).

Is the max criterion also applicable in studies of *tri-sensory integration*? A naïve application of this criterion would require the tri-sensory response to exceed each of the unisensory responses. A possible activation profile of such a tri-sensory

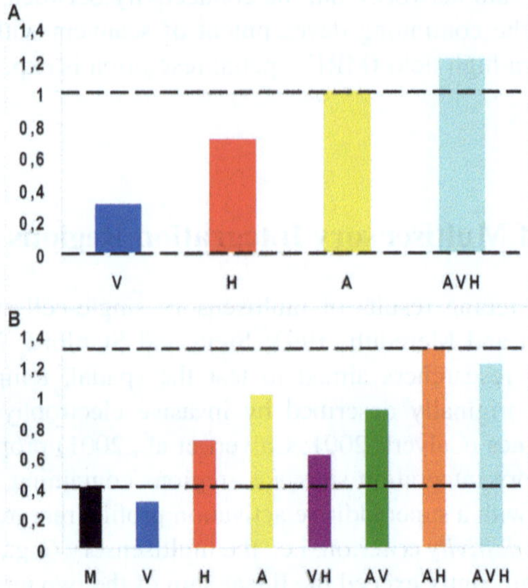

Fig. 6.2 Extending the max criterion to detect tri-sensory integration. According to the max criterion brain sites of tri-sensory (e.g., audio-visuo-haptic) integration should show a similar activation pattern as shown in (**a**) with the audio-visuo-haptic (AVH) response (in *light blue*) exceeding the respective auditory (A, in *yellow*), visual (V, in *blue*), and haptic (H, in *red*) responses. Panel (**c**), however, illustrates that this is a necessary but not a sufficient condition. Such a tri-sensory candidate region should be further characterized based on additional comparisons between the AVH response and the respective bi-sensory responses (i.e., audio-visual, AV, in *green*; audio-haptic, AH, in *orange*; and visuo-haptic, VH, in *lilac*). If one of these bi-sensory responses (e.g., AH) exceeds the tri-sensory response, this would clearly indicate a region of bi-sensory instead of tri-sensory integration

(e.g., audio-visuo-haptic, AVH > max [A, V, H]) candidate region is illustrated in Fig. 6.2a. Under a particular type of bimodal (e.g., audio-haptic) stimulation, however, this same region might respond more strongly to audio-haptic rather than tri-modal input (Fig. 6.2b; Polony et al., 2007). Thus, it appears necessary to include appropriate bimodal control conditions, and testing for tri-sensory integration should be based on an extended max criterion, namely AVH > max (AV, AH, VH).

Earlier fMRI studies on multisensory object perception have often determined integration regions based on *conjunctions of unimodal contrasts*, e.g., intact > scrambled objects (or textures) in both the visual and the haptic modalities (see Lacey et al., 2009 for a recent review). Since then, it became common practice to compare BOLD responses to bimodal and unimodal experimental conditions, e.g., when applying the max criterion. However, bimodal and unimodal experimental stimuli might systematically differ in stimulus complexity and might thus bias the respective statistical comparisons, as has recently been argued by Hocking and Price (2008).

One option to circumvent this problem is the computation of *statistical interaction effects*. Under the null hypothesis that different (e.g., auditory and visual) inputs are processed independently (i.e., are not integrated) this might be implemented in a simple 2*2 factorial design with *Auditory Input* (off vs. on) * *Visual Input* (off vs. on; see Noppeney, in press, for an excellent overview of this topic).

Another option is the direct comparison of two or more bimodal conditions often involving a *congruency manipulation*. This has been inspired by the so-called *spatial* and *temporal rules* of multisensory integration as originally described in single-cell electrophysiology. Such congruency manipulations have been successfully used on the where (e.g., Meienbrock et al., 2007), when (e.g., Noesselt et al., 2007), and what dimensions (e.g., Hein et al., 2007; see Doehrmann and Naumer, 2008 for a recent review) or even involved several of these dimensions (e.g., van Atteveldt et al., 2007; Chapter 10 of this volume by Lewis).

More recently, several groups have demonstrated that *fMRI adaptation* designs (Grill-Spector and Malach, 2001; Weigelt et al., 2008) can also be used to elucidate the neural basis of multisensory object perception. Usually involving repeated presentations of both identical and more or less differing stimuli such adaptation designs allow determining whether a particular brain region actually contains a sufficiently large proportion of MSI cells or only a mixture of diverse unisensory neurons. While two visuo-haptic studies reported either repetition enhancement (James et al., 2002) or repetition suppression (Tal and Amedi, 2009), a recent audio-visual study of our group revealed both repetition suppression and enhancement in response to repetition of identical stimuli in the preferred and non-preferred sensory modality, respectively (Doehrmann et al., 2010; see also Hasson et al., 2007; van Atteveldt et al., 2008).

Finally, another recent trend concerns the use of *multivariate pattern analysis (MVPA)*. Inspired by the earlier finding of distributed and largely overlapping visual object representations (Haxby et al., 2001), this novel approach has already been applied successfully to classification studies in the visuo-haptic (Pietrini et al., 2004) and sensory–motor domains (Etzel et al., 2008). Here the basic question is whether the spatial activation pattern *within* a particular brain region allows stimulus classification only in one or in several modalities (Beauchamp et al., 2009; Haynes and Rees, 2006).

6.4 Connectivity

One major distinction of perspectives in neuroimaging is that of functional specialization vs. functional integration. The former, more common approach attempts to define the function or functional occupation of specific brain modules, irrespective of each other. However, since multisensory integration deals with the interaction of representations of diverse senses, most of which may be localized according to this view, its study needs to take into account such interactions in its description of the brain. In neuroimaging, such interactions are mainly described by the concept of connectivity. Several types of connectivity can be distinguished: anatomical, functional, and effective connectivity.

6.4.1 Anatomical Connectivity

Measures of anatomical connectivity describe the physical connections in the brain enabling neural communication. These can be quantified according to their density, number, or structure. The most common of these techniques of in vivo neuroimaging is diffusion tensor imaging (DTI), which estimates the diffusivity of water molecules. The local degree of preference for a certain direction is called fractional anisotropy, which can be mapped or visualized in the form of so-called tracts. Because such anatomical connectivity only changes to a measurable degree over longer time intervals, it can only inform about a limited range of questions like between-subjects comparison (Niogi and McCandliss, 2006; Rouw and Scholte, 2007) or longitudinal studies of developmental changes or neurological pathology (Beauchamp and Ro, 2008; see also Naumer and van den Bosch, 2009). Another, indirect application constitutes the informing of functional models of networks with structural constraints.

6.4.2 Functional Connectivity

Functional connectivity, i.e., the correlation between remote neurophysiological events (David et al., 2004), indicates functional coupling of brain regions. For any interaction between regions cooperating to integrate, e.g., object information from different modalities, such coupling is likely to arise. Therefore, methods detecting functional connectivity have been used to map networks of multisensory integration. Taking into account the dependencies between the measured variables, multivariate analysis, as opposed to the common univariate approaches used in conventional fMRI data analysis, lends itself to connectivity analysis. Common multivariate strategies are often applicable in a data-driven way, requiring no prior knowledge. Independent Component Analysis (ICA) is probably the most popular of these methods. Functionally connected networks have been mapped for different combinations of modalities and different levels of stimulus complexity (e.g., Eckert et al., 2008). In a comparative study, we recently found that self-organizing group ICA identified the functional networks of audio-visual processing with increased statistical sensitivity as compared to whole-brain regression analysis (Naumer et al., 2009b).

6.4.3 Effective Connectivity

Functional connectivity may describe a coupling between brain regions, but does not discern causality or directed influence. Thus, the direction of connectivity and the hierarchical role of a region within a network of regions remain unknown. Effective connectivity refers to relationships between regions for which these aspects are known. One measure of effective connectivity applied in multisensory research is the so-called psychophysiological interaction (PPI; Friston et al., 1997), where an interaction between a psychophysical manipulation and the regression slope of the time courses of two regions indicates an influence of one region on the other (von Kriegstein and Giraud, 2006; von Kriegstein et al., 2005; Noppeney et al., 2008). Another measure is Granger causality, where the past development of a region's time course can improve the prediction of another region's time course as compared to the past of the target itself. One implementation of this measure is Granger causality mapping (GCM, Roebroeck et al., 2005). GCM has been successfully used to investigate effective connectivity underlying visuo-haptic shape perception (Deshpande et al., 2008) and to establish that speech perception in auditory cortex is modulated by a script representation of the same message via heteromodal superior temporal regions (van Atteveldt et al., 2009).

Another method of determining effective connectivity is dynamic causal modeling (DCM; Friston et al., 2003), where the activity of the underlying neural sources is modeled with existing information and fitted to the data through a hemodynamic forward model. One recent example of a model informed by invasive electrophysiological data is the study by Werner and Noppeney (2009) which confirmed direct influence from visual cortex to auditory processing in superior temporal regions. For a discussion of differences between GCM and DCM see Roebroeck et al. (2009a, b).

6.4.4 Representational Connectivity

A fourth type of connectivity is the new and promising method of representational similarity analysis (Kriegeskorte et al., 2008a, b) which can yield information on how different brain regions represent certain classes of stimuli in a similar pattern, making use of multivariate pattern analysis (Mur et al., 2009).

6.5 Conclusion and Outlook

The consequent utilization of the recent improvements in fMRI methodology opens up new and promising avenues for noninvasive research on the neural basis of multisensory object perception. Often, however, the predominantly investigated cerebral cortex provides only half of the story. Thus, future studies should fully exploit the potential of MRI to acquire functional data with whole-brain coverage to assess the potentially critical contributions of subcortical structures such as the cerebellum (Naumer et al., 2009a), basal ganglia, and thalamus (Cappe et al., 2009; Naumer and van den Bosch, 2009). In order to further disentangle effects of bottom-up and top-down processing the implementation of different experimental tasks (from passive via implicit to explicit tasks) appears to be especially important. This would also allow relating multisensory object perception performance to human brain activation more closely (Werner and Noppeney, 2009).

Acknowledgments This work was supported by the German Ministry of Education and Research (BMBF) and Frankfurt Medical School (Intramural Young Investigator Program to M.J.N.). We are grateful to Sarah Weigelt for helpful suggestions and Christoph Bledowski and Yavor Yalachkov for their helpful comments to an earlier version of this chapter.

References

Beauchamp MS (2005) Statistical criteria in FMRI studies of multisensory integration. Neuroinformatics 3:93–113

Beauchamp MS, Laconte S, Yasar N (2009) Distributed representation of single touches in somatosensory and visual cortex. Hum Brain Mapp 30:3163–3171

Beauchamp MS, Ro T (2008) Neural substrates of sound-touch synesthesia after a thalamic lesion. J Neurosci 28:13696–13702

Calvert GA (2001) Crossmodal processing in the human brain: insights from functional neuroimaging studies. Cereb Cortex 11:1110–1123

Calvert GA, Hansen PC, Iversen SD, Brammer MJ (2001) Detection of audio-visual integration sites in humans by application of electrophysiological criteria to the BOLD effect. Neuroimage 14:427–438

Cappe C, Morel A, Barone P, Rouiller EM (2009) The thalamocortical projection systems in primate: an anatomical support for multisensory and sensorimotor interplay. Cerebral Cortex, Jan 15 [Epub ahead of print]

David O, Cosmelli D, Friston KJ (2004) Evaluation of different measures of functional connectivity using a neural mass model. Neuroimage 21:659–673

Deshpande G, Hu X, Stilla R, Sathian K (2008) Effective connectivity during haptic perception: a study using Granger causality analysis of functional magnetic resonance imaging data. Neuroimage 40:1807–1814

Doehrmann O, Naumer MJ (2008) Semantics and the multisensory brain: how meaning modulates processes of audio-visual integration. Brain Res 1242:136–150

Doehrmann O, Weigelt S, Altmann CF, Kaiser J, Naumer MJ (2010) Audio-visual fMRI adaptation reveals multisensory integration effects in object-related sensory cortices J Neurosci 30: 3370–3379

Driver J, Noesselt T (2008) Multisensory interplay reveals crossmodal influences on 'sensory-specific' brain regions, neural responses, and judgments. Neuron 57:11–23

Eckert MA, Kamdar NV, Chang CE, Beckmann CF, Greicius MD, Menon V (2008) A cross-modal system linking primary auditory and visual cortices: evidence from intrinsic fMRI connectivity analysis. Hum Brain Mapp 29:848–857

Etzel JA, Gazzola V, Keysers C (2008) Testing simulation theory with cross-modal multivariate classification of fMRI data. PLoS ONE 3:e3690

Fischl B, Sereno MI, Tootel RBH, Dale AM (1999) High-resolution intersubject averaging and a coordinate system for the cortical surface. Hum Brain Mapp 8:272–284

Friston KJ, Buechel C, Fink GR, Morris J, Rolls E, Dolan RJ (1997) Psychophysiological and modulatory interactions in neuroimaging. Neuroimage 6:218–229

Friston KJ, Harrison L, Penny W (2003) Dynamic causal modelling. Neuroimage 19: 1273–1302

Goebel R, Esposito F, Formisano E (2006) Analysis of Functional Image Analysis Contest (FIAC) data with BrainVoyager QX: from single-subject to cortically aligned group general linear model analysis and self-organizing group independent component analysis. Hum Brain Mapp 27:392–402

Goebel R, van Atteveldt N (2009) Multisensory functional magnetic resonance imaging: a future perspective. Exp Brain Res 198:153–164

Grill-Spector K, Malach R (2001) fMR-adaptation: a tool for studying the functional properties of human cortical neurons. Acta Psychol (Amst) 107:293–321

Haller S, Wetzel SG, Radue EW, Bilecen D (2006) Mapping continuous neuronal activation without an ON–OFF paradigm: initial results of BOLD ceiling fMRI. Eur J Neurosci 24:2672–2678

Hasson U, Skipper JI, Nusbaum HC, Small SL (2007) Abstract coding of audiovisual speech: beyond sensory representation. Neuron 56:1116–1126

Haynes JD, Rees G (2006) Decoding mental states from brain activity in humans. Nat Rev Neurosci 7:523–534

Haxby JV, Gobbini MI, Furey ML, Ishai A, Schouten JL, Pietrini P (2001) Distributed and overlapping representations of faces and objects in ventral temporal cortex. Science 293: 2425–2430

Hein G, Doehrmann O, Muller NG, Kaiser J, Muckli L, Naumer MJ (2007) Object familiarity and semantic congruency modulate responses in cortical audiovisual integration areas. J Neurosci 27:7881–7887

Hocking J, Price CJ (2008) The role of the posterior superior temporal sulcus in audiovisual processing. Cereb Cortex 18:2439–2449

James TW, Humphrey GK, Gati JS, Servos P, Menon RS, Goodale MA (2002) Haptic study of three-dimensional objects activates extrastriate visual areas. Neuropsychologia 40:1706–1714

Kim S, James TW (2009) Enhanced effectiveness in visuo-haptic object-selective brain regions with increasing stimulus salience. Hum Brain Mapp, Oct 14 [Epub ahead of print]

Kriegeskorte N, Mur M, Bandettini P (2008a) Representational similarity analysis - connecting the branches of systems neuroscience. Front Syst Neurosci 2:4

Kriegeskorte N, Mur M, Ruff DA, Kiani R, Bodurka J, Esteky H, Tanaka K, Bandettini PA (2008b) Matching categorical object representations in inferior temporal cortex of man and monkey. Neuron 60:1126–1141

Lacey S, Tal N, Amedi A, Sathian K (2009) A putative model of multisensory object representation. Brain Topogr 21:269–274

Laurienti PJ, Perrault TJ, Stanford TR, Wallace MT, Stein BE (2005) On the use of superadditivity as a metric for characterizing multisensory integration in functional neuroimaging studies. Exp Brain Res 166:289–297

Logothetis NK (2008) What we can do and what we cannot do with fMRI. Nature 453:869–878

Meienbrock A, Naumer MJ, Doehrmann O, Singer W, Muckli L (2007) Retinotopic effects during spatial audio-visual integration. Neuropsychologia 45:531–539

Mur M, Bandettini PA, Kriegeskorte N (2009) Revealing representational content with pattern-information fMRI--an introductory guide. Soc Cogn Affect Neurosci 4:101–109

Naumer MJ, Ratz L, Yalachkov Y, Polony A, Doehrmann O, Müller NG, Kaiser J, Hein G (2010) Visuo-haptic convergence in a cortico-cerebellar network. Eur J Neurosci (in press)

Naumer MJ, van den Bosch JJF (2009) Touching sounds: thalamo-cortical plasticity and the neural basis of multisensory integration. J Neurophysiol 102:7–8

Naumer MJ, van den Bosch JJF, Wibral M, Kohler A, Singer W, Kaiser J, van de Ven V, Muckli L (2009b) Audio-visual integration in the human brain: data-driven detection and independent validation.

Niogi SN, McCandliss BD (2006) Left lateralized white matter microstructure accounts for individual differences in reading ability and disability. Neuropsychologia 44:2178–2188

Noesselt T, Rieger JW, Schoenfeld MA, Kanowski M, Hinrichs H, Heinze HJ, Driver J (2007) Audiovisual temporal correspondence modulates human multisensory superior temporal sulcus plus primary sensory cortices. J Neurosci 27:11431–11441

Noppeney U Characterization of multisensory integration with fMRI - experimental design, statistical analysis and interpretation. In: Wallace M, Murray M (eds) Frontiers in the neural bases of multisensory processes. Taylor and Francis Group, London (in press)

Noppeney U, Josephs O, Hocking J, Price CJ, Friston KJ (2008) The Effect of Prior Visual Information on Recognition of Speech and Sounds. Cereb Cortex 18:598–609

Pietrini P, Furey ML, Ricciardi E, Gobbini MI, Wu W-HC, Cohen L, Guazzelli M, Haxby JV (2004) Beyond sensory images: objectbased representation in the human ventral pathway. Proc Natl Acad Sci USA 101:5658–5663

Polony A, Ratz L, Doehrmann O, Kaiser J, Naumer MJ (2007) Audio-tactile integration of meaningful objects in the human brain. In: Annual Meeting of the International Multisensory Research Forum, Sydney, Australia

Roebroeck A, Formisano E, Goebel R (2005) Mapping directed influence over the brain using Granger causality and fMRI. Neuroimage 25:230–242

Roebroeck A, Formisano E, Goebel R (2009a) The identification of interacting networks in the brain using fMRI: Model selection, causality and deconvolution. Neuroimage, Sep 25 [Epub ahead of print]

Roebroeck A, Formisano E, Goebel R (2009b) Reply to Friston and David After comments on: The identification of interacting networks in the brain using fMRI: Model selection, causality and deconvolution. Neuroimage, Oct 31 [Epub ahead of print]

Rouw R, Scholte HS (2007) Increased structural connectivity in grapheme-color synesthesia. Nat Neurosci 10:792–797

Stein BE, Meredith MA (1993) The merging of the senses. Cambridge, Massachussetts: MIT Press

Stein BE, Stanford TR (2008) Multisensory integration: current issues from the perspective of the single neuron. Nat Rev Neurosci 9:255–266

Stevenson RA, Geoghegan ML, James TW (2007) Superadditive BOLD activation in superior temporal sulcus with threshold nonspeech objects. Exp Brain Res 179:85–95

Stevenson RA, James TW (2009) Audiovisual integration in human superior temporal sulcus: Inverse effectiveness and the neural processing of speech and object recognition. Neuroimage 44:1210–1223

Stevenson RA, Kim S, James TW (2009) An additive-factors design to disambiguate neuronal and areal convergence: measuring multisensory interactions between audio, visual, and haptic sensory streams using fMRI. Exp Brain Res 198:183–194

Tal N, Amedi A (2009) Multisensory visual-tactile object related network in humans: insights gained using a novel crossmodal adaptation approach. Exp Brain Res 198:165–182

van Atteveldt N, Blau V, Blomert L, Goebel R (2008) fMR-adaptation reveals multisensory integration in human superior temporal cortex. In: Annual Meeting of the International Multisensory Research Forum, Hamburg, Germany

van Atteveldt NM, Formisano E, Blomert L, Goebel R (2007) The effect of temporal asynchrony on the multisensory integration of letters and speech sounds. Cereb Cortex 17: 962–974

van Atteveldt N, Roebroeck A, Goebel R (2009) Interaction of speech and script in human auditory cortex: Insights from neuroimaging and effective connectivity. Hear Res 258: 152–164

van Essen DC, Drury HA, Joshi S, Miller MI (1998) Functional and structural mapping of human cerebral cortex: solutions are in the surfaces. Proc Natl Acad Sci U S A 95:788–95

von Kriegstein K, Giraud AL (2006) Implicit multisensory associations influence voice recognition. PLoS Biol 4:e326

von Kriegstein K, Kleinschmidt A, Sterzer P, Giraud AL (2005) Interaction of face and voice areas during speaker recognition. J Cogn Neurosci 17:367–376

Weigelt S, Muckli L, Kohler A (2008) Functional magnetic resonance adaptation in visual neuroscience. Rev Neurosci 19:363–380

Werner S, Noppeney U (2009) Superadditive responses in superior temporal sulcus predict audiovisual benefits in object categorization. Nov 18 [Epub ahead of print]

Part II
Audio-Visual Integration

Chapter 7
Audiovisual Temporal Integration for Complex Speech, Object-Action, Animal Call, and Musical Stimuli

Argiro Vatakis and Charles Spence

7.1 Introduction

One important, but as yet unresolved, issue in the field of cognitive science concerns the perception of synchrony for the various sensory cues relating to specific environmental events, especially when those cues occur in different sensory modalities. Scientists are well aware of the fact that processing the information that may be available to each of our senses requires differing amounts of time (Spence and Squire, 2003; Zeki, 1993). As yet, however, researchers still do not know how and where in the brain the integration of those incoming inputs occurs in order to yield the percept of synchronous and unified (i.e., perceived as referring to the same event) multisensory events/objects (Efron, 1963; King, 2005; Spence and Squire, 2003). For instance, a particular environmental event (such as an individual talking) may consist of two sensory inputs (in this case, the auditory- and visual-speech signals) that, even though they are produced simultaneously from a particular source (i.e., the speaker), may or may not be received at the human sensory receptor surfaces at the same time. Naturally, therefore, one may wonder how the human perceptual system 'decides' whether or not the two sensory inputs belong to the same perceptual event/object or not?

Understanding the mechanisms underlying our ability to perceive the synchrony of multisensory events is thus of great interest to the cognitive sciences. Until very recently, the majority of the research interest in the multisensory perception of synchrony (e.g., King, 2005; Spence, in press; Spence and Squire, 2003) has tended to focus on the perception of simple stimuli (such as brief noise bursts and light flashes; i.e., Hirsh and Sherrick, 1961; Spence et al., 2001; Zampini et al., 2003b). While such studies have certainly advanced our knowledge of synchrony perception, and the factors influencing people's temporal perception for simple stimuli (see

---

A. Vatakis (✉)
Institute for Language and Speech Processing, Research Centers "Athena", Artemidos 6 & Epidavrou, Athens 151 25, Greece
e-mail: argiro.vatakis@gmail.com

M.J. Naumer, J. Kaiser (eds.), *Multisensory Object Perception in the Primate Brain*,
DOI 10.1007/978-1-4419-5615-6_7,

Spence and Squire, 2003; Spence et al., 2001, for reviews), it now seems increasingly appropriate that this research be expanded in order to investigate the extent to which these factors also influence our perception of more ecologically valid multisensory stimuli (e.g., see de Gelder and Bertelson, 2003; Mauk and Buonomano, 2004; Spence, in press). In this chapter, we introduce the issue of audiovisual temporal perception by presenting the theories and methodologies that have been used to measure audiovisual temporal perception. We first review those studies of temporal perception that have used simple stimuli and then move on to looking at those studies that have investigated the temporal perception of more complex audiovisual stimuli. We close by summarizing our own research findings on the audiovisual temporal perception of complex stimuli (looking at musical events and animal calls for the first time) and the factors that appear to modulate the multisensory perception of synchrony.

7.2 Multisensory Integration and Temporal Synchrony

In their daily interaction with the environment, humans (and other animals) are exposed to multiple sensory inputs from the same and/or different sensory modalities. The appropriate combination of those sensory inputs can promote an organism's ability to discriminate the occurrence of salient events, to localize those objects/events in the environment, and to minimize the time needed by the organism in order to react to those events if need be (e.g., Calvert et al., 2004). How does one, however, know how to separate and combine the various sensory inputs that are attributed to one object/event from those that are related to another environmental object/event? This can be thought of as a crossmodal form of the correspondence problem (Fujisaki and Nishida, 2007). To date, research has shown that the binding of multiple sensory inputs referring to the same distal event (or object) depends on both temporal and spatial constraints (Parise and Spence, 2009; Slutsky and Recanzone, 2001; Spence, 2007). Specifically, spatial coincidence (i.e., when both sensory inputs originate from the same spatial location; e.g., Soto-Faraco et al., 2002; Spence and Driver, 1997; Spence and Squire, 2003; Zampini et al., 2003a; although see also Fujisaki and Nishida, 2005; Noesselt et al., 2005; Recanzone, 2003; Teder-Sälejärvi et al., 2005; Vroomen and Keetels, 2006) and temporal synchrony (i.e., when the two sensory inputs occur at more or less the same time) of multiple sensory inputs comprise two of the key factors that determine whether or not multisensory integration will take place in order to yield a unified percept of a given event or object (e.g., Calvert et al., 2004; de Gelder and Bertelson, 2003; Driver and Spence, 2000; Kallinen and Ravaja, 2007; Sekuler et al., 1997; Slutsky and Recanzone, 2001; Stein and Meredith, 1993; though see Spence, 2007). An additional factor that is increasingly being recognized as being important in terms of modulating audiovisual multisensory integration is the semantic congruency of the auditory and visual stimuli (see Chen and Spence, 2010; Doehrmann and Naumer, 2008; Hein et al., 2007; Spence, 2007).

Although we experience synchrony for the vast majority of proximal events around us, the generation of such percepts is by no means a simple process. The complexity here comes from the fact that both neuronal and non-neuronal factors (see Fig. 7.1 for an overview of those factors) influence the arrival and processing time of two or more sensory signals to the brain, even though those signals may have emanated simultaneously from a given distal source (e.g., King, 2005; King and Palmer, 1985; Pöppel et al., 1990; Spence and Squire, 2003). One of the classic demonstrations of the unified nature of our perceptual experience comes from research on the well-known ventriloquist effect (Alais and Burr, 2004; Choe et al., 1975; de Gelder and Bertelson, 2003; Howard and Templeton, 1966), whereby auditory signals appear to originate from (or close to) the location of simultaneously presented visual signals.

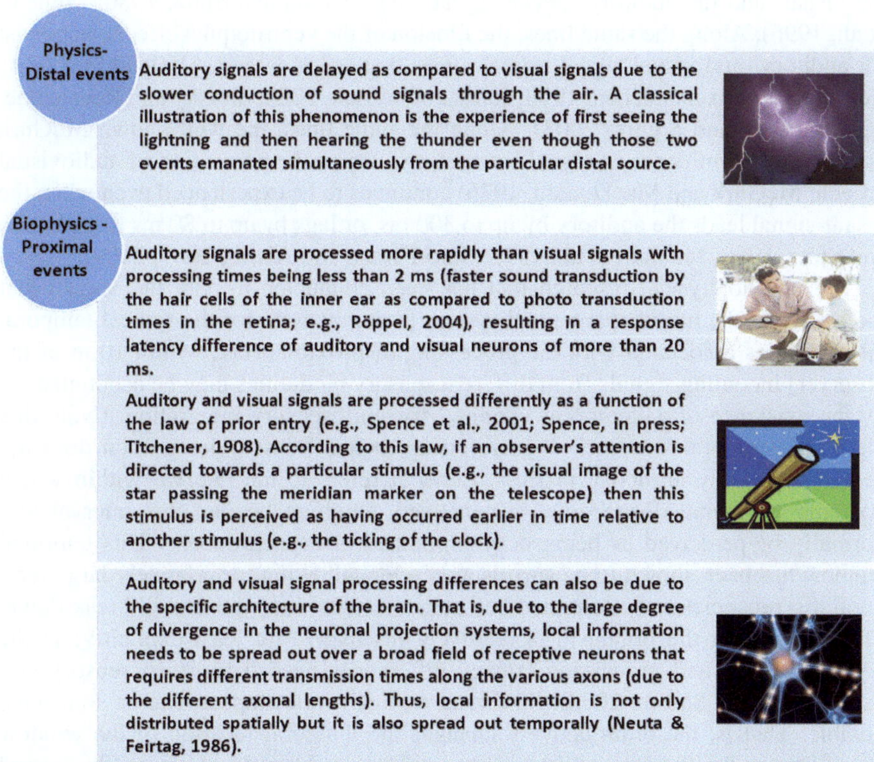

Fig. 7.1 An outline and examples of some of the neuronal (i.e., biophysical) and non-neuronal (i.e., physical) factors responsible for the arrival and processing differences of the auditory and visual signals of a multisensory event

7.3 The Mechanisms Underlying Multisensory Temporal Perception

The human perceptual system appears to promote the perception of events that are both temporally coherent and perceptually unified. Many studies utilizing both simple (e.g., light flashes and sound bursts) and more complex (e.g., speech, object-actions) stimuli, however, have now shown that even though multisensory integration is frequently enhanced when multiple sensory signals are synchronized (e.g., see Calvert et al., 2004; de Gelder and Bertelson, 2003), precise temporal coincidence is by no means mandatory for the human perceptual system to create a unified perceptual representation of a multisensory event (e.g., Dixon and Spitz, 1980; Engel and Doherty, 1971; Grant et al., 2004; Kopinska and Harris, 2004; Morein-Zamir et al., 2003; Navarra et al., 2005; Rihs, 1995; Soto-Faraco and Alsius, 2007, 2009; Sugita and Suzuki, 2003; Thorne and Debner, 2008). For example, it has been shown that the intelligibility of audiovisual speech stimuli remains high even when a temporal asynchrony of as much as 250 ms is introduced between the visual- and the auditory-speech signals (e.g., Dixon and Spitz, 1980; Munhall et al., 1996). Along the same lines, the illusion of the ventriloquist effect can persist for auditory- and visual-signal asynchronies that are as large as 250–300 ms (e.g., Bertelson and Aschersleben, 1998; Jones and Jarrick, 2006; Slutsky and Recanzone, 2001; Spence and Squire, 2003). Along the same lines, the well-known McGurk effect (i.e., the influence of visual lip-reading cues on the perception of audiovisual speech; McGurk and MacDonald, 1976) continues to be experienced even when the visual-signal leads the auditory by up to 300 ms, or lags by up to 80 ms (e.g., Dixon and Spitz, 1980; Munhall et al., 1996; Soto-Faraco and Alsius, 2007, 2009).

The psychophysical research that has been conducted to date has shown that the ability of the human perceptual system to compensate for the typical temporal discrepancies associated with the processing of proximal (i.e., within 10 m of the observer) incoming signals from different sensory modalities may be accounted for by the existence of a 'temporal window' for multisensory integration. (Note that the term 'temporal window' when referring to multisensory integration does not necessarily imply an active process, rather it refers to the interval within which no temporal signal discrepancy is perceived, anything beyond this interval will normally be perceived as being desynchronized or asynchronous.) This temporal window has been shown to be surprisingly wide, allowing for relatively large temporal discrepancies to be tolerated before audiovisual integration will break down. This means that the organism is relatively insensitive (at least explicitly) to the possible arrival and processing latency differences needed for each sensory signal (Spence and Squire, 2003). Furthermore, this window appears to be somewhat flexible. That is, the brain actively changes the temporal location of the window depending on the distance of the visible sound event from the observer (Sugita and Suzuki, 2003). Thus, this 'movable temporal window' would presumably have a narrow width for proximal events since the perceptual system has to deal mainly with the biophysical factors related to the neuronal processing of the events that may lead to the introduction of temporal discrepancies. However, as the distance of the event from the observer increases (especially above 10 m), the window appears to become moveable in order to allow for larger temporal discrepancies to be tolerated, given that increasing distance results in larger differences in the arrival and processing times of the relevant sensory signals due to physical differences in transduction latencies (i.e., visual signals travel faster through air than auditory signals). It should, however, be noted that this moveable temporal window appears to break down for events occurring at distances larger than 10 m from the observer (Sugita and Suzuki, 2003), thus accounting for the fact that the perception of synchrony for events occurring at greater distances eventually breaks down (e.g., think, for example, of first seeing the fireworks and then hearing the explosive sound that they produced).

The temporal window for multisensory integration therefore appears to be reasonably adaptable depending on the particular conditions under which the sensory signals are presented (Arnold et al., 2005; Engel and Dougherty, 1971; King, 2005; King and Palmer, 1985; Kopinska and Harris, 2004; Spence and Squire, 2003; Sugita and Suzuki, 2003; but see also Lewald and Guski, 2004; see Keetels and Vroomen, submitted, for a recent review). The existence of the phenomenon of 'temporal ventriloquism' is considered as another compensatory mechanism. That is, the auditory and visual signals associated with a particular event that, for whatever reason, are temporally misaligned are 'pulled' back into approximate temporal register (Fendrich and Corballis, 2001; Morein-Zamir et al., 2003; Navarra et al., 2009; Scheier et al., 1999; Spence and Squire, 2003; Vroomen and de Gelder, 2004; Vroomen and Keetels, 2006; but see also Kopinska and Harris, 2004).

7.4 Measuring Temporal Perception

This section focuses on three of the most popular psychophysical paradigms that have been used over the years in order to measure temporal perception: they are the temporal order judgment (TOJ) task, the simultaneity judgment (SJ) task, and the three-response (or ternary response) task (see Spence, in press). In a typical TOJ task, participants are presented with a pair of stimuli (e.g., one auditory and the other visual) at various SOAs and they are asked to make a judgment as to the order of stimulus presentation (i.e., 'Which stimulus was presented first?'; e.g., Bald et al., 1942; Hirsh and Sherrick, 1961; Spence et al., 2001; Sternberg et al., 1971; or 'Which was presented second?'; e.g., Parise and Spence, 2009). The data obtained from such TOJ studies allows for the calculation of two measures, the just noticeable difference (JND) and the point of subjective simultaneity (PSS). The JND provides a standardized measure of the sensitivity with which participants can judge the temporal order of the two stimuli that have been presented at a given performance threshold (typically 75% correct). The PSS provides an estimate of the time interval by which the stimulus in one sensory modality has to lead or lag the stimulus in the other modality in order for the two to be perceived as having been presented synchronously (or rather, for the two responses to be chosen equally often; see

Coren et al., 2004, for further details). The two measures derived from the TOJ task, the JND and the PSS, allow for the calculation of the temporal window described above. Specifically, the width of the temporal window is conventionally taken to equal the PSS±JND (see Koppen and Spence, 2007; Soto-Faraco and Alsius, 2007, 2009).

In an SJ task, just as in the TOJ task, participants are presented with pairs of stimuli at various SOAs and are asked to judge whether the two stimuli were presented simultaneously or successively (e.g., Fraisse, 1984; Stone et al., 2001; Zampini et al., 2005). The data from an SJ task allows one to derive three measures: the mean of the distribution (which corresponds to the PSS), the peak height of the distribution (i.e., the peak probability with which participants made the simultaneous response), and the standard deviation of the distribution (SD); this measure provides an estimate of the spread, or variance, of the distribution and therefore gives an indication of how difficult participants found the task across the range of SOAs tested; comparable to the JND, smaller SD values indicate steeper functions and more precise discriminative performance.

Finally, the three-response task represents a combination of the TOJ and SJ tasks in which participants have to decide whether or not the two stimuli were presented simultaneously and if not which stimulus was presented first (e.g., Allan, 1975; van de Par et al., 1999; Zampini et al., 2007).

7.5 Audiovisual Temporal Perception for Simple Stimuli

As mentioned earlier, the initial research studies on the topic of temporal perception all tended to use simple stimuli, such as light flashes and sound bursts, using either a TOJ or an SJ task. The results of these studies, however, exhibited great differences in terms of the JND and PSS values reported. These differences therefore led to quite some debate regarding the source(s) of these inconsistencies in participants' sensitivity to audiovisual temporal order for simple stimuli and the possible factors that may have promoted those differences (e.g., Kohlrausch and van de Par, 2005; van Eijk et al., 2008; Spence, in press). In subsequent TOJ studies using simple transitory stimuli, researchers have shown that the auditory and visual stimuli must be separated by a minimum of 20–60 ms in order for people to be able to judge correctly which modality came first on 75% of the trials (e.g., Hirsh, 1959; Hirsh and Sherrick, 1961). According to the results of Hirsh and Sherrick's classic early work, this 20 ms value remained relatively constant across a number of different intramodal and crossmodal combinations of auditory, visual, and tactile stimuli.

It is important to note, however, that the auditory and visual stimuli utilized in Hirsh and Sherrick's (1961) research (and in many of the early TOJ studies; e.g., Bald et al., 1942; Bushara et al., 2001; Jaśkowski et al., 1990; Rutschmann and Link, 1964) were presented from different spatial locations (e.g., the auditory stimuli were presented over headphones, while the visual stimuli were presented from a screen or from LEDs placed directly in front of the participants). It has now been demonstrated that the use of experimental setups that present pairs of stimuli from different spatial locations can introduce potential confounds, since the spatial separation of the stimulus sources can sometimes impair multisensory integration (e.g., Soto-Faraco et al., 2002; Spence and Driver, 1997; Spence and Squire, 2003; Zampini et al., 2003a, b, 2005). It is possible, therefore, that the participants in these early studies could have used redundant spatial cues in order to facilitate their TOJ responses (i.e., the participants in Hirsh and Sherrick's studies may have judged which *location* came first rather than which *sensory modality* came first; see Spence et al., 2001; Zampini et al., 2003b, on this point). Thus, the presence of this spatial confound may imply that the 20 ms value reported by Hirsh and Sherrick actually reflects an *overestimation* (or possibly even an underestimation) of people's sensitivity to the temporal order in which simple auditory and visual stimuli are presented. Note also that participants in Hirsh and Sherrick's studies were highly experienced in performing temporal tasks and, what is more, were also given feedback concerning the 'correctness' of their responses.

Subsequent research, in which such spatial confounds have been removed, have revealed that discrete pairs of auditory and visual stimuli actually need to be separated by at least 60–70 ms in order for naïve participants to judge accurately (i.e., 75% correct) which modality was presented first (Zampini et al., 2003a; see Fig. 7.2). Given that environmental events giving rise to multisensory stimulation typically originate from a single environmental location, this value of

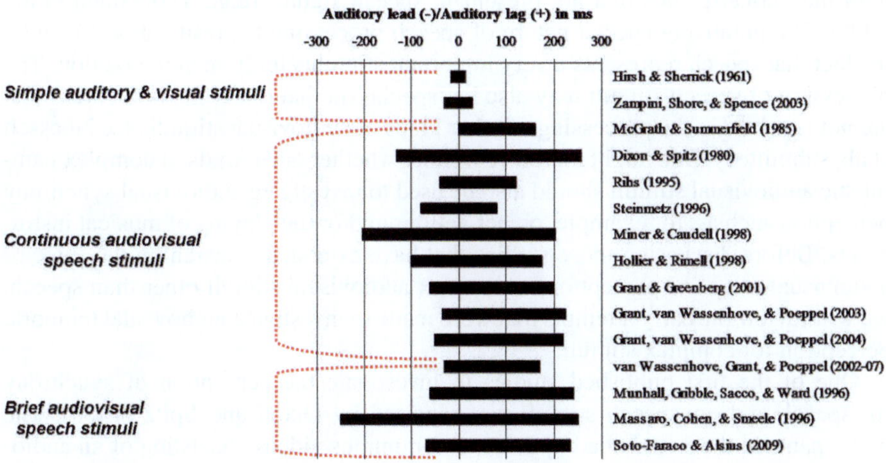

Fig. 7.2 The temporal window of integration (i.e., the interval in which no signal discrepancy is perceived, anything beyond this interval will normally be perceived as being desynchronized or asynchronous) for simple auditory and visual stimuli, continuous audiovisual speech stimuli, and brief speech stimuli. Note that a mean temporal window (derived from all the previous studies) for the studies by van Wassenhove and colleagues (van Wassenhove et al., 2002–2007) is represented here

60–70 ms likely provides a more representative estimate of the temporal window for simple multisensory stimuli than the 20 ms value proposed by Hirsh and Sherrick (1961).

7.6 Audiovisual Temporal Perception for Complex Stimuli

In order to investigate the temporal constraints on the multisensory perception of synchrony under more realistic conditions, however, one has to move away from the study of simple transitory stimuli of low informational content (e.g., Hirsh and Sherrick, 1961; Zampini et al., 2003b) toward the use of more ecologically valid and complex stimuli (such as speech, musical stimuli, or object-actions; see de Gelder and Bertelson, 2003; Lee and Noppeney, 2009; Mauk and Buonomano, 2004; McGrath and Summerfield, 1985). Until very recently, however, such research has primarily been focused on the perception of synchrony for audiovisual speech stimuli (e.g., Dixon and Spitz, 1980; Grant and Greenberg, 2001; Steinmetz, 1996). It is, however, important to note that speech represents a highly overlearned class of stimuli for most people. Furthermore, it has even been argued by some researchers that it may represent a special type of sensory event (e.g., Bernstein et al., 2004; Eskelund and Andersen, 2009; Massaro, 2004; Munhall and Vatikiotis-Bateson, 2004; Spence, 2007; Tuomainen et al., 2005; Vatakis et al., 2008). In particular, it has been claimed that the special nature of speech may lie in the existence of a 'specific mode of perception' that refers to the structural and functional processes related to the articulatory gestures of speech and/or to the perceptual processes associated with the phonetic cues that are present in speech signals (e.g., Tuomainen et al., 2005). The putatively special nature of speech processing is presumably driven by the fact that speech represents a very important stimulus for human interaction. The processing of speech stimuli may also be 'special' in that it recruits brain areas that are not involved in the processing of other kinds of audiovisual stimuli (see Noesselt et al., submitted). One might therefore wonder whether other kinds of complex naturalistic audiovisual stimuli should also be used to investigate audiovisual synchrony perception such as, for example, object-actions and/or the playing of musical instruments. Before discussing those studies that have extended research on the topic of audiovisual temporal perception to complex audiovisual stimuli other than speech, we will review the early attempts that were made to investigate audiovisual temporal perception for complex stimuli.

One of the first published studies to investigate the perception of synchrony for speech and nonspeech stimuli was reported by Dixon and Spitz (1980). The participants in their study had to monitor continuous videos consisting of an audiovisual speech stream or an object-action event consisting of a hammer repeatedly hitting a peg. The videos started out in synchrony and were then gradually desynchronized at a constant rate of 51 ms/s up to a maximum asynchrony of 500 ms. Participants were instructed to respond as soon as they noticed the asynchrony in the video that they were monitoring. Dixon and Spitz's results showed that the auditory stream had to lag the visual stream by 258 ms or else to lead by 131 ms before the temporal discrepancy was detected when participants monitored the continuous audiovisual speech stream (see also Conrey and Pisoni, 2006). By contrast, an auditory lag of only 188 ms or a lead of 75 ms was sufficient for participants to report the asynchrony in the object-action video. That is, the participants in Dixon and Spitz's study were significantly *more* sensitive to the asynchrony present in the object-action video than in the continuous speech video (see Fig. 7.3).

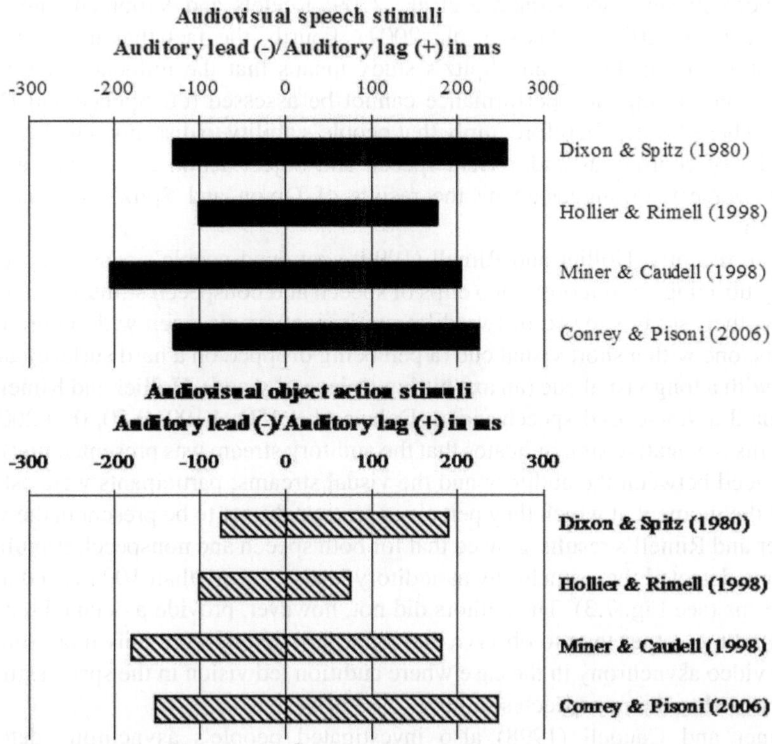

Fig. 7.3 A comparison of the temporal window of audiovisual integration reported in those studies that have compared continuous speech to nonspeech (object-action) stimuli. It is evident from this figure that the temporal window varies from study to study and that in some cases the temporal window for nonspeech stimuli is narrower than that for speech stimuli (e.g., see Hollier and Rimell, 1998)

It is important to note, however, that, for a number of reasons, the results of Dixon and Spitz's (1980) study may not provide a particularly accurate estimate of people's sensitivity to asynchrony for complex audiovisual stimuli. First, the auditory stimuli in their study were presented over headphones while the visual stimuli were presented from directly in front of the participants. As mentioned already, research has shown that the spatial separation of the stimulus sources can sometimes impair multisensory integration (e.g., Soto-Faraco et al., 2002; Spence and Driver, 1997; Spence and Squire, 2003; though see Zampini et al., 2003a). Second, the gradual desynchronization of the auditory and visual streams in Dixon and Spitz's study might inadvertently have presented participants with subtle auditory pitch-shifting cues that could also have facilitated their performance (Reeves and Voelker, 1993). Third, it should be noted that the magnitude of the asynchrony in Dixon and Spitz's study was always proportional to the length of time for which the video had been running in a given trial, thus potentially confounding participants' sensitivity to asynchrony with any temporal adaptation effects that may have been present (see Asakawa et al., 2009; Keetels and Vroomen, submitted; Navarra et al., 2005; Vatakis et al., 2007). Fourth, the fact that no catch trials were presented in Dixon and Spitz's study means that the influence of criterion shifting on participants' performance cannot be assessed (cf. Spence and Driver, 1997). These factors therefore imply that people's ability to discriminate the temporal order of continuous audiovisual speech and object-action events may actually be far worse than suggested by the results of Dixon and Spitz's seminal early research.

More recently, Hollier and Rimell (1998) examined people's sensitivity to synchrony utilizing continuous video clips of speech and nonspeech stimuli. The videos used in their study showed of two different impact events, each with a duration of 500 ms, one with a short visual cue (a pen being dropped on a hard surface) and the other with a long visual cue (an axe hitting a piece of wood). Hollier and Rimell also presented a 4.5-second speech video. Delays of ±150, ±100, ±50, 0, +200, and +300 ms (a negative sign indicates that the auditory stream was presented first) were introduced between the auditory and the visual streams; participants were asked to report the moment at which they perceived an asynchrony to be present in the video. Hollier and Rimell's results showed that for both speech and nonspeech stimuli, participants detected the asynchrony at auditory leads of more than 100 ms and at lags of 175 ms (see Fig. 7.3). The authors did not, however, provide a detailed report of their findings, other than to observe that their participants were only more sensitive to the video asynchrony in the case where audition led vision in the speech stimulus (as opposed to the nonspeech stimuli).

Miner and Caudell (1998) also investigated people's asynchrony detection thresholds by presenting a series of complex audiovisual stimuli. In this study, the participants were presented with one speech video, three single impact events (a hammer striking a block, two drumsticks colliding, and two wine glasses colliding), and one repeated impact event (two silver balls colliding 12 times). The participants made a vocal response (yes or no) concerning whether the auditory- and visual-stimulus streams were synchronous or not. Overall, the participants perceived the audiovisual stimuli as being desynchronized when the auditory/visual streams were delayed by more than 184 ms. The results showed that the average threshold was lowest (indicating superior discriminative performance) for the repetitive impact event (auditory/visual delays of 172 ms) and highest for the speech event (auditory/visual delays of 203 ms). The threshold of the single impact events ranged from auditory/visual delays of 177–191 ms (a hammer striking a block and two drumsticks colliding: 177 ms for both; two wine glasses colliding: 191 ms; see Fig. 7.3).

Based on the studies reported above (i.e., Dixon and Spitz, 1980; Hollier and Rimell, 1998; Miner and Caudell, 1998), it would appear that people are more sensitive to temporal asynchrony for object-action events as compared with speech events (see Fig. 7.3). As pointed out earlier, however, these early studies all had a number of potential shortcomings (e.g., the lack of spatial coincidence between the auditory and the visual information sources, the potential presence of pitch-shifting cues, and the possibility of criterion shifting effects) that, until very recently, had not been addressed. Until recently, therefore, there was no clear evidence that the temporal window of audiovisual integration reported in these previous studies for speech and nonspeech stimuli should be taken as providing an accurate representation of human perception of synchrony for different kinds of complex events. However, the experiments that we have conducted over the last few years (see below) were designed to provide more convincing evidence with regard to the temporal constraints modulating participants' sensitivity to temporal order for both speech and nonspeech stimuli (while eliminating the confounds identified in previous research).

The studies reviewed thus far suggest that a continuous speech signal remains perceptually unified under a wide range of asynchronies. It is not clear, however, what the exact audiovisual time range, or temporal window, is since a high level of variability has been observed between the results of previous research. This variability motivated us to conduct a series of studies to investigate the temporal perception of synchrony by utilizing audiovisual speech as the experimental stimulus (e.g., Vatakis and Spence, submitted, 2007e).

Many of the studies reported so far have focused on continuous audiovisual speech stimuli. For example, in one such study, Rihs (1995) measured the influence of the desynchronization of an audiovisual speech signal on people's perception of scenes presented in a television program. Specifically, he utilized a video extracted from a television talk show consisting of several speakers. The participants were presented with 50-s segments from the talk show with a fixed audiovisual delay. The participants had to report their perception of decreased video quality on a 5-point impairment scale (a decrease of 0.5 point on this scale was defined as the detectability threshold). According to the participants' reports, an auditory lead of about 40 ms and a lag of 120 ms constituted the interval needed for a noticeable scene impairment to be experienced (see Fig. 7.2). Rihs also measured the participants' acceptability thresholds (which were defined by a quality decrease of 1.5 points on the 5-point scale) finding that participants were willing to tolerate an auditory lead of as much as 90 ms and a lag of as much as 180 ms. The results of this study served to develop the recommendations for the ITU 1998 (Rec. ITU-R BT.1359) in terms of the relative timing of auditory and visual streams for broadcasting. Interestingly, the values reported by Rihs were much lower than those observed in Dixon and Spitz's (1980) study reported earlier.

Steinmetz (1996) conducted another study that focused on the effect of audiovisual stream delays on the temporal perception of segments of continuous speech. In this study, video segments of television news reports were presented at one of three different viewing angles (comprising head, shoulder, and body views). The auditory-

and visual-speech streams were desynchronized at intervals of 40 ms, resulting in audiovisual stream delays of ±120, ±40, ±80, and 0 ms. At the end of each video, the participants had to complete a questionnaire regarding the quality of the video presented and which of the two streams (auditory or visual) they had perceived as either leading or lagging. Steinmetz reported that the participants exhibited a lower sensitivity to the desynchronization of speech events occurring far away (i.e., whole-body view) than for events occurring closer to the observer (i.e., consisting of a view of the head and shoulders). Overall, the breakdown of the perception of synchrony for the news report videos was found to occur at auditory or visual lags exceeding ± 80 ms (see Fig. 7.2).

More recently, researchers have tended to focus on the use of single sentences (instead of passages composed of several sentences) in looking at the audiovisual temporal perception of speech in human observers. For instance, Grant et al. (2004) used audiovisual sentences in a two-interval forced choice adaptive procedure. They found asynchrony discrimination thresholds of approximately 35 ms for auditory leads and 160 ms for visual leads for unfiltered speech and auditory leads of 35 ms and visual leads of 225 ms for bandpass-filtered speech. Similarly, Grant et al. (2003) reported that participants only noticed the asynchrony in a continuous stream of audiovisual speech when the auditory-speech led the visual lip movements by at least 50 ms or else lagged by 220 ms or more (see also Grant and Greenberg, 2001; Grant and Seitz, 1998; Soto-Faraco and Alsius, 2007, 2009). Meanwhile, McGrath and Summerfield (1985) reported that the intelligibility of audiovisual sentences in auditory white noise deteriorated at much lower visual leads (160 ms; see also Pandey et al., 1986) than those observed in the studies of both Dixon and Spitz (1980) and Grant and colleagues (Grant and Greenberg, 2001; Grant and Seitz, 1998; see Fig. 7.2).

In addition to the studies using continuous speech stimuli (i.e., those including passages and/or sentences), audiovisual synchrony perception has also been evaluated for brief speech tokens using the McGurk effect (McGurk and MacDonald, 1976). For instance, Massaro et al. (1996) evaluated the temporal perception of synthetic and natural CV syllables (e.g., /ba, da, ða, and va/) dubbed onto either congruent or incongruent synthetic animated facial point-light displays. In this study, two groups of participants were tested on their ability to identify the audiovisual speech tokens that had been presented. The first group of participants was tested with video clips including stream delays of ±67, ±167, and ±267 ms, while the second group was tested with stream asynchronies of ±133, ±267, and ±500 ms. The results showed that the identification of congruent audiovisual speech pairs was overall less affected by the introduction of audiovisual asynchrony than was the identification of incongruent speech pairs. Massaro et al. proposed that the boundary of audiovisual integration for congruent speech occurs at an auditory lead/lag of about ±250 ms (see also Massaro and Cohen, 1993; see Fig. 7.2).

Munhall et al. (1996), however, described results that were quite different from those observed by Massaro et al. (1996). Specifically, they reported a series of experiments looking at the effect of asynchrony and vowel context on the McGurk effect. Their stimuli consisted of the auditory utterances /aga and ibi/ dubbed onto a video of an individual's face whose lip movements were articulating /aba/. The range of auditory- and visual-stream asynchronies extended over the range from ±360 ms in 60 ms steps. The participants in this study had to complete an identification task. Overall, the results showed that responses driven by the auditory stream dominated for the majority of the asynchronies presented, while responses that were driven by the visual stream (i.e., /g/) predominated for auditory leads of 60 ms to lags of 240 ms (see Fig. 7.2).

van Wassenhove et al. (2007) recently extended Munhall et al.'s (1996) study by testing more asynchronies under two different experimental conditions, one using a syllable identification task and the other using an SJ task. In these experiments, audio recordings of /pa and ba/ were dubbed onto video recordings of /ka or ga/, respectively (which usually produces the illusory percept /ta or da/). Their first experiment investigated the effect of audiovisual asynchrony (stream asynchronies in the range from ±467 ms were used) on the identification of incongruent (i.e., McGurk) audiovisual speech stimuli. Their second experiment focused on SJs for congruent and incongruent audiovisual speech stimuli tested at the same asynchronies. The major finding to emerge from van Wassenhove et al.'s study was that both the identification experiment and the SJ experiment revealed temporal windows of audiovisual integration having a width of about 200 ms. McGurk fusion responses were prevalent when the temporal asynchrony was in the range from –30 ms to +170 ms (once again, where negative values indicate that the auditory stream was presented first) and was more robust under those conditions where the auditory signal lagged. Overall, the data from this study was consistent with Munhall et al.'s (1996) observations and extended those findings by sampling a greater range of temporal asynchronies, using smaller temporal step sizes, and determining the boundaries for subjective audiovisual simultaneity. van Wassenhove et al.'s findings were also similar to those reported in other studies previously conducted by her group (e.g., see van Wassenhove et al., 2003, 2005; see Fig. 7.2).

7.7 Factors Affecting the Audiovisual Temporal Perception of Complex Speech and Nonspeech Stimuli

On the whole, the results of previous studies concerning the temporal perception of audiovisual speech stimuli have shown that the intelligibility of human speech signals remains high over a wide range of audiovisual temporal asynchronies. This time range (i.e., temporal window), however, demonstrates great variability across different studies (see Figs. 7.2 and 7.3). Additionally, the limitations identified in the previous section, and the fact that only a small number of studies have looked at the perception of more complex stimuli such as music, naturally leads one to wonder what the temporal window would look like once the limitations were addressed and other kinds of experimental stimuli were tested. The question therefore arises as to what accounts for such dramatic differences in sensitivity in temporal order discrimination for simple versus more complex audiovisual stimuli.

Despite the differences in the type of stimuli, response tasks, and statistical procedures used in previous studies, three characteristics of the audiovisual temporal window of multisensory integration have proved relatively consistent across the majority of studies that have been reported using complex stimuli. First, the temporal window for synchrony perception for audiovisual stimuli has a width on the order of several hundred milliseconds. Second, the temporal window of synchrony is asymmetrical, that is, people generally find it more difficult to detect the asynchrony under conditions where the visual signal leads, as compared to conditions where the auditory signal leads. This trend has typically been explained in terms of the inherent adaptation of the nervous system to time differences in the speed of light and sound signals (e.g., Massaro, 1996). Third, the width of the temporal window exhibits great variability across experimental setups and stimuli (Spence, in press).

Given the importance of understanding time perception and the fact that the great variability observed in the findings of the previous studies on audiovisual temporal perception, we have spent the last few years investigating some of the factors that affect audiovisual temporal perception for complex speech and nonspeech stimuli. Specifically, our work has focused on how audiovisual temporal perception is affected by the type of stimulus used, the physical characteristics of the stimulus, the medium used to present this stimulus, and the unity effect (Vatakis and Spence, 2006a, b, c, 2007d, e, 2008a).

7.7.1 How Does Stimulus Type Affect Audiovisual Temporal Perception?

In order to address this question, we have conducted a series of experiments in which we have explored temporal perception in normal adult human observers for complex speech and nonspeech stimuli, while eliminating the confounds identified in previous studies (e.g., the lack of spatial coincidence between the auditory and the visual information sources, the potential presence of pitch-shifting cues). (Note that complex stimuli are defined here as stimuli of higher information content and having a continuously changing audiovisual temporal profile; Vatakis and Spence, 2006a, c.) The aim of these experiments was to define the temporal limits for complex audiovisual stimuli more accurately. In particular, the experiments conducted utilized complex speech (i.e., continuous and brief speech tokens; e.g., sentences, words, and syllables) and nonspeech stimuli (i.e., object-actions and musical stimuli; e.g., single impact events, such as the smashing of a block of ice with a hammer, and the playing of musical pieces with single and double notes) in a TOJ task. Interestingly, the results of these experiments were similar to those reported by Dixon and Spitz (1980). That is, people found it significantly easier to detect the temporal asynchrony present in desynchronized audiovisual object-action events than to detect the asynchrony in speech events (see Fig. 7.4a). However, the overall pattern of results demonstrated that participants were generally more sensitive when reporting the

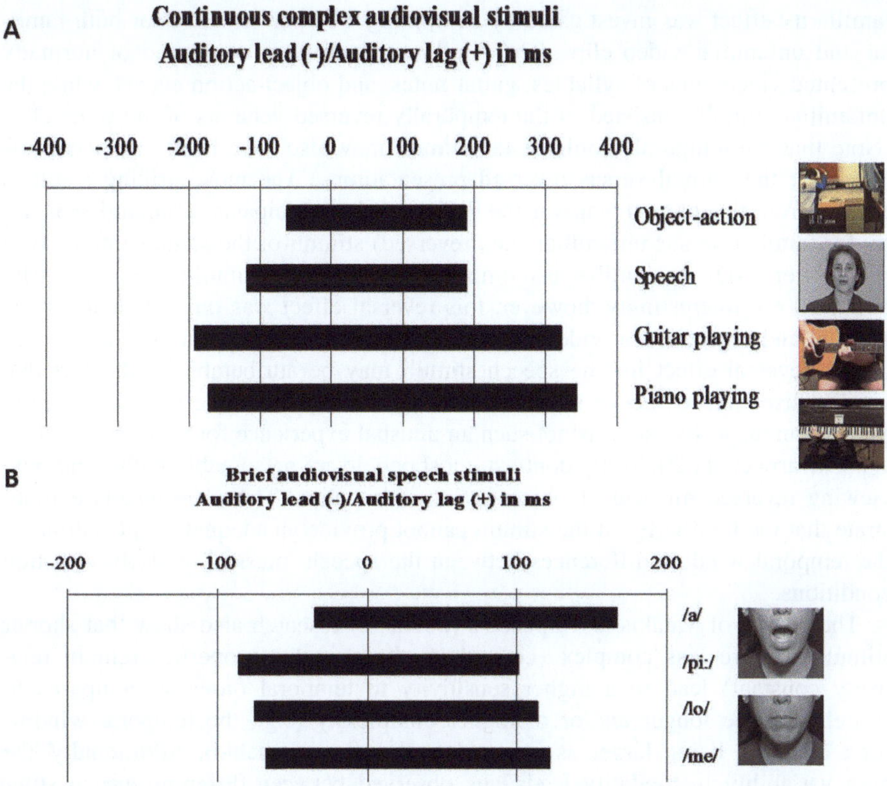

Fig. 7.4 (**a**) The temporal window of audiovisual integration for continuous audiovisual object-action, speech, and guitar and piano playing stimuli. (**b**) The temporal window of integration for brief audiovisual speech tokens (/a, pi:, lo, me/). Sample still images for each type of video are presented on the right side of the figure

temporal order of an audiovisual event than has been suggested by Dixon and Spitz's previous research (possibly due to the use of continuous speech streams as experimental stimuli in their study which, as we noted earlier, may have led to temporal adaptation effects; that is, prolonged viewing of the asynchronous stimulus could have somehow widened the temporal window of integration leading participant's to experience smaller discrepancies than those delays that were actually presented to the participants; cf. Navarra et al., 2005, 2009; Vatakis et al., 2007).

The results obtained in Vatakis and Spence's experiments also demonstrated, for the first time, that people are less sensitive to the asynchrony present when viewing audiovisual musical events than when viewing either speech or object-action events. Most importantly, however, their results showed that the temporal window of multisensory integration is modulated by the properties and complexity of a given stimulus as well as by the level of familiarity that a participant has with that particular stimulus (cf. Petrini et al., 2009; Schutz and Kubovy, 2009). The familiarity effect was investigated by comparing TOJ performance for both familiar and unfamiliar video clips. The familiar stimuli were composed of normally presented video clips of syllables, guitar notes, and object-action events, while the unfamiliar stimuli consisted of the temporally reversed versions of the same clips (Note that the temporal profile of the stimuli may also have been somewhat different for the normal versus reversed presentations.) The most striking result to emerge from this experiment was the difference in participants' temporal sensitivity for familiar versus unfamiliar (i.e., reversed) stimuli of the same stimulus type (i.e., lower JNDs for familiar as compared to unfamiliar stimuli; see also Petrini et al., 2009). Interestingly, however, this reversal effect was only evident for the musical and object-action video clips but not for the speech stimuli. The absence of any reversal effect for the speech stimuli may be attributable to the fact that viewing/listening to speech segments that are incomprehensible (i.e., listening to foreign language speakers) is not such an unusual experience for many people (see also Navarra et al., 2010). By contrast, most people are presumably unfamiliar with viewing reversed musical or object-action video clips. Thus, the results demonstrate that the familiarity of the stimuli cannot provide an adequate explanation for the temporal window differences between the speech, musical, and object-action conditions.

The results of Vatakis and Spence's (2006a, c) research also show that shorter stimuli that are less complex (i.e., where the stimulus properties remain relatively constant) lead to a higher sensitivity to temporal order as compared to stimuli that are longer and/or of higher complexity (e.g., the temporal window for a sentence being larger as opposed to that for a syllable). Additionally, the high variability in modality leads/lags observed between different speech stimuli could have been driven by the fact that the phonetic and physical properties involved in the production of speech sounds vary as a function of the particular speech sound being uttered (see Fig. 7.4b; e.g., Kent, 1997; van Wassenhove et al., 2005).

7.7.2 How Do the Physical Characteristics in the Articulation of a Speech Stimulus Affect Audiovisual Temporal Perception?

Having a clearer picture (compared to previous studies) of the actual limits of temporal perception for speech and nonspeech stimuli (and how to measure them), a new line of experiments was designed, focusing on the perception of speech stimuli and how physical differences present in the articulation of various speech tokens affect people's temporal sensitivity (Vatakis and Spence, 2007e, submitted). Specifically, the experiments were designed to investigate the possible effects that physical changes occurring during the articulation of different consonants (i.e., varying as a function of the place and manner of articulation and voicing) and vowels

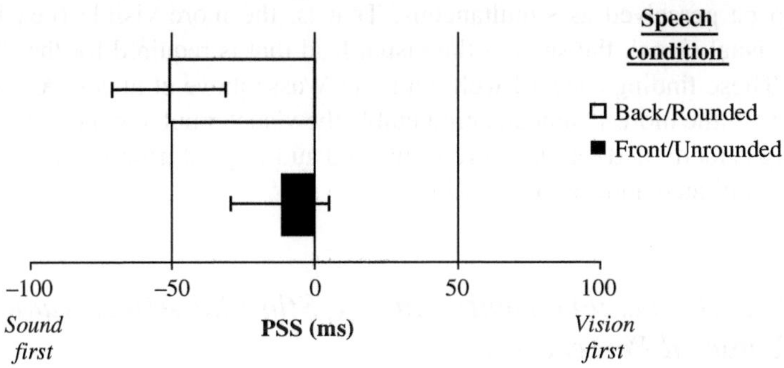

Fig. 7.5 PSSs for the backness/roundedness of the vowel stimuli (two levels: front/unrounded, /i, ɛ, æ/ and back/rounded, /u, ɔ, ɒ/) presented in Experiment 3 in Vatakis and Spence (submitted, 2007e). The error bars represent the standard errors of the mean

(i.e., varying as a function of the height and backness of the tongue and roundedness of the lips; see Kent, 1997) might have on the temporal window of audiovisual integration for speech stimuli in normal observers. The results of the experiments showed that visual speech had to lead auditory-speech in order for the PSS to be attained (except in the case of vowels; see Fig. 7.5). Specifically for vowels, larger auditory leads were observed for the highly visible rounded vowels as compared to the less-visible unrounded vowels (e.g., see Massaro and Cohen, 1993, for a comparison of /i/ and /u/ vowels and the /ui/ cluster; Traunmüller and Öhrström, 2007). Additionally, the participants were also more sensitive to the temporal order of the rounded vowels as compared to the unrounded vowels. It should, however, be noted that differences in sensitivity to the temporal order of the audiovisual speech stimuli were only found as a function of roundedness/backness, while no such differences were observed as a function of the height of the tongue positions (which happens to be a highly auditory-dominant feature).

Visual speech leads were generally larger for lower saliency visual-speech signals (e.g., alveolar tokens) as compared to the smaller visual leads observed for speech signals that were higher in visibility (e.g., such as bilabial tokens). Vatakis and Spence's (2007e, submitted) findings therefore replicated previous research showing that the visual-speech signal typically precedes the onset of the auditory-speech signal in the perception of audiovisual speech (e.g., Munhall et al., 1996). More importantly, this study extended previous research by using multiple speakers and by demonstrating that the precedence of the visual-speech signal changes as a function of the physical characteristics in the articulation of the particular speech signal that is uttered.

The results obtained in Vatakis and Spence's experiments lend support to the 'information reliability hypothesis,' with perception being dominated by the modality stream that provides the more reliable information (e.g., place versus manner of articulation of consonants; Schwartz et al., 1998; cf. Wada et al., 2003). Additionally, Vatakis and Spence's results also support the idea that the degree of visibility of the visual-speech signal can modulate the visual lead required for two stimuli to be perceived as simultaneous. That is, the more visible (i.e., informative) the visual signal, the smaller the visual lead that is required for the PSS to be reached. These findings accord well with van Wassenhove et al.'s (2005, p. 1183) claim that '. . .the more salient and predictable the visual input, the more the auditory processing is facilitated (or, the more visual and auditory information are redundant, the more facilitated auditory processing.'

7.7.3 Does the Visual Orientation of a Stimulus Affect Audiovisual Temporal Perception?

Audiovisual temporal perception can be modulated by any inherent differences in the properties of a complex audiovisual stimulus (such as in terms of physical differences attributable to the articulation of a particular speech sound). However, in an otherwise constant (in terms of its properties) stimulus, changes in the orientation of that stimulus (such as shifts in the orientation of a speaker's head during conversation) may also result in changes in sensitivity in the temporal perception of the stimulus. In order to evaluate this possibility, dynamic complex speech and nonspeech stimuli (short video clips) were presented in an upright or inverted orientation (Vatakis and Spence, 2008a). The results of these experiments revealed that the inversion of a dynamic visual-speech stream did not have a significant effect on the sensitivity of participants' TOJs concerning the auditory- and visual-speech and nonspeech stimuli (i.e., the JNDs were unchanged). The perception of synchrony was, however, affected in terms of a significant shift of the PSS being observed when the speech stimuli were inverted. Specifically, inversion of the speech stimulus resulted in the visual stream having to lead the auditory stream by a greater interval in order for the PSS to be reached. This result agrees with the findings of previous research on the face inversion effect (FIE), that is, the inversion of a visual display has been shown to lead to the loss of configural information (thus leading to slower face processing when compared to faces presented in an upright orientation) and to the recruitment of additional processes for the processing of a face but not for the processing of nonspeech events (cf. Bentin et al., 1996).

7.7.4 What Role Does the 'Unity Effect' Play in Audiovisual Temporal Perception?

The differences observed in the temporal window of audiovisual integration for simple versus complex audiovisual stimuli suggests that the perception of synchrony may be affected by the complexity of the particular stimuli being judged by participants. One possible account of how complexity might modulate temporal perception is in terms of the idea that a high level of stimulus complexity may promote the perception of synchrony (even for objectively slightly asynchronous stimuli). This could be due to an increased likelihood of binding that may be attributable to the operation of the unity assumption (i.e., the assumption that a perceiver has as to whether he/she is observing a single multisensory event versus multiple separate unimodal events, a decision that is based, at least in part, on the consistency of the information available to each sensory modality, e.g., Spence, 2007; Vatakis and Spence, 2007c; Welch and Warren, 1980).

In order to investigate the impact of the unity effect on the temporal perception of complex audiovisual stimuli, matching and mismatching auditory- and visual-speech streams consisting of syllables and words and nonspeech stimuli consisting of object-action, musical, and monkey call stimuli (see Fig. 7.6) were presented (Vatakis and Spence, 2007a, b, c, 2008a, b; Vatakis et al., 2008). The results of 11 TOJ experiments provided psychophysical evidence in support of the conclusion that the unity effect can modulate the crossmodal binding of multisensory information at a perceptual level of information processing. This modulation was shown to be robust in the case of audiovisual speech events, while no such effect was reported for audiovisual nonspeech or animal call events (but see Petrini et al., 2009; Schutz and Kubovy, 2009, for musicians; see Fig. 7.7).

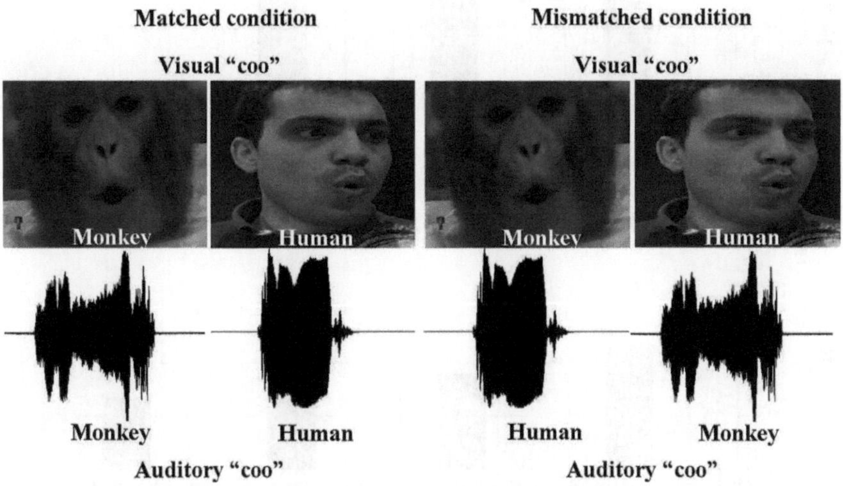

Fig. 7.6 Still frames taken from the matching and mismatching audiovisual 'coo' vocalization video clips of the rhesus monkey and human used. The frames highlight the most visible feature of the 'coo' vocalization. The *bottom panel* shows the entire acoustic 'coo' signal (see Vatakis et al., 2008)

Specifically, the results of the experiments (Vatakis and Spence, 2007c) showed that people were significantly more sensitive to the temporal order of the auditory- and visual-speech streams when they were mismatched (e.g., when the female voice was paired with a male face, or vice versa) than when they were matched. No such matching effect was found for audiovisual nonspeech and animal call stimuli, where no sensitivity differences between matched and mismatched conditions

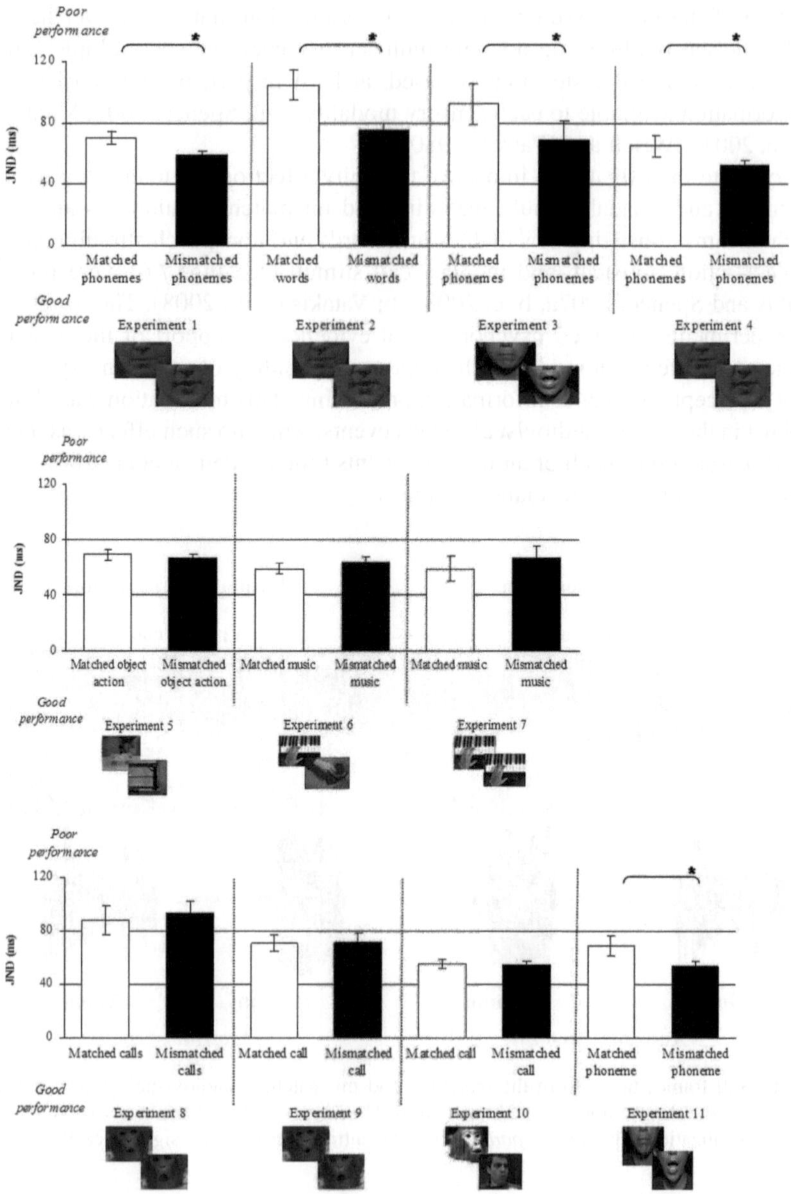

Fig. 7.7 Average JNDs for the matched and mismatched audiovisual stimuli speech (Experiments 1–4, 11), object-action (Experiments 5–7), and nonspeech calls (Experiments 8–10). The error bars represent the standard errors of the means. Significant differences ($P < 0.05$) are highlighted by an *asterisk* (Experiments 8–11 in Vatakis et al., 2008; Experiments 1–4 and 5–7 in Vatakis and Spence, 2007b, c, 2008b, respectively)

were observed. Moreover, it was shown that this modulatory effect was not present for all human vocalizations (e.g., for humans imitating the cooing call of a monkey) but that it was specific either to the integration of the auditory- and visual-speech signals or perhaps to the presence of the auditory-speech signal itself (though see also Parise and Spence, 2009, for a recent demonstration of the unity effect for synesthetically congruent simple auditory and visual stimuli).

7.8 Conclusions

To date, research on audiovisual temporal perception has shown that the temporal window of audiovisual synchrony perception has a width in the order of several hundred milliseconds and that this window is asymmetrical, being larger when the visual-stimulus leads than when it lags, and highly variable across studies and experimental stimuli. Over the last few years in our research at the Crossmodal Research Laboratory at Oxford University, we have sought to identify those factors driving the variability noted in the previously published research on temporal perception for both speech and nonspeech stimuli (Vatakis and Spence, 2007d). Taken together, the empirical research outlined here has demonstrated that the temporal window of integration for complex audiovisual stimuli is modulated by the type, complexity, and properties of the particular experimental stimuli used, the degree of unity of the auditory- and visual-stimulus streams (for the case of speech stimuli; see also Parise and Spence, 2009, for the case of nonspeech stimuli), the orientation of the visual stimulus (again for the case of speech, Vatakis and Spence, 2008a), and the medium used to present the desired experimental stimuli (Vatakis and Spence, 2006b).

As far as the special nature of speech is concerned, the issue still stands for further research (see Eskelund and Andersen, 2009; Maier et al., submitted). It should be noted that our experimental research was not designed to investigate the special nature of speech or to evaluate the existing theories that have been put forward to explain the audiovisual integration of speech perception. During experimental testing and data collection, however, an exploration of how certain of the speech data we obtained fitted with the major speech theories was considered. Thus, in our evaluation of how the physical differences in the articulation of speech affected temporal perception, we have shown that for the majority of speech tokens, the visual-speech signal had to precede the corresponding auditory-speech signal for synchrony to be perceived. Additionally, our research has also shown that highly visible visual-speech signals (such as bilabial stimuli) require less of a lead over auditory-speech signals than visual-speech signals that are less visible (such as velar stimuli). Our results therefore appear to be consistent with the 'analysis-by-synthesis' model, whereby the precedence of the visual-signal leads the speech-processing system to form a prediction regarding the auditory signal, which, in turn, is directly dependent on the saliency of the visual-signal (with higher saliency signals resulting in a better prediction concerning the auditory signal, e.g., van Wassenhove et al., 2005). It would therefore be interesting to further evaluate the validity of this theory and to examine what the specific properties that make a signal salient are and why is it that vowels appear not to fit the pattern described in this model.

Acknowledgments A.V. was supported by a Newton Abraham Studentship from the Medical Sciences Division, University of Oxford. Correspondence regarding this article should be addressed to Argiro Vatakis, Institute for Language and Speech Processing, Artemidos 6 & Epidavrou, Athens, 151 25, Greece. E-mail: argiro.vatakis@gmail.com.

References

Alais D, Burr D (2004) The ventriloquist effect results from near-optimal bimodal integration. Curr Biol 14:257–262

Allan LG (1975) The relationship between judgments of successiveness and judgments of order. Percept Psychophys 18:29–36

Arnold DH, Johnston A, Nishida S (2005) Timing sight and sound. Vision Research 45: 1275–1284

Asakawa K, Tanaka A, Imai H (2009). Temporal recalibration in audio-visual speech integration using a simultaneity judgment task and the McGurk identification task. Cognitive Science Meeting, Amsterdam, Netherlands

Bald L, Berrien FK, Price JB, Sprague RO (1942) Errors in perceiving the temporal order of auditory and visual stimuli. J Appl Psychol 26:382–388

Bentin S, Allison T, Puce A, Perez E, McCarthy G (1996) Electrophysiological studies of face perception in humans. J Cogn Neurosci 8:551–565

Bernstein LE, Auer ET, Moore JK (2004) Audiovisual speech binding: convergence or association? In: Calvert GA, Spence C, Stein BE (eds) The handbook of multisensory processing. MIT Press, Cambridge, MA, pp 203–223

Bertelson P, Aschersleben G (1998) Automatic visual bias of perceived auditory location. Psychonom Bull Rev 5:482–489

Bushara KO, Grafman J, Hallett M (2001). Neural correlates of auditory-visual stimulus onset asynchrony detection. J Neurosci 21:300–304

Calvert GA, Spence C, Stein BE (eds) (2004) The handbook of multisensory processing. MIT Press, Cambridge, MA

Chen Y-C, Spence C (2010). When hearing the bark helps to identify the dog: Semantically-congruent sounds modulate the identification of masked pictures. Cognition 114:389–404

Choe CS, Welch RB, Gilford RM, Juola JF (1975) The 'ventriloquist effect': visual dominance or response bias? Percept Psychophys 18:55–60

Conrey BL, Pisoni DB (2006) Auditory-visual speech perception and synchrony detection for speech and nonspeech signals. J Acoust Soc Am 119:4065–4073

Coren S, Ward LM, Enns JT (2004) Sensation perception, 6th edn. Harcourt Brace, Fort Worth

De Gelder B, Bertelson P (2003) Multisensory integration, perception and ecological validity. Trends Cogn Sci 7:460–467

Dixon NF, Spitz L (1980) The detection of auditory visual desynchrony. Perception 9: 719–721

Doehrmann O, Naumer MJ (2008) Semantics and the multisensory brain: how meaning modulates processes of audio-visual integration. Brain Res 1242:136–150

Driver J, Spence C (2000) Multisensory perception: beyond modularity and convergence. Curr Biol 10:R731–R735

Efron R (1963) The effect of handedness on the perception of simultaneity and temporal order. Brain 86:261–284

Engel GR, Dougherty WG (1971) Visual-auditory distance constancy. Nature 234:308

Eskelund K, Andersen TS (2009) Specialization in audiovisual speech perception: a replication study. Poster presented at the 10th Annual Meeting of the International Multisensory Research Forum (IMRF), New York City, 29th June–2nd July

Fendrich R, Corballis PM (2001) The temporal cross-capture of audition and vision. Percept Psychophys 63:719–725

Fraisse P (1984) Perception and estimation of time. Annu Rev Psychol 35:1–36

Fujisaki W, Nishida S (2005) Temporal frequency characteristics of synchrony-asynchrony discrimination of audio-visual signals. Exp Brain Res 166:455–464

Fujisaki W, Nishida S (2007) Feature-based processing of audio-visual synchrony perception revealed by random pulse trains. Vis Res 47:1075–1093

Grant KW, Greenberg S (2001) Speech intelligibility derived from asynchronous processing of auditory-visual speech information. Proceedings of the Workshop on Audio Visual Speech Processing, Scheelsminde, Denmark, September 7–9, pp 132–137

Grant KW, Seitz PF (1998) The use of visible speech cues (speechreading) for directing auditory attention: reducing temporal and spectral uncertainty in auditory detection of spoken sentences. In: Kuhl PK, Crum LA (eds) Proceedings of the 16th international congress on acoustics and the 135th meeting of the acoustical society of America, vol. 3. ASA, New York, pp 2335–2336

Grant KW, van Wassenhove V, Poeppel D (2003) Detection of auditory (cross-spectral) and auditory-visual (cross-modal) synchrony. Speech Commun 44:43–53

Grant KW, van Wassenhove V, Poeppel D (2004) Detection of auditory (cross-spectral) and auditory-visual (cross-modal) synchrony. J Acoust Soc Am 108:1197–1208

Hein G, Doehrmann O, Müller NG, Kaiser J, Muckli L, Naumer MJ (2007) Object familiarity and semantic congruency modulate responses in cortical audiovisual integration areas. J Neurosci 27:7881–7887

Hirsh IJ (1959) Auditory perception of temporal order. J Acoust Soc Am 31:759–767

Hirsh IJ, Sherrick CE Jr (1961) Perceived order in different sense modalities. J Exp Psychol 62:424–432

Hollier MP, Rimell AN (1998) An experimental investigation into multi-modal synchronisation sensitivity for perceptual model development. 105th AES Convention, Preprint No. 4790

Howard IP, Templeton WB (1966) Human spatial orientation. Wiley, New York

Jaśkowski P, Jaroszyk F, Hojan-Jesierska D (1990) Temporal-order judgments and reaction time for stimuli of different modalities. Psycholog Res 52:35–38

Jones JA, Jarick M (2006) Multisensory integration of speech signals: the relationship between space and time. Exp Brain Res 174:588–594

Kallinen K, Ravaja N (2007) Comparing speakers versus headphones in listening to news from a computer - individual differences and psychophysiological responses. Comp Human Behav 23:303–317

Keetels M, Vroomen J (in press). Perception of synchrony between the senses. In: Murrary MM, Wallace MT (eds.) Frontiers in the neural basis of multisensory processes

Kent RD (1997) The speech sciences. Singular, San Diego, CA

King AJ (2005) Multisensory integration: strategies for synchronization. Curr Biol 15: R339–R341

King AJ, Palmer AR (1985) Integration of visual and auditory information in bimodal neurones in the guinea-pig superior colliculus. Exp Brain Res 60:492–500

Kohlrausch A, van de Par S (2005) Audio-visual interaction in the context of multi-media applications. In: Blauert J (ed.) Communication acoustics. Springer, Berlin, pp 109–138

Kopinska A, Harris LR (2004) Simultaneity constancy. Perception 33:1049–1060

Koppen C, Spence C (2007). Audiovisual asynchrony modulates the Colavita visual dominance effect. Brain Res 1186:224–232

Lee H-L, Noppeney U (2009) Audiovisual synchrony detection for speech and music signals. Poster presented at the 10th annual meeting of the international multisensory research forum (IMRF), New York City, 29th June–2nd July

Lewald J, Guski R (2004) Auditory-visual temporal integration as a function of distance: no compensation of sound-transmission time in human perception. Neurosci Lett 357: 119–122
Maier JX, Di Luca M, Ghazanfar AA (2009; submitted). Auditory-visual asynchrony detection in humans. J Exp Psychol: Human Percept Perform
Massaro DW (1996) Integration of multiple sources of information in language processing. In: Inui T, McClelland JL (eds) Attention and performance XVI: information integration in perception and communication. MIT Press, New York, pp 397–432
Massaro DW (2004) From multisensory integration to talking heads and language learning. In: Calvert GA, Spence C, Stein BE (eds) The handbook of multisensory processing. MIT Press, Cambridge, MA, pp 153–176
Massaro DW, Cohen MM (1993) Perceiving asynchronous bimodal speech in consonant-vowel and vowel syllables. Speech Commun 13:127–134
Massaro DW, Cohen MM, Smeele PMT (1996) Perception of asynchronous and conflicting visual and auditory speech. J Acoust Soc Am 100:1777–1786
Mauk MD, Buonomano DV (2004) The neural basis of temporal processing. Annu Rev Neurosci 27:307–340
McGrath M, Summerfield Q (1985) Intermodal timing relations and audiovisual speech recognition by normal hearing adults. J Acoust Soc Am 77:678–685
McGurk H, MacDonald J (1976) Hearing lips and seeing voices. Nature 264:746–748
Miner N, Caudell T (1998) Computational requirements and synchronization issues of virtual acoustic displays. Presence: Teleop Virt Environ 7:396–409
Morein-Zamir S, Soto-Faraco S, Kingstone A (2003) Auditory capture of vision: examining temporal ventriloquism. Cogn Brain Res 17:154–163
Munhall KG, Gribble P, Sacco L, Ward M (1996) Temporal constraints on the McGurk effect. Percept Psychophys 58:351–362
Munhall KG, Vatikiotis-Bateson E (2004) Spatial and temporal constraints on audiovisual speech perception. In: Calvert GA, Spence C, Stein BE (eds) The handbook of multisensory processing. MIT Press, Cambridge, MA, pp 177–188
Navarra J, Alsius A, Velasco I, Soto-Faraco S, Spence C (2010) Perception of audiovisual speech synchrony for native and non-native speech. Brain Res 1323:84–93
Navarra J, Hartcher-O'Brien J, Piazza E, Spence C (2009) Adaptation to audiovisual asynchrony modulates the speeded detection of sound. Proc Natl Acad Sci USA 106:9169–9173
Navarra J, Vatakis A, Zampini M, Soto-Faraco S, Humphreys W, Spence C (2005) Exposure to asynchronous audiovisual speech extends the temporal window for audiovisual integration. Cogn Brain Res 25:499–507
Neuta W, Feirtag M (1986) Fundamental neuroanatomy. Freeman Co, New York
Noesselt T, Bergmann D, Heinze H-J., Münte T, Spence C (submitted) Spatial coding of multisensory temporal relations in human superior temporal sulcus. PLoS ONE
Noesselt T, Fendrich R, Bonath B, Tyll S, Heinze H-J. (2005) Closer in time when farther in space - Spatial factors in audiovisual temporal integration. Cogn Brain Res 25:443–458
Pandey CP, Kunov H, Abel MS (1986) Disruptive effects of auditory signal delay on speech perception with lip-reading. J Audit Res 26:27–41
Parise C, Spence C (2009) 'When birds of a feather flock together': synesthetic correspondences modulate audiovisual integration in non-synesthetes. PLoS ONE 4(5):e5664. doi:10.1371/journal.pone.0005664
Petrini K, Russell M, Pollick F (2009) When knowing can replace seeing in audiovisual integration of actions. Cognition 110:432–439
Pöppel E, Schill K, von Steinbüchel N (1990) Sensory integration within temporally neutral system states: a hypothesis. Naturwissenschaften 77:89–91
Recanzone GH (2003) Auditory influences on visual temporal rate perception. J Neurophysiol 89:1078–1093
Reeves B, Voelker D (1993) Effects of audio-video asynchrony on viewer's memory, evaluation of content and detection ability. Research report prepared for Pixel Instruments. Los Gatos, California

Rihs S (1995) The influence of audio on perceived picture quality and subjective audio-visual delay tolerance. In: Hamberg R, de Ridder H (eds) Proceedings of the MOSAIC workshop: advanced methods for the evaluation of television picture quality, Eindhoven, September 18th–19th, pp 133–137

Rutschmann J, Link R (1964) Perception of temporal order of stimuli differing in sense mode and simple reaction time. Percept Motor Skills 18:345–352

Scheier CR, Nijhawan R, Shimojo S (1999) Sound alters visual temporal resolution. Invest Opthalmol Vis Sci 40:S792

Schutz M, Kubovy M (2009) Causality in audio-visual sensory integration. J Exp Psychol: Human Percept Perform 35:1791–1810

Schwartz J-L, Robert-Ribes J, Escudier P (1998) Ten years after Summerfield: a taxonomy of models for audio-visual fusion in speech perception. In: Burnham D (ed.) Hearing by eye II: advances in the psychology of speechreading and auditory-visual speech. Psychology Press, Hove, UK, pp 85–108

Sekuler R, Sekuler AB, Lau R (1997) Sound alters visual motion perception. Nature 385:308

Slutsky DA, Recanzone GH (2001) Temporal and spatial dependency of the ventriloquism effect. Neuroreport 12:7–10

Soto-Faraco S, Alsius A (2007) Access to the uni-sensory components in a cross-modal illusion. Neuroreport 18:347–350

Soto-Faraco S, Alsius A (2009) Deconstructing the McGurk-MacDonald illusion. J Exp Psychol: Human Percept Perform 35:580–587

Soto-Faraco S, Lyons J, Gazzaniga M, Spence C, Kingstone A (2002) The ventriloquist in motion: illusory capture of dynamic information across sensory modalities. Cogn Brain Res 14: 139–146

Spence C (2007) Audiovisual multisensory integration. J Acoust Soc Jpn: Acoust Sci Technol 28:61–70

Spence C. Prior entry: attention and temporal perception. In: Nobre AC, Coull JT (eds) Attention and time. Oxford University Press, Oxford (in press)

Spence C, Driver J (1997) On measuring selective attention to a specific sensory modality. Percept Psychophys 59:389–403

Spence C, Shore DI, Klein RM (2001) Multisensory prior entry. J Exp Psychol: Gen 130: 799–832

Spence C, Squire SB (2003) Multisensory integration: maintaining the perception of synchrony. Curr Biol 13:R519–R521

Stein BE, Meredith MA (1993) The merging of the senses. MIT Press, Cambridge, MA

Steinmetz R (1996) Human perception of jitter and media synchronization. IEEE J Select Areas Commun 14:61–72

Sternberg S, Knoll RL, Gates BA (1971) Prior entry reexamined: Effect of attentional bias on order perception. Paper presented at the meeting of the Psychonomic Society, St. Louis, Missouri

Stone JV, Hunkin NM, Porrill J, Wood R, Keeler V, Beanland M, Port M, Porter NR (2001) When is now? Perception of simultaneity. Proc R Soc Lond B, Biol Sci 268:31–38

Sugita Y, Suzuki Y (2003) Implicit estimation of sound-arrival time. Nature 421:911

Teder-Sälejärvi WA, Di Russo F, McDonald JJ, Hillyard SA (2005) Effects of spatial congruity on audio-visual multimodal integration. J Cogn Neurosci 17:1396–1409

Thorne JD, Debner S (2008) Irrelevant visual stimuli improve auditory task performance. Neuroreport 19:553–557

Titchener- EB (1908) Lecture on the elementary psychology of feeling and attention. Macmillan, New York

Traunmüller H, Öhrström N (2007) Audiovisual perception of openness and lip rounding in front vowels. J Phonet 35:244–258

Tuomainen J, Andersen TS, Tiippana K, Sams M (2005) Audio-visual speech is special. Cognition 96:B13–B22

van de Par S, Kohlrausch A, Juola JF (1999) Judged synchrony/asynchrony for light-tone pairs. Poster presented at the 40th Annual Meeting of the Psychonomic Society, Los Angeles, CA
van Eijk RL J., Kohlrausch A, Juola JF, van de Par S (2008) Audiovisual synchrony and temporal order judgments: effects of experimental method and stimulus type. Percept Psychophys 70:955–968
van Wassenhove V., Grant KW., Poeppel D (2003) Electrophysiology of auditory-visual speech integration. International conference on auditory-visual speech processing (AVSP), St Jorioz, France, pp 31–35
van Wassenhove V, Grant KW, Poeppel D (2005) Visual speech speeds up the neural processing of auditory speech. Proc Natl Acad Sci USA 102:1181–1186
van Wassenhove V, Grant KW, Poeppel D (2007) Temporal window of integration in auditory-visual speech perception. Neuropsychologia 45:598–607
Vatakis A, Ghazanfar AA, Spence C (2008) Facilitation of multisensory integration by the "unity effect" reveals that speech is special. J Vis 8(9):14:1–11
Vatakis A, Navarra J, Soto-Faraco S, Spence C (2007) Temporal recalibration during asynchronous audiovisual speech perception. Exp Brain Res 181:173–181
Vatakis A, Spence C (2006a) Audiovisual synchrony perception for music, speech, and object actions. Brain Res 1111:134–142
Vatakis A, Spence C (2006b) Evaluating the influence of frame rate on the temporal aspects of audiovisual speech perception. Neurosci Lett 405:132–136
Vatakis A, Spence C (2006c) Audiovisual synchrony perception for speech and music using a temporal order judgment task. Neurosci Lett 393:40–44
Vatakis A, Spence C (2007a) How 'special' is the human face? Evidence from an audiovisual temporal order judgment task. Neuroreport 18:1807–1811
Vatakis A, Spence C (2007b) Crossmodal binding: evaluating the 'unity assumption' using complex audiovisual stimuli. Proceedings of the 19th international congress on acoustics (ICA), Madrid, Spain
Vatakis A, Spence C (2007c) Crossmodal binding: evaluating the 'unity assumption' using audiovisual speech stimuli. Percept Psychophys 69:744–756
Vatakis A, Spence C (2007d) Investigating the factors that influence the temporal perception of complex audiovisual events. Proc Eur Cogn Sci 2007 (EuroCogSci07):389–394
Vatakis A, Spence C (2007e) An assessment of the effect of physical differences in the articulation of consonants and vowels on audiovisual temporal perception. Poster presented at the one-day meeting for young speech researchers, University College London, London, UK
Vatakis A, Spence C (2008a). Investigating the effects of inversion on configural processing using an audiovisual temporal order judgment task. Perception 37:143–160
Vatakis A, Spence C (2008b). Evaluating the influence of the 'unity assumption' on the temporal perception of realistic audiovisual stimuli. Acta Psychol 127:12–23
Vatakis A, Spence C (submitted). Assessing the effect of physical differences in the articulation of consonants and vowels on audiovisual temporal perception. J Speech Lang Hear Res
Vroomen J, de Gelder B (2004) Temporal ventriloquism: sound modulates the flash-lag effect. J Exp Psychol: Human Percept Perform 30:513–518
Vroomen J, Keetels M (2006) The spatial constraint in intersensory pairing: no role in temporal ventriloquism. J Exp Psychol: Human Percept Perform 32:1063–1071
Wada Y, Kitagawa N, Noguchi K (2003) Audio-visual integration in temporal perception. Int J Psychophysiol 50:117–124
Welch RB, Warren DH (1980) Immediate perceptual response to intersensory discrepancy. Psychol Bull 88:638–667
Zampini M, Bird KJ, Bentley DE, Watson A, Barrett G, Jones AK, Spence C (2007) Prior entry for pain and vision: attention speeds the perceptual processing of painful stimuli. Neurosci Lett 414:75–79
Zampini M, Guest S, Shore DI, Spence C (2005) Audio-visual simultaneity judgments. Percept Psychophys 67:531–544

Zampini M, Shore DI, Spence C (2003a) Multisensory temporal order judgments: the role of hemispheric redundancy. Int J Psychophysiol 50:165–180
Zampini M, Shore DI, Spence C (2003b) Audiovisual temporal order judgments. Exp Brain Res 152:198–210
Zeki E (1993) A vision of the brain. New York: Oxford University Press

Chapter 8
Imaging Cross-Modal Influences in Auditory Cortex

Christoph Kayser, Christopher I. Petkov, and Nikos K. Logothetis

8.1 Introduction

During everyday interactions we benefit from the synergistic interplay of our different sensory modalities. In many situations evidence from one modality can complement or even input to another, often resulting in an improved percept or faster reactions (Driver and Spence, 1998; Hershenson, 1962; Lehmann and Murray, 2005; McDonald et al., 2000; Seitz et al., 2006; Vroomen and de Gelder, 2000). In the auditory domain, for example, our ability to recognize sounds or to understand speech profits considerably from visual information. A well-known example for such multisensory benefits of hearing is the cocktail party: with loud music playing and people chatting we can much better understand somebody speaking when we watch the movements of their lips. In such situations, the additional visual information can boost hearing capabilities by an equivalent of about 15–20 dB sound intensity (Sumby and Polack, 1954). Another frequently encountered example of audio-visual interaction is the ventriloquist illusion (Alais and Burr, 2004), which we experience, for example, when we watch a movie and attribute the sound of a voice to the actor on the screen, although it may be coming from a loudspeaker quite some distance from the screen. Together we can say that cross-modal influences play an important role in enhancing our sensory capabilities, even if we are not explicitly aware of this.

This influence of cross-modal input on our acoustic perception and the ability to hear begs the questions of where and how the brain combines the acoustic and non-acoustic information. Classical neuroanatomical studies proposed that sensory information converges only in higher association cortices of the superior temporal or intra-parietal regions and in the frontal lobe (Benevento et al., 1977; Bruce et al., 1981; Felleman and Van Essen, 1991; Hikosaka et al., 1988; Hyvarinen and

---

C. Kayser (✉)
Max Planck Institute for Biological Cybernetics, Spemannstrasse 38, 72076 Tübingen, Germany
e-mail: christoph.kayser@tuebingen.mpg.de

CP is now at the Institute of Neuroscience, University of Newcastle, Newcastle upon Tyne, NE2 4HH, UK

M.J. Naumer, J. Kaiser (eds.), *Multisensory Object Perception in the Primate Brain*,
DOI 10.1007/978-1-4419-5615-6_8,

Shelepin, 1979; Jones and Powell, 1970). Mostly, this conclusion was drawn based on anatomical studies overlooking the interconnections between early sensory processing stages of the neocortex (e.g., the primary sensory cortices); the earlier work mainly reported anatomical connectivity between higher level association cortices and neurons that responded to multiple sensory modalities mainly in such higher association regions. More recent evidence suggests that this picture is incomplete (Driver and Noesselt, 2008; Ghazanfar and Schroeder, 2006; Kayser and Logothetis, 2007; Schroeder et al., 2008). Results from functional imaging, electrophysiology, and anatomy suggest that cross-modal interactions appear to occur much earlier in the cortical processing hierarchy than previously appreciated, for example, in primary sensory cortices (Foxe and Schroeder, 2005; Ghazanfar and Schroeder, 2006; Lakatos et al., 2007, 2008; Schroeder et al., 2004) or even earlier. In the following we will review the evidence for cross-modal influences in auditory cortex and place particular emphasis on functional imaging studies and their ability to localize cross-modal influences in particular auditory fields.

8.2 Cross-Modal Influences in Auditory Cortex Revealed by fMRI

Functional magnetic resonance imaging (fMRI) contributes the largest body of evidence showing that visual or somatosensory stimuli can influence activity in auditory cortex. Usually, 'activity' in imaging studies is assessed using the blood-oxygen level-dependent (BOLD) signal, which provides a hemodynamic signal that is indirectly coupled to neuronal activity (Logothetis, 2008; Logothetis et al., 2001). To detect and localize cross-modal influences, imaging studies usually compare the sensory-evoked response following the presentation of an acoustic stimulus to the response when the same stimulus is paired with, for instance, a visual stimulus. Using this approach, several imaging studies demonstrated that visual or somatosensory stimuli can enhance (or reduce) the BOLD response in regions of auditory cortex and in a wide variety of sensory paradigms. Especially, visual influences on responses in auditory cortex were found in the broad context of simplistic, natural, and speech stimuli, hence covering a wide range of behaviorally relevant settings, as examples see (Bernstein et al., 2002; Calvert and Campbell, 2003; Calvert et al., 1997, 1999; Foxe et al., 2002; Lehmann et al., 2006; Martuzzi et al., 2006; Pekkola et al., 2005a; Schurmann et al., 2006; van Atteveldt et al., 2004; van Wassenhove et al., 2005). In addition, BOLD responses in auditory cortex were also reported as a result of plain visual stimulation, providing additional support for a persistent visual drive in this region (Calvert, 2001; Calvert et al., 1997). While these studies leave little doubt as to the existence of visual influences on BOLD responses in auditory cortex, they provide only limited insight into the exact localization of these cross-modal influences with respect to individual auditory fields, the main reason being the absence of an exact map of human auditory cortex. In the following we will discuss some of the underlying problems and approaches to overcome these.

Auditory cortex is subdivided into a number of different fields that can be distinguished based on anatomical and functional properties. For example, cytoarchitectonic features such as the koniocortical appearance of cells, the density of myelin, or the expression of histological markers such as acetylcholinesterase, cytochrome oxidase, and parvalbumin allow the differentiation of so-called core, belt, and parabelt regions (Fullerton and Pandya, 2007; Hackett et al., 1998, 2001; Morel et al., 1993). The core region receives prominent afferents from the auditory thalamus, hence constituting a primary-like region. The belt and parabelt in contrast receive input from the core region and constitute higher stages of auditory processing. However, anatomical studies further demonstrate that both core and (para-) belt regions can be further subdivided into a number of distinct auditory fields each. In non-human primates such as the macaque monkey, for example, the core is divided into three fields, one called the primary field A1 and two the rostral and primary-like fields (abbreviated R and RT). The belt, in addition, is divided into about eight fields, and the parabelt into at least two, resulting in a total of more than a dozen auditory fields that can be characterized based on anatomical properties. While this division of auditory cortex has been extensively characterized in animals, anatomical studies demonstrate that a parcellation of auditory cortex can be made of the human brain (Fullerton and Pandya, 2007; Rivier and Clarke, 1997), although how this corresponds to the animal mapping of fields is still an open question.

Several features of this division of auditory cortex are important with regard to functional imaging studies. Many of these regions are rather small and even in the human brain often have a diameter of only a few millimeters. This small scale as well as their complex spatial arrangement makes it difficult to 'see' individual fields using conventional human imaging studies with a resolution of 2–3 mm or more. In addition, the exact position of individual auditory fields with respect to anatomical landmarks such as Heschel's gyrus varies considerably in individual subjects (Chiry et al., 2003; Clarke and Rivier, 1998). As a result, group averaging techniques are likely to blur over individual fields, leading to a miss-localization of the observed activations (Crivello et al., 2002; Desai et al., 2005; Rademacher et al., 1993). The scale and organization of auditory cortex hence pose considerable limitations on our possibility to localize cross-modal influences with high fidelity to particular auditory fields.

One possibility to overcome these difficulties would be to localize individual auditory fields on a subject by subject basis and to analyze functional data with respect to these functionally defined fields. Studies in the visual system frequently follow this strategy by exploiting the retinotopic organization of early visual areas: localizer stimuli can be used to map the boundaries and spatial layout of early visual areas in individual brains (Engel et al., 1994; Warnking et al., 2002). Such a functional localization however requires a certain selectivity of the responses in those regions to be localized, which in the visual cortex is given by the retinotopic organization. Incidentally, auditory cortex possesses a similar response property that allows a functional identification of individual auditory fields: Electrophysiological recordings have demonstrated that many neurons in auditory cortex respond selectively to sounds of particular frequencies. Importantly, this response preference is

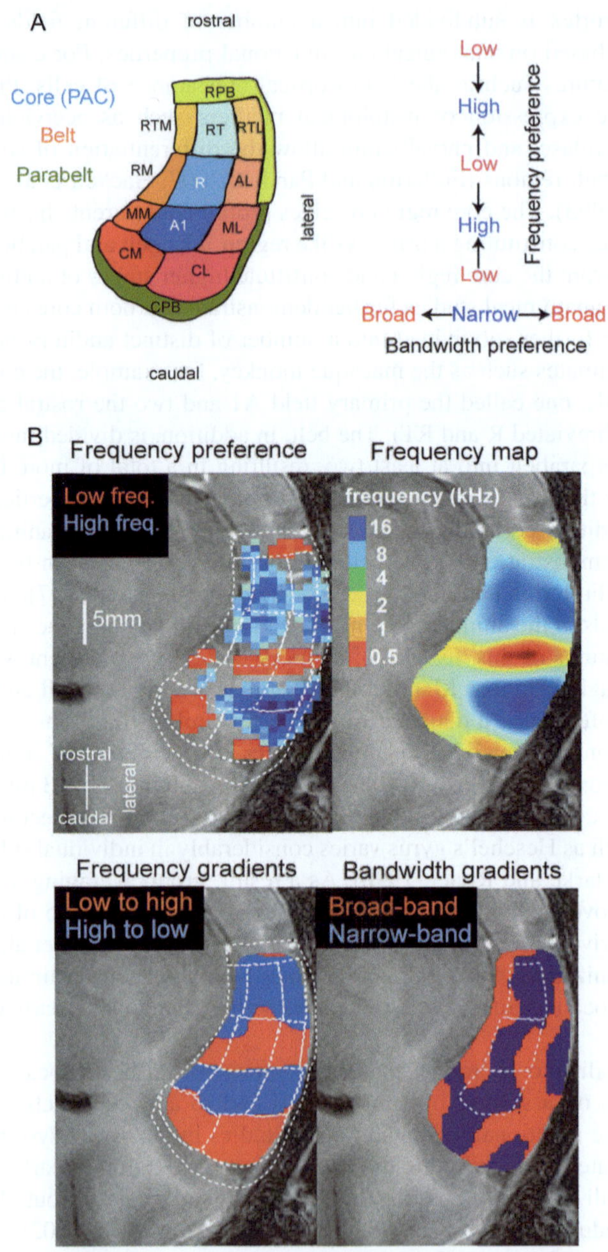

Fig. 8.1 Functional parcellation of auditory cortex using fMRI. (**a**) Functional parcellation of auditory cortex. Three primary auditory fields (the core region) are surrounded by the secondary fields (the belt region) as well as higher association areas (the parabelt). Electrophysiological studies have shown that several of these fields contain an ordered representation of sound frequency (tonotopic map, indicated on the *left*) and that core and belt fields prefer narrow and broadband sounds, respectively. These two functional properties can be exploited to map the layout of these

not distributed randomly, but neurons preferring similar sound frequencies are spatially clustered resulting in an ordered distribution of sound frequency selectivity throughout auditory cortex. More specifically, many anatomically defined fields, such as the core fields A1 or R, for example, contain a full representation of the entire frequency range. As a result, many tonotopic maps exist in auditory cortex, with each individual map belonging to one auditory field. Localizing such a map allows one not only to determine the position within each field, but can also be used to distinguish borders between individual fields, since the direction of the frequency-gradient changes between fields. In addition, neurons in the primary-like fields of auditory cortex, the core region, often respond stronger to narrowband sounds, while neurons in the higher auditory fields (the belt region) respond more to broadband or complex sounds (Rauschecker, 1998; Rauschecker and Tian, 2004; Rauschecker et al., 1997). This selectivity to sound bandwidth allows a further differentiation of auditory fields and provides one means to differentiate core and belt regions. Together these response properties allow the functional localization of individual fields and a parcellation of auditory cortex into a number of core and belt fields (Fig. 8.1a). These properties are routinely employed in electrophysiological studies to localize particular auditory fields and, as we will see next, can also be exploited in functional imaging studies.

8.3 A Functional Parcellation of Auditory Cortex

Several human functional imaging studies have exploited this functional organization to map individual auditory fields in the human brain (Formisano et al., 2003; Talavage et al., 2004; Wessinger et al., 1997). However, although anatomical studies suggest the existence of nearly a dozen fields, these imaging studies have not been able to elucidate more than a few individual fields. Partly this might be attributable to the small scale of many fields, which as noted above, can easily be at the limits of the resolution achieved using current fMRI (but see (Formisano et al., 2003) for a human high-resolution approach). However, partly this might also relate

Fig. 8.1 (continued) auditory fields in individual subjects using functional imaging. Abbreviations: A1, primary auditory cortex; R, rostral core field; CM, caudo-medial field; CL, caudo-lateral field) indicate individual auditory fields. See, e.g., Petkov et al. (2006) for details. (**b**) Single-slice fMRI data showing frequency selective BOLD responses to low and high tones (*upper left panel*) and a complete (smoothed) frequency map obtained from stimulation using six frequency bands (*upper right panel*). From this frequency map borders between regions of opposite frequency gradients can be determined (*lower left panel*). These borders delineate different auditory fields in the antero-posterior direction. In addition, a smoothed bandwidth map can be obtained, from which borders between regions preferring narrowband and broadband sounds are obtained (*lower right panel*). Combining the evidence from frequency and bandwidth gradients finally allows constructing the full functional parcellation of auditory cortex in individual subjects (as shown in **a**) to differences in the response selectivity as seen by fMRI and electrophysiology. Most of our knowledge regarding the tonotopic organization is based on single-unit recordings, which might provide a more selective signal than the BOLD signal, which reflects neuronal responses only indirectly and on a larger scale (Logothetis, 2008). As a result, it is not a priori clear whether the limited success of the above studies in functionally mapping auditory cortex reflects only the limitations of the used imaging paradigms (lower spatial resolution, partly unknown organization of human auditory cortex) or a true limitation of the functional imaging technique per se.

To determine whether an extensive functional parcellation of auditory cortex into several fields is indeed possible, we exploited high-resolution imaging facilities in combination with a non-human model system for which there exists considerably prior knowledge about the organization of auditory cortex. Several decades of neuroanatomical and electrophysiological studies have provided insights into the organization of the auditory cortex in the macaque monkey. In fact, many studies carefully delineated a number of distinct auditory fields and demonstrated the tonotopic organization of these (Kosaki et al., 1997; Merzenich and Brugge, 1973; Morel et al., 1993; Recanzone et al., 2000). As a result, considerable a priori knowledge existed about the expected functional organization of auditory cortex for this model system which guided the fMRI.

We exploited this a priori anatomical and functional knowledge to constrain models for the organization of auditory cortex that were derived from functional imaging experiments. Exploiting the high resolution offered by high-field (4.7 and 7 T) imaging, we were able to obtain reliable maps of the tonotopic functional parcellation in individual animals (Petkov et al., 2006). By comparing the activation to stimulation with sounds of different frequency composition, we obtained a frequency preference map, which allowed determining the anterior–posterior borders of potential fields (Fig. 8.1b). In more detail, borders were defined as known from electrophysiology, and between regions featuring tonotopic maps with spatially opponent frequency gradients, i.e., regions were defined where the tonotopic map switched from increasing to decreasing frequency or vice versa. In addition, the preference to sounds of different bandwidth allowed a segregation of core and belt fields, hence providing borders in medial–lateral directions. When combined with the known organization of auditory cortex the evidence from these activation patterns allowed a complete parcellation into distinct core and belt fields and provided constraints for the localization of the parabelt regions. We were able to statistically evaluate the localized fields and to reliably functionally parcellate the core and belt regions of the auditory cortex in several animals. This now serves as a routine tool, similar as retinotopic mapping, in the analysis of experimental data, such as that related to cross-modal interactions or higher auditory function (Kayser et al., 2007; Kikuchi et al., 2008; Petkov et al., 2008a, b). Overall, these findings suggest that a functional mapping of auditory cortex should also be feasible in the human brain, yet high-resolution imaging would be needed in case the fields are as small and numerous as in monkeys.

8.4 Localizing Cross-Modal Influences to Individual Fields

Having the possibility to map individual auditory fields in single subjects, we then exploited this approach to localize cross-modal influences to particular fields. Following prior human cross-modal studies, we combined both visual and tactile stimuli with various sounds and determined regions of cross-modal enhancement using conventional criteria (Kayser et al., 2005, 2007) (Fig. 8.2). To probe the interaction of acoustic and somatosensory stimulation we combined auditory broadband stimuli with touch stimulation of the hand and foot. Measuring BOLD responses in anesthetized animals we reliably found voxels exhibiting enhanced responses to the multisensory stimulation in caudal auditory region. Across animals, these voxels were consistently found in the caudo-medial and caudo-lateral belt fields (CM and CL), but not in primary auditory cortex. These findings are in good concordance with previous results from human imaging studies (Foxe et al., 2002; Lutkenhoner et al., 2002; Murray et al., 2005; Schurmann et al., 2006) in that they demonstrate an influence of somatosensory stimulation on auditory responses in (secondary) auditory cortices, although some human studies also reported influences directly in A1 (Pekkola et al., 2005b).

To rule out non-specific projections (e.g., attention related neuromodulators) as the source of these cross-modal influences, we employed two important functional criteria for sensory integration (Stein and Meredith, 1993): the principles of temporal coincidence and inverse effectiveness. The former principle posits that cross-modal interactions should be stronger when the stimuli in both modalities are presented at the same time, while the later posits that the benefit (e.g., enhancement) of the interaction should be stronger under conditions where the unisensory stimuli are themselves little effective. In our data we indeed found that for many voxels the audio–tactile interaction obeyed both these principles, confirming that the discovered cross-modal influence at least functionally resembles sensory integration (Kayser et al., 2005).

In addition to the influence of tactile stimuli on auditory responses, we tested for a similar influence of visual stimulation on auditory fields. Measuring the fMRI BOLD response to naturalistic sounds, we quantified whether the corresponding naturalistic movies enhance the responses when presented simultaneously with the sound. Indeed, we found such regions of response enhancement in the caudo-medial and caudo-lateral fields (CM, CL), portions of the medial belt (MM), and the caudal parabelt (Fig. 8.2). These cross-modal interactions in secondary and higher auditory regions occurred reliably in both anesthetized and alert animals. In addition, we found cross-modal interactions in the primary region A1, but only in the alert animal, indicating that these very early interactions could be dependent on the vigilance of the animal, perhaps involving cognitive or top-down influences, or might depend on other projections that are silenced during anesthesia.

In addition to these cross-modal influences in primary and secondary auditory cortices, we also found modulation of auditory responses in higher auditory association regions of the parabelt and superior temporal gyrus and in classical

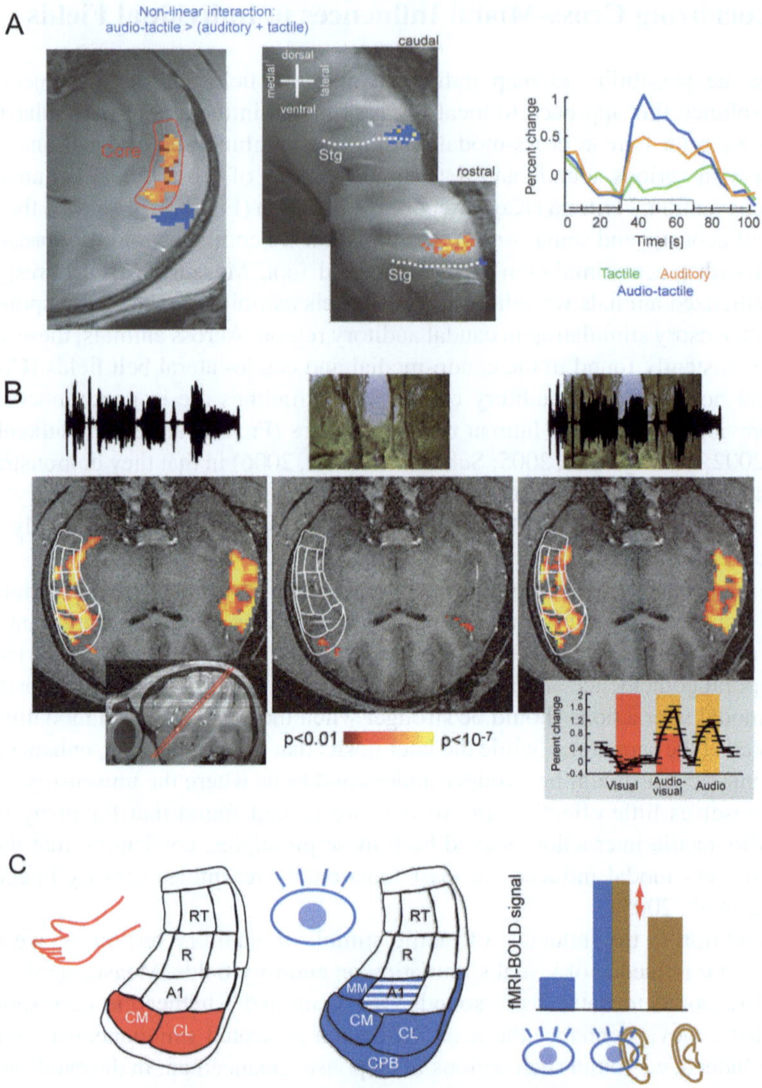

Fig. 8.2 High-resolution functional imaging of cross-modal influences in auditory cortex. (**a**) Experiments using combinations of auditory and tactile stimuli. The auditory stimulus consisted of *white* noise while the tactile stimuli consisted of a brush rotating in the hand. Regions with multisensory (supralinear) enhancement were found in caudal auditory cortex and are shown in *blue* for two example experiments. In addition, the auditory core region was localized using narrowband sounds and is shown in *red*. The time course on the right indicates the response enhancement of auditory responses in the absence of a response to just tactile stimulation. (**b**) Experiments using combinations of auditory and visual stimuli. The auditory stimuli consisted of natural sounds and the visual stimuli of the corresponding natural movies. The image slices display responses to all three sensory conditions for a single experiment. Image slices were aligned to the lateral sulcus, as indicated in the *lower left inset*. The functional parcellation of auditory cortex is indicted in *white* on each slice for one hemisphere. The time course indicated on the *lower right* displays the

multisensory cortices in the superior temporal sulcus. Exploiting the full advantages offered by the large spatial coverage by functional imaging studies, we systematically quantified the influence of visual stimuli on regions along the auditory cortical processing pathway. It turned out that visual influences increase systematically at higher processing stages, being minimal in primary auditory cortex (A1) and strongest in the well-known multisensory association regions of the superior temporal sulcus (Fig. 8.3).

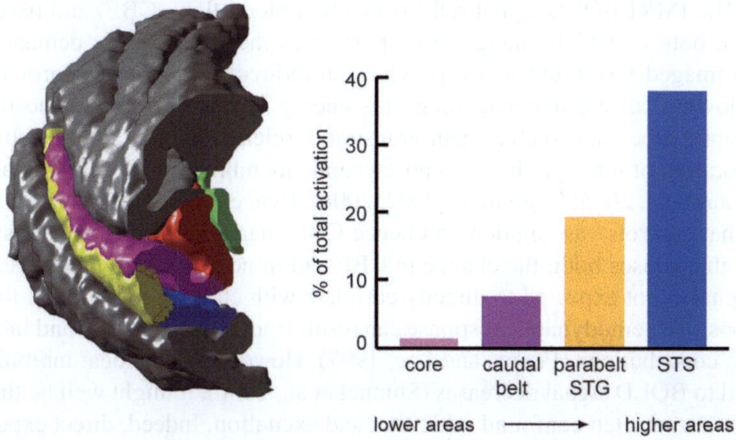

Fig. 8.3 Audio-visual influences along auditory processing streams. The *left panel* displays a three-dimensional rendering of auditory areas obtained from an anatomical MR image. *Pink*: auditory cortex; orange: superior temporal gyrus; *blue*: superior temporal sulcus; *red*: insula cortex. The *right panel* displays the BOLD response to just visual stimuli, normalized with respect to the total activation obtained in all paradigms. The visual influence increases from lower to higher auditory processing stages

Altogether, our results agree with those of previous human studies in that non-acoustic influences indeed occur in auditory cortex when quantified using the BOLD signal. The high resolution of our data and the functional localization of individual auditory fields allowed us to localize cross-modal influences to a network of caudal auditory fields, mostly comprising secondary stages but partly including primary auditory cortex. This network seems to be susceptible to cross-modal influences from a variety of sensory modalities and might serve to complement the processing of acoustic information with evidence from the other senses.

Fig. 8.2 (continued) response enhancement in the absence of a response to just visual stimulation. (**c**) Summary of cross-modal influences in auditory cortex. In our experiments we found enhancement of auditory responses by touch stimulation reliably in the medial and lateral caudal belt fields (CM, CL, *red shading*). Enhancement of auditory responses by visual stimulation was reliably found in the caudal belt and parabelt, the medial belt field (CM, CL, CPB caudal parabelt, and medial field MM *blue shading*), and (only in alert animals) in primary auditory field A1 (*white/blue shading*). The bars on the right depict the typical activations found in caudal belt: weak response to visual stimuli and stronger responses to multisensory audio-visual stimuli than to auditory stimuli

8.5 Extrapolation from Functional Imaging to Neuronal Responses

It is difficult to extrapolate from functional imaging results to the level of neuronal activity. Predicting neuronal responses from the fMRI signal would require an accurate understanding of the coupling of hemodynamic and neuronal processes – hence the neural correlate of the imaging signal, which is still not resolved (Logothetis, 2008). The fMRI BOLD signal reflects cerebral blood flow (CBF) and tissue oxygenation, both of which change in proportion to the local energy demands in or near an imaged voxel and hence provides an indirect measure of neuronal activity. Following current understanding, this energy demand originates mostly from perisynaptic processes, such as neurotransmitter release, uptake, and recycling, and the restoration of ionic gradients in postsynaptic membranes (Attwell and Iadecola, 2002; Lauritzen, 2005; Logothetis, 2002, 2008). Hence it is not the neuronal spiking per se that controls vasodilation and hence CBF chances, but it is the perisynaptic activity that causes both, the change in CBF and in neuronal firing. As a result the CBF signal is not expected to directly correlate with changes in neuronal firing. In fact, a positive hemodynamic response can result from both excitatory and inhibitory synaptic contributions (Fergus and Lee, 1997). However, since local inhibition can also lead to BOLD signal decreases (Shmuel et al., 2006), it might well be that CBF and BOLD can often confound inhibition and excitation. Indeed, direct experimental evidence demonstrates powerful dissociations of CBF and spiking activity, i.e. increases in CBF in the absence of spiking activity (Mathiesen et al., 1998, 2000; Thomsen et al., 2004), or decoupling of CBF and afferent input (Norup Nielsen and Lauritzen, 2001). Importantly, such dissociations of CBF and neuronal firing occur not only during artificially induced situations but also during typical sensory stimulation protocols (Goense and Logothetis, 2008; Logothetis et al., 2001).

The indirect coupling of neuronal and hemodynamic responses, however, does not imply that the BOLD signal is usually unrelated to neuronal firing. Under many conditions neuronal firing is well proportional to the local synaptic input and both the local field potentials (LFP) characterizing the aggregate synaptic activity in a local region and the firing rates correlate with the BOLD signal (Logothetis et al., 2001; Mukamel et al., 2005; Niessing et al., 2005; Nir et al., 2007). Yet, this is mostly the case in those situations where both LFPs and firing rates correlate with each other, and hence the (input–output) system is in a linear state. In other circumstances, the BOLD signal and neuronal firing can dissociate, while signals characterizing perisynaptic events such as LFPs still correlate with BOLD (Lauritzen and Gold, 2003).

In addition to the indirect coupling of neuronal and hemodynamic responses, the interpretation of imaging results is further complicated by the relatively coarse resolution of functional imaging (Bartels et al., 2008; Laurienti et al., 2005). In most cases, a given BOLD response allows multiple interpretations with respect to the properties of individual neurons. For example, an activation to audio-visual stimulation that matches the sum of the activations to auditory and visual stimuli, $AV = A + V$, allows multiple interpretations: This response could either result from pooling the responses of two distinct and unisensory groups of neurons, one responding only to acoustic and one only to visual stimuli. But the same response could also result from one pool of unisensory acoustic neurons and one group of multisensory neurons that respond to both the auditory and visual stimuli. While in the former case no individual neuron exhibits multisensory responses, some neurons in the latter interpretation respond to both sensory modalities. As this example shows, a definite interpretation of functional imaging data in the context of neuronal responses is limited by our current understanding of the neuronal basis underlying the imaging signal. As a result, only direct measurements of neuronal responses allow a definite localization of cross-modal interactions to particular auditory fields.

8.6 Conclusions

The notion that auditory fields in or near primary auditory cortex receive inputs from other modalities and possibly integrate this with the acoustic information has become increasingly popular over the last years (Ghazanfar and Schroeder, 2006). A good deal of this evidence comes from functional imaging experiments. Indeed, combined evidence from studies with human subjects and high-resolution imaging studies with animals provide consistent evidence that cross-modal influences are strongest in caudal auditory cortex and increase along the auditory processing pathway. All in all this suggests that early sensory cortices receive some but weak cross-modal influences, while higher association regions receive prominent afferents from several modalities. However, as we also argue here, functional imaging provides good means to localize regions of interest, but does not warrant strong conclusions about the properties of individual or populations of neurons in those regions. As a result, direct electrophysiological recordings are required to settle the questions of whether individual neurons in auditory cortex indeed integrate information from multiple modalities and hence whether the representation of the acoustic environment indeed benefits from the cross-modal input. Future work along these directions could be guided by functional imaging studies and, for example, benefit from systematic analysis of the neuronal information coding properties which might provide a more intuitive way of quantifying cross-modal influences rather than reporting the absolute changes in response strength. Clearly, much remains to be learned in this field and it will be exciting to combine such investigations focused purely on neuronal response properties with behavioral paradigms that allow a direct comparison of the benefits of sensory integration at the level of neurons and behavior.

Acknowledgements This work was supported by the Max Planck Society and the Alexander von Humboldt Foundation.

References

Alais D, Burr D (2004) The ventriloquist effect results from near-optimal bimodal integration. Curr Biol 14:257–262

Attwell D, Iadecola C (2002) The neural basis of functional brain imaging signals. Trends Neurosci 25:621–625

Bartels A, Logothetis NK, Moutoussis K (2008) fMRI and its interpretations: an illustration on directional selectivity in area V5/MT. Trends Neurosci 31:444–453

Benevento LA, Fallon J, Davis BJ, Rezak M (1977) Auditory--visual interaction in single cells in the cortex of the superior temporal sulcus and the orbital frontal cortex of the macaque monkey. Exp Neurol 57:849–872

Bernstein LE, Auer ET Jr, Moore JK, Ponton CW, Don M, Singh M (2002) Visual speech perception without primary auditory cortex activation. Neuroreport 13:311–315

Bruce C, Desimone R, Gross CG (1981) Visual properties of neurons in a polysensory area in superior temporal sulcus of the macaque. J Neurophysiol 46:369–384

Calvert GA (2001) Crossmodal processing in the human brain: insights from functional neuroimaging studies. Cereb Cortex 11:1110–1123

Calvert GA, Campbell R (2003) Reading speech from still and moving faces: the neural substrates of visible speech. J Cogn Neurosci 15:57–70

Calvert GA, Brammer MJ, Bullmore ET, Campbell R, Iversen SD, David AS (1999) Response amplification in sensory-specific cortices during crossmodal binding. Neuroreport 10: 2619–2623

Calvert GA, Bullmore ET, Brammer MJ, Campbell R, Williams SC, McGuire PK, Woodruff PW, Iversen SD, David AS (1997) Activation of auditory cortex during silent lipreading. Science 276:593–596

Chiry O, Tardif E, Magistretti PJ, Clarke S (2003) Patterns of calcium-binding proteins support parallel and hierarchical organization of human auditory areas. Eur J Neurosci 17:397–410

Clarke S, Rivier F (1998) Compartments within human primary auditory cortex: evidence from cytochrome oxidase and acetylcholinesterase staining. Eur J Neurosci 10:741–745

Crivello F, Schormann T, Tzourio-Mazoyer N, Roland PE, Zilles K, Mazoyer BM (2002) Comparison of spatial normalization procedures and their impact on functional maps. Hum Brain Mapp 16:228–250

Desai R, Liebenthal E, Possing ET, Waldron E, Binder JR (2005) Volumetric vs. surface-based alignment for localization of auditory cortex activation. Neuroimage 26:1019–1029

Driver J, Spence C (1998) Crossmodal attention. Curr Opin Neurobiol 8:245–253

Driver J, Noesselt T (2008) Multisensory interplay reveals crossmodal influences on ′sensory-specific′ brain regions, neural responses, and judgments. Neuron 57:11–23

Engel SA, Rumelhart DE, Wandell BA, Lee AT, Glover GH, Chichilnisky EJ, Shadlen MN (1994) fMRI of human visual cortex. Nature 369:525

Felleman DJ, van Essen DC (1991) Distributed hierarchical processing in the primate cerebral cortex. Cereb Cortex 1:1–47

Fergus A, Lee KS (1997) GABAergic regulation of cerebral microvascular tone in the rat. J Cereb Blood Flow Metab 17:992–1003

Formisano E, Kim DS, Di Salle F, van de Moortele PF, Ugurbil K, Goebel R (2003) Mirror-symmetric tonotopic maps in human primary auditory cortex. Neuron 40:859–869

Foxe JJ, Schroeder CE (2005) The case for feedforward multisensory convergence during early cortical processing. Neuroreport 16:419–423

Foxe JJ, Wylie GR, Martinez A, Schroeder CE, Javitt DC, Guilfoyle D, Ritter W, Murray MM (2002) Auditory-somatosensory multisensory processing in auditory association cortex: an fMRI study. J Neurophysiol 88:540–543

Fullerton BC, Pandya DN (2007) Architectonic analysis of the auditory-related areas of the superior temporal region in human brain. J Comp Neurol 504:470–498

Ghazanfar AA, Schroeder CE (2006) Is neocortex essentially multisensory? Trends Cogn Sci 10:278–285

Goense J, Logothetis N (2008) Neurophysiology of the BOLD fMRI signal in awake monkeys. Curr Biol In Press

Hackett TA, Stepniewska I, Kaas JH (1998) Subdivisions of auditory cortex and ipsilateral cortical connections of the parabelt auditory cortex in macaque monkeys. J Comp Neurol 394: 475–495

Hackett TA, Preuss TM, Kaas JH (2001) Architectonic identification of the core region in auditory cortex of macaques, chimpanzees, and humans. J Comp Neurol 441:197–222

Hershenson M (1962) Reaction time as a measure of intersensory facilitation. J Exp Psychol 63:289–293

Hikosaka K, Iwai E, Saito H, Tanaka K (1988) Polysensory properties of neurons in the anterior bank of the caudal superior temporal sulcus of the macaque monkey. J Neurophysiol 60: 1615–1637

Hyvarinen J, Shelepin Y (1979) Distribution of visual and somatic functions in the parietal associative area 7 of the monkey. Brain Res 169:561–564

Jones EG, Powell TP (1970) An anatomical study of converging sensory pathways within the cerebral cortex of the monkey. Brain 93:793–820

Kayser C, Logothetis NK (2007) Do early sensory cortices integrate cross-modal information? Brain Struct Funct 212

Kayser C, Petkov CI, Augath M, Logothetis NK (2005) Integration of touch and sound in auditory cortex. Neuron 48:373–384

Kayser C, Petkov CI, Augath M, Logothetis NK (2007) Functional imaging reveals visual modulation of specific fields in auditory cortex. J Neurosci 27:1824–1835

Kikuchi Y, Rauschecker JP, Mishkin M, Augath M, Logothetis N, Petkov C (2008) Voice region connectivity in the monkey assessed with microstimulation and functional imaging. In: Program No. 850.2. 2008 Neuroscience meeting planner. Society for Neuroscience, 2008, Washington, DC. Online

Kosaki H, Hashikawa T, He J, Jones EG (1997) Tonotopic organization of auditory cortical fields delineated by parvalbumin immunoreactivity in macaque monkeys. J Comp Neurol 386: 304–316

Lakatos P, Chen CM, O'Connell MN, Mills A, Schroeder CE (2007) Neuronal oscillations and multisensory interaction in primary auditory cortex. Neuron 53:279–292

Lakatos P, Karmos G, Mehta AD, Ulbert I, Schroeder CE (2008) Entrainment of neuronal oscillations as a mechanism of attentional selection. Science 320:110–113

Laurienti PJ, Perrault TJ, Stanford TR, Wallace MT, Stein BE (2005) On the use of superadditivity as a metric for characterizing multisensory integration in functional neuroimaging studies. Exp Brain Res 166:298–297

Lauritzen M (2005) Reading vascular changes in brain imaging: is dendritic calcium the key? Nat Rev Neurosci 6:77–85

Lauritzen M, Gold L (2003) Brain function and neurophysiological correlates of signals used in functional neuroimaging. J Neurosci 23:3972–3980

Lehmann C, Herdener M, Esposito F, Hubl D, di Salle F, Scheffler K, Bach DR, Federspiel A, Kretz R, Dierks T, Seifritz E (2006) Differential patterns of multisensory interactions in core and belt areas of human auditory cortex. Neuroimage 31:294–300

Lehmann S, Murray MM (2005) The role of multisensory memories in unisensory object discrimination. Brain Res Cogn Brain Res 24:326–334

Logothetis NK (2002) The neural basis of the blood-oxygen-level-dependent functional magnetic resonance imaging signal. Philos Trans R Soc Lond B Biol Sci 357:1003–1037

Logothetis NK (2008) What we can do and what we cannot do with fMRI. Nature 453:869–878

Logothetis NK, Pauls J, Augath M, Trinath T, Oeltermann A (2001) Neurophysiological investigation of the basis of the fMRI signal. Nature 412:150–157

Lutkenhoner B, Lammertmann C, Simoes C, Hari R (2002) Magnetoencephalographic correlates of audiotactile interaction. Neuroimage 15:509–522

Martuzzi R, Murray MM, Michel CM, Thiran JP, Maeder PP, Clarke S, Meuli RA (2006) Multisensory interactions within human primary cortices revealed by BOLD dynamics. Cereb Cortex 17:1672–1679

Mathiesen C, Caesar K, Lauritzen M (2000) Temporal coupling between neuronal activity and blood flow in rat cerebellar cortex as indicated by field potential analysis. J Physiol 523 Pt 1:235–246

Mathiesen C, Caesar K, Akgoren N, Lauritzen M (1998) Modification of activity-dependent increases of cerebral blood flow by excitatory synaptic activity and spikes in rat cerebellar cortex. J Physiol 512(Pt 2):555–566
McDonald JJ, Teder-Salejarvi WA, Hillyard SA (2000) Involuntary orienting to sound improves visual perception. Nature 407:906–908
Merzenich MM, Brugge JF (1973) Representation of the cochlear partition of the superior temporal plane of the macaque monkey. Brain Res 50:275–296
Morel A, Garraghty PE, Kaas JH (1993) Tonotopic organization, architectonic fields, and connections of auditory cortex in macaque monkeys. J Comp Neurol 335:437–459
Mukamel R, Gelbard H, Arieli A, Hasson U, Fried I, Malach R (2005) Coupling between neuronal firing, field potentials, and FMRI in human auditory cortex. Science 309:951–954
Murray MM, Molholm S, Michel CM, Heslenfeld DJ, Ritter W, Javitt DC, Schroeder CE, Foxe JJ (2005) Grabbing your ear: rapid auditory-somatosensory multisensory interactions in low-level sensory cortices are not constrained by stimulus alignment. Cereb Cortex 15:963–974
Niessing J, Ebisch B, Schmidt KE, Niessing M, Singer W, Galuske RA (2005) Hemodynamic signals correlate tightly with synchronized gamma oscillations. Science 309:948–951
Nir Y, Fisch L, Mukamel R, Gelbard-Sagiv H, Arieli A, Fried I, Malach R (2007) Coupling between neuronal firing rate, gamma LFP, and BOLD fMRI is related to interneuronal correlations. Curr Biol 17:1275–1285
Norup Nielsen A, Lauritzen M (2001) Coupling and uncoupling of activity-dependent increases of neuronal activity and blood flow in rat somatosensory cortex. J Physiol 533:773–785
Pekkola J, Ojanen V, Autti T, Jaaskelainen IP, Mottonen R, Sams M (2005a) Attention to visual speech gestures enhances hemodynamic activity in the left planum temporale. Hum Brain Mapp 27:471–477
Pekkola J, Ojanen V, Autti T, Jaaskelainen IP, Mottonen R, Tarkiainen A, Sams M (2005b) Primary auditory cortex activation by visual speech: an fMRI study at 3 T. Neuroreport 16:125–128
Petkov C, Kayser C, Ghazanfar AA, Patterson RD, Logothetis N (2008a) Functional imaging of sensitivity to components of the voice in monkey auditory cortex. In: Program No. 851.19. 2008 Neuroscience meeting planner. Society for Neuroscience, Washington, DC. Online
Petkov C, Kayser C, Steudel T, Whittingstall K, Augath M, Logothetis N (2008b) A voice region in the monkey brain. Nat Neurosci 11:367–374
Petkov CI, Kayser C, Augath M, Logothetis NK (2006) Functional imaging reveals numerous fields in the monkey auditory cortex. PLOS Biol 4:e215
Rademacher J, Caviness VS, Jr., Steinmetz H, Galaburda AM (1993) Topographical variation of the human primary cortices: implications for neuroimaging, brain mapping, and neurobiology. Cereb Cortex 3:313–329
Rauschecker JP (1998) Cortical processing of complex sounds. Curr Opin Neurobiol 8:516–521
Rauschecker JP, Tian B (2004) Processing of band-passed noise in the lateral auditory belt cortex of the rhesus monkey. J Neurophysiol 91:2578–2589
Rauschecker JP, Tian B, Pons T, Mishkin M (1997) Serial and parallel processing in rhesus monkey auditory cortex. J Comp Neurol 382:89–103
Recanzone GH, Guard DC, Phan ML (2000) Frequency and intensity response properties of single neurons in the auditory cortex of the behaving macaque monkey. J Neurophysiol 83: 2315–2331
Rivier F, Clarke S (1997) Cytochrome oxidase, acetylcholinesterase, and NADPH-diaphorase staining in human supratemporal and insular cortex: evidence for multiple auditory areas. Neuroimage 6:288–304
Schroeder CE, Molholm S, Lakatos P, Ritter W, Foxe JJ (2004) Human–simian correspondence in the early cortical processing of multisensory cues. Cogn Process 5:140–151
Schroeder CE, Lakatos P, Kajikawa Y, Partan S, Puce A (2008) Neuronal oscillations and visual amplification of speech. Trends Cogn Sci 12:106–113
Schurmann M, Caetano G, Hlushchuk Y, Jousmaki V, Hari R (2006) Touch activates human auditory cortex. Neuroimage 30:1325–1331

Seitz AR, Kim R, Shams L (2006) Sound facilitates visual learning. Curr Biol 16:1422–1427
Shmuel A, Augath M, Oeltermann A, Logothetis NK (2006) Negative functional MRI response correlates with decreases in neuronal activity in monkey visual area V1. Nat Neurosci 9: 569–577
Stein BE, Meredith MA (1993) Merging of the senses. MIT Press, Cambridge
Sumby WH, Polack I (1954) Visual contribution to speech intelligibility in noise. J Acoust Soc Am 26:212–215
Talavage TM, Sereno MI, Melcher JR, Ledden PJ, Rosen BR, Dale AM (2004) Tonotopic organization in human auditory cortex revealed by progressions of frequency sensitivity. J Neurophysiol 91:1282–1296
Thomsen K, Offenhauser N, Lauritzen M (2004) Principal neuron spiking: neither necessary nor sufficient for cerebral blood flow in rat cerebellum. J Physiol 560:181–189
van Atteveldt N, Formisano E, Goebel R, Blomert L (2004) Integration of letters and speech sounds in the human brain. Neuron 43:271–282
van Wassenhove V, Grant KW, Poeppel D (2005) Visual speech speeds up the neural processing of auditory speech. Proc Natl Acad Sci U S A 102:1181–1186
Vroomen J, de Gelder B (2000) Sound enhances visual perception: cross-modal effects of auditory organization on vision. J Exp Psychol Hum Percept Perform 26:1583–1590
Warnking J, Dojat M, Guerin-Dugue A, Delon-Martin C, Olympieff S, Richard N, Chehikian A, Segebarth C (2002) fMRI retinotopic mapping--step by step. Neuroimage 17:1665–1683
Wessinger CM, Buonocore MH, Kussmaul CL, Mangun GR (1997) Tonotopy in human auditory cortex examined with functional magnetic resonance imaging. Neuroimage 5:18–25

Chapter 9
The Default Mode of Primate Vocal Communication and Its Neural Correlates

Asif A. Ghazanfar

9.1 Introduction

It's been argued that the integration of the visual and auditory channels during human speech perception is the default mode of speech processing (Rosenblum, 2005). That is, speech perception is not a capacity that is 'piggybacked' on to auditory-only speech perception. Visual information from the mouth and other parts of the face is used by all perceivers and readily integrates with auditory speech. This integration is ubiquitous and automatic (McGurk and MacDonald, 1976) and is similar across all sighted individuals across all cultures (Rosenblum, 2008). The two modalities seem to be integrated even at the earliest stages of human cognitive development (Gogate et al., 2001; Patterson and Werker, 2003).

If multisensory speech is the default mode of perception, then this should be reflected both in the evolution of vocal communication and in the organization of neural processes related to communication. The purpose of this chapter is (1) to briefly describe the data that reveal that human speech is not uniquely multisensory, that in fact, the default mode of communication is multisensory in nonhuman primates as well and (2) to suggest that this mode of communication is reflected in the organization of the neocortex. By focusing on the properties of a presumptive unisensory region – the auditory cortex – I will argue that multisensory associations are not mediated solely through association areas, but are instead mediated through large-scale networks that include both 'lower' and 'higher' sensory areas.

9.2 Faces and Voices Are Inextricably Linked in Primates

Human and primate vocalizations are produced by coordinated movements of the lungs, larynx (vocal folds), and the supralaryngeal vocal tract (Fitch and Hauser,

A.A. Ghazanfar (✉)
Departments of Psychology and Ecology & Evolutionary Biology, Neuroscience Institute, Princeton University, Princeton, NJ 08540, USA
e-mail: asifg@princeton.edu

M.J. Naumer, J. Kaiser (eds.), *Multisensory Object Perception in the Primate Brain*,
DOI 10.1007/978-1-4419-5615-6_9, © Springer Science+Business Media, LLC 2010

1995; Ghazanfar and Rendall, 2008). The vocal tract consists of the column of air derived from the pharynx, mouth, and nasal cavity. In humans, speech-related vocal tract motion results in the predictable deformation of the face around the oral aperture and other parts of the face (Jiang et al., 2002; Yehia et al., 1998, 2002). Thus, facial motion is inextricably linked to the production of vocal sounds. For example, human adults automatically link high-pitched sounds to facial postures producing an /i/ sound and low-pitched sounds to faces producing an /a/ sound (Kuhl et al., 1991). In nonhuman primate vocal production, there is a similar link between acoustic output and facial dynamics. Different macaque monkey vocalizations are produced with unique lip configurations and mandibular positions and the motion of such articulators influences the acoustics of the signal (Hauser and Ybarra, 1994; Hauser et al., 1993). Coo calls, like /u/ in speech, are produced with the lips protruded, while screams, like the /i/ in speech, are produced with the lips retracted (Fig. 9.1). Thus, it is likely that many of the facial motion cues that humans use for speech-reading are present in other primates as well.

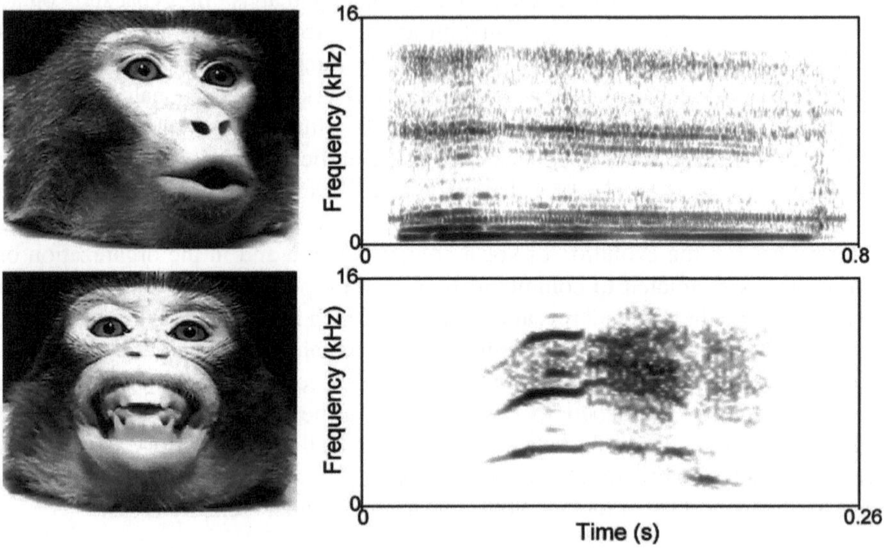

Fig. 9.1 Exemplars of the facial expressions produced concomitantly with vocalizations. Rhesus monkey coo and scream calls taken at the midpoint of the expressions with their corresponding spectrograms

The link between facial motion and vocalizations presents an obvious opportunity to exploit the concordance of both channels. Thus, it is not surprising that many primates other than humans recognize the correspondence between the visual and the auditory components of vocal signals. Rhesus and Japanese macaques (*Macaca mulatta and Macaca fuscata*), capuchins (*Cebus apella*), and chimpanzees (*Pan troglodytes*) (the only nonhuman primates tested thus far) all recognize auditory–visual correspondences between their various vocalizations (Adachi et al., 2006; Evans et al., 2005; Ghazanfar and Logothetis, 2003; Izumi and Kojima, 2004; Parr, 2004). For example, rhesus monkeys readily match the facial expressions of 'coo' and 'threat' calls with their associated vocal components (Ghazanfar and Logothetis, 2003). Perhaps more pertinent, rhesus monkeys can also segregate competing voices in a chorus of coos, much as humans might with speech in a cocktail party scenario, and match them to the correct number of individuals seen cooing on a video screen (Jordan et al., 2005) (Fig. 9.2a). Finally, macaque monkeys use

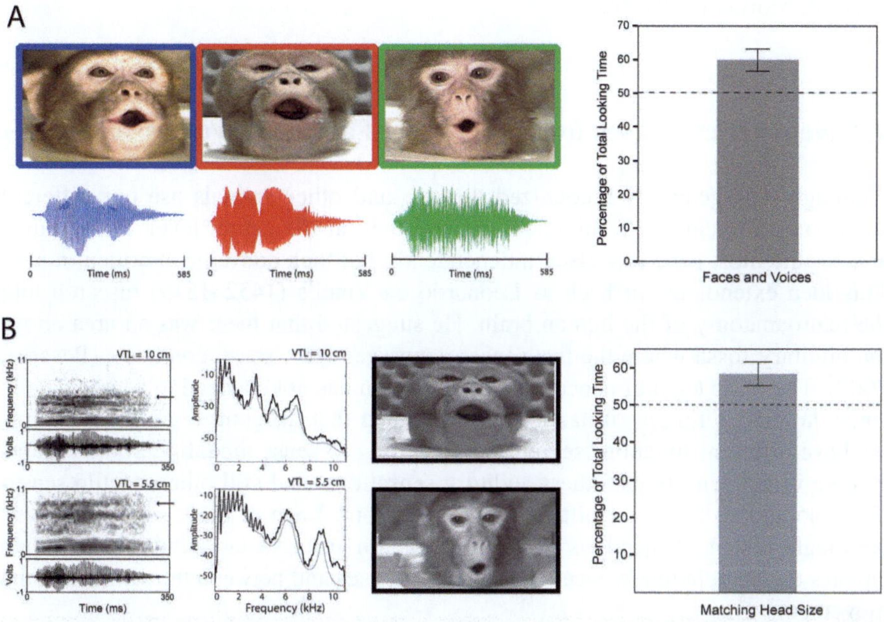

Fig. 9.2 Monkeys can match across modalities. To test this, we adopted the preferential-looking paradigm which does not require training or reward. In the paradigm, subjects are seated in front of two LCD monitors and shown two side-by-side digital videos, only one of which corresponds to the sound track heard through a centrally located speaker. A trial consists of two videos played in a continuous loop with one of the two sound tracks also played in a loop through the speaker. The dependent measure is percentage of total looking time to the match video. (**a**) Monkeys segregate coo vocalizations from different individuals and look to correct number of conspecific individuals displayed on the screen. Still frames extracted from a stimulus set along with their acoustic counterparts below. The bar graph shows the mean percentage (±SEM) of total looking time to the matching video display; chance is 50%. (**b**) A single coo vocalizations were synthesized to mimic large and small sounding individuals. Diagrams in the *left panels* show the spectrograms and waveforms of a coo vocalization re-synthesized with two different vocal tract lengths. The *arrow* in the spectrogram indicates the position of an individual formant which increases in frequency as the apparent vocal tract length decreases. In the *middle panels*, power spectra (*black line*) and linear predictive coding spectra (*gray lines*) for the long vocal tract length (10 cm, *top panel*) and short vocal tract length (5.5 cm, *bottom panel*). Still frames show the visual components of a large and small monkey. The bar graph shows the mean percentage of total time spent looking at the matching video display; the *dotted line* indicates chance expectation. Error bars are SEM

formants (i.e., vocal tract resonances) as acoustic cues to assess age-related body size differences among conspecifics (Ghazanfar et al., 2007) (Fig. 9.2b). They do so by linking across modalities the body size information embedded in the formant spacing of vocalizations (Fitch, 1997) with the visual size of animals who are likely to produce such vocalizations (Ghazanfar et al., 2007). Taken together these data suggest that humans are not at all unique in their ability to communicate across modalities. Indeed, as will be described below, vocal communication is a fully integrated *multi-sensorimotor* system with numerous similarities between humans and monkeys and in which auditory cortex may serve as a key node in a larger neocortical network.

9.3 Neocortical Bases for Integrating Faces and Voices

Although it is generally recognized that we and other animals use our different senses in an integrated fashion, we assume that, at the neural level, these senses are, for the most part, processed independently but then converge at critical nodes. This idea extends as far back as Leonardo da Vinci's (1452–1519) research into the neuroanatomy of the human brain. He suggested that there was an area above the pituitary fossa where the five senses converged (the 'sensu comune') (Pevsner, 2002). The basic tenet of neocortical organization has not changed to a large degree since da Vinci's time as it has long been argued that different regions of the cortex have different functions segregated according to sense modality. Some regions receive visual sensations, others auditory sensations, and still others tactile sensations (so and so forth, for olfaction and gustation). Each of these sensory regions is thought to send projections which converge on an 'association area' which then enables the association between the different senses and between the senses and the movement.

According to this line of thinking, the linking of vision with audition in the multisensory vocal perception described above would be attributed to the functions of association areas such as the superior temporal sulcus in the temporal lobe or the principal and intraparietal sulci located in the frontal and parietal lobes, respectively. Although these regions may certainly play important roles (see below), they are certainly not necessary for all types of multisensory behaviors (Ettlinger and Wilson, 1990), nor are they the sole regions for multisensory convergence (Driver and Noesselt, 2008; Ghazanfar and Schroeder, 2006). The auditory cortex, in particular, has many potential sources of visual inputs (Ghazanfar and Schroeder, 2006) and this is borne out in the increasing number of studies demonstrating visual modulation of auditory cortical activity (Bizley et al., 2007; Ghazanfar et al., 2005, 2008; Kayser et al., 2007, 2008; Schroeder and Foxe, 2002). Here I focus on those auditory cortical studies investigating face/voice integration specifically.

Neural activity in both the primary and the lateral belt regions of auditory cortex is influenced by the presence of a dynamic face (Ghazanfar et al., 2005, 2008). Monkey subjects viewing unimodal and bimodal versions of two different species-typical vocalizations ('coos' and 'grunts') show both enhanced and suppressed local field potential (LFP) responses in the bimodal condition relative to the unimodal auditory condition (Ghazanfar et al., 2005). Consistent with evoked potential studies in humans (Besle et al., 2004; van Wassenhove et al., 2005), the combination of faces and voices led to integrative responses (significantly different from unimodal responses) in the vast majority of auditory cortical sites – both in the primary auditory cortex and in the lateral belt auditory cortex. The data demonstrated that LFP signals in the auditory cortex are capable of multisensory integration of facial and vocal signals in monkeys (Ghazanfar et al., 2005) and have subsequently been confirmed at the single-unit level in the lateral belt cortex as well (Ghazanfar et al., 2008) (Fig. 9.3).

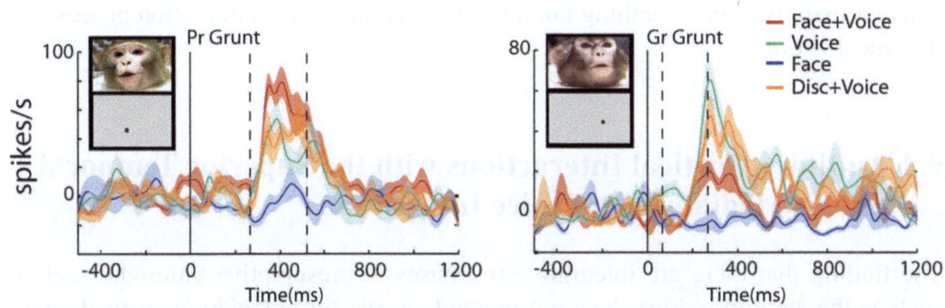

Fig. 9.3 Single neuron examples of multisensory integration of face+voice stimuli compared with disk+voice stimuli in the lateral belt area. The *left panel* shows an enhanced response when voices are coupled with faces, but no similar modulation when coupled with disks. The *right panel* shows similar effects for a suppressed response. *x*-axes show time aligned to onset of the face (*solid line*). *Dashed lines* indicate the onset and offset of the voice signal. *y*-axes depict the firing rate of the neuron in spikes per second. *Shaded regions* denote the SEM

To test the specificity of face/voice integrative responses, the dynamic faces were replaced with dynamic discs which mimicked the aperture and displacement of the mouth. In human psychophysical experiments, such artificial dynamic stimuli can still lead to enhanced speech detection, but not to the same degree as a real face (Bernstein et al., 2004; Schwartz et al., 2004). When cortical sites or single units were tested with dynamic discs, far less integration was seen when compared to the real monkey faces (Ghazanfar et al., 2005, 2008) (Fig. 9.3). This was true primarily for the lateral belt auditory cortex (LFPs and single units) and was observed to a lesser extent in the primary auditory cortex (LFPs only).

A comparison of grunt calls versus coo calls revealed that the former seemed to be over-represented relative to the latter. That is, grunt vocalizations elicited more enhanced multisensory LFP responses than did coo calls (Ghazanfar et al., 2005). As both coos and grunts are produced frequently in a variety of affiliative contexts and are broadband spectrally, the differential representation cannot be attributed to experience, valence, or the frequency tuning of neurons. One remaining possibility is that this differential representation may reflect a behaviorally relevant distinction, as coos and grunts differ in their direction of expression and range. Coos are generally contact calls rarely directed toward any particular individual. In contrast, grunts are often directed toward individuals in one-on-one situations, often during social approaches as in baboons and vervet monkeys (Cheney and Seyfarth, 1982; Palombit et al., 1999). Given their production at close range and context, grunts may produce a stronger face/voice association than coo calls. This distinction appeared to be reflected in the pattern of significant multisensory responses in auditory cortex; that is, this multisensory bias toward grunt calls may be related to the fact that the grunts (relative to coos) are often produced during intimate, one-to-one social interactions. That said, much more work needs to be done to explore whether these multisensory differences are simply due to greater numbers of auditory neurons responding to grunts in general (something that the LFP signal cannot tell us) or whether there is truly something unique about the face/voice integration process for this vocalization.

9.4 Auditory Cortical Interactions with the Superior Temporal Sulcus Mediates Face/Voice Integration

The finding that there are integrative responses in presumptive unimodal regions such as the auditory cortex does not preclude a role for association cortical areas. The face-specific visual influence on the lateral belt auditory cortex begs the question as to its anatomical source and the likely possibilities for such a source include the superior temporal sulcus (STS), the prefrontal cortex, and the amygdala – regions which have abundant face-sensitive neurons. The STS is likely to be a prominent source for the following reasons. First, there are reciprocal connections between the STS and the lateral belt and other parts of auditory cortex (Barnes and Pandya, 1992; Seltzer and Pandya, 1994). Second, neurons in the STS are sensitive to both faces and biological motion (Harries and Perrett, 1991; Oram and Perrett, 1994). Finally, the STS is known to be multisensory (Barraclough et al., 2005; Benevento et al., 1977; Bruce et al., 1981; Chandrasekaran and Ghazanfar, 2009; Schroeder and Foxe, 2002).

One mechanism for establishing whether auditory cortex and the STS interact at the functional level is to measure their temporal correlations as a function of stimulus condition. Concurrent recordings of LFPs and spiking activity in the lateral belt of auditory cortex and the upper bank of the STS revealed that functional interactions, in the form of gamma band correlations, between these two regions increased in strength during presentations of faces and voices together relative to the unimodal conditions (Ghazanfar et al., 2008) (Fig. 9.4a). Furthermore, these interactions were not solely modulations of response strength, as phase relationships were significantly less variable (tighter) in the multisensory conditions (Fig. 9.4b).

The influence of the STS on auditory cortex was not merely on its gamma oscillations. Spiking activity seems to be *modulated*, but not 'driven', by an ongoing

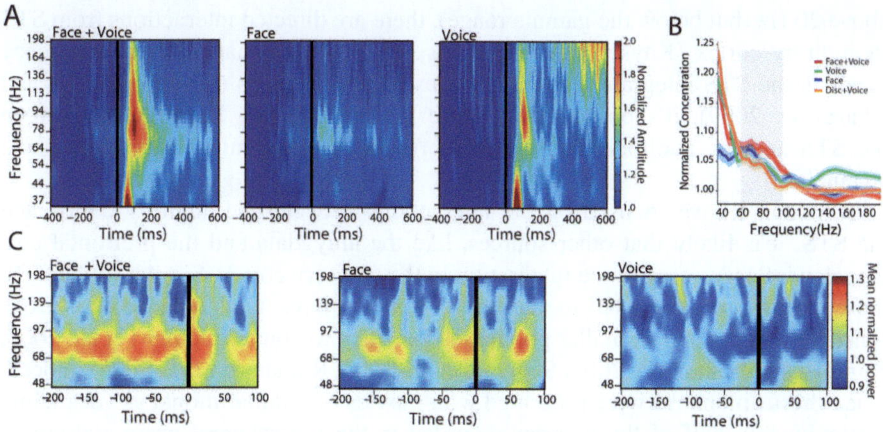

Fig. 9.4 (**a**) Time–frequency plots (cross-spectrograms) illustrate the modulation of functional interactions (as a function of stimulus condition) between the lateral belt auditory cortex and the STS for a population of cortical sites. *x*-axes depict the time in milliseconds as a function of onset of the auditory signal (*solid black line*). *y*-axes depict the frequency of the oscillations in Hz. *Color bar* indicates the amplitude of these signals normalized by the baseline mean. (**b**) Population phase concentration from 0 to 300 ms after voice onset. *x*-axes depict frequency in Hz. *y*-axes depict the average normalized phase concentration. *Shaded regions* denote the SEM across all electrode pairs and calls. All values are normalized by the baseline mean for different frequency bands. *Right panel* shows the phase concentration across all calls and electrode pairs in the gamma band for the four conditions. (**c**) Spike-field cross-spectrogram illustrates the relationship between the spiking activity of auditory cortical neurons and the STS local field potential across the population of cortical sites. *x*-axes depict time in milliseconds as a function of the onset of the multisensory response in the auditory neuron (*solid black line*). *y*-axes depict the frequency in Hz. *Color bar* denotes the cross-spectral power normalized by the baseline mean for different frequencies

activity arising from the STS. Three lines of evidence suggest this scenario. First, visual influences on single neurons were most robust when in the form of dynamic faces and were only apparent when neurons had a significant response to a vocalization (i.e., there were no overt responses to faces alone). Second, these integrative responses were often 'face specific' and had a wide distribution of latencies, which suggested that the face signal was an ongoing signal that influenced auditory responses (Ghazanfar et al., 2008). Finally, this hypothesis for an ongoing signal is supported by the sustained gamma band activity between auditory cortex and STS and by a spike-field coherence analysis of the relationship between auditory cortical spiking activity and gamma oscillations from the STS (Ghazanfar et al., 2008) (Fig. 9.4c).

Both the auditory cortex and the STS have multiple bands of oscillatory activity generated in response to stimuli that may mediate different functions (Chandrasekaran and Ghazanfar, 2009; Lakatos et al., 2005). Thus, interactions between the auditory cortex and the STS are not limited to spiking activity and high-frequency gamma oscillations. Below 20 Hz, and in response to naturalistic audiovisual stimuli, there are directed interactions from auditory cortex to STS, while above 20 Hz (but below the gamma range), there are directed interactions from STS to auditory cortex (Kayser and Logothetis, 2009). Given that different frequency bands in the STS integrate faces and voices in distinct ways (Chandrasekaran and Ghazanfar, 2009), it's possible that these lower frequency interactions between the STS and the auditory cortex also represent distinct multisensory processing channels.

Although I have focused on the interactions between the auditory cortex and the STS, it is likely that other sources, like the amygdala and the prefrontal cortex, also influence face/voice integration in the auditory cortex. Similar to the STS, these regions are connected to auditory cortex and have an abundance of neurons sensitive to faces and a smaller population sensitive to voices (Gothard et al., 2007; Kuraoka and Nakamura, 2007; Sugihara et al., 2006; Romanski et al., 2005). Indeed, when rhesus monkeys were presented with movies of familiar monkeys vocalizing, approximately half of the neurons recorded in the ventrolateral prefrontal cortex were bimodal in the sense that they responded to both unimodal auditory and visual stimuli or responded differently to bimodal stimuli than to either unimodal stimulus (Sugihara et al., 2006). As in the STS and auditory cortex, prefrontal neurons exhibited enhancement or suppression, and, like the STS but unlike the auditory cortex, suppression (73% of neurons) was more commonly observed than enhancement (27% of neurons).

9.5 Beyond Visual Influences in Auditory Cortex

The auditory cortex is also responsive to other modalities besides vision and these other modalities could also be important for vocal communication. For example, both humans and monkeys tend to look at the eyes more than mouth when viewing vocalizing conspecifics (Ghazanfar et al., 2006; Klin et al., 2002; Vatikiotis-Bateson et al., 1998). When they do fixate on the mouth, it is highly correlated with the onset of mouth movement (Ghazanfar et al., 2006; Lansing and McConkie, 2003). Surprisingly, auditory cortex is likely sensitive to such eye movement patterns: activity in both primary auditory cortex and belt areas is influenced by eye position. For example, when the spatial tuning of primary auditory cortical neurons is measured with the eyes gazing in different directions, ~30% of the neurons are affected by the position of the eyes (Werner-Reiss et al., 2003). Similarly, when LFP-derived current-source density activity was measured from auditory cortex (both primary auditory cortex and caudal belt regions), eye position significantly modulated auditory-evoked amplitude in about 80% of sites (Fu et al., 2004). These eye-position (proprioceptive) effects occurred mainly in the upper cortical layers, suggesting that the signal is fed-back from another cortical area. A possible source includes the frontal eye field (FEF) located in the frontal lobes, the medial portion of which generates relatively long saccades (Robinson and Fuchs, 1969), is interconnected with both the STS (Schall et al., 1995; Seltzer and Pandya, 1989) and multiple regions of the auditory cortex (Hackett et al., 1999; Schall et al., 1995; Romanski et al., 1999).

The auditory cortex is also sensitive to tactile inputs. Numerous lines of both physiological and anatomical evidence demonstrate that at least some regions of the auditory cortex respond to touch as well as sound (Fu et al., 2003; Hackett et al., 2007a, b; Kayser et al., 2005; Lakatos et al., 2007; Schroeder and Foxe, 2002; Smiley et al., 2007). How might tactile signals be involved in vocal communication? Oddly enough, kinesthetic feedback from one's own speech movements also integrates with heard speech (Sams et al., 2005). More directly, if a robotic device is used to artificially deform the facial skin of subjects in a way that mimics the deformation seen during speech production, then subjects actually hear speech differently (Ito et al., 2009). Surprisingly, there is a systematic perceptual variation with speech-like patterns of skin deformation that implicate a robust somatosensory influence on auditory processes under normal conditions (Ito et al., 2009). While the substrates for these somatosensory–auditory effects have not been explored, interactions between the somatosensory system and the auditory cortex seem like a likely source for the phenomena described above for the following reasons. First, many auditory cortical fields respond to, or are modulated by, tactile inputs (Fu et al., 2003; Kayser et al., 2005; Schroeder et al., 2001). Second, there are intercortical connections between somatosensory areas and the auditory cortex (Cappe and Barone, 2005; de la Mothe et al., 2006; Smiley et al., 2007). Third, auditory area CM, where many auditory-tactile responses seem to converge, is directly connected to somatosensory areas in the retroinsular cortex and the granular insula (de la Mothe et al., 2006; Smiley et al., 2006). Finally, the tactile receptive fields of neurons in auditory cortical area CM are confined to the upper body, primarily the face and neck regions (areas consisting of, or covering, the vocal tract) (Fu et al., 2003) and the primary somatosensory cortical (area 3b) representation for the tongue (a vocal tract articulator) projects to auditory areas in the lower bank of the lateral sulcus (Iyengar et al., 2007). All of these facts lend further credibility to the putative role of somatosensory–auditory interactions during vocal production and perception.

9.6 The Development of Multisensory Systems for Communication

While it appears that monkeys and humans share numerous behavioral and neural phenotypes related to multisensory integration of communication signals, how these systems emerge may not be identical. Given that monkeys and humans develop at different rates, it is important to know how this difference in developmental timing (or *heterochrony*) might influence the behavior and neural circuitry underlying multisensory communication. One line of investigation suggests that an interaction between developmental timing (*heterochrony*) and social experience may shape the neural circuits underlying both human and primate vocal communication (Lewkowicz and Ghazanfar, 2006; Lewkowicz et al., 2008; Zangehenpour et al., 2008).

The rate of neural development in Old World monkeys is faster than in humans and, as a result, they are neurologically precocial relative to humans. For example, in terms of overall brain size at birth, Old World monkeys are among the most precocial of all mammals (Sacher and Staffeldt, 1974), possessing ~65% of their brain size at birth compared to only ~25% for human infants (Malkova et al., 2006; Sacher and Staffeldt, 1974). Second, fiber pathways in the developing monkey brain are more heavily myelinated than in the human brain at the same postnatal age (Gibson, 1991), suggesting that postnatal myelination in the rhesus monkey brain is about three to four times faster than in the human brain (Gibson, 1991; Malkova et al., 2006). All sensorimotor tracts are heavily myelinated by 2–3 months after birth in rhesus monkeys, but not until 8–12 months after birth in human infants. Finally, at the behavioral level, the differential patterns of brain growth in the two species lead to differential timing in the emergence of species-specific motor, socio-emotional, and cognitive abilities (Antinucci, 1989; Konner, 1991).

The differences in the timing of neural and behavioral development across different primate species raise the possibility that the development of intersensory integration may be different in monkeys relative to humans. The slow rate of neural development in human infants (relative to monkeys) may actually be advantageous because their altricial brains may provide them with greater functional plasticity and better correspondence with their postnatal environment. As a consequence, however, this may make human infants initially more sensitive to a broader range of sensory stimulation and to the relations among multisensory inputs. This theoretical observation has received empirical support from studies showing that infants go through a process of 'perceptual narrowing' in their processing of unisensory as well as multisensory information; that is, where initially they exhibit broad sensory tuning, they later exhibit narrower tuning. For example, 4- to 6-month-old human infants can match rhesus monkey faces and voices, but 8- to 10-month-old infants no longer do so (Lewkowicz and Ghazanfar, 2006). These findings suggest that as human infants acquire increasingly greater experience with conspecific human faces and vocalizations – but none with heterospecific faces and vocalizations – their sensory tuning (and their neural systems) narrows to match their early experience.

If a relatively immature state of neural development leaves a developing human infant more 'open' to the effects of early sensory experience, then it stands to reason that the more advanced state of neural development in monkeys might result in a different outcome. In support of this, a study of infant vervet monkeys that was identical in design to the human infant study of cross-species multisensory matching (Lewkowicz and Ghazanfar, 2006) revealed that, unlike human infants, they exhibit no evidence of perceptual narrowing (Zangehenpour et al., 2008). That is, the infant vervet monkeys could match faces and voices of rhesus monkeys despite the fact that they had no prior experience with macaque monkeys and that they continued to do so well beyond the ages where such matching ability declines in human infants (Zangehenpour et al., 2008). The reason for this lack of perceptual narrowing may lie in the precocial neurological development of this Old World monkey species.

These comparative developmental data reveal that while monkeys and humans may appear to share similarities at the behavioral and neural levels, their different developmental trajectories are likely to reveal important differences. It is important to keep this in mind when making claims about homologies at either of these levels.

9.7 Conclusion

Communication is, by default, a multisensory phenomenon. This is evident in the automatic integration of the senses during vocal perception in humans and monkeys, the evidence of such integration early in development, and most importantly, by the organization of the neocortex. The overwhelming evidence from the studies reviewed here, and numerous other studies from different domains of neuroscience, all converge on the idea that the neocortex is fundamentally multisensory (Ghazanfar and Schroeder, 2006). It is not confined to a few 'sensu comune' in the association cortices. It is all over. This does not mean, however, that every cortical area is uniformly multisensory, but rather that cortical areas maybe weighted differently by 'extra'-modal inputs depending on the task at hand and its context.

Acknowledgments The author gratefully acknowledges the scientific contributions and numerous discussions with the following people: Chand Chandrasekaran, Kari Hoffman, David Lewkowicz, Joost Maier, and Hjalmar Turesson. This work was supported by NIH R01NS054898 and NSF BCS-0547760 CAREER Award.

References

Adachi I, Kuwahata H, Fujita K, Tomonaga M, Matsuzawa T (2006) Japanese macaques form a cross-modal representation of their own species in their first year of life. Primates 47: 350–354

Antinucci F (1989) Systematic comparison of early sensorimotor development. In: Antinucci F (ed) Cognitive structure and development in nonhuman primates.: Lawrence Erlbaum Associates, Hillsdale, NJ, pp 67–85

Barnes CL, Pandya DN (1992) Efferent cortical connections of multimodal cortex of the superior temporal sulcus in the rhesus-monkey. J Comp Neurol 318:222–244

Barraclough NE, Xiao D, Baker CI, Oram MW, Perrett DI (2005) Integration of visual and auditory information by superior temporal sulcus neurons responsive to the sight of actions. J Cogn Neurosci 17:377–391

Benevento LA, Fallon J, Davis BJ, Rezak M (1977) Auditory-visual interactions in single cells in the cortex of the superior temporal sulcus and the orbital frontal cortex of the macaque monkey. Exp Neurol 57:849–872

Bernstein LE, Auer ET, Takayanagi S (2004) Auditory speech detection in noise enhanced by lipreading. Speech Commun 44:5–18

Besle J, Fort A, Delpuech C, Giard MH (2004) Bimodal speech: early suppressive visual effects in human auditory cortex. Eur J Neurosci 20:2225–2234

Bizley JK, Nodal FR, Bajo VM, Nelken I, King AJ (2007) Physiological and anatomical evidence for multisensory interactions in auditory cortex. Cereb Cortex 17:2172–2189

Bruce C, Desimone R, Gross CG (1981) Visual properties of neurons in a polysensory area in superior temporal sulcus of the macaque. J Neurophysiol 46:369–384

Cappe C, Barone P (2005) Heteromodal connections supporting multisensory integration at low levels of cortical processing in the monkey. Eur J Neurosci 22:2886–2902

Chandrasekaran C, Ghazanfar AA (2009) Different neural frequency bands integrate faces and voices differently in the superior temporal sulcus. J Neurophysiol 101:773–788
Cheney DL, Seyfarth RM (1982) How vervet monkeys perceive their grunts – field playback experiments. Animal Behav 30:739–751
de la Mothe LA, Blumell S, Kajikawa Y, Hackett TA (2006) Cortical connections of the auditory cortex in marmoset monkeys: Core and medial belt regions. J Comp Neurol 496:27–71
Driver J, Noesselt T (2008) Multisensory interplay reveals crossmodal influences on 'sensory-specific' brain regions, neural responses, and judgments. Neuron 57:11–23
Ettlinger G, Wilson WA (1990) Cross-modal performance: behavioural processes, phylogenetic considerations and neural mechanisms. Behav Brain Res 40:169–192
Evans TA, Howell S, Westergaard GC (2005) Auditory-visual cross-modal perception of communicative stimuli in tufted capuchin monkeys (Cebus apella). J Exp Psychol-Anim Behav Proc 31:399–406
Fitch WT (1997) Vocal tract length and formant frequency dispersion correlate with body size in rhesus macaques. J Acoust Soc Am 102:1213–1222
Fitch WT, Hauser MD (1995) Vocal production in nonhuman-primates - acoustics, physiology, and functional constraints on honest advertisement. Am J Primatol 37:191–219
Fu KMG, Johnston TA, Shah AS, Arnold L, Smiley J, Hackett TA, Garraghty PE, Schroeder CE (2003) Auditory cortical neurons respond to somatosensory stimulation. J Neurosci 23: 7510–7515
Fu KMG, Shah AS, O'Connell MN, McGinnis T, Eckholdt H, Lakatos P, Smiley J, Schroeder CE (2004) Timing and laminar profile of eye-position effects on auditory responses in primate auditory cortex. J Neurophysiol 92:3522–3531
Ghazanfar AA, Logothetis NK (2003) Facial expressions linked to monkey calls. Nature 423: 937–938
Ghazanfar AA, Schroeder CE (2006) Is neocortex essentially multisensory? Trends Cogn Sci 10:278–285
Ghazanfar AA, Rendall D (2008) Evolution of human vocal production. Curr Biol 18:R457–R460
Ghazanfar AA, Nielsen K, Logothetis NK (2006) Eye movements of monkeys viewing vocalizing conspecifics. Cognition 101:515–529
Ghazanfar AA, Chandrasekaran C, Logothetis NK (2008) Interactions between the superior temporal sulcus and auditory cortex mediate dynamic face/voice integration in rhesus monkeys. J Neurosci 28:4457–4469
Ghazanfar AA, Maier JX, Hoffman KL, Logothetis NK (2005) Multisensory integration of dynamic faces and voices in rhesus monkey auditory cortex. J Neurosci 25:5004–5012
Ghazanfar AA, Turesson HK, Maier JX, van Dinther R, Patterson RD, Logothetis NK (2007) Vocal tract resonances as indexical cues in rhesus monkeys. Curr Biol 17:425–430
Gibson KR (1991) Myelination and behavioral development: A comparative perspective on questions of neoteny, altriciality and intelligence. In: Gibson KR, Petersen AC (eds) Brain maturation and cognitive development: comparative and cross-cultural perspective. Aldine de Gruyter, New York, pp 29–63
Gogate LJ, Walker-Andrews AS, Bahrick LE (2001) The intersensory origins of word comprehension: an ecological-dynamic systems view. Develop Sci 4:1–18
Gothard KM, Battaglia FP, Erickson CA, Spitler KM, Amaral DG (2007) Neural responses to facial expression and face identity in the monkey amygdala. J Neurophysiol 97:1671–1683
Hackett TA, Stepniewska I, Kaas JH (1999) Prefrontal connections of the parabelt auditory cortex in macaque monkeys. Brain Res 817:45–58
Hackett TA, De La Mothe LA, Ulbert I, Karmos G, Smiley J, Schroeder CE (2007a) Multisensory convergence in auditory cortex, II. Thalamocortical connections of the caudal superior temporal plane. J Comp Neurol 502:924–952
Hackett TA, Smiley JF, Ulbert I, Karmos G, Lakatos P, de la Mothe LA, Schroeder CE (2007b) Sources of somatosensory input to the caudal belt areas of auditory cortex. Perception 36: 1419–1430

Harries MH, Perrett DI (1991) Visual processing of faces in temporal cortex - physiological evidence for a modular organization and possible anatomical correlates. J Cogn Neurosci 3:9–24

Hauser MD, Ybarra MS (1994) The role of lip configuration in monkey vocalizations - experiments using xylocaine as a nerve block. Brain Lang 46:232–244

Hauser MD, Evans CS, Marler P (1993) The role of articulation in the production of rhesus-monkey, Macaca-Mulatta, vocalizations. Anim Behav 45:423–433

Ito T, Tiede M, Ostry DJ (2009) Somatosensory function in speech perception. Proc Natl Acad Sci U S A 106:1245–1248

Iyengar S, Qi H, Jain N, Kaas JH (2007) Cortical and thalamic connections of the representations of the teeth and tongue in somatosensory cortex of new world monkeys. J Comp Neurol 501: 95–120

Izumi A, Kojima S (2004) Matching vocalizations to vocalizing faces in a chimpanzee (Pan troglodytes). Anim Cogn 7:179–184

Jiang JT, Alwan A, Keating PA, Auer ET, Bernstein LE (2002) On the relationship between face movements, tongue movements, and speech acoustics. Eurasip J Appl Sig Proc 2002: 1174–1188

Jordan KE, Brannon EM, Logothetis NK, Ghazanfar AA (2005) Monkeys match the number of voices they hear with the number of faces they see. Curr Biol 15:1034–1038

Kayser C, Logothetis NK (2009) Directed interactions between auditory and superior temporal cortices and their role in sensory integration. Front Integr Neurosci 3:7

Kayser C, Petkov CI, Logothetis NK (2008) Visual modulation of neurons in auditory cortex. Cereb Cortex 18:1560–1574

Kayser C, Petkov CI, Augath M, Logothetis NK (2005) Integration of touch and sound in auditory cortex. Neuron 48:373–384

Kayser C, Petkov CI, Augath M, Logothetis NK (2007) Functional imaging reveals visual modulation of specific fields in auditory cortex. J Neurosci 27:1824–1835

Klin A, Jones W, Schultz R, Volkmar F, Cohen D (2002) Visual fixation patterns during viewing of naturalistic social situations as predictors of social competence in individuals with autism. Archiv Gen Psychiatry 59:809–816

Konner M (1991) Universals of behavioral development in relation to brain myelination. In: Gibson KR, Petersen AC (eds) Brain maturation and cognitive development: comparative and cross-cultural perspectives. Aldine de Gruyter, New York, pp 181–223

Kuhl PK, Williams KA, Meltzoff AN (1991) Cross-modal speech perception in adults and infants using nonspeech auditory stimuli. J Exp Psychol: Human Percept Perform 17:829–840

Kuraoka K, Nakamura K (2007) Responses of single neurons in monkey amygdala to facial and vocal emotions. J Neurophysiol 97:1379–1387

Lakatos P, Chen C-M, O'Connell MN, Mills A, Schroeder CE (2007) Neuronal oscillations and multisensory interaction in primary auditory cortex. Neuron 53:279–292

Lakatos P, Shah AS, Knuth KH, Ulbert I, Karmos G, Schroeder CE (2005) An oscillatory hierarchy controlling neuronal excitability and stimulus processing in the auditory cortex. J Neurophysiol 94:1904–1911

Lansing IR, McConkie GW (2003) Word identification and eye fixation locations in visual and visual-plus-auditory presentations of spoken sentences. Percept Psychophys 65: 536–552

Lewkowicz DJ, Ghazanfar AA (2006) The decline of cross-species intersensory perception in human infants. Proc Natl Acad Sci U S A 103:6771–6774

Lewkowicz DJ, Sowinski R, Place S (2008) The decline of cross-species intersensory perception in human infants: underlying mechanisms and its developmental persistence. Brain Res 1242:291–302

Malkova L, Heuer E, Saunders RC (2006) Longitudinal magnetic resonance imaging study of rhesus monkey brain development. Eur J Neurosci 24:3204–3212

McGurk H, MacDonald J (1976) Hearing lips and seeing voices. Nature 264:229–239

Oram MW, Perrett DI (1994) Responses of anterior superior temporal polysensory (Stpa) neurons to biological motion stimuli. J Cogn Neurosci 6:99–116
Palombit RA, Cheney DL, Seyfarth RM (1999) Male grunts as mediators of social interaction with females in wild chacma baboons (Papio cynocephalus ursinus). Behaviour 136: 221–242
Parr LA (2004) Perceptual biases for multimodal cues in chimpanzee (Pan troglodytes) affect recognition. Anim Cogn 7:171–178
Patterson ML, Werker JF (2003) Two-month-old infants match phonetic information in lips and voice. Develop Sci 6:191–196
Pevsner J (2002) Leonardo da Vinci's contributions to neuroscience. Trends Neurosci 25: 217–220
Robinson DA, Fuchs AF (1969) Eye movements evoked by stimulation of frontal eye fields. J Neurophysiol 32:637–648
Romanski LM, Bates JF, Goldman-Rakic PS (1999) Auditory belt and parabelt projections to the prefrontal cortex in the rhesus monkey. J Comp Neurol 403:141–157
Romanski LM, Averbeck BB, Diltz M (2005) Neural representation of vocalizations in the primate ventrolateral prefrontal cortex. J Neurophysiol 93:734–747
Rosenblum LD (2005) Primacy of multimodal speech perception. In: Pisoni DB, Remez RE (eds) Handbook of speech perception. Blackwell, Malden, MA, pp 51–78
Rosenblum LD (2008) Speech perception as a multimodal phenomenon. Curr Direct Psychol Sci 17:405–409
Sacher GA, Staffeldt EF (1974) Relation of gestation time to brain weight for placental mammals: implications for the theory of vertebrate growth. Am Naturalist 108:593–615
Sams M, Mottonen R, Sihvonen T (2005) Seeing and hearing others and oneself talk. Cogn Brain Res 23:429–435
Schall JD, Morel A, King DJ, Bullier J (1995) Topography of visual cortex connections with frontal eye field in macaque: convergence and segregation of processing streams. J Neurosci 15: 4464–4487
Schroeder CE, Foxe JJ (2002) The timing and laminar profile of converging inputs to multisensory areas of the macaque neocortex. Cogn Brain Res 14:187–198
Schroeder CE, Lindsley RW, Specht C, Marcovici A, Smiley JF, Javitt DC (2001) Somatosensory input to auditory association cortex in the macaque monkey. J Neurophysiol 85:1322–1327
Schwartz J-L, Berthommier F, Savariaux C (2004) Seeing to hear better: evidence for early audiovisual interactions in speech identification. Cognition 93:B69–B78
Seltzer B, Pandya DN (1989) Frontal-lobe connections of the superior temporal sulcus in the rhesus-monkey. J Comp Neurol 281:97–113
Seltzer B, Pandya DN (1994) Parietal, temporal, and occipital projections to cortex of the superior temporal sulcus in the rhesus monkey: a retrograde tracer study. J Comp Neurol 343: 445–463
Smiley JF, Hackett TA, Ulbert I, Karmas G, Lakatos P, Javitt DC, Schroeder CE (2007) Multisensory convergence in auditory cortex, I. Cortical connections of the caudal superior temporal plane in macaque monkeys. J Comp Neurol 502:894–923
Sugihara T, Diltz MD, Averbeck BB, Romanski LM (2006) Integration of auditory and visual communication information in the primate ventrolateral prefrontal cortex. J Neurosci 26: 11138–11147
van Wassenhove V, Grant KW, Poeppel D (2005) Visual speech speeds up the neural processing of auditory speech. Proc Natl Acad Sci US A 102:1181–1186
Vatikiotis-Bateson E, Eigsti IM, Yano S, Munhall KG (1998) Eye movement of perceivers during audiovisual speech perception. Percept Psychophys 60:926–940
Werner-Reiss U, Kelly KA, Trause AS, Underhill AM, Groh JM (2003) Eye position affects activity in primary auditory cortex of primates. Curr Biol 13:554–562
Yehia H, Rubin P, Vatikiotis-Bateson E (1998) Quantitative association of vocal-tract and facial behavior. Speech Commun 26:23–43

Yehia HC, Kuratate T, Vatikiotis-Bateson E (2002) Linking facial animation, head motion and speech acoustics. J Phonet 30:555–568

Zangehenpour S, Ghazanfar AA, Lewkowicz DJ, Zatorre RJ (2008) Heterochrony and cross-species intersensory matching by infant vervet monkeys. PLoS ONE 4:e4302

Chapter 10
Audio-Visual Perception of Everyday Natural Objects – Hemodynamic Studies in Humans

James W. Lewis

10.1 Introduction

In human and nonhuman primates, the ability to integrate auditory and visual sensory input features to detect and recognize objects, other individuals, and communicative gestures can be a critical cognitive function for survival. For hearing and sighted people, information from these separate modalities provides both complementary and redundant information, which in many circumstances may improve the accuracy and speed of recognition and identification. From an early age, the ability to bind auditory and visual inputs coming from natural objects and people becomes effortless. This process of integration becomes so habituated that we have no problem with accepting the illusion of perceiving object actions and characters seen and heard while watching television or a movie. How do these two separate sources of sensory inputs integrate in the brain to provide a unified percept of objects and people in our environment?

Physical attributes of visual and auditory signal inputs (changes in energy) that share direct commonalities of an object or event are termed *intermodal invariant* features (Lewkowicz, 2000). This includes, for instance, localizing acoustic and visual information emanating from the same location in space relative to the observer. Another intermodal invariant feature is temporal coherence, wherein the timing of changes in intensity or energy conveyed via the visual and auditory systems are in synchrony, such as when viewing and hearing a basketball bouncing to rest (which is robust even when depicted on television). Representations along early stages of the visual system may include the encoding of features such as changes in local luminance and changes in motion direction (e.g., a basketball bouncing along the ground). Representations along early auditory processing stages may include encoding changes in intensity, changes in pitch and motion cues such as Doppler shifts, spectral attenuation, and changes in distance (e.g., the loud to quiet "thuds"

J.W. Lewis (✉)
Department of Physiology and Pharmacology, Sensory Neuroscience Research Center, and Center for Advanced Imaging, West Virginia University, Morgantown, WV 26506, USA
e-mail: jlewis@hsc.wvu.edu

M.J. Naumer, J. Kaiser (eds.), *Multisensory Object Perception in the Primate Brain*,
DOI 10.1007/978-1-4419-5615-6_10,

correlated with a basketball bouncing away from you). Because these two separate sensory inputs can be naturally correlated in time and/or spatial location, they consequently can be encoded in neural systems as representing the same "multisensory" event (via Hebbian mechanisms). The notion of matching intermodal invariant features will be an important theme for interpreting results regarding audio-visual processing pathways revealed through the series of meta-analyses presented in this chapter.

Separately, each sense also provides qualitatively distinct subjective impressions of objects and events in the world. For instance, visual color and object shapes have no direct counterparts in audition (e.g., an orange, spherical-shaped object), while acoustic attributes such as pitch, timbre, and harmonicity have no direct counterparts in vision (e.g., the distinctive frequencies and echoes associated with a basketball bouncing in a large gymnasium). Despite some of these fundamental differences in the physical nature of signal energy and information conveyed by the visual and auditory systems, we nevertheless are able to maintain a unified and stable multisensory perception of moving objects and people in our surroundings.

This chapter reviews human brain regions, networks, and parallel hierarchical processing pathways underlying audio-visual interactions, based predominantly on hemodynamic neuroimaging studies, using functional magnetic resonance imaging (fMRI) and positron emission tomography (PET). Although the focus of this chapter is on perception of everyday "natural objects," results from a wide variety of audio-visual interaction studies from over roughly the past decade have been compiled to provide a broader context for interpretation. This includes meta-analysis results from 49 neuroimaging paradigms, illustrating a substantial, though not exhaustive, portion of the fMRI and PET literature on this topic to date.

Based on animal studies, methods for identifying whether neurons or specific brain regions are truly integrating multisensory information are formally addressed elsewhere (Beauchamp, 2005; Calvert and Lewis, 2004; Calvert et al., 2001; Ethofer et al., 2006). Notwithstanding, precise criteria for identifying audio-visual "*integration*" remain somewhat open to interpretation with regard to how rigorously one can apply principles from single-unit studies of superadditivity, subadditivity, and inverse effectiveness to hemodynamic studies that measure changes in blood flow. Consequently, this chapter includes results from a relatively broad range of studies that examined audio-visual "*interactions,*" which includes findings that were derived using potentially more liberal criteria for revealing multisensory associations.

As background, the Section 10.2 provides an illustrated overview of the global sum of cortical networks reported to be involved in various aspects of multisensory interactions and perception. Section 10.3 highlights a few theories regarding cortical organizational principles that will influence interpretation of the meta-analysis results. Sections 10.4.1–10.4.5 summarize the meta-analysis results divided into five of the most commonly used audio-visual interaction paradigms, including cortical networks involved in (1) the integration of spatially and/or temporally matched auditory and visual stimuli, (2–3) the processing of semantically congruent audio-visual objects (videos versus static pictures), (4) the processing of learned arbitrary audio-visual pairings, including written languages and novel objects/sounds, and (5) the processing of auditory and visual pairs that are semantically incongruent or non-matching. Results from each meta-analysis highlight some of the proposed functional roles of distinct brain regions or networks, which are addressed under various sub-headings. In a final section, the illustrated parallel and hierarchical pathways implicated in audio-visual object perception are further addressed in the context of theories regarding conceptual systems for object knowledge representations, which are necessary for providing a sense of meaning behind the perceived audio-visual events.

10.2 Overview of Audio-Visual Interaction Sites in Human Cortex

Figure 10.1 illustrates a composite functional map from the neuroimaging paradigms examined in this chapter that directly assessed brain regions involved in some aspect of congruent audio-visual interactions. These studies incorporated a wide range of stimuli, from moving dots on a screen paired with pure tones, to videos of natural objects or people in action – here people are regarded as one category of natural objects. For most studies, the "centroids" (centers of mass) of activated brain regions were reported in one of two standardized x, y, z coordinate systems, derived using software packages such as AFNI (Analysis of Functional NeuroImages), BrainVoyager, and SPM (Statistical Parametric Mapping). These coordinate systems allow for more direct comparisons of data across individuals and across studies by accounting for variations in the size and shape of different people's brains. For the present meta-analyses, the reported group-averaged data from each paradigm were all converted to AFNI-Talairach coordinate space (Cox, 1996). Using meta-analysis mapping methods described previously (Lewis, 2006), activation volumes were approximated by spheres, which were projected into a brain volume space using AFNI software. These volumetric data (roughly marble to ping-pong ball-sized spheres) were then projected onto the PALS (Population-Average, Landmark and Surface-based) atlas cortical surface models (left and right hemispheres), using freely available Caret software (http://brainmap.wustl.edu). The PALS surface models represent averaged cortical surfaces of 12 individuals (van Essen, 2005; van Essen et al., 2001).

Figure 10.1A shows the sum of meta-analysis results projected onto a cortical surface model from one individual (approximating layer 2 of the cortical mantle). Figure 10.1B depicts lateral and medial views of an inflated rendering of the PALS surface models (approximating layer 4) and Fig. 10.1C depicts the corresponding "flat map" representations. The inflated and unfolded models reveal activation foci contained deep within sulcal folds, thereby facilitating visualization as well as quantitative cross-study comparisons of the data. Most activation foci appear roughly as circular disks on the cortical surface maps, depending on how they intersected with the underlying three-dimensional spherical volumes. The grayscale hues in the "heat maps" depict cortical regions showing an increasing number of paradigms that reported audio-visual (AV) interactions to congruent AV stimuli (refer to Tables 10.1, 10.2, 10.3, and 10.4). This global meta-analysis revealed roughly 10 functionally distinct regions of interest (ROIs; labeled brain regions or domains) in the left hemisphere, most of which are mirrored in the right hemisphere.

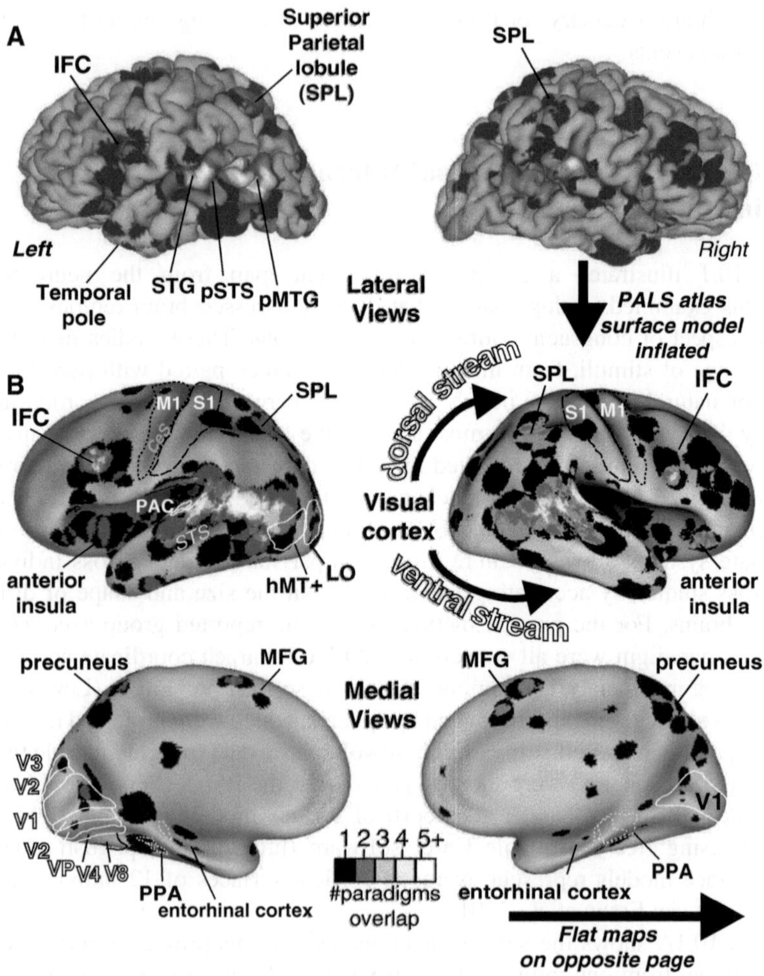

Fig. 10.1 Summary of audio-visual interaction sites in human cortex, from 49 neuroimaging paradigms. Meta-analysis data are illustrated on (**a**) an individual brain model depicting cortical layer 2 surface, (**b**) inflated renditions of the PALS surface model, and (**c**) flat maps (on opposite page) of the PALS surface model. White outlines and labels refer to early sensory areas and landmarks from the PALS atlas database. CeS=central sulcus, IPS=intraparietal sulcus, MFG=medial frontal gyrus, pMTG=posterior middle temporal gyrus, STS=superior temporal sulcus, refer to text Section 10.3 for other mapping details and Tables 10.1, 10.2, 10.3, and 10.4 for descriptions of the studies included

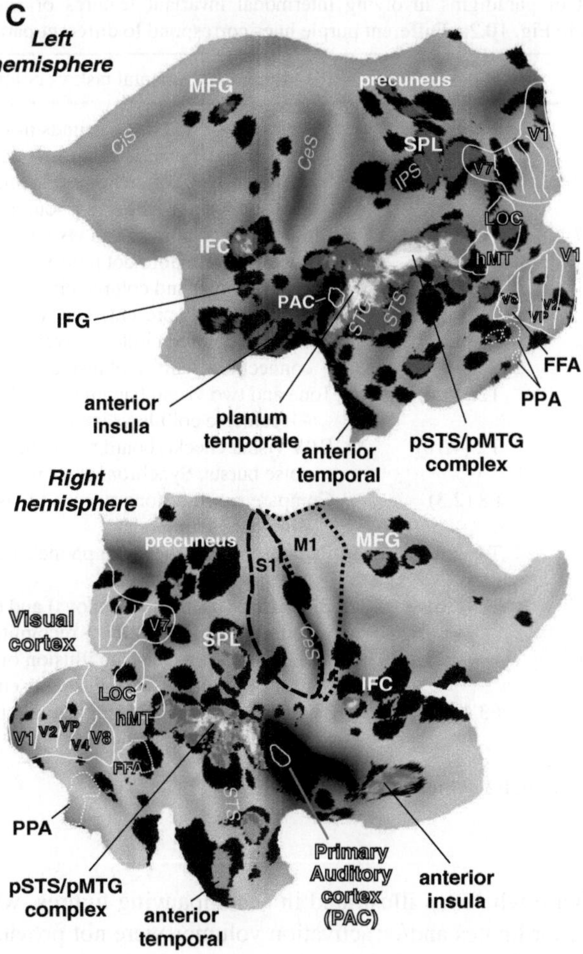

Fig. 10.1 (continued)

Though this approach to defining functionally distinct brain foci will necessarily be biased by the particular studies included, it nonetheless serves to highlight the major cortical regions germane to audio-visual interactions that will be referred to throughout this chapter.

All reported brain activation foci were equally weighted in the meta-analyses. The extent of cortex reported to be activated was dependent on statistical threshold settings adopted by each study paradigm, which varied across studies. Thus, this source of variability places limits on detailed analyses of the functions of subdivisions within ROIs, such as different subdivisions of the inferior frontal cortex (IFC). Also the reported centroid locations of some reported foci were slightly modified so that the spherical volumes would project onto the PALS cortical surface models and

Table 10.1 List of paradigms involving intermodal invariant features of auditory and visual stimuli, depicted in Fig. 10.2a. Different purple hues correspond to different paradigms

Study	#Foci (L,R)	Paradigm: experimental task vs control condition
Alink (2008)	T1c (4,6)	Visual spheres and drum sounds moving: crossmodal dynamic capture vs conflicting motion
Baumann (2007)	T1b (2,1)	Visual dots 16% coherent motion and in-phase acoustic noise > stationary sound
Baumann (2007)	T2b (15,14)	Moving acoustic noise and visual dots 16% in-phase coherent > random dot motion
Bushara (2001)	F2 (0,3)	Tones (100 ms) and colored circles: detect if synchronously presented or not
Bushara (2001)	F4 (1,3)	Tones and colored circles: correlated functional connections with right insula
Bushara (2003)	T2a (4,2)	Tone and two visual bars moving: Tone synchronous → perceive collide vs pass
Calvert (2001)	T2 (4,10)	B/W visual checkerboard reversing and 100 ms noise bursts: Synchronous vs not
Lewis (2000)	F8 (2,3)	Compare speed of tone sweeps to visual dot sweeps: Bimodal vs unimodal
Meyer (2007)	T4 (2,6)	Paired screen flashes with phone ring: View flashes after post-conditioned
Sestieri (2006)	T1 (2,6)	B/W images (animal, weapons) and environmental sounds: Match location > recognition
Watkins (2006)	F4 (0,2)	Two audio bleeps leads to illusion of two screen flashes when only one flash present
Watkins (2007)	F3 (0,1)	Two visual flashes and single audio bleep leads to the illusion of a single flash

T1b = Table 1, part b; F2 = Figure 2

be consistent with what was illustrated in accompanying figures, when present. In some studies, coordinates and/or activation volumes were not provided, so the data were approximated based on visual inspection of illustrated figures.

Functionally defined cortical landmarks from the PALS atlas database are also illustrated (Fig. 10.1, outlines), including visuotopically defined visual areas (V1, V2, V3, V4, hMT, V7, V8, solid white outlines), the primary motor cortices (Brodmann area 4 or "M1"; dotted black outline), and somatosensory cortex ("S1": Brodmann areas 3a, 3b, 1 and 2; dashed black outline). The boundary estimates for the primary auditory cortices (PAC, thick white outlines) were defined based on the anatomical location of the medial two-thirds of Heschl's gyrus (Rademacher et al., 2001), which are boundaries that generally receive support from fMRI tonotopic mapping studies (Formisano et al., 2003; Lewis et al., 2009; Talavage et al., 2004). Also illustrated are outlines estimating the location of the parahippocampal place area (PPA) (Epstein and Kanwisher, 1998; Gron et al., 2000). Before addressing the meta-analysis results, however, a brief account of theories underlying or potentially impacting an interpretation of multisensory interaction brain networks are considered first.

Table 10.2 List of paradigms involving common semantic features and intermodal invariant features of auditory (A) and visual (V) stimuli using videos of natural objects and faces, depicted in Figs. 10.2b and 10.3a, b (yellow hues)

Study	#Foci (L,R)	Paradigm: experimental task vs control condition
Beauchamp (2004a)	F3 (2,1)	Hear and see videos of realistic hand tools in motion vs unimodal (AV>A,V)
Beauchamp (2004b)	F2 (1,1)	High-resolution version of above study: AV tool videos vs unimodal (AV>A,V)
Calvert (1999)	F1 (3,4)	Lower face and hear numbers 1 through 10 vs unimodal (AVvideo>AVpic,A)
Calvert (2000)	F2 (1,0)	Speech and lower face: supra-additive plus sub-additive (AVcong>A,V>AVincong)
Kreifelts (2007)	F2 (1,1)	Facial expression and intonated spoken words, judge emotion expressed (AV>A,V)
Olson (2002)	F3 (2,0)	Whole face video and heard words: Synchronized vs not
Reale (2007)	F11	Electrode array: Speech and lower face video: AVcong>AVincong (left STG)
Robins (2009)	T2 (2,1)	Face speaking sentences: angry, fearful, happy, neutral (AV>A,V)
Robins (2009)	T4a (1,3)	AV faces expressing fear or neutral valence vs unimodal conditions (AV>A,V)
Stevenson (2009)	T1b (1,1)	Hand tools in use video: inverse effectiveness (degraded ΔAV>ΔA, ΔV)
Stevenson (2009)	T1c (1,1)	Face and words video: inverse effectiveness (degraded ΔAV>ΔA, ΔV)

Triangle = inverse effectiveness (see Chapter 13)

Table 10.3 List of paradigms involving common semantic features of auditory (A) and visual (V) stimuli using static pictures of natural objects and faces, depicted in Fig. 10.3a, d (red hues, and overlap color orange)

Study	#Foci (L,R)	Paradigm: experimental task vs control condition
Belardinelli (2004)	T1 (4,1)	Colored images and sounds of tools, animals, humans: matched vs mis-matched
Hein (2007)	F2,3 (3,3)	B/W images of animals and vocalizations versus unimodal A, V
Hocking (2008)	(pSTS outlines)	AV, VV, AA conceptually matched pictures and sounds → no differential activation
Naghavi (2007)	F1c (0,1)	B/W pictures (animals, tools, instruments, vehicles) and their sounds: Cong vs Incong
Taylor (2006)	F1 (2,0)	Color photos (V), sounds (A), spoken words (W): Cong AV vs Incong (living objects)
Taylor (2009)	T3a (5,3)	Subset of above study stimuli: crossmodal integration vs Unimodal (VW or AW)
Taylor (2009)	T3b (5,5)	Above study stimuli: AV crossmodal integration of living > nonliving things

Table 10.4 List of paradigms involving artificially paired acoustic stimuli and visual objects or characters, depicted in Fig. 10.3b, d (green hues)

Study	#Foci (L,R)	Paradigm: experimental task vs control condition
Light greens (familiar letters)		
VanAtteveldt (2004)	F2 (3,1)	Familiar letters and their speech sounds: Congruent vs not and Bimodal vs Unimodal
VanAtteveldt (2007a)	T2b (2,1)	Familiar letters and their speech sounds: Passive perception, event-related design
VanAtteveldt (2007b)	T1a (2,2)	Familiar letters and their speech sounds: temporal synchrony, Congruent vs incongruent
VanAtteveldt (2007b)	T1c (1,0)	Familiar letters and their speech sounds: AV>A,V>baseline and temporal synchrony effect
Dark greens (unfamiliar objects/characters):		
Gonzalo (2000)	Ex1 (1,1)	Novel Kanji characters and musical chords, learn consistent (vs inconsistent) pairings
Hashimoto (2004)	F4b (2,1)	Unfamiliar Hangul letters and nonsense words, learn speech vs tone/noise pairings
Hein (2007)	F2-3 (3,1)	Visual "Fribbles" and backward/underwater distorted animal sounds, learn pairings
James (2003)	F2 (0,1)	Visual "Greebles" and verbally cued auditory features (e.g., buzzes), learn pairings
McNamara (2008)	F3 (2,2)	Videos of meaningless hand gestures and FM tone sounds: Increases with learning
Naumer (2008)	F2 (8,6)	Images of "Fribbles" and learned corresponding artificial sounds: Post- vs Pre-training
Scheef (2009)	T1 (1,1)	Video of cartoon person jumping and "sonification" of a tone, learn correlated pairings
Tanabe (2005)	T1a (4,3)	2D amorphous texture patterns and modulated noises: Activation during learning delay

10.3 Background of System-Level Mechanisms for Multisensory Integration

This section examines dorsal versus ventral cortical pathways ("where versus what") for visual processing, and for acoustic processing, followed by a consideration of how motor systems appear to interact with audio-visual systems. Ideas regarding the extent to which "nature versus nurture" mechanisms may impact cortical networks for processing or representing audio-visual interactions are also considered.

Dorsal versus ventral processing streams for primate vision. Primate visual cortex includes a highly diverse and hierarchically structured system (Felleman and van Essen, 1991) that largely originates from early visual areas (e.g., Fig. 10.1, outlines of V1, V2, V3). The dorsally directed stream routes information to higher order processing centers in parietal and frontal cortices (see Fig. 10.1b, upward curved arrow). The dorsal stream is involved in transforming visual stimulus locations to motor coordinate spaces ("where is it" information), such as localizing objects relative to the body and for mediating eye–hand motor coordination (Andersen and Zipser, 1988; Creem and Proffitt, 2001), and has also been more generally described as processing *vision for action* (Goodale et al., 1994). The ventrally directed visual processing stream propagates information from occipital cortex to the temporal lobes (Fig. 10.1b, downward curved arrow) (Ungerleider and Haxby, 1994; Ungerleider et al., 1982) and is involved more in analyzing features of objects ("what is it" information) and the processing of *vision for perception* (Goodale et al., 1994). At high-level processing stages, the ventral stream (and to some extent the dorsal stream) is characterized by cortical regions and networks that show selectivity for processing different "conceptual" categories of real-world objects and other complex stimuli (Kanwisher et al., 2001; Martin, 2001). This includes, for example, temporal lobe regions sensitive or selective for object categories such as faces (Allison et al., 1994; Kanwisher et al., 1997; McCarthy et al., 1997), scenes or places (Epstein and Kanwisher, 1998; Gron et al., 2000), human body parts (Downing et al., 2001), buildings (Hasson et al., 2003), animals versus tools (Beauchamp et al., 2002; Chao and Martin, 2000), letters (Polk et al., 2002), or visual word forms (McCandliss et al., 2003). How these object representation "hot spots" are, or become, organized is an ongoing matter of research (Caramazza and Mahon, 2003; Gauthier et al., 2000; Grill-Spector and Malach, 2004; Haxby et al., 2001; Martin, 2007; Tranel et al., 1997; Tootell et al., 2003).

Dorsal versus ventral processing streams for primate hearing. Early stages of the auditory system are organized based on tonotopic representations, which is evident in primary auditory cortex (PAC) and some of the immediately surrounding cortical areas (Talavage et al., 2004). In a manner analogous to the visual system, the auditory system appears to be further organized, at least roughly, based on dorsal and ventral systems that are respectively involved in *hearing for action* and *hearing for perception* (Arnott et al., 2004; Belin and Zatorre, 2000; Kaas and Hackett, 1999; Rauschecker and Scott, 2009; Rauschecker and Tian, 2000; Recanzone and Cohen, 2009; Romanski et al., 1999). For instance, as with visual spatial attention, processing for sound localization and motion perception involves activation of dorsal stream frontal and parietal networks related to spatial attention (Lewis et al., 2000; Mesulam, 1998; Warren et al., 2002).

Regarding sound recognition, systems for identifying "what" an acoustic event is have more commonly been explored using conspecific (same-species) vocalizations, which involve pathways along the temporal lobes that may be, or may become, specialized for vocal communication (Altmann et al., 2007; Belin et al., 2000; Fecteau et al., 2004; Lewis et al., 2009; Rauschecker and Scott, 2009). However, the presence of sound necessarily implies some form of action or motion. Thus, many *action* sounds (defined here as excluding vocalizations) are strongly associated with dynamic visual motions. For instance, sounds produced by human actions lead to activation along motor-related networks and presumed mirror-neuron systems (Aziz-Zadeh et al., 2004; Doehrmann et al., 2008; Engel et al., 2009; Gazzola et al., 2006; Lewis et al., 2005, 2006). A recent study that examined four conceptually distinct categories of natural action sounds (nonvocal) revealed category-specific representations across various cortical regions, distinguishing sounds depicting living versus nonliving sources (Engel et al., 2009). Other categories of action sound sources that were preferentially represented included sounds produced by nonhuman animal actions (e.g., flight, galloping), by mechanical sources (e.g., clocks, washing machine), and by environmental sources (e.g., wind, rain, fire, rivers) (Engel et al., 2009). The extent or degree to which these category-specific representations of auditory objects and actions mesh with category-specific visual representations awaits future study. Nonetheless, together the above studies indicate that different categories of visually defined objects along with different categories of sound sources should be taken into consideration when addressing audio-visual interactions. These interactions may not only involve processing within and between the broadly described dorsal and ventral processing streams but also appear to be strongly influenced by motor associations, as addressed next.

Influence of motor systems on audio-visual processing. Throughout development, we typically learn about audio-visual objects in the context of motor and tactile interactions. Thus, not too surprisingly, cortical networks implicated in auditory and visual sensory perception are strongly influenced by, and interact with, sensory-motor systems (see Part III of this book). One salient "category" of object–action for both the visual and the auditory modalities includes biological motion (Johansson, 1973). For instance, viewing and/or hearing a person dribbling a basketball is thought to immediately and involuntarily lead to activation of the observer's own networks related to motor production (Aglioti et al., 2008; Barsalou et al., 2003; Buccino et al., 2001; Corballis, 1992; Cross et al., 2008; Kilner et al., 2004; Liberman and Mattingly, 1985; Nishitani and Hari, 2000; Norman and Shallice, 1986; Wheaton et al., 2001). This may be mediated, at least in part, by mirror-neuron systems (Keysers et al., 2003; Kohler et al., 2002), involving left lateralized frontal and parietal cortical network processing via dorsal stream pathways (Engel et al., 2009; Galati et al., 2008; Gazzola et al., 2006; Iacoboni et al., 2005; Lewis et al., 2006; Pizzamiglio et al., 2005; Rizzolatti and Craighero, 2004). Thus, aspects of object recognition may be thought of as a "selfish" process, in that our nervous system utilizes networks that function to assess the behavioral relevance of object actions (especially biological actions) relative to the growing repertoire of actions that we ourselves have produced. From this perspective, some of the systems for processing audition, vision, and consequently for audio-visual integration may ultimately be organized to interact with motor systems to help convey a sense of meaning behind the observed action or event as well as to interact with the material object, if desired. In sum, audio-visual object representations may develop based in part on a complex combination of characteristic visual form features, characteristic action sounds, and motor-related associations, which in many instances manifest, or appear to manifest, as category-specific object representations (Martin, 2007; McClelland and Rogers, 2003; Miller et al., 2003; Rosch, 1973).

Multisensory object perception and a metamodal organization of cortex. Inferred largely from animal studies, a mechanism for crossmodal object perception involves the recruitment of various multisensory convergence zones, wherein the specific zones activated, the degree to which they are activated, and synchrony of activity is what mediates perception, and that this depends on both the information content and the dominant modality providing the most relevant information (Amedi et al., 2005; Bulkin and Groh, 2006; Ettlinger and Wilson, 1990; Gray et al., 1989). However, it is important to point out that some brain regions that appear to function as audio-visual "integration sites" may actually represent metamodal operations (Pascual-Leone and Hamilton, 2001). For example, some local neural networks may be innately predisposed to develop to compete for the ability to perform particular types of operations or computations with sensory information regardless of the modality of the sensory input – whether or not a person had grown up with hearing and/or visual ability. Thus, the organization of the brain may be influenced more by internal influences than by external sensory experiences. Consequently, some cortical regions may be better thought of in terms of their "amodal" roles in extracting certain types of information from the senses, if those senses happen to be present (Amedi et al., 2007; Burton et al., 2004; Patterson et al., 2007). This concept will be important to consider in latter sections (and in Chapter 18) when accounting for how object knowledge and perception may be mediated in the brains of people who are born deaf, blind, or both (e.g., Helen Keller), yet nonetheless learn to recognize objects and associated conceptual knowledge about objects (Mahon et al., 2009; Wheeler and Griffin, 1997).

For the purposes of this chapter, an *object* is loosely defined as "a thing, person, or matter to which thought or action is directed" (Random House dictionary) and "something material that may be perceived by the senses" (Merriam-Webster dictionary). However, the concepts of object and action *perception* will remain as poorly defined. To understand how object and action meaning may be represented in the nervous system, and to what end the associated sensory information is ultimately processed, requires examination of cognitive models and models of conceptual systems, which will be briefly addressed in later sections to facilitate interpretation of the meta-analysis results addressed herein and to help inspire future directions. Despite the numerous complexities to consider when attempting to understand the nature of object and action perception, and how perceptual and conceptual systems may interact, recent fMRI and PET studies are beginning to reveal some of the gross-level cortical organizations that subserve audio-visual object perception.

10.4 Meta-analyses of Brain Networks Involved in Audio-Visual Interactions

10.4.1 Networks That Reflect Intermodal Invariant Audio-Visual Attributes

Both spatial and temporal congruence of multisensory stimuli, or intermodal invariant features, are known to lead to more robust integration in the neurons of both sub-cortical and cortical brain regions (Bulkin and Groh, 2006; Stein and Meredith, 1993). Figure 10.2a depicts a composite map from 12 paradigms (coded by different purple hues) that had an emphasis on examining brain networks activated while processing intermodal invariant features from relatively simple nonnatural auditory and/or visual objects, such as dots moving on a computer screen or use of pure tone stimuli (Alink et al., 2008; Baumann and Greenlee, 2007; Bushara et al., 2001; Calvert et al., 2001; Lewis et al., 2000; Meyer et al., 2007; Sestieri et al., 2006; Watkins et al., 2006 , 2007). The audio and visual inputs were either perceived to be coming from roughly the same region of space, were moving in the same direction, and/or had common temporal synchrony – any of which served to lead to the perception (illusion) of a unified multisensory event. Table 10.1, lists each study reference (where a given study included one or more paradigms), the figure or table depicted from that study, the number of reported cortical foci in each hemisphere, and a brief description of the task versus control condition(s) used.

At first glance, the network activation patterns in Fig. 10.2a (purple hues) spanned a fairly distributed range of regions. Most cortical foci outside of early visual areas (e.g., V1, V2) and early auditory areas (e.g., PAC) were located in dorsal brain regions (above the green dotted line). Four of the most prominent regions and/or networks that emerged from this meta-analysis are addressed below, while a discussion of other ROIs and networks will be considered after presenting results from the other meta-analyses.

Early visual and auditory cortical interactions. Although multisensory integration is classically assigned to higher hierarchical cortical regions, both early visual areas and early auditory processing stages immediately surrounding PAC are reported to show crossmodal enhancement effects. For instance, a single brief flash paired with two auditory bleeps can lead to the illusion of two flashes having occurred, with a corresponding and disproportionately greater increase of activation in visual areas V1, V2, and V3 (Watkins et al., 2006). Similarly, two flashes paired with a single audio bleep can render the illusion of a single flash and a corresponding decrease in activation in early visual cortices (Watkins et al., 2007). Together with activation of the superior colliculi and posterior temporal regions, enhanced crossmodal activation in these early stages of visual-related cortex correlated with the individual's perception of the illusory event, rather than simply reflecting the actual physical properties of the visual stimuli. Conversely, early auditory areas are also reported to show activation to purely visual stimuli. For instance, after learning to associate screen flashes with a complex auditory event such as a phone ringing, the flashes alone could lead to greater activation of auditory cortices (Meyer et al., 2007). The results from this study may have reflected top-down feedback leading to some form of acoustic imagery rather than a true integration per se. Nonetheless, such studies highlight the point that visual stimuli can influence activation in cortical regions that are traditionally considered to be early auditory areas (Recanzone and Sutter, 2008).

The findings that human primary sensory areas show crossmodal interactions have been corroborated using more invasive techniques in awake behaving monkeys, as addressed in Chapter 7. Importantly, single neurons in V1 can integrate acoustic information as fast as ~61 ms from stimulus onset, indicating that some forms of audio-visual interactions must be occurring via direct projections between early sensory areas (Kayser and Logothetis, 2007) versus feedback projections from higher level association or polymodal areas (Wang et al., 2008). Additionally, anatomical tracing studies have demonstrated connections from the core and belt auditory regions to peripheral field representations of primary visual areas in the macaque (Falchier et al., 2002; Rockland and Ojima, 2003), and possibly in humans (Catani et al., 2003). Thus, a relatively small, yet significant, amount of direct cross talk between early auditory and visual sensory stages appears to exist in humans, which mediate some aspects of audio-visual interaction. However, cross talk between these early processing stages may not reflect interactions that convey meaningful or semantic representations of object features. Rather, they are thought to support low-level processing such as enhancing sensory perception in noisy environments and increasing attentional focus in order to enhance streaming and/or convergence of information to higher processing stages (Macaluso, 2006).

The cerebellum and optimization of audio-visual sensory input acquisition. Though not illustrated, the cerebellum was more frequently reported to be activated in tasks involving simple intermodal invariant matching paradigms than in the semantic matching paradigms that are presented later (Baumann and Greenlee, 2007; Bushara et al., 2001; Calvert et al., 2001). In addition to tactile and motor-related functions, portions of the cerebellum are known to have roles in the processing of acoustic sensory input, showing increased activity to increased sensory processing demand, and relatively independent of task (Petacchi et al., 2005). In general, this structure appears to play a role in optimizing the active, dynamic acquisition of sensory input. Thus, the cerebellum may be extracting or comparing intermodal invariant audio-visual features, perhaps facilitating the streaming of correlated crossmodal information from noisy signal inputs.

The anterior insulae and the ventriloquist effect. Activation of the bilateral anterior insulae (Fig. 10.2a, purple) was largely unique to those paradigms emphasizing relatively simple audio-visual spatial and temporal matching, which is in contrast to meta-analyses presented in the following sections (e.g., Fig. 10.2b). In several paradigms, the superior colliculi and posterior thalamus (not shown) were co-activated with the anterior insulae. Consistent with known anatomical connections, this led to the proposal that crossmodal processing of temporally matching auditory and visual features is in part mediated along an input pathway from the superior colliculus to the thalamus and then to the anterior insulae, especially in the right hemisphere (Bushara et al., 2001; Calvert et al., 2001; Meyer et al., 2007). Portions of this processing pathway, termed the tectal-cortical pathway, showed a direct association between the degree of activity and the subjects' perceptual experience of crossmodal interactions (Bushara et al., 2003).

The tectal-cortical processing pathway entails relatively early processing stages, and thus parallels the processing of information flowing from thalamus to early visual and auditory cortices. These parallel processing streams are thought to later feed into a common network of association cortices, wherein an interplay between these two sources of audio-visual input may have a significant role in well-documented illusions such as the ventriloquist effect (Bushara et al., 2001; Calvert and Campbell, 2003) and McGurk effect (McGurk and MacDonald, 1976). In both illusions spatial and temporal voice and lip movement cues are combined (i.e., along tectal-cortical pathways) before unimodal acoustic and visual information are assigned to a phoneme or word category level of semantic recognition and before the processes of attending to specific spatial locations are complete (Driver, 1996; Skipper et al., 2007). This allows, for example, for the perception of speech sounds as being captured by the temporally congruent visual movements of a puppet's mouth, leading to the illusion that the puppet is doing the talking. In this regard, audio-visual illusions have provided invaluable clues as to the underlying cortical mechanisms for integrating multisensory inputs, and their limitations for mediating our sense of perception.

Fronto-parietal networks and multisensory spatial attention. In Fig. 10.2a, the activation foci in dorsal cortical regions included frontal and parietal cortices. In frontal regions, this included the inferior frontal cortex (IFC) and medial frontal gyri (MFG) in both hemispheres (bilaterally). These regions are implicated in functions related to spatial attention and attention allocation (Corbetta et al., 1993; Mesulam, 1998). Portions of the IFC are also involved in working memory functions, in motor-related functions, and various cognitive functions, as will be further addressed in a later section after considering networks for processing abstract and non-matching crossmodal interactions.

In parietal cortex, the most prominent ROIs included the bilateral precuneus regions (medial parietal cortex) bilaterally plus the *right* superior parietal lobule (SPL). The functions of these parietal regions generally include aspects of visuospatial and motor coordinate processing (Grefkes et al., 2004). For instance, the SPL regions, representing end stages of the dorsal processing stream (Section 10.3), have roles in converting representations of object or target locations from eye-centered space (retina) into to head-centered and body-centered coordinates, and conversely, converting acoustic target locations from head-centered into body-centered and/or eye-centered coordinates, as mentioned previously (Andersen, 1997; Avillac et al., 2005; Creem and Proffitt, 2001). These transformations allow one to track the spatial locations of salient objects in the environment, whether heard or viewed, relative to one's body and limbs in preparation for potential motor interaction.

Patients with bilateral injury to parietal cortex often have severe deficits in processing and interacting with objects in space (Balint's syndrome) and impaired spatial audio-visual interactions (Valenza et al., 2004). Moreover, patients with lesions restricted to the right parietal lobe (i.e., the SPL), sparing the left SPL, are also reported to have severe impairments of visual spatial perception, spatial attention, and neglect syndromes (De Renzi et al., 1977; Irving-Bell et al., 1999; Krumbholz et al., 2005). For instance, they may be unaware of objects in their left visual hemifield and have deficits in visual, auditory, and tactile spatial processing, including a failure to recognize their own left arm and hand! Thus, much of the right lateralized fronto-parietal network for audio-visual interactions (Fig. 10.2a, purple) appears to be consistent with its proposed role in processing spatial-related information, including establishing and maintaining crossmodal or "supramodal" representations of objects in space relative to the observer (Sestieri et al., 2006). The proposed functional roles of the left SPL will be addressed in Section 10.4.4, after addressing artificial audio-visual interactions.

10.4.2 Networks for Audio-Visual Integration of Natural Objects and Faces in Motion

In striking contrast to the cortical networks for spatial and temporal audio-visual matching addressed above, networks associated with the perception of natural and semantically congruent pairings of audio-visual motion prominently activated more ventral brain regions (Fig. 10.2b, yellow hues below the green dotted line), including posterior temporal cortex (Beauchamp et al., 2004a, b; Calvert et al., 1999, 2000; Kreifelts et al., 2007; Olson et al., 2002; Robins et al., 2009; Stevenson and James, 2009). All of these paradigms depicted complex natural actions that were produced by humans (or implied human agency), together with emphasis on semantic congruence and effectively recognizing *what* the multisensory action was. These paradigms included talking faces and hand tools in use (see Table 10.2), contrasting matched audio-visual presentations with each of the unimodal conditions (AV>A,V), while other paradigms contrasted congruent (matched) versus incongruent (mis-matched) audio-visual pairings (e.g., $AV_{cong} > AV_{incong}$). Also depicted are results from a human electrophysiology study (Reale et al., 2007), wherein electrode arrays were placed over the posterior superior temporal gyri (pSTG) of epilepsy patients slated for surgical intervention (Fig. 10.2b, array of small orange circles). These authors reported enhanced neural activation along portions of the STG to video clips of a lower face speaking while the patient heard congruent versus incongruent spoken phonemes.

Ostensibly, the temporal dynamics of the visual motion in the video matched the temporal and amplitude energy dynamics of the acoustic information, and thus these paradigms also involved matching of intermodal invariant features, as addressed in the previous section. Yet the use of these more natural and highly familiar audio-visual actions depicting human conspecifics led to a dramatically different network of enhanced activation. This difference in network involvement may have been due to a combination of either the greater degree of life-long familiarity and experience with human actions (especially exposure during early development), the greater behavioral relevance typically associated with human actions (e.g., for social cognition functions), the richer semantic information conveyed by humans and natural object actions, or possibly unintended differences in the task(s) that the participants were performing. The relative influences of these factors remain to be further elucidated.

The regions preferentially activated by congruent human audio-visual interactions largely involved a swath of cortex along posterior temporal regions in both hemispheres, including the pSTS and posterior middle temporal gyri (pMTG), herein referred to as the pSTS/pMTG complex. In some paradigms, reported activation foci also included anterior temporal regions and the claustrum (not shown). The proposed functions of the above three regions will be addressed after first considering results from the other meta-analyses, illustrated in Fig. 10.3, which incorporated auditory and visual stimuli that had little or no intermodal invariant features in common, including audio-visual matching paradigms using pictures instead of motion

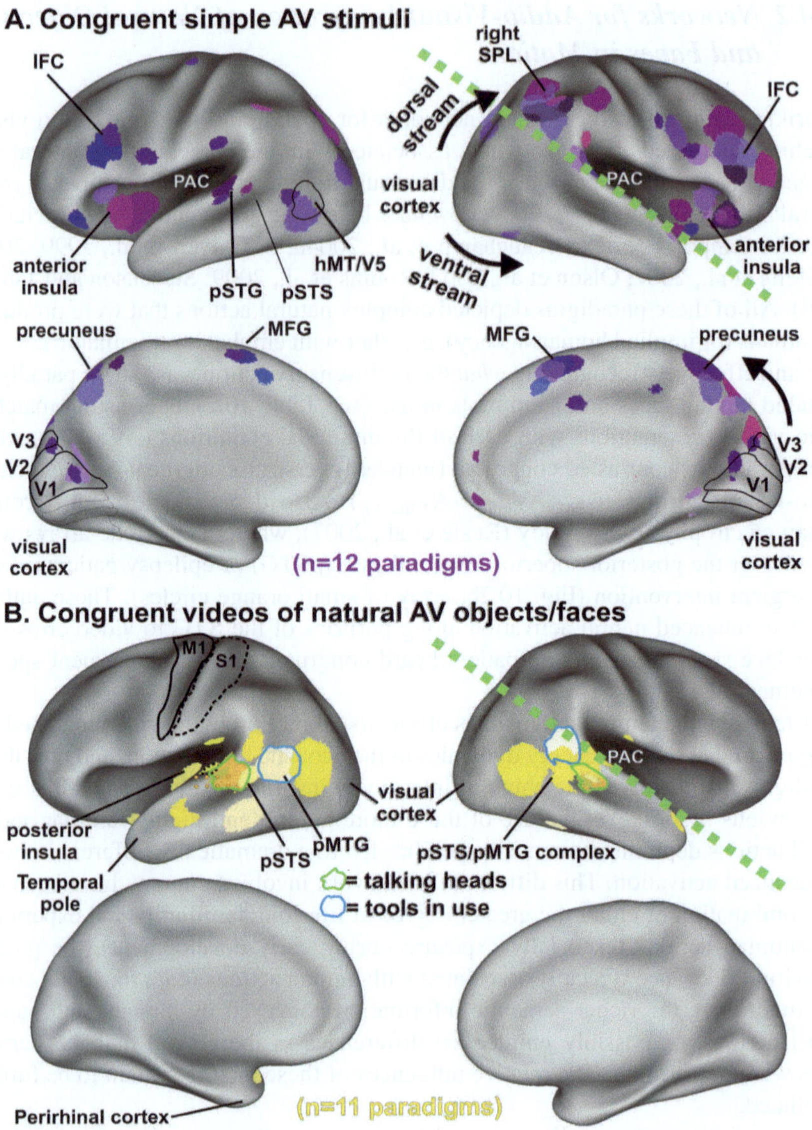

Fig. 10.2 (**a**) Brain networks sensitive to audio-visual matching based on spatially and/or temporally matched intermodal invariant features of artificial stimuli (each purple hue corresponds to a given paradigm; see Table 10.1). (**b**) Networks involved in audio-visual interactions for semantically congruent auditory stimuli when using videos of natural objects and/or faces in motion (each *yellow* hue corresponds to a given paradigm; Table 10.2). Refer to Fig. 10.1 and text for other details

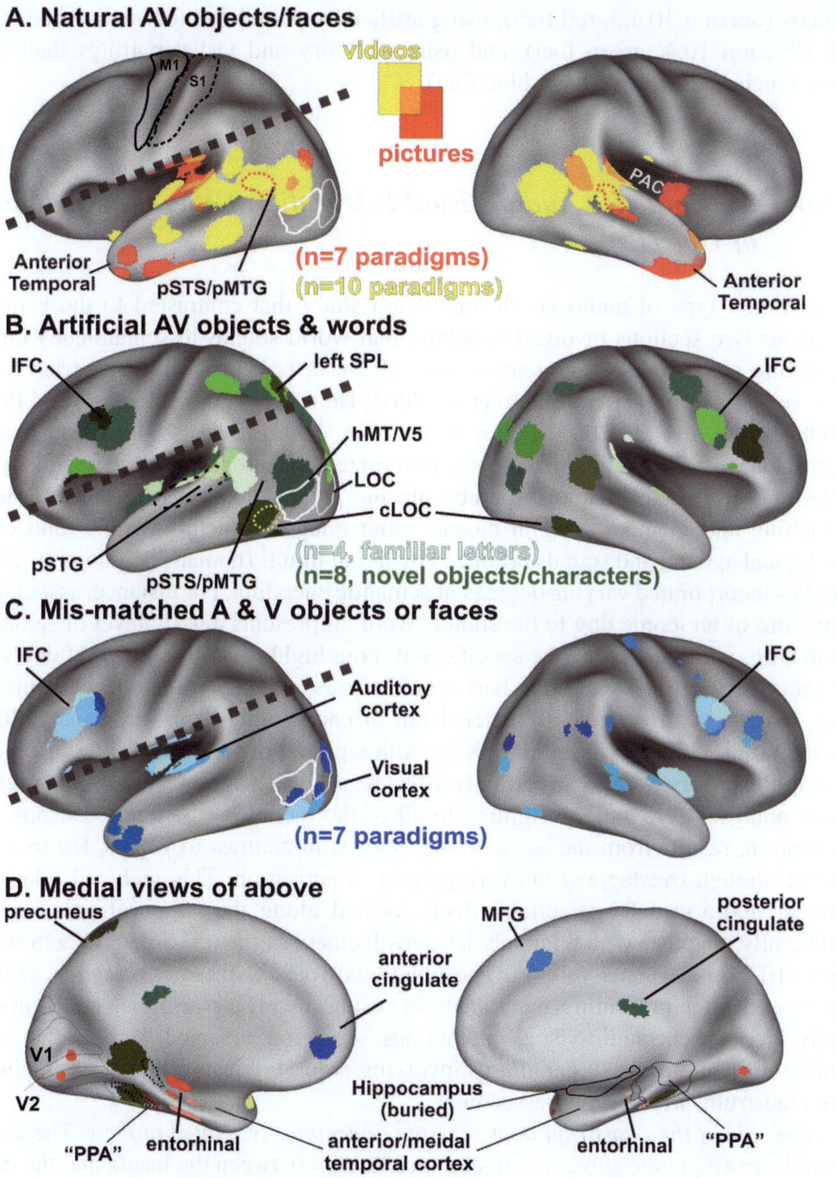

Fig. 10.3 (**a**) Cortical networks involved in the perception of natural objects and faces using still pictures (*red*, see Table 10.3) superimposed on data using videos (*single yellow hue* combined from results in Fig. 10.2B), showing overlap (*orange*). (**b**) Cortical networks for interactions involving learned abstract semantic pairings of auditory and visual stimuli (each *green hue* corresponds to a given paradigm, Table 10.4). (**c**) Incongruent versus congruent auditory and visual objects (each *blue hue* corresponds to a given paradigm, Table 10.5). (**d**) Medial view of brain showing meta-analysis results from **a** to **c**. Refer to text for other details

videos (Section 10.4.3, red foci), using artificially paired auditory and visual stimuli (Section 10.4, green foci), and using auditory and visual pairings that were mis-matched (Section 10.4.5, blue foci).

10.4.3 Networks for Audio-Visual Integration Using Static Pictures of Objects

A common type of audio-visual interaction study that contrasted to those in the previous two sections involved matching real-world sounds to semantically corresponding static images or pictures (Fig. 10.3a, d, red, plus orange overlap hues; also see Table 10.3) (Belardinelli et al., 2004; Hein et al., 2007; Hocking and Price, 2008; Naghavi et al., 2007; Taylor et al., 2006, 2009). Because the visual images were static, these paradigms greatly minimized the processing and matching of intermodal invariant features, thereby placing greater emphasis on semantic-level matching and memory – again mostly using images depicting people (and other biological agents) and sounds produced by living things (humans or animals). These studies incorporated varying degrees of semantic encoding. For instance, associating a picture of an iconic dog to the sound "woof" represents a basic level of semantic matching, while matching the specific and more highly familiar image of one's pet Tibetan terrier to her particular bark would represent a subordinate level of matching, which usually entails a greater depth of encoding (Adams and Janata, 2002; Tranel et al., 2003; Tyler et al., 2004). Although the depth of encoding of objects and multisensory events can certainly affect resulting activation patterns, the present meta-analysis gave equal weighting to all of the included paradigms. Despite this limitation, results from the use of static images, in contrast to videos, led to a different, though overlapping network pattern of activation. This included relatively greater audio-visual interaction effects located along the temporal pole regions bilaterally, together with relatively less involvement of the pSTS/pMTG complexes (Fig. 10.3a, red versus yellow). The functional roles of these two regions will be addressed after presenting meta-analysis results involving artificial and semantically incongruent audio-visual interactions. A region activated by both the static and motion audio-visual paradigms involving natural objects additionally included the claustrum, which is addressed next.

The role of the claustrum in processing conceptually related objects. The claustrum is an enigmatic sub-cortical structure located between the insula and the basal ganglia in both hemispheres and has reciprocal connections with nearly all regions of cortex (Ettlinger and Wilson, 1990). This region was activated by a few of the above-mentioned paradigms (data not shown) including studies that incorporated audio-visual speech integration (Calvert and Brammer, 1999; Olson et al., 2002) and also semantically congruent pictures of animals or objects and their corresponding characteristic sounds (Naghavi et al., 2007). The claustrum (and nearby insula) regions have also been implicated in other types of multisensory integration, such as tactile-visual matching of object shapes and audio-visual onset synchrony (Bushara et al., 2001; Calvert et al., 1999; Hadjikhani and Roland, 1998). Thus, the claustrum appears to have a role in integrative processes that require the analysis of the *content* of the stimuli, coordinating the rapid integration of object attributes across different modalities, leading to coherent conscious percepts (Crick and Koch, 2005; Naghavi et al., 2007). How the high-level multisensory object representations of the claustrum relate to those high-level functions of the anterior temporal regions, which are addressed later, remains unclear. Before further addressing issues of semantic representations, the next section illustrates networks associated with representing artificial, abstract, or "nonnatural" audio-visual pairings.

10.4.4 Interactions When Using Artificial or Abstract Audio-Visual Pairings

One of the main features that set humans apart from other primates is our more advanced use of symbolism that permits a more complex and precise means for mediating communication and social learning (Corballis, 1992; Rilling, 2008; Whiten et al., 2004). This notably involves the use of abstract acoustic and visual representations to convey conceptual knowledge and information. In particular, written languages represent a major category of behaviorally relevant visual forms that are artificially or arbitrarily paired with acoustic (spoken) representations (Arbib, 2005; Corballis, 2003; van Atteveldt et al., 2004). At an early age one typically learns to speak, and subsequently learns to read and pronounce written text, which are processes that are known to be associated with critical periods in childhood development (Bates and Dick, 2002; Mayberry et al., 2002). After those critical periods, the acquisition of language representations may be mediated by different cortical mechanisms (Castro-Caldas et al., 1998; Hashimoto and Sakai, 2004). Thus, the central nervous system may similarly utilize different mechanisms for associating arbitrary audio-visual object pairs (linguistic or otherwise) at different stages in maturation, further complicating the interpretation of the crossmodal studies addressed herein. Nonetheless, differences in cortical organization for different types of arbitrarily associated audio-visual pairing are becoming apparent, as revealed through neuroimaging studies that involved highly familiar objects (learned early in life) versus audio-visual pairings recently learned (in adulthood).

Familiar arbitrary audio-visual pairings. A composite meta-analysis map of cortical regions activated by highly familiar letter and phoneme pairings (such as one's native language) is illustrated in Fig. 10.3b, d (light green hues, Table 10.4) (van Atteveldt et al., 2004, 2007a, b). Although a more thorough account of networks subserving audio-visual language associations is beyond the scope of the present chapter, in general, the presentation of familiar written and spoken pairings activated regions along the pSTG bilaterally, especially in the left hemisphere.

Interestingly, in a study that examined congenitally deaf individuals a correlation was found between the skill of speech reading (lip reading) and the degree of activity along the left pSTG (Fig. 10.3b, black dashed outline) (Capek et al.,

2008), which overlapped with some of the other light green foci. Thus, in the complete absence of hearing, the left STG region apparently still functions to perform linguistic- or prelinguistic-related operations with available sensory input, namely visual lip reading, and possibly even tactile lip reading (e.g., methods used by Anne Sullivan to teach Helen Keller to detect and comprehend articulated language). These seemingly paradoxical results are consistent with the idea of a metamodal brain organization (Section 10.3), wherein the STG may typically develop to process acoustic-related inputs, but in the absence of cochlear input it can be recruited or adapted for performing operations with information derived from vision and possibly tactile senses.

Unfamiliar arbitrary audio-visual pairings. In contrast to highly familiar arbitrary audio-visual pairings, the focus of several other studies emphasized learning novel acoustic associations with novel complex visual objects and/or object actions (Fig. 10.3b, dark green hues) (Gonzalo et al., 2000; Hashimoto et al., 2006; Hein et al., 2007; James and Gauthier, 2003; McNamara et al., 2008; Neal et al., 1988; Scheef et al., 2009; Tanabe et al., 2005). For instance, some studies used novel visual objects such as "Greebles" (James and Gauthier, 2003), and artificial 3D object images called "Fribbles" (Hein et al., 2007; Naumer et al., 2009), either of which could become newly associated with novel sounds, such as buzzes or modified animal sounds. Other studies involved pairing written characters from an unfamiliar written language with different sounds – though not to the extent of actually learning a foreign language. Together, the above paradigms led to a relatively widespread network of activation involving posterior temporal regions, bilateral IFC cortices, and the left SPL.

In posterior temporal cortices, activation foci involved predominantly visual related regions, including those for object motion and object shape processing. One study involved teaching participants to pair artificial "sonifications" (tones rising up and down) that could be correlated to the cartoon video of a human leaping (Scheef et al., 2009). Practice with the dynamic audio-visual pairings led to enhanced effects in the visual motion area hMT or V5 (Fig. 10.3b; white bold outline). Regarding object shape processing, the caudal lateral occipital cortex (cLOC, also termed LOC) is a region implicated in visual shape processing (Amedi et al., 2007). This region showed enhanced activation that was correlated with learning of novel Hangul letters paired with nonsense words (i.e., 2-dimensional shapes) (Hashimoto et al., 2006). In another study, portions of the cLOC region (Fig. 10.3b, dotted yellow outline) showed activation by congenitally blind participants after they had learned to associate object surface features with artificial "soundscapes" that conveyed object shape information (Amedi et al., 2007). Thus, the LOC/cLOC region appears to be representing shape features, independent of whether or not one has even experienced visual sensory input (also see Chapter 17). This represents yet another example that was consistent with the concept of some cortical regions perform metamodal operations for sensory perception, whether or not a person had grown up with sight or hearing ability.

Left versus right parietal cortical functions in AV interactions. While the right superior parietal lobe (SPL) was involved more with audio-visual interactions depicting nonnatural objects that had intermodal invariant features in common to bind the two sensory inputs (Fig. 10.2a, purple, Section 10.4.1), the left SPL was more prominently involved in artificial audio-visual pairings that linked at a more abstract or symbolic level (cf Fig. 10.3b, dark green). The left parietal cortex has been associated with mathematic skills: In particular, number processing and mental arithmetic appears to utilize similar circuits for attention to external space and internal representations of quantity or numbers (Hubbard et al., 2005; Nieder and Dehaene, 2009). These symbolic representations and comparisons may also relate to left-lateralized language functions, which also represent symbolic representations that can develop independent of whether or not a person has vision and/or hearing ability (Neville et al., 1998; Roder et al., 2002). More precisely determining what differences are driving these functional lateralization biases in parietal cortices will be a matter of future research.

10.4.5 Networks for Semantically Mis-matched Auditory and Visual Pairs

Typically as a control condition, several of the above-mentioned paradigms included contrast conditions wherein either sequentially or simultaneously presented acoustic and visual stimuli were semantically incongruent with one another (Fig. 10.3c, d, blue hues, Table 10.5) (Belardinelli et al., 2004; Gonzalo et al., 2000; Hein et al., 2007; Hocking and Price, 2008; Naumer et al., 2008; Taylor et al., 2009; van Atteveldt et al., 2007). This included studies in which participants either already knew incorrect audio-visual pairings from life experience or had been recently trained to identify novel abstract audio-visual pairings and discriminate those from mis-matched pairings. These meta-analysis results support the idea that attending

Table 10.5 List of paradigms involving incongruent or mis-matching auditory (A) and visual (V) objects or characters, depicted in Fig. 10.3c, d (blue hues)

Study	#Foci (L,R)	Paradigm: familiar or learned AV pairs: incorrect vs correct matches
Belardinelli (2004)	T3 (2,3)	Colored images of tools, animals, humans and incongruent vs congruent sounds
Gonzalo (2000)	F4 (4,3)	Novel Kanji characters and musical chords, activity increases to inconsistent pairings
Hein (2007)	F2-3 (6,2)	Familiar animal images and incorrect vocalizations (dog: meow) vs correct pairs
Hocking (2008)	T3 (6,10)	Incongruent sequential AV pairs (see drum, hear bagpipes) vs congruent pairs
Naumer (2008)	F4 (1,1)	Learn of "Freebles" and distorted sounds as incongruent vs congruent pairs
Taylor (2009)	F3c (3,0)	Familiar incongruent AV pairs (see bunny, hear rabbit) vs congruent pairs
VanAtteveldt (2007a)	T4 (1,6)	Familiar letters and their spoken (Dutch) phoneme sounds: Incongruent vs congruent

to an auditory stimulus and a mis-matched visual stimulus implies that the nervous system effectively has to deal with recognition of two different sensory events, leading to an increased working memory load and/or increased attentional demands (Belardinelli et al., 2004; Doehrmann and Naumer, 2008; Taylor et al., 2006). Thus, early auditory and early visual areas were activated, but there was a relative lack of naturally congruent features to drive the pSTS/pMTG, and a relative lack of symbolic or intermodal invariant spatio-temporal features to drive activation in parietal (SPL) cortices (cf Figs. 10.2 and 10.3, blue versus green, yellow, and purple). Other regions showing positive activation to the mis-matched conditions included a few anterior midline structures, the lateral portions of anterior temporal cortex, and foci along the bilateral IFC regions. Now having presented all five audio-visual interaction meta-analyses, discussions of the proposed functions of the IFC, pSTS/pMTG complex, and anterior temporal regions are presented next.

The bilateral IFC and monitoring of conflicting A and V stimuli. Interestingly, the bilateral IFC regions (Figs. 10.2 and 10.3) showed interaction effects to all of the nonnatural audio-visual paradigms, including mis-matched pairings (blue), artificially matched audio-visual objects (green), and relatively simple audio-visual objects bound by intermodal invariant features (purple). However, these regions showed very little activation to the processing of congruent natural audio-visual stimuli (yellow and red). This meta-analysis finding was largely consistent with an earlier study that hypothesized that frontal cortical regions are involved more in processing incongruent than congruent audio-visual objects or speech, due to the requirement of greater cognitive control and conflict monitoring (Doehrmann and Naumer, 2008). However, the present meta-analysis results further suggest that frontal regions may be involved when congruent *nonnatural* and *artificial* audio-visual matches are being made, which may similarly require a relatively greater degree of cognitive control to bind or conceptually match features of nonnatural object events.

The pSTS/pMTG complex and complex motion encoding. When comparing the results across all five meta-analyses, the pSTS/pMTG complexes were most prominently activated when subjects viewed motion videos of congruent, natural audio-visual interactions depicting human-produced actions (i.e., Figs. 10.2b and 10.3b, yellow), which represented highly familiar, life-long behaviorally relevant actions. Anatomically, the pSTS/pMTG complexes were situated between early visual and early auditory cortices, which presumably minimizes cortical "wiring" (Van Essen, 1997) for efficiently associating complex motion information, or abstractions of action representations, between these two modalities. The closer proximity of these regions to primary auditory cortices (see Fig. 10.1c, flat maps) may be due to the greater number of acoustic processing stages that occur in sub-cortical regions prior to reaching primary auditory cortex (King and Nelken, 2009). Regardless, these meta-analysis results support the hypothesis that the lateral temporal cortices are the primary loci for complex natural motion processing (Martin, 2007), though with some regional variations based on object or sound-source category membership.

In many studies in addition to those illustrated in this chapter, the pSTS/pMTG complexes showed greater or preferential activation to actions produced by living things, most notably including human actions as a distinct category, whether viewed (Beauchamp et al., 2002; Grossman and Blake, 2002; Thompson et al., 2005) or heard (Bidet-Caulet et al., 2005; Engel et al., 2009; Gazzola et al., 2006, 2008; Lewis et al., 2006) or both (Campbell, 2008; Campanella and Belin, 2007). Some of the studies illustrated in this chapter similarly reported category-related differences for audio-visual interactions. For instance, one study found that audio-visual interactions to videos of talking heads activated anterior regions overlapping the STG (Fig. 10.2b, green outlines) while more posterior subdivisions (blue outlines) were preferentially activated by videos of humans using tools (Stevenson and James, 2009). These results were demonstrated using a novel fMRI method for assessing integration, termed "enhanced effectiveness," addressed further in Chapter 13. Similarly, note that the audio-visual speech regions were located closer to auditory cortices (along the STG), consistent with forming associations with vocalization content, while tools in use (by humans) activated regions closer to visual cortex proper, perhaps consistent with representing visual motion and form.

Tools represent an interesting object category from the perspective of audio-visual interactions. As a visual stimulus, a tool (i.e., hand tool or utensil) that is presented in the absence of a biological agent manipulating it is generally regarded as a nonliving thing. However, when sounds are produced by a hand tool it implies an action by a living thing or agent. Thus tool action sounds tend to yield activation of widespread motor-related systems associated with the dominant hand (Lewis et al., 2005, 2006). Hence, tools, as a seemingly distinct category of object (Section 10.3), can effectively cross the living versus nonliving conceptual category boundary, depending on how they are presented (Lewis, 2006). Given this caveat, videos of people in motion have been reported to preferentially activate the pSTS subdivision, while puppeted hand tool actions (which may imply actions of a human agent) preferentially activate the pMTG region (Beauchamp et al., 2002). From the perspective of audition, sounds produced by automated machinery or tools that were judged to *not* be directly associated with a living agent instigating the action, as opposed to human action sounds, led to substantially less activation in the pSTS/pMTG regions (Engel et al., 2009). Rather, this non-human category of action sound preferentially activated the anterior STG and parahippocampal regions – both well outside the pSTS/pMTG complexes. Thus, audio-visual actions involving hand tools, which tend to have strong motor affordances and associations, in some situations may be represented in cortical networks more as an extension of one's hand and arm motoric associations rather than as the form features of the tool per se.

The pSTS/pMTG complexes appear to play a prominent perceptual role in transforming the spatially and temporally dynamic features of natural auditory and visual action information together into a common neural code (if/when both inputs are present), which may then facilitate or confer crossmodal integration (Beauchamp et al., 2004a, b; Calvert et al., 2000; Lewis et al., 2004; Taylor et al., 2006, 2009). Similar to proposed mechanisms of parietal regions (Avillac et al., 2005; von Kriegstein and Giraud, 2006), one possibility is that the pSTS/pMTG complexes may additionally form an elaborate temporal reference frame for probabilistically comparing the predicted or expected incoming auditory and/or visual information based on what actions have already occurred. The greater degree of experience and life-long familiarity with everyday motor actions produced by human conspecifics, including facial movements and hand tool use, may thus be the primary reason why stimuli depicting human conspecifics most robustly activate those cortical regions and possibly why human actions and vocalizations apparently manifest as distinct object or event categories. Consistent with this interpretation, the pSTS/pMTG regions are also reported to have prominent roles in social cognition (Jellema and Perrett, 2006; Pelphrey et al., 2004; Zilbovicius et al., 2006), wherein reading subtleties of human expressions and body language can be highly relevant for conveying information and guiding social interactions.

Another view of the function of the pSTS/pMTG is that they serve a more conceptual role in matching across stimulus features independent of sensory modality (Hocking and Price, 2008; Tanabe et al., 2005). For instance, when attention and stimuli were well controlled, an analysis based on pre-defined bilateral pSTS regions (Fig. 10.3a, within red dotted outlines) was comparable when matching static natural images of pictures to their corresponding sounds as to when matching visual–visual and audio–audio semantically matched pairs (Hocking and Price, 2008). This indicates that the pSTS/pMTG regions may be generally involved in establishing paired associations, including audio–audio and visual–visual associations, and not just functioning for purposes of audio-visual integration. Note, however, that the above paradigms did not involve real-time visual motion information that could naturally be bound to real-time acoustic information (e.g., use of pictures or delay period processing), which, based on the present meta-analyses, appear to be non-optimal stimuli for activating the lateral temporal regions. Notwithstanding, human lesions studies indicate that the pSTS/pMTG regions may not even be necessary for conceptually binding auditory and visual congruence of inputs (Taylor et al., 2009). Rather, this more conceptual role of semantic binding is subserved by anterior temporal regions (addressed below). So how are the above seemingly conflicting findings reconciled?

The present results are consistent with the pSTS/pMTG complexes principally functioning at a perceptual input level for representing complex actions, and that they convey symbolic associations of naturally matched audio-visual features (if hearing and vision are present). This information may then be fed into anterior temporal regions, facilitating relatively more semantic or cognitive-level binding of object features. Words or phrases that depict human actions, and even imagining complex actions, can lead to activation of these lateral temporal regions (Kellenbach et al., 2003; Kiefer et al., 2008; Tettamanti et al., 2005; Noppeney et al., 2008). This may reflect re-activation of the pSTS/pMTG networks (especially in the left hemisphere) due to top-down influences, wherein dynamic action sequences, or symbolic representations of them, are being accessed or re-enacted through associative knowledge representations that had been formed over past experience. This hypothesis is in line with the Hierarchy of Associativity model linking the pSTS/pMTG and anterior temporal regions, as addressed next.

Anterior temporal cortices as "master binders" of crossmodal inputs. In macaques and humans, the anterior temporal regions, including perirhinal, entorhinal, and nearby anterior-medial temporal cortices, are proposed to serve as "master binders" of semantic information, integrating or associating polymodal inputs from each sensory modality, including auditory, visual, tactile, and olfactory information (Belardinelli et al., 2004; Murray and Richmond, 2001; Murray et al., 2000; Taylor et al., 2006, 2009; Patterson et al., 2007). Lesion studies suggest that these regions are critical for crossmodal integration of objects as meaningful representations to the observer (Taylor et al., 2009). So why didn't more of the audio-visual interaction paradigms emphasizing semantic associations report activity in these regions? This may in part have been due to the considerably poorer fMRI signal quality generally obtained from ventral temporal regions (fMRI susceptibility artifacts), and thus a relative lack of data reaching statistical significance. Alternatively, only those tasks that demand a high degree of fine-grained discrimination may adequately drive anterior temporal regions (to reveal significant blood flow changes), such as matching features at subordinate as opposed to basic levels of recognition (Adams and Janata, 2002; Tyler et al., 2004). For instance, living things, as a general category, tend to have more features in common with one another than do nonliving things, requiring a greater degree of fine-grained discriminations to distinguish them – a concept termed "visual crowding" (Gale et al., 2001). The antero-medial temporal cortex (left > right hemisphere) is highly involved in multimodal representations of person-related information, and is necessary for conscious recognition of animals and living things (Moss et al., 2005). This is consistent with the greater activation revealed in the meta-analysis of studies using static pictures (i.e., Fig. 10.3a, red versus yellow), which may have required a greater degree of processing relative to motion videos: Motion videos can more completely convey features leading to crossmodal binding via the pSTS/pMTG complexes, thereby effectively reducing the amount of integration or computation required at the level of anterior temporal regions to determine if there is a semantic match (i.e., the pSTS/pMTG regions already "did the work" of binding features).

Given the wide range of parallel hierarchical pathways and interaction zones that may contribute to audio-visual integration, how does all of this information ultimately combine and become represented as meaningful events? Theories such as the Hierarchy of Associativity model help to provide a possible account for how information may flow along the parallel audio-visual processing pathways and mediate perception (Davachi, 2006; Lavenex and Amaral, 2000; Taylor et al., 2006, 2009). For instance, in its simplest form, the anterior temporal regions include two divisions, perirhinal and entorhinal cortex. The perirhinal regions predominantly receive input from unisensory areas, where some forms of audio-visual integration can occur, such as integrating static pictures and sounds (i.e., Fig. 10.3a, red). The integrated sensory information in perirhinal cortex, together with higher level audio-visual association cortices, then feed into entorhinal cortex. The entorhinal cortex is thought to additionally receive input from association areas, such as the pSTS/pMTG complexes (i.e., Fig. 10.3a, yellow), which could convey representations of crossmodal actions that have already been bound or integrated due to their congruent matching of dynamic, natural intermodal invariant motion attributes. Additionally the entorhinal cortex may receive input from parietal cortices, where the left SPL, for instance, may convey symbolic matches between learned artificial audio-visual pairings. The entorhinal regions then synthesize and feed highly integrated audio-visual associated information into the hippocampal system, supporting the formation of episodic memories.

The Hierarchy of Associativity model also accounts for results from the paradigms where auditory and visual stimuli did not match (Fig. 10.3c, d, blue). In particular, the more lateral anterior temporal activations were thought to be recruited due to the fine discrimination processing required to determine whether the stimuli presented were semantically congruent, comparing representations of the object actions with networks that have encoded previously learned crossmodal associations and semantic-level object memories (Taylor et al., 2009). Additionally, words depicting action events might preferentially re-activate portions of hierarchically organized pathways, such as the pSTS/pMTG complexes, due to learned symbolic associations between the word and the object–events that they represent, as mentioned earlier. The idea that conceptual-level representations of action events might be encoded along the same networks that mediate perception is consistent with grounded cognition theories for knowledge representations, which are addressed next in the final section.

10.5 Category-Specific Audio-Visual Object Processing and Knowledge Representations

As alluded to in the previous sections, different pathways and networks appear to preferentially or selectively process objects that belong to different categories. Accounting for these semantic effects on perceptual object processing has inspired the development of cognitive models for semantic object memory. In particular, the Conceptual Structure Account model (Tyler and Moss, 2001) proposes that "objects in different categories can be characterized by the number and statistical properties of their constituent features" (Taylor et al., 2009). This would suggest, for instance, that the diverse range of reported category-specific networks for different visual object categories and for different auditory object categories (Section 10.3) may well influence one another and thereby impact how knowledge representations may become organized in the brain.

Two currently debated theoretical frameworks for how the brain becomes organized for object knowledge have influenced our understanding of the mechanisms that may mediate audio-visual perception. One is a domain-specific framework, wherein some cortical regions may function to perform metamodal or amodal operations independent of the presence or absence of a sensory modality input (as addressed earlier). An extreme form of this theory posits that evolutionary pressures have produced specialized, and functionally dissociable, neuronal circuits that are predisposed or effectively "hard wired" for processing perceptually and/or conceptually different categories of objects, such as living versus nonliving things, and possibly even more specific categories such as faces, tools, fruits/vegetables, animals, body parts (Caramazza and Shelton, 1998; Damasio et al., 1996; Mahon and Caramazza, 2005; Mahon et al., 2009; Pascual-Leone and Hamilton, 2001). This would enable rapid and efficient identification of particular objects (unisensory or multisensory) and would have obvious survival and reproductive advantages. A second theoretical framework includes sensory-motor property-based models. According to this theory object knowledge and category-specific representations become organized based on one's experiences with sensory and sensory-motor features, including distinctive visual, auditory, and/or tactile attributes of an object together with representations of any corresponding motor properties associated with object interactions (e.g., pounding with a hammer) (Lissauer, 1890/1988; Martin, 2007). In the hypothetical extreme of this theory, the brain basically bootstraps its organization based on sensory and multisensory experiences. Combinations of these two mechanisms are likely to mediate the formation of knowledge representations.

Grounded cognition models further provide a mechanistic account for the encoding of knowledge representations, in that our thoughts and concepts may ultimately be encoded ("grounded") in the very same networks and pathways that are utilized for perception (Barsalou, 2008; Barsalou et al., 2003). Human development theories have long promulgated the idea that our knowledge, and hence what we perceive when experiencing multisensory objects and object events, builds incrementally on our past experiences through symbolic mediation (Vygotsky, 1978). As to exactly how we are able to house concepts that allow us to contemplate and introspect on our own thoughts remains as an exciting area of research that will impact models of sensory perception (Craig, 2009). Ostensibly, this capacity for housing concepts derived from sensory and multisensory information, in conjunction with our developing conceptual systems, appears to co-construct our sense of material reality and an apparent sense of awareness of the world around us.

10.6 Conclusions

This chapter reviewed evidence derived from meta-analyses across 49 paradigms, revealing several parallel and hierarchical processing pathways for purposes of audio-visual perception of objects. This included dissociations of pathways and networks for the binding of audio-visual features based either on (1) intermodal invariant attributes of nonnatural, simple objects; (2) semantically congruent features of natural objects, especially life-long highly familiar human conspecific actions; and (3) semantically congruent features of nonnatural or artificial audio-visual pairings. A dorsally directed pathway preferentially processes intermodal invariant features of nonnatural object–actions and was largely consistent with *processing for action*, including functions related to spatial attention and for establishing reference frames of presented objects and actions relative to one's body representations and motor systems. The right parietal cortices showed robust activation for spatio-temporal binding of intermodal invariant audio-visual features of nonnatural objects, while left parietal cortex showed prevalence for binding more symbolic attributes of learned auditory and visual pairings associated with complex objects. Bilateral inferior frontal networks were recruited in paradigms where audio-visual stimuli were either artificially or nonnaturally paired or were altogether mis-matched, consistent with their proposed roles associated with cognitive control and conflict monitoring. Natural audio-visual object actions preferentially activated the temporal lobes. Congruent and highly familiar action events, especially those produced by humans and other living things, activated the pSTS/pMTG complexes. These regions appear to have prominent perceptual-level functions in establishing reference frames for representing dynamic actions, whether heard, viewed, or both, consistent with their proposed functions in perception for social cognition. The anterior temporal cortices had a prevalent role in high-level semantic associations, serving as a master binders of information across all modalities. Together, these meta-analyses not only highlight several parallel hierarchical pathways for audio-visual interactions but also underscore the importance of addressing the organization of conceptual systems related to object knowledge representations in order to develop more complete models of multisensory perception.

Acknowledgements Thanks to Mr. Chris Frum for assistance with preparation of figures. Thanks also to Dr. David Van Essen, Donna Hanlon, and John Harwell for continual development of cortical data analysis and presentation with CARET software, and William J. Talkington, Mary Pettit, and two anonymous reviewers for helpful comments on earlier versions of the text. This work was supported by the NCRR/NIH COBRE grant P20 RR15574 (to the Sensory Neuroscience Research Center of West Virginia University) and subproject to JWL.

References

Adams RB, Janata P (2002) A comparison of neural circuits underlying auditory and visual object categorization. Neuroimage 16:361–377

Aglioti SM, Cesari P, Romani M, Urgesi C (2008) Action anticipation and motor resonance in elite basketball players. Nat Neurosci 11:1109–1116

Alink A, Singer W, Muckli L (2008) Capture of auditory motion by vision is represented by an activation shift from auditory to visual motion cortex. J Neurosci 28:2690–2697

Allison T, McCarthy G, Nobre A, Puce A, Belger A (1994) Human extrastriate visual cortex and the perception of faces, words, numbers, and colors. Cereb Cortex 5:544–554

Altmann CF, Doehrmann O, Kaiser J (2007) Selectivity for animal vocalizations in the human auditory cortex. Cereb Cortex 17:2601–2608

Amedi A, von Kriegstein K, van Atteveldt NM, Beauchamp MS, Naumer MJ (2005) Functional imaging of human crossmodal identification and object recognition. Exp Brain Res 166:559–571

Amedi A, Stern WM, Camprodon JA, Bermpohl F, Merabet L, Rotman S, Hemond C, Meijer P, Pascual-Leone A (2007) Shape conveyed by visual-to-auditory sensory substitution activates the lateral occipital complex. Nat Neurosci 10:687–689

Andersen RA (1997) Multimodal representation of space in the posterior parietal cortex and its use in planning movements. Annu Rev Neurosci 20:303–330

Andersen RA, Zipser D (1988) The role of the posterior parietal cortex in coordinate transformations for visual-motor integration. Can J Physiol Pharmacol 66:488–501

Arbib MA (2005) From monkey-like action recognition to human language: an evolutionary framework for neurolinguistics. Behav Brain Sci 28:105–124; discussion 125–167

Arnott SR, Binns MA, Grady CL, Alain C (2004) Assessing the auditory dual-pathway model in humans. Neuroimage 22:401–408

Avillac M, Deneve S, Olivier E, Pouget A, Duhamel JR (2005) Reference frames for representing visual and tactile locations in parietal cortex. Nat Neurosci 8:941–949

Aziz-Zadeh L, Iacoboni M, Zaidel E, Wilson S, Mazziotta J (2004) Left hemisphere motor facilitation in response to manual action sounds. Eur J Neurosci 19:2609–2612

Barsalou LW (2008) Grounded cognition. Annu Rev Psychol 59:617–645

Barsalou LW, Kyle Simmons W, Barbey AK, Wilson CD (2003) Grounding conceptual knowledge in modality-specific systems. Trends Cogn Sci 7:84–91

Bates E, Dick F (2002) Language, gesture, and the developing brain. Dev Psychobiol 40:293–310

Baumann O, Greenlee MW (2007) Neural correlates of coherent audiovisual motion perception. Cereb Cortex 17:1433–1443

Beauchamp M, Lee K, Haxby J, Martin A (2002) Parallel visual motion processing streams for manipulable objects and human movements. Neuron 34:149–159

Beauchamp MS (2005) Statistical criteria in FMRI studies of multisensory integration. Neuroinformatics 3:93–113

Beauchamp MS, Argall BD, Bodurka J, Duyn JH, Martin A (2004) Unraveling multisensory integration: patchy organization within human STS multisensory cortex. Nat Neurosci 7:1190–1192

Beauchamp MS, Lee KM, Argall BD, Martin A (2004) Integration of auditory and visual information about objects in superior temporal sulcus. Neuron 41:809–823

Belardinelli M, Sestieri C, Di Matteo R, Delogu F, Del Gratta C, Ferretti A, Caulo M, Tartaro A, Romani G (2004) Audio-visual crossmodal interactions in environmental perception: an fMRI investigation. Cogn Process 5:167–174

Belin P, Zatorre R (2000) 'What', 'where' and 'how' in auditory cortex. Nat Neurosci 3:965–966

Belin P, Zatorre RJ, Lafaille P, Ahad P, Pike B (2000) Voice-selective areas in human auditory cortex. Nature 403:309–312

Bidet-Caulet A, Voisin J, Bertrand O, Fonlupt P (2005) Listening to a walking human activates the temporal biological motion area. Neuroimage 28:132–139

Buccino G, Binkofski F, Fink GR, Fadiga L, Fogassi L, Gallese V, Seitz RJ, Zilles K, Rizzolatti G, Freund H-J (2001) Action observation activates premotor and parietal areas in a somatotopic manner: an fMRI study. Eur J Neurosci 13:400–404

Bulkin DA, Groh JM (2006) Seeing sounds: visual and auditory interactions in the brain. Curr Opin Neurobiol 16:415–419

Burton H, Snyder AZ, Raichle ME (2004) Default brain functionality in blind people. Proc Natl Acad Sci USA 101:15500–15505

Bushara KO, Grafman J, Hallett M (2001) Neural correlates of auditory-visual stimulus onset asynchrony detection. J Neurosci 21:300–304

Bushara KO, Hanakawa T, Immisch I, Toma K, Kansaku K, Hallett M (2003) Neural correlates of cross-modal binding. Nat Neurosci 6:190–195

Calvert GA, Brammer MJ (1999) FMRI evidence of a multimodal response in human superior temporal sulcus. Neuroimage 9:(S1038)

Calvert GA, Campbell R (2003) Reading speech from still and moving faces: the neural substrates of visible speech. J Cogn Neurosci 15:57–70

Calvert GA, Lewis JW (2004) Hemodynamic studies of audio-visual interactions. In: Calvert GA, Spence C, Stein B (eds) Handbook of multisensory processing. MIT Press, Cambridge, MA, pp 483–502

Calvert GA, Campbell R, Brammer MJ (2000) Evidence from functional magnetic resonance imaging of crossmodal binding in the human heteromodal cortex. Curr Biol 10:649–657

Calvert GA, Hansen PC, Iversen SD, Brammer MJ (2001) Detection of audio-visual integration sites in humans by application of electrophysiological criteria to the BOLD effect. Neuroimage 14:427–438

Calvert GA, Brammer MJ, Bullmore ET, Campbell R, Iversen SD, David AS (1999) Response amplification in sensory-specific cortices during crossmodal binding. Neuroreport 10:2619–2623

Campanella S, Belin P (2007) Integrating face and voice in person perception. Trends Cogn Sci 11:535–543

Campbell R (2008) The processing of audio-visual speech: empirical and neural bases. Philos Trans R Soc Lond B Biol Sci 363:1001–1010

Capek CM, Macsweeney M, Woll B, Waters D, McGuire PK, David AS, Brammer MJ, Campbell R (2008) Cortical circuits for silent speechreading in deaf and hearing people. Neuropsychologia 46:1233–1241

Caramazza A, Shelton JR (1998) Domain-specific knowledge systems in the brain the animate-inanimate distinction. J Cogn Neurosci 10:1–34

Caramazza A, Mahon BZ (2003) The organization of conceptual knowledge: the evidence from category-specific semantic deficits. Trends Cogn Sci 7:354–361

Castro-Caldas A, Petersson KM, Reis A, Stone-Elander S, Ingvar M (1998) The illiterate brain. Learning to read and write during childhood influences the functional organization of the adult brain. Brain 121 (Pt 6):1053–1063

Catani M, Jones DK, Donato R, Ffytche DH (2003) Occipito-temporal connections in the human brain. Brain 126:2093–2107

Chao LL, Martin A (2000) Representation of manipulable man-made objects in the dorsal stream. Neuroimage 12:478–484

Corballis MC (1992) On the evolution of language and generativity. Cognition 44:197–126

Corballis MC (2003) From mouth to hand: gesture, speech, and the evolution of right-handedness. Behav Brain Sci 26:199–208; discussion 208–160

Corbetta M, Miezin FM, Shulman GL, Petersen SE (1993) A PET study of visuospatial attention. J Neurosci 13:1202–1226

Cox RW (1996) AFNI: Software for analysis and visualization of functional magnetic resonance neuroimages. Comput Biomed Res 29:162–173

Craig AD (2009) How do you feel--now? The anterior insula and human awareness. Nat Rev Neurosci 10:59–70

Creem SH, Proffitt DR (2001) Defining the cortical visual systems: "what", "where", and "how". Acta Psychol (Amst) 107:43–68

Crick FC, Koch C (2005) What is the function of the claustrum? Philos Trans R Soc Lond B Biol Sci 360:1271–1279

Cross ES, Kraemer DJ, Hamilton AF, Kelley WM, Grafton ST (2008) Sensitivity of the action observation network to physical and observational learning. Cereb Cortex 19(2):315–326

Damasio H, Grabowski TJ, Tranel D, Hichwa RD, Damasio RD (1996) A neural basis for lexical retrieval. Nature 380:499–505

Davachi L (2006) Item, context and relational episodic encoding in humans. Curr Opin Neurobiol 16:693–700

De Renzi E, Faglioni P, Previdi P (1977) Spatial memory and hemispheric locus of lesion. Cortex 13:424–433

Doehrmann O, Naumer MJ (2008) Semantics and the multisensory brain: how meaning modulates processes of audio-visual integration. Brain Res 1242:136–150

Doehrmann O, Naumer MJ, Volz S, Kaiser J, Altmann CF (2008) Probing category selectivity for environmental sounds in the human auditory brain. Neuropsychologia 46:2776–2786

Downing PE, Jiang Y, Shuman M, Kanwisher N (2001) A cortical area selective for visual processing of the human body. Science 293:2470–2473

Driver J (1996) Enhancement of selective listening by illusory mislocation of speech sounds due to lip-reading. Nature 381:66–68

Engel LR, Frum C, Puce A, Walker NA, Lewis JW (2009) Different categories of living and non-living sound-sources activate distinct cortical networks. Neuroimage 47:1778–1791

Epstein R, Kanwisher N (1998) A cortical representation of the local visual environment. Nature 392:598–601

Ethofer T, Pourtois G, Wildgruber D (2006) Investigating audiovisual integration of emotional signals in the human brain. Prog Brain Res 156:345–361

Ettlinger G, Wilson WA (1990) Cross-modal performance: behavioural processes, phylogenetic considerations and neural mechanisms. Behav Brain Res 40:169–192

Falchier A, Clavagnier S, Barone P, Kennedy H (2002) Anatomical evidence of multimodal integration in primate striate cortex. J Neurosci 22:5749–5759

Fecteau S, Armony JL, Joanette Y, Belin P (2004) Is voice processing species-specific in human auditory cortex? An fMRI study. Neuroimage 23:840–848

Felleman DJ, van Essen DC (1991) Distributed hierarchical processing in the primate cerebral cortex. Cerebral Cortex 1:1–47

Formisano E, Kim DS, Di Salle F, van de Moortele PF, Ugurbil K, Goebel R (2003) Mirror-symmetric tonotopic maps in human primary auditory cortex. Neuron 40:859–869

Galati G, Committeri G, Spitoni G, Aprile T, Di Russo F, Pitzalis S, Pizzamiglio L (2008) A selective representation of the meaning of actions in the auditory mirror system. Neuroimage 40:1274–1286

Gale TM, Done DJ, Frank RJ (2001) Visual crowding and category specific deficits for pictorial stimuli: a neural network model. Cogn Neuropsychol 18:509–550

Gauthier I, Skudlarski P, Gore JC, Anderson AW (2000) Expertise for cars and birds recruits brain areas involved in face recognition. Nat Neurosci 3:191–197

Gazzola V, Aziz-Zadeh L, Keysers C (2006) Empathy and the somatotopic auditory mirror system in humans. Curr Biol 16:1824–1829

Gonzalo D, Shallice T, Dolan R (2000) Time-dependent changes in learning audiovisual associations: a single-trial fMRI study. Neuroimage 11:243–255

Goodale MA, Meenan JP, Bulthoff HH, Nicolle DA, Murphy KJ, Racicot CI (1994) Separate neural pathways for the visual analysis of object shape in perception and prehension. Curr Biol 4:604–610

Gray CM, Konig P, Engel AK, Singer W (1989) Oscillatory responses in cat visual cortex exhibit inter-columnar synchronization which reflects global stimulus properties. Nature 338:334–337

Grefkes C, Ritzl A, Zilles K, Fink GR (2004) Human medial intraparietal cortex subserves visuomotor coordinate transformation. Neuroimage 23:1494–1506

Grill-Spector K, Malach R (2004) The human visual cortex. Annu Rev Neurosci 27:649–677

Gron G, Wunderlich AP, Spitzer M, Tomczak R, Riepe MW (2000) Brain activation during human navigation: gender-different neural networks as substrate of performance. Nat Neurosci 3:404–408

Grossman ED, Blake R (2002) Brain areas active during visual perception of biological motion. Neuron 35:1167–1175

Hadjikhani N, Roland PE (1998) Cross-modal transfer of information between the tactile and the visual representations in the human brain: a positron emission tomographic study. J Neurosci 18:1072–1084

Hashimoto R, Sakai KL (2004) Learning letters in adulthood: direct visualization of cortical plasticity for forming a new link between orthography and phonology. Neuron 42:311–322

Hashimoto T, Usui N, Taira M, Nose I, Haji T, Kojima S (2006) The neural mechanism associated with the processing of onomatopoeic sounds. Neuroimage 31:1762–1770

Hasson U, Harel M, Levy I, Malach R (2003) Large-scale mirror-symmetry organization of human occipito-temporal object areas. Neuron 37:1027–1041

Haxby JV, Gobbini MI, Furey ML, Ishai A, Schouten JL, Pietrini P (2001) Distributed and overlapping representations of faces and objects in ventral temporal cortex. Science 293:2425–2430

Hein G, Doehrmann O, Muller NG, Kaiser J, Muckli L, Naumer MJ (2007) Object familiarity and semantic congruency modulate responses in cortical audiovisual integration areas. J Neurosci 27:7881–7887

Hocking J, Price CJ (2008) The role of the posterior superior temporal sulcus in audiovisual processing. Cereb Cortex 18:2439–2449
Hubbard EM, Piazza M, Pinel P, Dehaene S (2005) Interactions between number and space in parietal cortex. Nat Rev Neurosci 6:435–448
Iacoboni M, Molnar-Szakacs I, Gallese V, Buccino G, Mazziotta JC, Rizzolatti G (2005) Grasping the intentions of others with one's own mirror neuron system. PLoS Biol 3:529–535
Irving-Bell L, Small M, Cowey A (1999) A distortion of perceived space in patients with right-hemisphere lesions and visual hemineglect. Neuropsychologia 37:919–925
James TW, Gauthier I (2003) Auditory and action semantic features activate sensory-specific perceptual brain regions. Curr Biol 13:1792–1796
Jellema T, Perrett DI (2006) Neural representations of perceived bodily actions using a categorical frame of reference. Neuropsychologia 44:1535–1546
Johansson G (1973) Visual perception of biological motion and a model for its analysis. Percept Psychophys 14:201–211
Kaas JH, Hackett TA (1999) 'What' and 'where' processing in auditory cortex. Nat Neurosci 2:1045–1047
Kanwisher N, McDermott J, Chun MM (1997) The fusiform face area: a module in human extrastriate cortex specialized for face perception. J Neurosci 17:4302–4311
Kanwisher N, Downing P, Epstein R, Kourtzi Z (2001) Functional neuroimaging of visual recognition. In: Cabeza R, Kingstone A (eds) Handbook of functional neuroimaging of cognition. MIT Press, Cambridge, MA, pp 109–152
Kayser C, Logothetis NK (2007) Do early sensory cortices integrate cross-modal information? Brain Struct Funct 212:121–132
Kellenbach ML, Brett M, Patterson K (2003) Actions speak louder than functions: the importance of manipulability and action in tool representation. J Cogn Neurosci 15:30–46
Keysers C, Kohler E, Umilta A, Nanetti L, Fogassi L, Gallese V (2003) Audiovisual mirror neurons and action recognition. Exp Brain Res 153:628–636
Kiefer M, Sim EJ, Herrnberger B, Grothe J, Hoenig K (2008) The sound of concepts: four markers for a link between auditory and conceptual brain systems. J Neurosci 28:12224–12230
Kilner JM, Vargas C, Duval S, Blakemore SJ, Sirigu A (2004) Motor activation prior to observation of a predicted movement. Nat Neurosci 7:1299–1301
King AJ, Nelken I (2009) Unraveling the principles of auditory cortical processing: can we learn from the visual system? Nat Neurosci 12:698–701
Kohler E, Keysers C, Umilta A, Fogassi L, Gallese V, Rizzolatti G (2002) Hearing sounds, understanding actions: action representation in mirror neurons. Science 297:846–848
Kreifelts B, Ethofer T, Grodd W, Erb M, Wildgruber D (2007) Audiovisual integration of emotional signals in voice and face: an event-related fMRI study. Neuroimage 37:1445–1456
Krumbholz K, Schonwiesner M, von Cramon DY, Rubsamen R, Shah NJ, Zilles K, Fink GR (2005) Representation of interaural temporal information from left and right auditory space in the human planum temporale and inferior parietal lobe. Cereb Cortex 15:317–324
Lavenex P, Amaral DG (2000) Hippocampal-neocortical interaction: a hierarchy of associativity. Hippocampus 10:420–430
Lewis JW (2006) Cortical networks related to human use of tools. Neuroscientist 12:211–231
Lewis JW, Beauchamp MS, DeYoe EA (2000) A comparison of visual and auditory motion processing in human cerebral cortex. Cereb Cortex 10:873–888
Lewis JW, Phinney RE, Brefczynski-Lewis JA, DeYoe EA (2006) Lefties get it "right" when hearing tool sounds. J Cogn Neurosci 18(8):1314–1330
Lewis JW, Brefczynski JA, Phinney RE, Janik JJ, DeYoe EA (2005) Distinct cortical pathways for processing tool versus animal sounds. J Neurosci 25:5148–5158
Lewis JW, Wightman FL, Brefczynski JA, Phinney RE, Binder JR, DeYoe EA (2004) Human brain regions involved in recognizing environmental sounds. Cereb Cortex 14:1008–1021
Lewis JW, Talkington WJ, Walker NA, Spirou GA, Jajosky A, Frum C, Brefczynski-Lewis JA (2009) Human cortical organization for processing vocalizations indicates representation of harmonic structure as a signal attribute. J Neurosci 29:2283–2296

Lewkowicz DJ (2000) The development of intersensory temporal perception: an epigenetic systems/limitations view. Psycholog Bull 126:281–308

Liberman AM, Mattingly IG (1985) The motor theory of speech perception revised. Cognition 21:1–36

Lissauer H (1890/1988) A case of visual agnosia with a contribution to theory. Cogn Neuropsychol 5:157–192

Macaluso E (2006) Multisensory processing in sensory-specific cortical areas. Neuroscientist 12:327–338

Mahon BZ, Caramazza A (2005) The orchestration of the sensory-motor systems: clues from neuropsychology. Cogn Neuropsychol 22:480–494

Mahon BZ, Anzellotti S, Schwarzbach J, Zampini M, Caramazza A (2009) Category-specific organization in the human brain does not require visual experience. Neuron 63:397–405

Martin A (2001) Functional neuroimaging of semantic memory. In: Cabeza R, Kingstone A (eds) Handbook of functional neuroimaging of cognition. The MIT Press, Cambridge, MA, pp 153–186

Martin A (2007) The representation of object concepts in the brain. Annu Rev Psychol 58:25–45

Mayberry RI, Lock E, Kazmi H (2002) Linguistic ability and early language exposure. Nature 417:38

McCandliss BD, Cohen L, Dehaene S (2003) The visual word form area: expertise for reading in the fusiform gyrus. Trends Cogn Sci 7:293–299

McCarthy G, Puce A, Gore JC, Allison T (1997) Face-specific processing in the human fusiform gyrus. J Cogn Neurosci 9:605–610

McClelland JL, Rogers TT (2003) The parallel distributed processing approach to semantic cognition. Nat Rev Neurosci 4:310–322

McGurk H, MacDonald J (1976) Hearing lips and seeing voices. Nature 264:746–748

McNamara A, Buccino G, Menz MM, Glascher J, Wolbers T, Baumgartner A, Binkofski F (2008) Neural dynamics of learning sound-action associations. PLoS ONE 3:e3845

Mesulam MM (1998) From sensation to cognition. Brain 121:1013–1052

Meyer M, Baumann S, Marchina S, Jancke L (2007) Hemodynamic responses in human multisensory and auditory association cortex to purely visual stimulation. BMC Neurosci 8:14

Miller EK, Nieder A, Freedman DJ, Wallis JD (2003) Neural correlates of categories and concepts. Curr Opin Neurobiol 13:198–203

Moss HE, Rodd JM, Stamatakis EA, Bright P, Tyler LK (2005) Anteromedial temporal cortex supports fine-grained differentiation among objects. Cereb Cortex 15:616–627

Murray EA, Richmond BJ (2001) Role of perirhinal cortex in object perception, memory, and associations. Curr Opin Neurobiol 11:188–193

Murray EA, Bussey TJ, Hampton RR, Saksida LM (2000) The parahippocampal region and object identification. Ann N Y Acad Sci 911:166–174

Naghavi HR, Eriksson J, Larsson A, Nyberg L (2007) The claustrum/insula region integrates conceptually related sounds and pictures. Neurosci Lett 422:77–80

Naumer MJ, Doehrmann O, Muller NG, Muckli L, Kaiser J, Hein G (2008) Cortical plasticity of audio-visual object representations. Cereb Cortex 19:1641–1653

Neal JW, Pearson RCA, Powell TPS (1988) The cortico-cortical connections within the parieto-temporal lobe of area PG,7a in the monkey. Brain Res 438:343–350

Neville H, Bavelier D, Corina D, Rauschecker J, Karni A, Lalwani A, Braun A, Clark V, Jezzard P, Turner R (1998) Cerebral organization for language in deaf and hearing subjects: biological constraints and effects of experience. Proc Natl Acad Sci USA 95:922–929

Nieder A, Dehaene S (2009) Representation of number in the brain. Annu Rev Neurosci 32:185–208

Nishitani N, Hari R (2000) Temporal dynamics of cortical representation for action. Proc Natl Acad Sci USA 97:913–918

Noppeney U, Josephs O, Hocking J, Price CJ, Friston KJ (2008) The effect of prior visual information on recognition of speech and sounds. Cereb Cortex 18:598–609

Norman D, Shallice T (1986) Attention to action: willed and automatic control of behaviour. In: Davidson RJ, Schwartz GE, Shapiro D (eds) Consciousness and self regulation. Plenum, New York, pp 1–18

Olson IR, Gatenby JC, Gore JC (2002) A comparison of bound and unbound audio-visual information processing in the human cerebral cortex. Brain Res Cogn Brain Res 14:129–138

Pascual-Leone A, Hamilton R (2001) The metamodal organization of the brain. Prog Brain Res 134:427–445

Patterson K, Nestor PJ, Rogers TT (2007) Where do you know what you know? The representation of semantic knowledge in the human brain. Nat Rev Neurosci 8:976–987

Pelphrey KA, Morris JP, McCarthy G (2004) Grasping the intentions of others: the perceived intentionality of an action influences activity in the superior temporal sulcus during social perception. J Cogn Neurosci 16:1706–1716

Petacchi A, Laird AR, Fox PT, Bower JM (2005) Cerebellum and auditory function: an ALE meta-analysis of functional neuroimaging studies. Hum Brain Mapp 25:118–128

Pizzamiglio L, Aprile T, Spitoni G, Pitzalis S, Bates E, D'Amico S, Di Russo F (2005) Separate neural systems for processing action- or non-action-related sounds. Neuroimage 24:852–861

Polk TA, Stallcup M, Aguirre GK, Alsop DC, D'Esposito M, Detre JA, Farah MJ (2002) Neural specialization for letter recognition. J Cogn Neurosci 14:145–159

Rademacher J, Morosan P, Schormann T, Schleicher A, Werner C, Freund HJ, Zilles K (2001) Probabilistic mapping and volume measurement of human primary auditory cortex. Neuroimage 13:669–683

Rauschecker JP, Tian B (2000) Mechanisms and streams for processing of "what" and "where" in auditory cortex. Proc Natl Acad Sci USA 97:11800–11806

Rauschecker JP, Scott SK (2009) Maps and streams in the auditory cortex: nonhuman primates illuminate human speech processing. Nat Neurosci 12:718–724

Reale RA, Calvert GA, Thesen T, Jenison RL, Kawasaki H, Oya H, Howard MA, Brugge JF (2007) Auditory-visual processing represented in the human superior temporal gyrus. Neuroscience 145:162–184

Recanzone GH, Sutter ML (2008) The biological basis of audition. Annu Rev Psychol 59:119–142

Recanzone GH, Cohen YE (2009) Serial and parallel processing in the primate auditory cortex revisited. Behav Brain Res 206:1–7

Rilling JK (2008) Neuroscientific approaches and applications within anthropology. Am J Phys Anthropol Suppl 47:2–32

Rizzolatti G, Craighero L (2004) The mirror-neuron system. Annu Rev Neurosci 27:169–192

Robins DL, Hunyadi E, Schultz RT (2009) Superior temporal activation in response to dynamic audio-visual emotional cues. Brain Cogn 69:269–278

Rockland KS, Ojima H (2003) Multisensory convergence in calcarine visual areas in macaque monkey. Int J Psychophysiol 50:19–26

Roder B, Stock O, Bien S, Neville H, Rosler F (2002) Speech processing activates visual cortex in congenitally blind humans. Eur J Neurosci 16:930–936

Romanski LM, Tian B, Fritz J, Mishkin M, Goldman-Rakic PS, Rauschecker JP (1999) Dual streams of auditory afferents target multiple domains in the primate prefrontal cortex. Nat Neurosci 2:1131–1136

Rosch EH (1973) Natural categories. Cogn Psychol 4:328–350

Scheef L, Boecker H, Daamen M, Fehse U, Landsberg MW, Granath DO, Mechling H, Effenberg AO (2009) Multimodal motion processing in area V5/MT: evidence from an artificial class of audio-visual events. Brain Res 1252:94–104

Sestieri C, Di Matteo R, Ferretti A, Del Gratta C, Caulo M, Tartaro A, Olivetti Belardinelli M, Romani GL (2006) "What" versus "where" in the audiovisual domain: an fMRI study. Neuroimage 33:672–680

Skipper JI, van Wassenhove V, Nusbaum HC, Small SL (2007) Hearing lips and seeing voices: how cortical areas supporting speech production mediate audiovisual speech perception. Cereb Cortex 17:2387–2399

Stein BE, Meredith MA (1993) The merging of the senses. MIT Press, Cambridge, MA
Stevenson RA, James TW (2009) Audiovisual integration in human superior temporal sulcus: inverse effectiveness and the neural processing of speech and object recognition. Neuroimage 44:1210–1223
Talavage TM, Sereno MI, Melcher JR, Ledden PJ, Rosen BR, Dale AM (2004) Tonotopic organization in human auditory cortex revealed by progressions of frequency sensitivity. J Neurophysiol 91:1282–1296
Tanabe HC, Honda M, Sadato N (2005) Functionally segregated neural substrates for arbitrary audiovisual paired-association learning. J Neurosci 25:6409–6418
Taylor KI, Stamatakis EA, Tyler LK (2009) Crossmodal integration of object features: voxel-based correlations in brain-damaged patients. Brain 132:671–683
Taylor KI, Moss HE, Stamatakis EA, Tyler LK (2006) Binding crossmodal object features in perirhinal cortex. Proc Natl Acad Sci USA 103:8239–8244
Tettamanti M, Buccino G, Saccuman MC, Gallese V, Danna M, Scifo P, Fazio F, Rizzolatti G, Cappa SF, Perani D (2005) Listening to action-related sentences activates fronto-parietal motor circuits. J Cogn Neurosci 17:273–281
Thompson JC, Clarke M, Stewart T, Puce A (2005) Configural processing of biological motion in human superior temporal sulcus. J Neurosci 25:9059–9066
Tootell RB, Tsao D, Vanduffel W (2003) Neuroimaging weighs in: humans meet macaques in "primate" visual cortex. J Neurosci 23:3981–3989
Tranel D, Logan CG, Frank RJ, Damasio AR (1997) Explaining category-related effects in the retrieval of conceptual and lexical knowledge for concrete entities: operationalization and analysis of factors. Neuropsychologia 35:1329–1339
Tranel D, Damasio H, Eichhorn GR, Grabowski TJ, Ponto LLB, Hichwa RD (2003) Neural correlates of naming animals from their characteristic sounds. Neuropsychologia 41:847–854
Tyler LK, Moss HE (2001) Towards a distributed account of conceptual knowledge. Trends Cogn Sci 5:244–252
Tyler LK, Stamatakis EA, Bright P, Acres K, Abdallah S, Rodd JM, Moss HE (2004) Processing objects at different levels of specificity. J Cogn Neurosci 16:351–362
Ungerleider LG, Haxby JV (1994) 'What' and 'where' in the human brain. Curr Opin Neurobiol 4:157–165
Ungerleider LG, Mishkin M, Goodale MA, Mansfield RJW (1982) Two cortical visual systems. In: Ingle DJ (ed.) Analysis of visual behavior. MIT Press, Cambridge, MA, pp 549–586
Valenza N, Murray MM, Ptak R, Vuilleumier P (2004) The space of senses: impaired crossmodal interactions in a patient with Balint syndrome after bilateral parietal damage. Neuropsychologia 42:1737–1748
van Atteveldt N, Formisano E, Goebel R, Blomert L (2004) Integration of letters and speech sounds in the human brain. Neuron 43:271–282
van Atteveldt NM, Formisano E, Blomert L, Goebel R (2007) The effect of temporal asynchrony on the multisensory integration of letters and speech sounds. Cereb Cortex 17:962–974
van Atteveldt NM, Formisano E, Goebel R, Blomert L (2007) Top-down task effects overrule automatic multisensory responses to letter-sound pairs in auditory association cortex. Neuroimage 36:1345–1360
van Essen DC (1997) A tension-based theory of morphogenesis and compact wiring in the central nervous system. Nature 385:313–318
van Essen DC (2005) A population-average, landmark- and surface-based (PALS) atlas of human cerebral cortex. Neuroimage 28:635–662
van Essen DC, Drury HA, Dickson J, Harwell J, Hanlon D, Anderson CH (2001) An integrated software suite for surface-based analyses of cerebral cortex. J Am Med Inform Assoc 8:443–459
von Kriegstein K, Giraud AL (2006) Implicit multisensory associations influence voice recognition. PLoS Biol 4:e326

Vygotsky L (1978) Mind in society: the development of higher psychological processes. Harvard University Press, Cambridge, MA

Wang Y, Celebrini S, Trotter Y, Barone P (2008) Visuo-auditory interactions in the primary visual cortex of the behaving monkey: electrophysiological evidence. BMC Neurosci 9:79

Warren J, Zielinski B, Green G, Rauschecker J, Griffiths T (2002) Perception of sound-source motion by the human brain. Neuron 34:139–148

Watkins S, Shams L, Josephs O, Rees G (2007) Activity in human V1 follows multisensory perception. Neuroimage 37:572–578

Watkins S, Shams L, Tanaka S, Haynes JD, Rees G (2006) Sound alters activity in human V1 in association with illusory visual perception. Neuroimage 31:1247–1256

Wheaton KJ, Pipingas A, Silberstein RB, Puce A (2001) Human neural responses elicited to observing the actions of others. Vis Neurosci 18:401–406

Wheeler L, Griffin HC (1997) A movement-based approach to language development in children who are deaf-blind. Am Ann Deaf 142:387–390

Whiten A, Horner V, Litchfield CA, Marshall-Pescini S (2004) How do apes ape? Learn Behav 32:36–52

Zilbovicius M, Meresse I, Chabane N, Brunelle F, Samson Y, Boddaert N (2006) Autism, the superior temporal sulcus and social perception. Trends Neurosci 29:359–366

Chapter 11
Single-Trial Multisensory Learning and Memory Retrieval

Micah M. Murray and Holger F. Sperdin

11.1 Background

Multisensory research on object processing has generally focused on how information from one sensory modality (e.g. audition) can impact the processing of simultaneously presented (and often co-localized) information from another sensory modality (e.g. vision) (e.g. Amedi et al., 2005). Substantially less consideration has been given to how multisensory information processing at one point in time might subsequently affect unisensory processing and behaviour. Yet, this kind of situation is commonplace. For example, you meet someone for the first time, speak and see him, and later recognize his face at a party. Similarly, acquiring fluent reading skills initially involves ascribing sounds to written letters, but later progresses to whole-word reading without need for vocalization.

Similarly, the neurophysiological mechanisms of learning and memory processes have almost exclusively been investigated under unisensory conditions. One example is found in studies of repetition priming and repetition suppression showing that behaviour and brain responses change with repeated exposure to the same or similar stimuli (e.g. Desimone, 1996; Grill-Spector et al., 2006; Murray et al., 2008a; Tulving and Schacter, 1990; Wiggs and Martin, 1998). Repetition suppression is typically considered as an index of plasticity in that it reflects experience-dependent modulation in neural activity that supports perceptual learning (e.g. Desimone, 1996). At present, the precise mechanism(s) leading to this reduced activity remain unresolved, and candidate mechanisms include (but are not limited to): fatigue, sharpening, and facilitation (reviewed in Grill-Spector et al., 2006). Despite the uncertainty concerning the precise mechanism, it is commonly agreed that repetition

M.M. Murray (✉)
Centre for Biomedical Imaging, Department of Clinical Neurosciences, Department of Radiology, Vaudois University Hospital Centre and University of Lausanne, BH08.078, Rue du Bugnon 46, 1011, Lausanne, Switzerland
Department of Hearing and Speech Sciences, Vanderbilt University Medical Center, Nashville, Tennessee, USA
e-mail: micah.murray@chuv.ch

M.J. Naumer, J. Kaiser (eds.), *Multisensory Object Perception in the Primate Brain*,
DOI 10.1007/978-1-4419-5615-6_11,

suppression is one indication of a change in the cortical representation of the stimulus/object (e.g. Desimone, 1996; Henson and Rugg, 2003; Wiggs and Martin, 1998).

Some investigations have examined how experiences in one or multiple senses alter later processing of stimuli of another sensory modality. These studies provide evidence that brain regions involved in an experience's encoding can also be involved during its subsequent active retrieval (e.g. James et al., 2002; Nyberg et al., 2000; Wheeler et al., 2000; see also von Kriegstein and Giraud, 2006). In these studies, subjects learned auditory–visual or visual–visual associations during a separate session and later classified visual stimuli according to the sensory modality with which it initially appeared (Nyberg et al., 2000; Wheeler et al., 2000). During a test session auditory regions were active in response to those visual stimuli that had been presented with sounds during study sessions. This activity was taken as support for the psychological postulate of 'redintegration' (Hamilton, 1859), wherein a component part is sufficient to (re)activate the whole experience's consolidated representation. That is, a visual stimulus that had been studied and thus associated with a sound (and presumably formed a consolidated representation with that sound) could elicit activity within auditory cortices when participants actively remembered the initial encoding context.

A similar line of evidence is found in intracranial microelectrode recordings in monkeys that performed a delayed match-to-sample task. These studies demonstrate that there is selective delay-period activity with visual–visual, somatosensory–visual, and auditory–visual paired associates (e.g. Colombo and Gross, 1994; Gibson and Maunsell, 1997; Haenny et al., 1988; Maunsell et al., 1991; see also Guo and Guo, 2005 for an example in drosophila). In these studies, responses were elicited in cortical areas involved in visual object recognition (i.e. areas V4 and IT) by non-visual stimuli and were selective for specific associations among the learned set. In these studies, like those above, extensive studying of the paired associations was performed prior to testing, though it is worth mentioning that the stimulus set did not include environmental objects. The effects manifested as a modulation in the response profile within the visual system (though it should be mentioned that no recordings were concurrently performed in the auditory cortices).

One implication of these collective data is that prior multisensory experiences can influence and be part of memory functions such that when an association is formed between sensory modalities for a given object, presentation of the stimulus in just one sensory modality can alter the activity in regions typically implicated in the processing of the other, non-stimulated sensory modality. That is, responses to an incoming stimulus may vary, either in terms of their pattern within a region or overall activated network, according to whether it is part of a multisensory or unisensory memory. However, the above-mentioned literature is ambiguous on several issues.

First, these studies do not demonstrate whether multisensory experiences influence subsequent behaviour with unisensory stimuli, leaving it unknown whether such multisensory interactions are behaviourally relevant. Second, because these studies all involved extensive training with or exposure to the multisensory stimuli, the requisite conditions for eliciting such effects are unknown. This issue is further exacerbated in those studies requiring overt discrimination of stimuli that had been learned in a multisensory context from those that were not, thereby adding the possibility for a role of mental imagery. That is, to complete the task participants could reasonably adopt a strategy of thinking of sounds that correspond to a given picture or could actively recall the context under which a given picture had been encountered. Similarly, it was not addressed whether any effects on memory retrieval were simply a consequence of altered encoding processes with multisensory vs. unisensory stimuli. Finally, these previous studies either lacked adequate temporal resolution or had limited spatial sampling to provide information concerning where or when such effects *first* occur (either in terms of time post-stimulus or in terms of levels of processing). Identification of the earliest effects can be used to place critical limits on mechanisms of multisensory memory retrieval.

In a series of experiments, we first examined the time course and initial locus of incidental effects of past multisensory experiences on current unisensory responses when subjects neither explicitly studied multisensory associations nor later classified stimuli according to these associations. We then examined the necessary conditions for eliciting such effects.

11.2 Findings

Our studies used a continuous recognition task that required participants to indicate whether the image presented on each trial was novel or had already been shown previously during the block of trials. In this way, trials could be subdivided at one level between initial and repeated presentations. In addition, the initial trials were further subdivided between those containing only visual information and those consisting of auditory–visual multisensory information (Fig. 11.1). Importantly, all repeated stimuli consisted only of images. In this way, there were repeated images that had been previously presented visually (hereafter termed V−) and repeated images that had been previously presented in a multisensory context (hereafter termed V+)[1]. Moreover, across our studies we manipulated whether multisensory presentations were always semantically congruent (i.e. an image with its corresponding environmental sound), varied in their semantic congruence, or were instead presented with a meaningless pure tone that was in turn paired with multiple images (see Fig. 11.2). In what follows, we first present and discuss the psychophysical findings from our studies and then the data concerning modulations in brain activity.

11.2.1 Multisensory Experiences Impact Subsequent Visual Memory Performance

In a first study, stimuli had been presented as a semantically congruent multisensory stimulus on initial trials (Murray et al., 2004). We observed that memory performance (i.e. correctly indicating that the image had been previously presented) was

[1] An alternative viewpoint is that the V+ condition is not a pure repetition, but rather a novel context (see Section 2.3 for discussion).

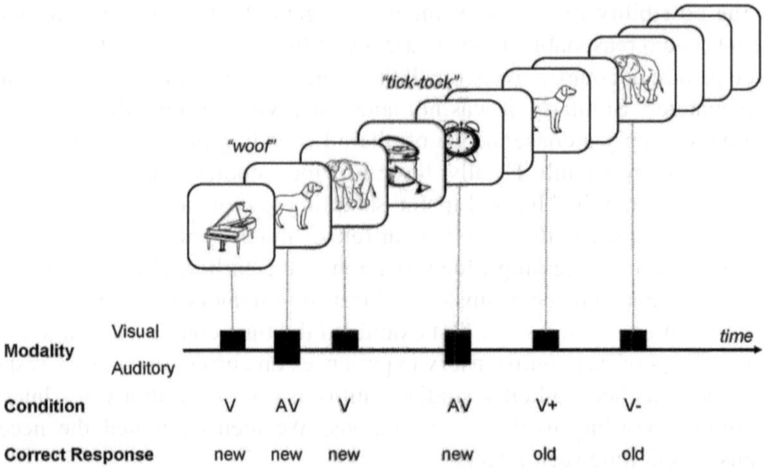

Fig. 11.1 Illustration of the continuous recognition task used in our work. In this paradigm participants indicate whether each image is being presented for the first or repeated time. Stimuli are presented for 500 ms. Initial presentations are divided between those containing only images (V condition) and those presented with sounds (AV condition). Repeated presentations consist only of images, but can be divided between those that had been initially presented as images only (V− condition) and those that had been initially presented with sounds (V+ condition). In this way, contrasting performance and/or brain activity from the V− and V+ conditions reveals effects of past multisensory experiences on current unisensory (visual) processing. In the example illustrated here, all sounds were semantically congruent environmental objects. See Fig. 11.2 for additional variations

significantly higher for V+ than V− trials ($t_{(10)} = 3.18$; $p = 0.010$; $\eta_p^2 = 0.503$). There was no evidence of differences in reaction times ($p > 0.85$), suggesting against an account based on alerting (see also Section 11.2.3 for a fuller discussion). This improved accuracy provides an indication that visual stimuli are differentially processed according to their past single-trial multisensory vs. unisensory experiences. Moreover, in this study, initial and repeated presentations were separated by 13 ± 3 trials (roughly equivalent to 25 s). In a subsequent study (Murray et al., 2005a) we replicated and extended these findings to show that memory performance was again enhanced for V+ trials even when participants completed the task within a noisy MR scanner environment and when the average temporal delay between initial and repeated presentations was nearly doubled (~50 s). As before, accuracy rates were significantly increased for V+ than V− trials ($t_{(7)} = 2.76$; $p = 0.028$; $\eta_p^2 = 0.521$), despite no difference in reaction times ($p > 0.45$). In a final study (Lehmann and Murray, 2005) we conducted two experiments to examine the requisite conditions for this enhanced memory performance. First, we examined whether purely episodic multisensory events would be sufficient by pairing half of the initial image presentations with an identical pure tone, rather than with a semantically congruent environmental sound as in our prior work. This manipulation led to a significant performance impairment on V+ relative to V− trials

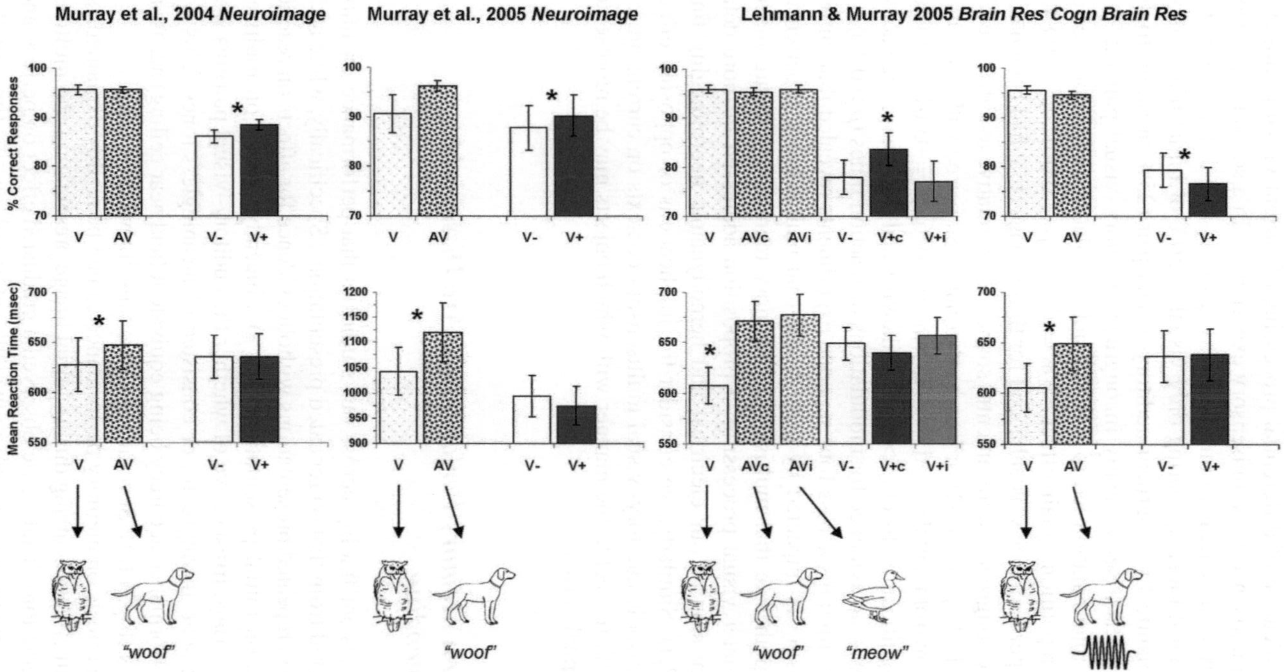

Fig. 11.2 Behavioural findings. The *top* set of bar graphs displays the mean (s.e.m. indicated) accuracy rates on the continuous recognition task for each experimental condition. The *bottom* set of bar graphs displays the mean (s.e.m. indicated) reaction times. Bars with *dotted fills* refer to initial image presentations under various conditions, whereas *solid* bars refer to repeated image presentations. An *asterisk* indicates a significant difference ($p < 0.05$; see Section 11.2 of the text for details). When the *asterisk* is positioned over the middle of a bar it indicates to have a main effect in the ANOVA as well as statistically reliable effects in the post-hoc contrasts

($t_{(15)} = 2.24$; $p = 0.041$; $\eta_p^2 = 0.250$) with no evidence of a reaction time difference ($p > 0.75$). However, it is important to note that this impairment nonetheless provides an indication of differential processing of current visual information according to past multisensory vs. unisensory experiences. In a second experiment we examined the role of semantic congruence by dividing initial image presentations into three groups: those appearing only visually (50% of initial trials), those appearing as a semantically congruent auditory–visual pair (25% of initial trials), and those appearing as a semantically incongruent auditory–visual pair (25% of initial trials). This manipulation led to a significant modulation in memory performance with image repetitions (main effect $F_{(2,9)} = 23.95$; $p < 0.001$; $\eta_p^2 = 0.842$). More specifically, performance was enhanced for those images that had been paired with a semantically congruent environmental sound (V+c) relative to either the V− condition ($t_{(10)} = 4.01$; $p = 0.002$) or images that had been paired with a semantically incongruent environmental sound ($V + i$; $t_{(10)} = 5.036$; $p = 0.001$); the latter of which did not significantly differ ($p > 0.65$). Likewise, and as we have seen across our studies, there was no evidence for modulations of reaction times ($p > 0.10$).

These collective findings across four separate experiments revealed that current unisensory processing is impacted by past, single-trial multisensory experiences. This led us to hypothesize that multisensory memory representations are indeed accessible during later visual processing; a proposition addressed in more detail below in our discussion of our electrical and hemodynamic neuroimaging findings (Section 11.2.4). Similarly, we consider the implications of opposing effects of semantic and episodic auditory–visual multisensory contexts on current unisensory processing with respect to the manner with which objects may be represented within lateral occipital cortices.

11.2.2 Effects on Memory Performance Are Dissociable from Encoding

Importantly, a consistent finding across our studies is that performance on initial presentations differed from that on repeated presentations. Specifically, while accuracy was affected on repeated presentations without evidence for effects on reaction time, performance on initial presentations was significantly slowed on multisensory vs. unisensory trials, irrespective of whether the auditory–visual pairings were semantically congruent, incongruent, or consisted of meaningless tones. This was the case despite performance accuracy being equivalent (and near ceiling) across all initial presentations (Fig. 11.2). Several points are thus noteworthy.

First, the above effects on memory performance cannot be readily explained as a direct transfer of an effect occurring during initial image presentation and multisensory encoding/interactions. That is, we observed a graduated effect on performance accuracy with image repetitions despite equivalent patterns of reaction times on their initial presentations. However, we would immediately add that we cannot exclude the possibility that equivalent performance measures are nonetheless leading to differential brain processes; something that additional brain imaging studies will need to address. Such being said, this pattern of results does provide evidence dissociating effects of multisensory interactions from memory performance with repetitions of unisensory components.

Second, participants were clearly unaware that the mere presence of sounds reliably indicated that the image was novel. Had this been the case, then faster reaction times would have been expected for initial presentations consisting of multisensory stimuli than unisensory (visual) stimuli. This pattern of results thus suggests that our memory effects are occurring not only orthogonally to task requirements (i.e. both V+ and V− trials required an 'old' response; see Fig. 11.1) but also incidentally and perhaps automatically.

11.2.3 The Role of Attention, Alerting, and Novelty

While we would propose that these effects follow from distinct neural representations of multisensory and unisensory experiences that are formed by single-trial exposures and later accessible during subsequent unisensory processing, it is worthwhile to also consider some alternative accounts. One possibility is that these effects are the consequence of selective attention to the auditory channel and/or novel contexts (e.g. Ranganath and Rainer, 2003; Tsivilis et al., 2001). We can exclude these possibilities because such accounts would have predicted faster and/or more accurate performance on initial multisensory presentations, particularly because the mere presence of non-visual information would have been a sufficient cue to indicate a novel image presentation (see Chen and Yeh, 2008 for an example of a sound-induced reduction in repetition blindness during a rapid serial visual presentation paradigm). That is, on the basis of selectively attending to audition, subjects would have been able to more accurately and rapidly indicate an image's initial presentation (for multisensory vs. unisensory trials). Such a pattern was not observed in any of our experiments. A similar argument applies to an explanation in terms of general alerting, wherein multisensory events would have been predicted to produce the fastest behaviour. Rather, the pattern of reaction times on initial stimulus presentations fits well with results suggesting that events in an unexpected modality during a discrimination task can lead to slowed reaction times (Spence et al., 2001). However, this variety of selective attention still would not account for the performance pattern observed with repeated image presentations, particularly those where the semantic congruence was varied (cf Experiment 2 in Lehmann and Murray, 2005). In addition, effects of general arousal and fatigue cannot readily account for our results, because the experimental design included a nearly homogenous distribution of the different stimulus conditions throughout blocks of trials. Thus, even if subjects were more engaged in the task during the beginning of a block of trials, this would have applied equally to all stimulus conditions.

11.2.4 Visual Stimuli Are Rapidly Discriminated Within Lateral Occipital Cortices According to Past Multisensory Experiences

In addition to the above behavioural effects, brain responses significantly differed between images that had been initially presented in a multisensory context and those that had only been presented visually, though to date we have only examined the situation where the initial multisensory exposure entailed a semantically congruent pairing. In a first study that included collection of 128-channel visual evoked potentials (Murray et al., 2004), we showed that brain responses to V+ and V− conditions first differed over the 60–136 ms post-stimulus period (Fig. 11.3). Importantly, the use of electrical neuroimaging analyses (detailed in Michel et al., 2004; Murray et al., 2008b) allowed us to determine that this response difference followed from a change in the electric field topography at the scalp, rather than a modulation in the strength of the response. That is, responses to the V+ and V− conditions differed in terms of the configuration of the generators active over this time period. In other words, different sets of brain regions were active at 60–136 ms

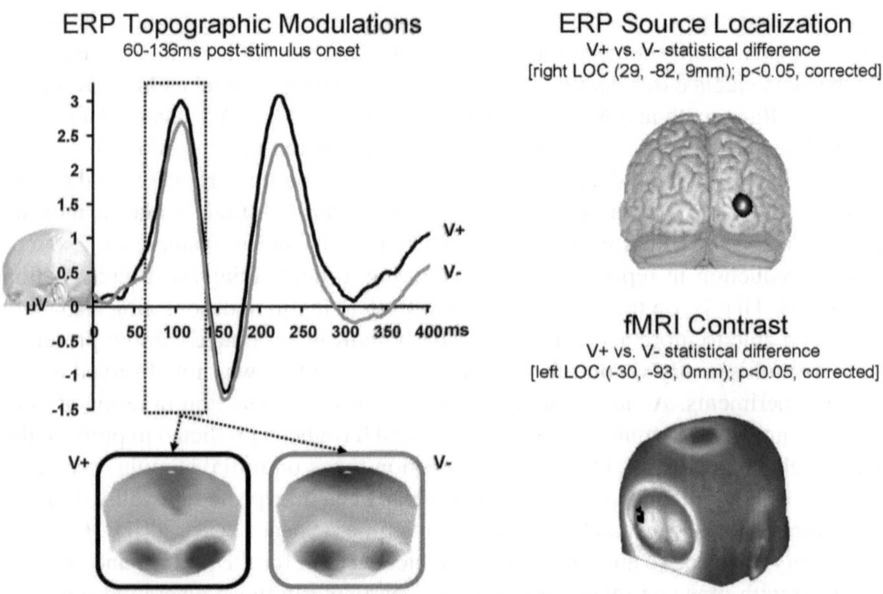

Fig. 11.3 Electrophysiologic and hemodynamic imaging results. The *left panel* displays group-averaged event-related potential waveforms from an exemplar posterior scalp site in response to the V+ and V− condition in our original study (Murray et al., 2004). The *dotted box* highlights the 60–136 ms post-stimulus period during which the topography significantly differed between conditions, indicative of configuration differences in the underlying sources. The *upper right panel* illustrates the locus of significant differences in the source estimations over this time period as being within the right lateral occipital cortex. In a subsequent fMRI study (Murray et al., 2005a), the V+ vs. V− contrast resulted in activation differences within the left lateral occipital cortex

post-stimulus onset depending on whether or not the incoming visual stimulus had been initially encountered in a multisensory context. Moreover, our source estimations and statistical analysis thereof indicated that distinct subsets of lateral occipital cortices mediated this early effect. We then conducted an event-related fMRI study at 1.5 T to both confirm the localization provided by our source estimations and also address discrepancies between our work and prior hemodynamic imaging studies (most notably those of Nyberg et al., 2000; Wheeler et al., 2000). As already detailed in Section 11.2.1 above, we were able to replicate our behavioural findings despite the modifications to the paradigm necessitated by fMRI constraints (i.e. the additional time between trials and the additional acoustic noise from the MR gradients). Additionally, we replicated our observation of response modulations within lateral occipital cortices between V+ and V− conditions. Both the electrical and the hemodynamic imaging results thus converge to reveal differential involvement of lateral occipital cortices in responding to images that had been presented previously as a multisensory pair. By contrast, there was no evidence in either study for distinct activity in response to those images that had been present previously in a visual-only context. Lastly, while the effects were lateralized (to the right hemisphere in the EEG study and to the left hemisphere in the fMRI study) with the statistical threshold applied, a more relaxed threshold revealed bilateral effects in both cases. Thus, we are hesitant to speculate regarding any possible lateralization.

Aside from effects at 60–136 ms, we observed additional later periods of response modulation at ∼210–260 and ∼318–390 ms that were each characterized by scalp topographic changes and, by extension, alterations in the configuration of the underlying active generators. The functional role of these later modulations is less clear. Given that the brain already distinguishes between stimuli having either unisensory or multisensory pasts over the 60–136 ms period and that this distinction is task irrelevant, these later periods of response modulation might reflect the brain's treatment of such incidental memory activations and their integration with current behavioural or task requirements. Indeed, their timing is consistent with those found in a number of recent ERP investigations of object recognition and identification processes (e.g. Doniger et al., 2000, 2001; Ritter et al., 1982; Vanni et al., 1996), recognition memory (e.g. Rugg et al., 1998; Tsivilis et al., 2001), as well as the discrimination of memories that pertain to ongoing reality from those that do not (Schnider et al., 2002). We would speculate that these later modulations might be serving these or similar functions.

11.3 Implications

Our principal finding across these studies is that past multisensory experiences influence both the ability to accurately discriminate image repetitions during a continuous recognition task and brain responses to image repetitions – thereby extending the effects of multisensory interactions across a substantially longer timescale than previously considered. This discrimination was according to past multisensory vs. unisensory experiences, during the task itself, and was influenced by both episodic (i.e. the simple co-occurrence of an unrelated, meaningless tone) and semantic (i.e. the co-occurrence of meaningful object stimuli) auditory–visual memory traces. Specifically, accuracy in indicating image repetitions (1) was significantly impaired for those images that had been presented with a 1000 Hz tone, (2) was not significantly affected for those images that had initially been presented with a semantically incongruent sound, and (3) was selectively improved for images initially presented with a semantically congruent sound. Such performance changes were relative to repetition discrimination accuracy with those images initially presented only visually. These effects provide some indications concerning the necessary conditions for multisensory perceptual/memory traces to be established and later accessed upon the repeated presentation of unisensory visual stimuli. The collective results reveal opposing effects of episodic and semantic contexts from auditory–visual multisensory events. Additionally, the effect of episodic multisensory contexts on later unisensory discrimination would appear to be limited to auditory–visual experiences. We have conducted some initial investigations into effects of somatosensory combinations, but have thus far obtained null results; cf Lehmann and Murray, 2005. Nonetheless, both unisensory visual and multisensory auditory–visual traces are accessible for processing incoming stimuli and indeed result in distinct treatment of incoming visual information during the initial stages of sensory-cognitive processing and within higher tier regions of visual cortex.

The incidental and single-trial underpinnings of our behavioural effects and differential brain activity within lateral occipital cortices sharply contrast with previous hemodynamic imaging studies of multisensory memory processes that observed activity within auditory (Nyberg et al., 2000; Wheeler et al., 2000) or olfactory (Gottfried et al., 2004) cortices in response to visual stimuli that had been learned during the experiment as multisensory pairs. For one, we observed that repetition discrimination was significantly improved for stimuli with (semantically congruent) multisensory, rather than unisensory, initial exposures. This behavioural improvement indicates that multisensory memory representations are established after single-trial exposure and are later accessible to facilitate memory. By contrast, the studies of Nyberg et al. (2000) and Wheeler et al. (2000) provide no behavioural evidence that past multisensory experiences were actually beneficial. Rather, the data of Nyberg and colleagues suggest relatively poorer performance for words that had been paired with sounds, with no condition-wise behavioural results reported by Wheeler and colleagues. One possibility, then, is that auditory areas were differentially (re)activated in these studies as a form of mental imagery (i.e. imagining sounds that correspond to the presented image) to either perform the active discrimination required of the participants and/or to compensate for the increased task difficulty. It is likewise noteworthy that the Nyberg et al. study limited their localization to regions defined by the statistical mask generated from the contrast between the auditory–visual and visual encoding conditions. Consequently, the full (spatial) extent of brain regions involved in their task is not readily apparent. Nonetheless, these authors considered the activation of auditory cortices in response to visual stimuli that had been learned under a multisensory context as neurophysiological support for the psychological construct of redintegration (Hamilton, 1859). Under this schema visual stimuli could reactivate associated sound representations within auditory cortex because the visual–auditory associations had been consolidated in memory.

Incorporating our findings into this framework would instead suggest that redintegration processes might also manifest without explicit consolidation of auditory–visual associations and first within regions involved in multisensory interactions. This is because the design of the continuous recognition task used in our work did not permit extensive studying of the multisensory associations. There were only single-trial exposures, and the initial and repeated presentations were pseudorandomly intermixed. More generally, the observed performance facilitation (and impairment) does not appear to be contingent upon extensive or explicit encoding.

We propose that the effects at 60–136 ms post-stimulus onset within regions of the lateral occipital cortex reflect the rapid reactivation of distinct multisensory and unisensory perceptual traces established during initial stimulus presentation. At least three prerequisites are satisfied that support this proposition, namely (1) auditory–visual convergence and interactions can occur within lateral occipital cortices, (2) multisensory memory representations are both localized and distinguishable from their unisensory counterparts within lateral occipital cortices, and (3) sensory responses can propagate to and differ within the LOC within the latency of the present effects.

Regarding the first prerequisite, hemodynamic and electrical neuroimaging studies of humans provide evidence of both auditory–visual (Calvert et al., 1999, 2000, 2001; Fort et al., 2002a, b; Giard and Perronet, 1999; MacSweeny et al., 2002; Molholm et al., 2002; Raij et al., 2000) and tactile–visual (Amedi et al., 2001, 2002; Deibert et al., 1999; James et al., 2002; Stoesz et al., 2003; Zangaladze et al., 1999) multisensory convergence in the LOC and other nearby visual cortices, including even primary visual cortices (e.g. Martuzzi et al., 2007; Romei et al., 2007; Wang et al., 2008). Additional support for the role of higher tier object recognition areas in multisensory interactions is found in microelectrode recordings from monkey inferotemporal (IT) cortex, for which the LOC is considered to be the human homologue, as well as visual area V4. In these studies, selective delay-period responses on a delayed match-to-sample task were observed for specific multisensory and unisensory paired associates (e.g. Colombo and Gross, 1994; Gibson and Maunsell, 1997; Haenny et al., 1988; Maunsell et al., 1991; see also Goulet and Murray, 2001). The selectivity of these responses indicates that neurons within these regions distinguish unisensory stimuli according to their learned association with another stimulus of the same or different sensory modality. Moreover, neurons showing selective responses for multisensory associations did not show selective responses to other unisensory associations (Gibson and Maunsell, 1997). The suggestion is that there are distinct neural responses to and perhaps also distinct representations of unisensory and multisensory associations within IT cortex, which would satisfy the second prerequisite described above. Finally, additional evidence speaks to the rapid time course of visual discrimination capabilities. During active discrimination, responses likely originating within object recognition areas modulate to specific classes of visual stimuli within the initial ~100 ms post-stimulus onset (e.g. Braeutigam et al., 2001; Debruille et al., 1998; George et al., 1997; Halgren et al., 2000; Landis et al., 1984; Mouchetant-Rostaing et al., 2000a, b; Seeck et al., 1997; Thierry et al., 2007), thereby supporting the third prerequisite described above.

Our results extend upon existing multisensory research to suggest that the multisensory representations established within these visual regions involved in neural response interactions are later accessible during subsequent visual stimulation for rapid stimulus discrimination, even though such is unnecessary for task completion. While such does not exclude additional brain regions from either multisensory or memory functions, the present data would indicate that the earliest discrimination of visual stimuli according to past experiences is within nominally visual cortices, rather than within other cortices typically attributed to auditory functions or other higher order associative, memory, or executive functions. While such awaits empirical support, we would speculate that effects of prior multisensory experience on auditory object repetition might result in effects within regions of the middle temporal cortex that our prior research has shown to be involved in early stages of auditory object discrimination (Murray et al., 2006, 2008a).

We propose that these traces reflect the consequences of object-based multisensory interactions, which may be distinct from interactions observed between more rudimentary stimuli (e.g. Besle et al., 2009). By the latter we refer to interactive, non-linear effects observed between sensory stimuli (e.g. visual flashes, auditory beeps, and somatosensory pulses), often during conditions of passive presentation or simple detection (e.g. Cappe et al., 2009; Foxe et al., 2000; Giard and Perronet, 1999; Murray et al., 2005b; Sperdin et al., 2009). These varieties of multisensory interactions are thought to depend on overlapping spatial receptive fields and temporal response profiles (reviewed in Stein and Meredith, 1993). In contrast to these effects, we would propose that object-based multisensory interactions are particularly sensitive to the identity and semantic attributes of stimuli (e.g. Doehrmann and Naumer, 2009; Laurienti et al., 2004), which in the present study are principally conveyed within the visual modality. In support of this distinction, recent studies show sensitivity to the semantic attributes of stimuli – i.e. whether information conveyed to the different senses derives from the same object – in generating multisensory effects. Some have shown that brain responses within sensory-related cortices modulate in strength as a function of such semantic congruence (e.g. Beauchamp et al., 2004; Calvert, 2001; Hein et al., 2007; Laurienti et al., 2003, 2004; Molholm et al., 2004). Others show that the timing of effects varies with semantic congruence (Fort et al., 2002a; Molholm et al., 2004). At present, however, the extent to which these modulations are directly linked to behaviour remains to be fully established, but will be important for interpreting psychophysical effects as reflecting perceptual vs. decision/response modulations.

Here, we propose the following model to account for the establishment of object-based perceptual/memory traces that yield the pattern of effects observed in the present study (Fig. 11.4). This proposition is predicated on the notion of distributed

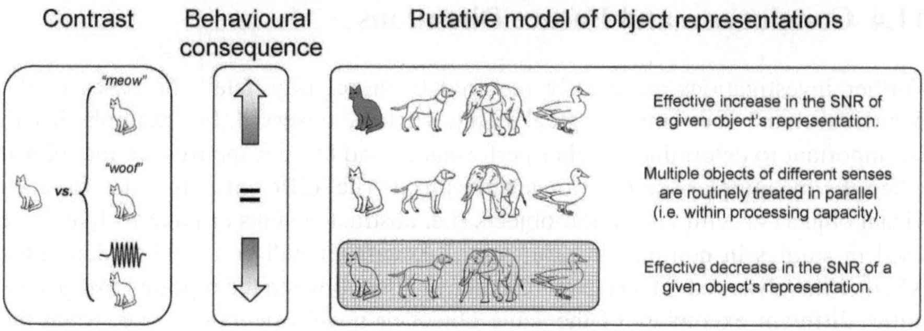

Fig. 11.4 A putative model of object representations to account for the behavioural findings. The *left panel* depicts various contrasts between unisensory (visual) and multisensory past experiences. The *centre panel* depicts the consequence of these past experiences on image repetition discrimination. The *right panel* depicts and describes how object representations might be affected by past multisensory experiences

object representations (e.g. Haxby et al., 2001; Rolls et al., 1997) and the hypothesis that semantic influences from the different senses modulate the fidelity with which these object representations can be (re)activated. In the case of semantically congruent auditory–visual objects, distinct perceptual/memory traces can be established that can be rapidly reactivated upon subsequent presentation of the visual component. This may arise through the enhanced activation of a singular object representation (e.g. 'cat') via multiple, redundant sources that in turn effectively yields a higher signal-to-noise ratio (in terms of the object's representation relative to other object representations) when the object system is confronted with repetition of just the visual component. That this reactivation has been observed to begin just 60 ms post-stimulus onset during the processing of image repetitions provides one indication of the particular fidelity that such perceptual/memory traces may have and that access to/influences from these traces need not be restricted to later, higher order processing stages. In the case of semantically incongruent pairs, no such enhanced trace is established relative to that established under unisensory conditions (and by extension no behavioral effect is observed). That is, the visual stimulus activates one object representation and the sound an altogether other. However, since multiple objects are routinely treated in parallel (i.e. visual scenes seldom include solitary objects; e.g. Rousselet et al., 2002) and since multiple objects can be simultaneously encoded via distributed neuronal representations (Rolls et al., 1997), no modulation in performance is observed relative to the visual-only condition. In the case of images paired with tones, we would propose that the association of a single sound with multiple visual objects over trials effectively leads to the introduction of 'noise' into the establishment of distinct multisensory perceptual/memory traces for individual objects. That is, the same sound has produced an interactive response with several different object representations. Consequently, the fidelity of perceptual/memory traces for these objects is diminished and comparatively impaired performance is obtained.

11.4 Conclusions and Future Directions

Further investigations are clearly required to more fully detail the bases of the behavioural and neurophysiological effects we have observed. For example, it will be important to determine whether performance and brain responses are modulated when the meaningless sound varies across trials (i.e. different pure tones for each visual object) or with visual non-objects (i.e. abstract designs or patterns like those used in studies in non-human primates). Likewise, it will be useful to determine whether the observed effects can be similarly elicited when the repeated images are either different exemplars of the same object or novel orientations (i.e. when the semantic attribute remains constant, but the physical features change), in order to better ascertain the precise nature of the representation accessed during retrieval. Nonetheless, irrespective of the direction of the change in performance accuracy, the present data suggest that unisensory percepts trigger the multisensory representations associated with them, which can be formed without explicit study or active classification. In this manner, our findings indicate how multisensory interactions can have long-term effects.

Of similar interest is whether homologous effects can be attained during object recognition within other sensory modalities (e.g. audition or touch). Along these lines, von Kriegstein and Giraud (2006) showed that voice recognition was improved when their subjects learned to associate faces with a voice relative to when they learned associations between names and voices. By contrast, there was no evidence that this benefit extended to other acoustic stimuli (ring tones of mobile telephones) that were either associated with images of the telephones or with brand names. Moreover, fMRI-based brain responses to voices that had been learned with faces resulted in enhanced activity within voice sensitive auditory cortices within the right middle temporal convexity as well as enhanced activity within the fusiform face area. These results, like our own findings, indicate that behaviour and brain activity in response to unisensory stimuli can be enhanced by past multisensory experiences. Given these results, however, it would be intriguing to assess whether larger effects than those we obtained with line drawings can be obtained using more realistic stimuli such as faces that had been previously presented either alone or with a voice. Another dimension that could be added would be to parametrically vary the emotional context of either the initial exposure (either visually and/or acoustically) or the repeated presentation. Finally, it would likewise be of interest to evaluate the impact of encoding conditions by parametrically varying the efficacy of the auditory and visual stimuli during initial presentations. One might anticipate there to be greater effects on subsequent memory performance if the multisensory interactions during encoding are themselves enhanced in a manner analogous to the principle of inverse effectiveness (Holmes, 2009; Stein and Meredith, 1993).

In conclusion, the findings reviewed in this chapter highlight the functional efficacy of multisensory learning on performance and brain activity not only when the multisensory associations are explicitly learned but also when such associations are formed incidentally after single-trial exposure. The growing interest in multisensory learning (e.g. Naumer et al., 2009; Shams and Seitz, 2008 for review) is not only opening new lines of basic research but also strategies for education and clinical rehabilitation.

References

Amedi A, von Kriegstein K, van Atteveldt NM, Beauchamp MS, Naumer MJ (2005) Functional imaging of human crossmodal identification and object recognition. Exp Brain Res 166: 559–71

Amedi A, Jacobson G, Hendler T, Malach R, Zohary E (2002) Convergence of visual and tactile shape processing in the human lateral occipital complex. Cereb Cortex 12:1202–1212

Amedi A, Malach R, Hendler T, Peled S, Zohary E (2001) Visuohaptic object-related activation in the ventral visual pathway. Nat Neurosci 4:324–330

Beauchamp MS, Argall BD, Bodurka J, Duyn JH, Martin A (2004) Unraveling multisensory integration: patchy organization within human STS multisensory cortex. Nat Neurosci 7:1190–1192

Besle J, Bertrand O, Giard MH (2009) Electrophysiological (EEG, sEEG, MEG) evidence for multiple audiovisual interactions in the human auditory cortex. Hear Res 258(1–2):143–151

Braeutigam S, Bailey AJ, Swithenby SJ (2001) Task-dependent early latency (30–60 ms) visual processing of human faces and other objects. Neuroreport 12:1531–36

Calvert GA, Brammer MJ, Bullmore ET, Campbell R, Iversen SD, David AS (1999) Response amplification in sensory-specific cortices during cross-modal binding. Neuroreport 10: 2619–23

Calvert GA, Campbell R, Brammer MJ (2000) Evidence from functional magnetic resonance imaging of crossmodal binding in the human heteromodal cortex. Curr Biol 10:649–57

Calvert GA, Hansen PC, Iversen SD, Brammer MJ (2001) Detection of audio–visual integration sites in humans by application of electrophysiological criteria to the BOLD effect. Neuroimage 14:427–38

Cappe C, Morel A, Barone P, Rouiller EM (2009) The thalamocortical projection systems in primate: an anatomical support for multisensory and sensorimotor integrations. Cereb Cortex. 19(9):2025–2037, doi:10.1093/cercor/bhn228

Chen YC, Yeh SL (2008) Visual events modulated by sound in repetition blindness. Psychonom Bull Rev 15:404–408

Colombo M, Gross CG (1994) Responses of inferior temporal cortex and hippocampal neurons during delayed matching to sample in monkeys (Macaca fascicularis). Behav Neurosci 108:443–55

Debruille JB, Guillem F, Renault B (1998) ERPs and chronometry of face recognition: following-up Seeck et al. and George et al. Neuroreport 9:3349–53

Deibert E, Kraut M, Kremen S, Hart JJ (1999) Neural pathways in tactile object recognition. Neurology 52:1413–7

Desimone, R (1996) Neural mechanisms for visual memory and their role in attention. Proc Natl Acad Sci USA 93:13494–9

Doehrmann O, Naumer MJ (2008) Semantics and the multisensory brain: how meaning modulates processes of audio-visual integration. Brain Res 1242:136–150

Doniger GM, Foxe JJ, Murray MM, Higgins BA, Snodgrass JG, Schroeder CE, Javitt DC (2000) Activation timecourse of ventral visual stream object-recognition areas: high density electrical mapping of perceptual closure processes. J Cogn Neurosci 12:615–621

Doniger GM, Foxe JJ, Schroeder CE, Murray MM, Higgins BA, Javitt DC (2001) Visual perceptual learning in human object recognition areas: a repetition priming study using high-density electrical mapping. Neuroimage 13:305–313

Fort A, Delpuech C, Pernier J, Giard MH (2002a) Dynamics of cortico-subcortical cross-modal operations involved in audio–visual object detection in humans. Cereb Cortex 12:1031–9

Fort A, Delpuech C, Pernier J, Giard MH (2002b) Early auditory–visual interactions in human cortex during nonredundant target identification. Brain Res Cogn Brain Res 14:20–30

Foxe JJ, Morocz IA, Murray MM, Higgins BA, Javitt DC, Schroeder CE (2000) Multisensory auditory–somatosensory interactions in early cortical processing revealed by high-density electrical mapping. Brain Res Cogn Brain Res 10:77–83

George N, Jemel B, Fiore N, Renault B (1997) Face and shape repetition effects in humans: a spatio-temporal ERP study. Neuroreport 8:1417–23

Giard MH, Peronnet, F (1999) Auditory–visual integration during multimodal object recognition in humans: a behavioral and electrophysiological study. J Cogn Neurosci 11:473–90

Gibson JR, Maunsell JHR (1997) Sensory modality specificity of neural activity related to memory in visual cortex. J Neurophysiol 78:1263–75

Gottfried JA, Smith APR, Rugg MD, Dolan RJ (2004) Remembrance of odors past: human olfactory cortex in cross-modal recognition memory. Neuron 42:687–95

Goulet S, Murray EA (2001) Neural substrates of crossmodal association memory in monkeys: the amygdala versus the anterior rhinal cortex. Behav Neurosci 115:271–284

Grill-Spector K, Henson RN, Martin A (2006) Repetition and the brain, neural models of stimulus specific effects. Trends Cogn Sci 10:14–23

Guo J, Guo A (2005) Crossmodal interactions between olfactory and visual learning in Drosophila. Science 309:307–310

Haenny PE, Maunsell JHR, Schiler PH (1988) State dependent activity in monkey visual cortex: II. Retinal and extraretinal factors in V4. Exp Brain Res 69:245–259

Halgren E, Raij T, Marinkovic K, Jousmaki V, Hari R (2000) Cognitive response profile of the human fusiform face area as determined by MEG. Cereb Cortex 10:69–81

Hamilton W (1859) Lectures on Metaphysics and Logic. Gould & Lincoln, Boston

Haxby JV, Gobbini MI, Furey ML, Ishai A, Schouten JL, Pietrini P (2001) Distributed and overlapping representations of faces and objects in ventral temporal cortex. Science 293: 2425–2430

Hein G, Doehrmann O, Müller NG, Kaiser J, Muckli L, Naumer MJ (2007) Object familiarity and semantic congruency modulate responses in cortical audiovisual integration areas. J Neurosci 27:7881–7887

Henson RN, Rugg MD (2003) Neural response suppression, haemodynamic repetition effects, and behavioural priming. Neuropsychologia 41:263–270

Holmes NP (2009) The principle of inverse effectiveness in multisensory integration: Some statistical considerations. Brain Topogr 21(3–4):168–176

James TW, Humphrey GK, Gati JS, Servos P, Menon RS, Goodale MA (2002) Haptic study of three-dimensional objects activates extrastriate visual areas. Neuropsychologia 40:1706–1714

Landis T, Lehmann D, Mita T, Skrandies W (1984) Evoked potential correlates of figure and ground Int J Psychophysiol 1:345–348

Laurienti PJ, Kraft RA, Maldjian JA, Burdette JH, Wallace MT (2004) Semantic congruence is a critical factor in multisensory behavioral performance. Exp Brain Res 158:405–414

Laurienti PJ, Wallace MT, Maldjian JA, Susi CM, Stein BE, Burdette JH (2003) Cross-modal sensory processing in the anterior cingulate and medial prefrontal cortices. Hum Brain Mapp 19:213–223

Lehmann S, Murray MM (2005) The role of multisensory memories in unisensory object discrimination. Brain Res Cogn Brain Res 24:326–334

MacSweeney M, Woll B, Campbell R, McGuire PK, David AS, Williams SC, Suckling J, Calvert GA, Brammer MJ (2002) Neural systems underlying British Sign Language and audio–visual English processing in native users. Brain 125:1583–93

Martuzzi R, Murray MM, Michel CM, Thiran JP, Maeder PP, Clarke S, Meuli RA (2007) Multisensory interactions within human primary cortices revealed by BOLD dynamics. Cereb Cortex 17:1672–1779

Maunsell JHR, Sclar G, Nealey TA, Depriest DD (1991) Extraretinal representations in area V4 in the macaque monkey. Vis Neurosci 7:561–573

Michel CM, Murray MM, Lantz G, Gonzalez S, Spinelli L, Grave de Peralta R (2004) EEG source imaging. Clin Neurophysiol 115:2195–222

Molholm S, Ritter W, Javitt DC, Foxe JJ (2004) Multisensory visual–auditory object recognition in humans: a high-density electrical mapping study. Cereb Cortex 14:452–465

Molholm S, Ritter W, Murray MM, Javitt DC, Schroeder CE, Foxe JJ (2002) Multisensory auditory–visual interactions during early sensory processing in humans: a high-density electrical mapping study. Brain Res Cogn Brain Res 14:115–128

Mouchetant-Rostaing Y, Giard MH, Delpuech C, Echallier JF, Pernier J (2000a) Early signs of visual categorization for biological and non-biological stimuli in humans. Neuroreport 11:2521–2525

Mouchetant-Rostaing Y, Giard MH, Bentin S, Aguera PE, Pernier J (2000b) Neurophysiological correlates of face gender processing in humans. Eur J Neurosci 12:303–310

Murray MM, Camen C, Spierer L, Clarke S (2008a) Plasticity in representations of environmental sounds revealed by electrical neuroimaging. Neuroimage 39:847–856

Murray MM, Brunet D, Michel CM (2008b) Topographic ERP analyses: a step-by-step tutorial review. Brain Topogr 20:249–264

Murray MM, Camen C, Gonzalez Andino SL, Bovet P, Clarke S (2006) Rapid brain discrimination of sounds of objects. J Neurosci 26:1293–1302

Murray MM, Foxe JJ, Wylie GR (2005a) The brain uses single-trial multisensory memories to discriminate without awareness. Neuroimage 27:473–478

Murray MM, Molholm S, Michel CM, Heslenfeld DJ, Ritter W, Javitt DC, Schroeder CE, Foxe JJ (2005b) Grabbing your ear: rapid auditory–somatosensory multisensory interactions in low-level sensory cortices are not constrained by stimulus alignment. Cereb Cortex 15: 963–974

Murray MM, Michel CM, Grave de Peralta R, Ortigue S, Brunet D, Andino SG, Schnider A (2004) Rapid discrimination of visual and multisensory memories revealed by electrical neuroimaging. Neuroimage 21:125–135

Naumer MJ, Doehrmann O, Müller NG, Muckli L, Kaiser J, Hein G (2009) Cortical plasticity of audio-visual object representations. Cereb Cortex. 19(7):1641–1653, doi: 10.1093/cercor/bhn200

Nyberg L, Habib R, McIntosh AR, Tulving E (2000) Reactivation of encoding-related brain activity during memory retrieval. Proc Natl Acad Sci USA 97:11120–11124

Raij T, Uutela K, Hari R (2000) Audiovisual integration of letters in the human brain. Neuron 28:617–625

Ranganath C, Rainer G (2003) Neural mechanisms for detecting and remembering novel events. Nat Rev Neurosci 4:193–202

Ritter W, Simson R, Vaughan Jr HG, Macht M (1982) Manipulation of event-related potential manifestations of information processing stages. Science 218:909–911

Rolls ET, Treves A, Tovee MJ (1997) The representational capacity of the distributed encoding of information provided by populations of neurons in primate temporal visual cortex. Exp Brain Res 114:149–162

Romei V, Murray MM, Merabet LB, Thut G (2007) Occipital transcranial magnetic stimulation has opposing effects on visual and auditory stimulus detection: implications for multisensory interactions. J Neurosci 27:11465–11472

Rousselet GA, Fabre-Thorpe M, Thorpe SJ (2002) Parallel processing in high-level categorization of natural images. Nat Neurosci 5:629–630

Rugg MD, Mark RE, Walla P, Schloerscheidt AM, Birch CS, Allan K (1998) Dissociation of the neural correlates of implicit and explicit memory. Nature 392:595–598

Schnider A, Valenza N, Morand S, Michel CM (2002) Early cortical distinction between memories that pertain to ongoing reality and memories that don't. Cereb Cortex 12:54–61

Seeck M, Michel CM, Mainwaring N, Cosgrove R, Blume H, Ives J, Landis T, Schomer DL (1997) Evidence for rapid face recognition from human scalp and intracranial electrodes. Neuroreport 8:2749–2754

Shams L, Seitz AR (2008) Benefits of multisensory learning. Trends Cogn Sci 12:411–417

Spence C, Nicholls ME, Driver J (2001) The cost of expecting events in the wrong sensory modality. Percept Psychophys 63:330–336

Sperdin HF, Cappe C, Foxe JJ, Murray MM (2009) Early, low-level auditory-somatosensory multisensory interactions impact reaction time speed. Front Integr Neurosci 3:2. doi:10.3389/neuro.07.002.2009

Stein BE, Meredith MA (1993) The merging of the senses. MIT Press, Cambridge, MA

Stoesz MR, Zhang M, Weisser VD, Prather SC, Mao H, Sathian K (2003) Neural networks active during tactile form perception: common and differential activity during macrospatial and microspatial tasks. Int J Psychophysiol 50:41–49

Thierry G, Martin CD, Downing P, Pegna AJ (2007) Controlling for interstimulus perceptual variance abolishes N170 face selectivity. Nat Neurosci 10:505–511

Tsivilis D, Otten LJ, Rugg MD (2001) Context effects on the neural correlates of recognition memory: an electrophysiological study. Neuron 31:497–505

Tulving E, Schacter DL (1990) Priming and human memory systems. Science 247:301–306

Vanni S, Revonsuo A, Saarinen J, Hari R (1996) Visual awareness of objects correlates with activity of right occipital cortex. Neuroreport 8:183–186

Von Kriegstein K, Giraud AL (2006) Implicit multisensory associations influence voice recognition. PLoS Biol 4:e326

Wang Y, Celebrini S, Trotter Y, Barone P (2008) Visuo-auditory interactions in the primary visual cortex of the behaving monkey: electrophysiological evidence. BMC Neurosci 9:79

Wheeler ME, Petersen SE, Buckner RL (2000) Memory's echo: vivid remembering reactivates sensory-specific cortex. Proc Natl Acad Sci USA 97:11125–11129

Wiggs CL, Martin A (1998) Properties and mechanisms of perceptual priming. Curr Opin Neurobiol 8:227–233

Zangaladze A, Epstein CM, Grafton ST, Sathian K (1999) Involvement of visual cortex in tactile discrimination of orientation. Nature 401:587–590

Part III
Visuo-Tactile Integration

Chapter 12
Multisensory Texture Perception

Roberta L. Klatzky and Susan J. Lederman

12.1 Introduction

The fine structural details of surfaces give rise to a perceptual property generally called *texture*. While any definition of texture will designate it as a surface property, as distinguished from the geometry of the object as a whole, beyond that point of consensus there is little agreement as to what constitutes texture. Indeed, the definition will vary with the sensory system that transduces the surface. The potential dimensions for texture are numerous, including shine/matte, coarse/fine, rough/smooth, sticky/smooth, or slippery/resistant. Some descriptors apply primarily to a particular modality, as "shine" does to vision, but others like "coarse" may be applied across modalities. As will be described below, there have been efforts to derive the underlying features of texture through behavioral techniques, particularly multidimensional scaling.

In this chapter, we consider the perception of texture in touch, vision, and audition, and how these senses interact. Within any modality, sensory mechanisms impose an unequivocal constraint on how a texture is perceived, producing intermodal differences in the periphery that extend further to influence attention and memory. What is just as clear is that the senses show commonalities as well as differences in responses to the same physical substrate.

As a starting point for this review, consider the paradigmatic case where a person sees and touches a textured surface while hearing the resulting sounds. Intuitively, we might think that a surface composed of punctate elements will look jittered, feel rough, and sound scratchy, whereas a glassy surface will look shiny, feel smooth, and emit little sound when touched. Our intuition tells us that the physical features of the surface are realized in different ways by the senses, yet reflect the common source. Given the inherent fascination of these phenomena, it is not surprising that texture perception has been the focus of a substantial body of research.

R.L. Klatzky (✉)
Department of Psychology, Carnegie Mellon University, Pittsburgh, PA 15213-3890, USA
e-mail: klatzky@cmu.edu

M.J. Naumer, J. Kaiser (eds.), *Multisensory Object Perception in the Primate Brain*,
DOI 10.1007/978-1-4419-5615-6_12,

Our chapter is based primarily on the psychological literature, but it includes important contributions from neuroscience and computational approaches. Research in these fields has dealt with such questions as the following: What information is computed from distributed surface elements and how? What are the perceptual properties that arise from these computations, and how do they compare across the senses? To what aspects of a surface texture are perceptual responses most responsive? How do perceptual responses vary across salient dimensions of the physical stimulus, with respect to perceived intensity and discriminability? What is the most salient multidimensional psychological texture space for unimodal and multisensory perception?

Many of these questions, and others, were first raised by the pioneering perceptual psychologist David Katz (1925; translated and edited by Krueger, 1989). He anticipated later interest in many of the topics of this chapter, for example, feeling textures through an intermediary device like a tool, the role of sounds, differences in processing fine vs. relatively coarse textures (by vibration and the "pressure sense," respectively), and the relative contributions of vision and touch.

12.2 Texture and Its Measurement

Fundamental to addressing the questions raised above are efforts to define and measure texture, and so we begin with this topic. Texture is predominantly thought of as a property that falls within the domain of touch, where it is most commonly designated by surface roughness. Haptically perceived textures may be labeled by other properties, such as sharpness, stickiness, or friction, or even by characteristics of the surface pattern, such as element width or spacing, to the extent that the pattern can be resolved by the somatosensory receptors.

Texture is, however, multisensory; it is not restricted to the sense of touch. As used in the context of vision, the word texture refers to a property arising from the pattern of brightness of elements across a surface. Adelson and Bergen (1991) referred to texture as "stuff" in an image, rather than "things." Visual texture can pertain to pattern features such as grain size, density, or regularity; alternatively, smoothly coated reflective surfaces can give rise to features of coarseness and glint (Kirchner et al., 2007). When it comes to audition, textural features arise from mechanical interactions with objects, such as rubbing or tapping. To our knowledge, there is no agreed-upon vocabulary for the family of contact sounds that reveal surface properties, but terms like crackliness, scratchiness, or rhythmicity might be applied. Auditory roughness has also been described in the context of tone perception, where it is related to the frequency difference in a dissonant interval (Plomp and Steeneken, 1968; Rasch and Plomp, 1999).

Just as texture is difficult to define as a concept, measures of perceived texture are elusive. When a homogeneous surface patch is considered, the size, height or depth, and spacing of surface elements can be measured objectively, as can visual surface properties such as element density. Auditory loudness can be scaled, and the spectral properties of a texture-induced sound can be analyzed. The perceptual concomitants of these physical entities, however, are more difficult to assess.

In psychophysical approaches to texture, two techniques have been commonly used to measure the perceptual outcome: magnitude estimation and discrimination. In a magnitude-estimation task, the participant gives a numerical response to indicate the intensity of a designated textural property, such as roughness. The typical finding is that perceived magnitude is related to physical value by a power function. This methodology can be used to perceptually scale the contributions of different physical parameters of the surface, with the exponent of the power function (or the slope in log/log space) being used to indicate relative differentiability along some physical surface dimension that is manipulated. Various versions of the task can be used, for example, by using free or constrained numerical scales.

Discrimination is also assessed by a variety of procedures. One measure is the just-noticeable difference (JND) along some dimension. The JND can be used to calculate a Weber fraction, which characterizes a general increment relative to a base value that is needed to barely detect a stimulus difference. Like magnitude estimation, measurement of the JND tells us about people's ability to differentiate surfaces, although the measures derived from the two approaches (magnitude-estimation slope and Weber fraction) for a given physical parameter do not always agree (Ross, 1997). Confusions among textured stimuli can also be used to calculate the amount of information transmitted by a marginally discriminable set of surfaces.

At the limit of discrimination, the absolute threshold, people are just able to detect a texture relative to a smooth surface. Haptic exploration has been shown to differentiate textured from smooth surfaces when the surface elements are below 1 μm (0.001 mm) in height (LaMotte and Srinivasan, 1991; Srinivasan et al., 1990). This ability is attributed to vibratory signals detected by the Pacinian corpuscles (PCs), mechanoreceptors lying deep beneath the skin surface. In vision, the threshold for texture could be measured by the limit on grating detection (i.e., highest resolvable spatial frequency), which depends on contrast. The resolution limit with high-contrast stripes is about 60 cycles per degree.

Another approach to the evaluation of perceived texture is multidimensional scaling (MDS), which converts judgments of similarity (or dissimilarity) to distances in a low-dimensional space. The dimensions of the space are then interpreted in terms of stimulus features that underlie the textural percept. A number of studies have taken this approach, using visual or haptic textures. A limitation of this method is that the solution derived from MDS depends on the population of textures that is judged. For example, Harvey and Gervais (1981) constructed visual textures by combining spatial frequencies with random amplitudes and found, perhaps not surprisingly, that the MDS solution corresponded to spatial frequency components rather than visual features. Rather different results were found by Rao and Lohse (1996), who had subjects rate a set of pictures on a set of Likert scales and, using MDS, recovered textural dimensions related to repetitiveness, contrast, and complexity.

Considering MDS approaches to haptic textures, again the solution will depend on the stimulus set. Raised-dot patterns were studied by Gescheider and colleagues

(2005), who found that three dimensions accounted for dissimilarity judgments, corresponding to blur, roughness, and clarity. Car-seat materials were used in a scaling study of Picard and colleagues (2003), where the outcome indicated dimensions of soft/harsh, thin/thick, relief, and hardness. Hollins and associates examined the perceptual structure of sets of natural stimuli, such as wood, sandpaper, and velvet. In an initial study (Hollins et al., 1993), a 3D solution was obtained. The two primary dimensions corresponded to roughness and hardness, and a third was tentatively attributed to elasticity. Using a different but related set of stimuli, Hollins and colleagues (2000) subsequently found that roughness and hardness were consistently obtained across subjects, but a third dimension, sticky/slippery, was salient only to a subset. The solution for a representative subject is shown in Fig. 12.1.

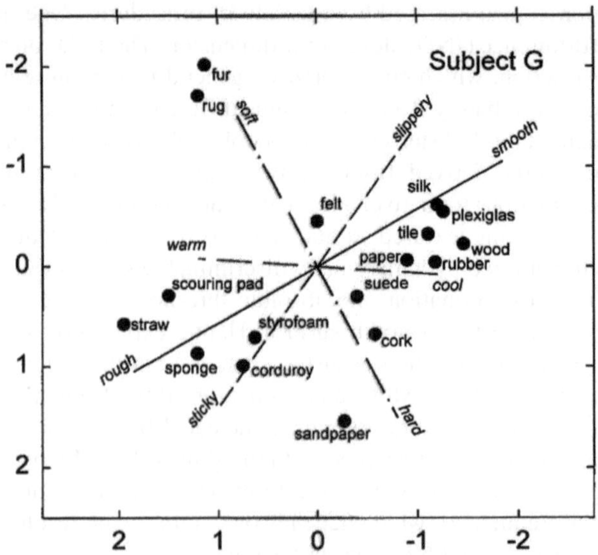

Fig. 12.1 2D MDS solution for a representative subject in Hollins et al. (2000). Adjective scales have been placed in the space according to their correlation with the dimensions (adapted from Fig. 2, with permission)

12.3 Haptic Roughness Perception

Since the largest body of work on texture perception is found in research on touch, we will review that work in detail (for a recent brief review, see Chapman and Smith, 2009). As was mentioned above, the most commonly assessed haptic feature related to touch is roughness, the perception of which arises when the skin or a handheld tool passes over a surface. Research on how people perceive roughness has been multi-pronged, including behavioral, neurophysiological, and computational approaches. Recently, it has become clear that to describe human roughness perception, it is necessary to distinguish surfaces at different levels of "grain size."

There is a change in the underlying processing giving rise to the roughness percept once the elements in the texture become very fine. Accordingly, we separately consider surfaces with spatial periods greater and less than ~200 μm (0.2 mm), called *macrotextures* and *microtextures*, respectively.

At the macrotextural scale, Lederman and colleagues (Lederman and Taylor, 1972; Taylor and Lederman, 1975) conducted seminal empirical work on the perception of roughness with the bare finger. These studies used various kinds of surfaces: sandpapers, manufactured plates with a rectangular-wave profile (gratings), and plates composed of randomly arranged conical elements. The parametric control permitted by the latter stimuli led to a number of basic findings. First, surface roughness appears to be primarily determined by the spacing between the elements that form the texture. Until spacing becomes sparse (~3.5 mm between element edges), roughness increases monotonically (generally, as a power function) with spacing. (Others have reported monotonicity beyond that range, e.g., Meftah et al., 2000.) In comparison to inter-element spacing, smaller effects are found of other variables, including the width of ridges in a grated plate or the force applied to the plate during exploration. Still smaller or negligible effects have been found for exploration speed and whether the surface is touched under active vs. passive control.

Based on this initial work, Lederman and Taylor developed a mechanical model of roughness perception (1972; Taylor and Lederman, 1975; see also Lederman, 1974, 1983). In this model, perceived roughness is determined by the total area of skin that is instantaneously indented from a resting position while in contact with a surface. Effects on perceived roughness described above were shown to be mediated by their impact on skin deformation. As the instantaneous deformation proved to be critical, it is not surprising that exploratory speed had little effect, although surfaces tended to be judged slightly less rough at higher speeds. This could be due to the smaller amount of skin displaced with higher speeds.

A critical point arising from this early behavioral work is that macrotexture perception is a spatial, rather than a temporal, phenomenon. Intuitively it may seem, to the contrary, that vibration would be involved, particularly because textured surfaces tend to be explored by a moving finger (or surfaces are rubbed against a stationary finger). However, the operative model's assumption that texture perception is independent of temporal cues was empirically supported by studies that directly addressed the role of vibration and found little relevance of temporal factors. As was noted above, speed has little effect on perceived roughness, in comparison to spatial parameters (Lederman, 1974, 1983). Moreover, when perceivers' fingers were preadapted to a vibration matched to the sensitivity of vibration-sensitive receptors in the skin, there was little effect on judgments of roughness (Lederman et al., 1982). More recently, others have shown evidence for small contributions of temporal frequency to perceived magnitude of macrotextures (Cascio and Sathian, 2001; Gamzu and Ahissar, 2001; Smith et al., 2002), but the predominant evidence supports a spatial mechanism.

An extensive program of research by Johnson and associates has pointed to the operative receptor population that underlies roughness perception of macrotextures (for review, see Johnson et al., 2002). This work supports the idea of a spatial code. Connor and colleagues (1990) measured neural responses from monkey SA, RA, and PC afferents and related them to roughness magnitudes for dotted textures varying in dot diameter and spacing. Mean impulse rate from any population of receptors failed to unambiguously predict the roughness data, whereas the spatial and temporal variabilities in SA1 impulse rates were highly correlated with roughness across the range of stimuli. Subsequent studies ruled out temporal variation in firing rate as the signal for roughness (Connor and Johnson, 1992) and implicated the *spatial* variation in the SA1 receptors (Blake et al., 1997).

We now turn to the perception of microtextures, those having spatial periods on the order of <200 μm or less. Katz (1925) suggested that very fine textures were perceived by vibration, whereas coarse textures were sensed by pressure. Recent work by Bensmaïa, Hollins, and colleagues supports a duplex model of roughness perception, which proposes a transition from spatial coding of macrotexture to vibratory coding at the micro-scale (Bensmaïa and Hollins, 2003, 2005; Bensmaïa et al., 2005; Hollins et al., 1998). Evidence for this proposal comes from several approaches. Vibrotactile adaptation has been found to affect the perception of microtextures, but not surfaces with spatial period > 200 μm (Hollins et al., 2001, 2006). Bensmaïa and Hollins (2005) found direct evidence that roughness of microtextures is mediated by responses from the PCs. Skin vibration measures (filtered by a PC) predicted psychophysical differentiation of fine textures.

As force-feedback devices have been developed to simulate textures, considerable interest has developed in how people perceive textures that they explore using a rigid tool as opposed to the bare skin. This situation, called *indirect touch*, is particularly relevant to the topic of multisensory texture perception, because the mechanical interactions between tool and surface can give rise to strong auditory cues. Figure 12.2 shows samples of rendered textures and spherical contact elements, like those used in research by Unger et al. (2008).

In initial studies of perception through a tool, Klatzky, Lederman, and associates investigated how people judged roughness when their fingers were covered with rigid sheaths or when they held a spherically tipped probe (Klatzky and Lederman, 1999; Klatzky et al. 2003; Lederman et al., 2000; see Klatzky and Lederman, 2008, for review). The underlying signal for roughness in this case must be vibratory, since the rigid intermediary eliminates spatial cues, in the form of the pressure array that would arise if the bare finger touched the surface. Vibratory coding is further supported by the finding that vibrotactile adaptation impairs roughness perception with a probe even at the macrotextural scale, where roughness coding with the bare skin is presumably spatial, as well as with very fine textures (Hollins et al., 2006).

Recall that bare-finger studies of perceived roughness under free exploration using magnitude estimation typically find a monotonic relation between roughness magnitude and the spacing between elements on a surface, up to spacing on the order of 3.5 mm. In contrast, Klatzky, Lederman, and associates found that when a probe was used to explore a surface, the monotonic relation between perceived roughness magnitude and inter-element spacing was violated well before this point. As shown in Fig. 12.3, instead of being monotonic over a wide range of spacing, the function

Fig. 12.2 Sample texture shapes and spherical probe tips rendered with a force-feedback device (figures from Bertram Unger, with permission)

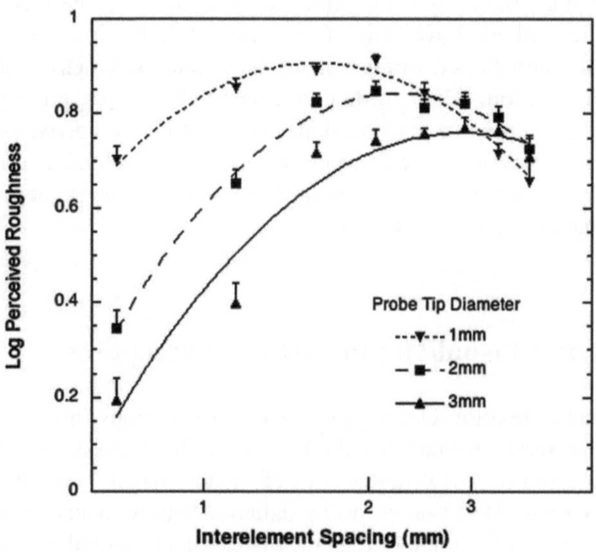

Fig. 12.3 Roughness magnitude as a function of inter-element spacing and probe tip size in Klatzky et al. (2003) (From Fig. 6, with permission)

relating roughness magnitude to spacing took the form of an inverted U. The spacing where the function peaked was found to be directly related to the size of the probe tip: The larger the tip, the further along the spacing dimension the function peaked. Klatzky, Lederman, and associates proposed that this reflected a critical geometric relation between probe and surface: Roughness peaked near the point where the surface elements became sufficiently widely spaced that the probe could drop between them and predominantly ride on the underlying substrate. Before this "drop point," the probe rode along the tops of the elements and was increasingly jarred by mechanical interactions as the spacing increased.

The static geometric model of texture perception with a probe, as proposed by Klatzky et al. (2003), has been extended by Unger to a dynamic model that takes into account detailed probe/surface interactions. This model appears to account well for the quadratic trend in the magnitude-estimation function (Unger, 2008; Unger et al., 2008). Further, the ability to discriminate textures on the basis of inter-element spacing, as measured by the JND, is greatest in the range of spacings where the roughness magnitude peaks, presumably reflecting the greater signal strength in that region (Unger et al., 2007).

Multidimensional scaling of haptic texture has been extended to exploration with a probe. Yoshioka and associates (Yoshioka et al., 2007) used MDS to compare perceptual spaces of natural textures (e.g., corduroy, paper, rubber) explored with a probe vs. the bare finger. They also had subjects rate the surfaces for roughness, hardness, and stickiness – the dimensions obtained in the studies of Hollins and associates described above. They found that while the roughness ratings were similar for probe and finger, ratings of hardness and stickiness varied according to mode of exploration. They further discovered that three physical quantities, vibratory power, compliance, and friction, predicted the perceived dissimilarity of textures felt with a probe. These were proposed to be the physical dimensions that constitute texture space, that is, that collectively underlie the perceptual properties of roughness, hardness, and stickiness.

12.4 Visual and Visual/Haptic Texture Perception

Measures of haptic texture tend to correspond to variations in magnitude along a single dimension and hence can be called *intensive*. In contrast, visual textures typically describe variations of brightness in 2D space, which constitute *pattern*. As Adelson and Bergen (1991) noted, to be called a texture, a visual display should exhibit variation on a scale smaller than the display itself; global gradients or shapes are not textures.

Early treatment of texture in studies of visual perception emphasized the role of the texture gradient as a depth cue (Gibson, 1950), rather than treating it as an object property. Subsequently, considerable effort in the vision literature has been directed at determining how different textural elements lead to segregation of regions in a 2D image (see Landy and Graham, 2004, for review). Julesz (1984;

Julesz and Bergen, 1983) proposed that the visual system pre-attentively extracts primitive features that he called *textons*, consisting of blobs, line ends, and crossings. Regions of common textons form textures, and texture boundaries arise where textons change. Early work of Treisman (1982) similarly treated texture segregation as the result of pre-attentive processing that extracted featural primitives.

Of greater interest in the present context is how visual textural variations give rise to the perception of surface properties, such as visual roughness. In a directly relevant study, Ho et al. (2006) asked subjects to make roughness comparisons of surfaces rendered with different lighting angles. Roughness judgments were not invariant with lighting angle, even when enhanced cues to lighting were added. This result suggested that the observers were relying on cues inherent to the texture, including shadows cast by the light. Ultimately, four cues were identified that were used to judge roughness: the proportion of image in shadow, the variability in luminance of pixels outside of shadow, the mean luminance of pixels outside of shadow, and the texture contrast (cf. Pont and Koenderink, 2005), a statistical measure responsive to the difference between high- and low-luminance regions. Failures in roughness constancy over lighting variations could be attributed to the weighted use of these cues, which vary as the lighting changes. The critical point here is that while other cues were possible, subjects were judging roughness based on shadows in the image, not on lighting-invariant cues such as binocular disparity. The authors suggested that the reliance on visual shading arises from everyday experience in which touch and vision are both present, and shadows from element depth become correlated with haptic roughness.

Several studies have made direct attempts to compare vision and touch with respect to textural sensitivity. In a very early study, Binns (1936) found no difference between the two modalities in the ordering of a small number of fabrics by softness and fineness. Björkman (1967) found that visual matching of sandpaper samples was less variable than matching by touch, but the numbers of subjects and samples were small. Lederman and Abbott (1981) found that surface roughness was judged equivalently whether people perceived the surfaces by vision alone, haptics, or both modalities. Similarity of visual and haptic roughness judgments was also found when the stimuli were virtual jittered-dot displays rendered by force feedback (Drewing et al., 2004). In an extensive comparison using natural surfaces, Bergmann Tiest and Kappers (2006) had subjects rank-order 96 samples of widely varying materials (wood, paper, ceramics, foams, etc.) according to their perceived roughness, using vision or haptics alone. Objective physical roughness measures were then used to benchmark perceptual ranking performance. Rank-order correlations of subjects' rankings with most physical measures were about equal under haptic and visual sorting, but there were variations across the individual subjects and the physical measures.

Another approach to comparing visual and haptic texture perception is to compare MDS solutions to a common set of stimuli, when similarity data are gathered using vision vs. touch. Previously, we noted that the scaled solution will depend on the stimulus set and that different dimensional solutions have been obtained for visual and haptic stimuli. When the same objects are used, it is possible to compare

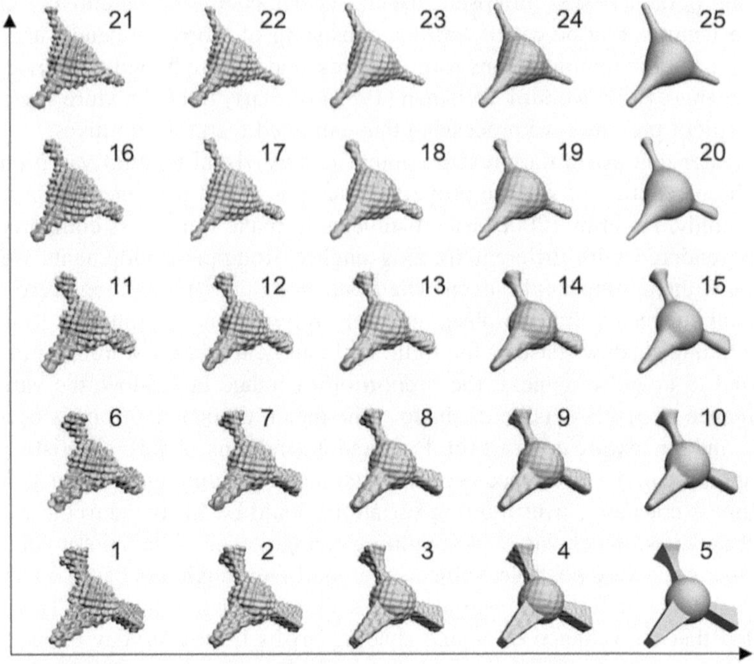

Fig. 12.4 Stimuli of Cooke et al. (2006), with microgeometry varying horizontally and macrogeometry varying vertically (adapted from Fig. 2, © 2006 ACM, Inc.; included here by permission)

spaces derived from unimodal vision, haptics, and bimodal judgments. With this goal, Cooke and associates constructed a set of stimuli varying parametrically in macrogeometry (angularity of protrusions around a central element) and microgeometry (smooth to bumpy) (see Fig. 12.4). A 3D printer was used to render the objects for haptic display. Physical similarities were computed by a number of measures, for purposes of comparing with the MDS outcome. The MDS computation produced a set of weighted dimensions, allowing the perceptual salience of shape vs. texture to be compared across the various perceptual conditions. Subjects who judged similarity by vision tended to weight shape more than texture, whereas those judging similarity by touch assigned the weights essentially equally, findings congruent with earlier results of Lederman and Abbott (1981) using a stimulus matching procedure. Subjects judging haptically also showed larger individual differences (Cooke et al., 2006, 2007). In the 2007 study, bimodal judgments were also used and found to resemble the haptic condition, suggesting that the presence of haptic cues mitigated against the perceptual concentration on shape.

Most commonly, textured surfaces are touched with vision present; they are not unimodal percepts. This gives rise to the question of how the two modalities interact to produce a textural percept. A general idea behind several theories of inter-sensory interaction is that modalities contribute to a common percept in some weighted combination (see Lederman and Klatzky, 2004, for review), reflecting *modality appropriateness*. In a maximum-likelihood integration model, the weights are assumed to be optimally derived so as to reflect the reliability of each modality (Ernst and Banks, 2002).

Under this model, since the spatial acuity of vision is greater than touch, judgments related to the pattern of textural elements should be given greater weight under vision. On the other hand, the spatial and temporal signals from cutaneous mechanoreceptors signal roughness as a magnitude or intensity, not pattern, and the greater weighting for vision may not pertain when roughness is treated intensively. Evidence for relatively greater contribution for touch than vision in texture perception has been provided by Heller (1982, 1989). In the 1982 study, bimodal visual/haptic input led to better discrimination performance than unimodal, but the contribution of vision could be attributed to sight of the exploring hand: Elimination of visual texture cues left bimodal performance unchanged, as long as the hand movements could be seen. The 1989 study showed equivalent discrimination for vision and touch with coarse textures, but haptic texture perception proved superior when the surfaces were fine.

Moreover, the sensitivity or reliability of perceptual modalities does not tell the whole story as to how they are weighted when multisensory information is present. It has also been suggested that people "bring to the table" long-term biases toward using one sense or another, depending on the perceptual property of interest. Such biases have been demonstrated in sorting tasks using multi-attribute objects. Sorting by one property means, de facto, that others must be combined; for example, sorting objects that vary in size and texture according to size means that the items called "small" will include a variety of textures. The extent of separation along a particular property is, then, an indication of the bias toward that property in the object representation. Using this approach, Klatzky, Lederman, and associates found that the tendency to sort by texture was greater when people felt objects, without sight, than when they could see the objects; conversely, the tendency to sort by shape was greater when people saw the objects than when they merely touched them (Klatzky et al., 1987; Lederman et al., 1996). Overall, this suggests that texture judgments would have a bias toward the haptic modality, which is particularly suited to yield information about intensive (cf. spatial) responses.

Lederman and colleagues pitted the spatial and intensive biases of vision and touch against one another in experiments using hybrid stimuli, created from discrepant visible vs. touched surfaces. In an experiment by Lederman and Abbott (1981, Experiment 1), subjects picked the best texture match for a target surface from a set of sample surfaces. In the bimodal condition, the "target" was actually two different surfaces that were seen and felt simultaneously. Bimodal matching led to a mean response that was halfway between the responses to the unimodal components, suggesting a process that averaged the inputs from the two channels.

Using a magnitude-estimation task, Lederman et al. (1986) further demonstrated that the weights given to the component modalities were labile and depended on attentional set. Subjects were asked to judge either the magnitude of spatial density or the roughness of surfaces with raised elements. Again, a discrepancy paradigm was used, where an apparently single bimodal surface was actually composed of different surfaces for vision and touch. Instructions to judge spatial density led to a higher weight for vision than touch (presumably because vision has such high spatial resolution), whereas the reverse held for judgments of roughness (for which spatial resolution is unnecessary).

A more specific mechanism for inter-modal interaction was tested by Guest and Spence (2003). The stimuli were textile samples, and the study assessed the interference generated by discrepant information from one modality as subjects did speeded discriminations in another. Discrepant haptic distractors affected visual discriminations, but not the reverse. This suggests that haptic inputs cannot be filtered under speeded assessment of roughness, whereas visual inputs can be gated from processing.

In general agreement with the inter-modal differences described here, a recent review by Whitaker et al. (2008) characterized the roles of vision and touch in texture perception as "independent, but complementary" (p. 59). The authors suggested that where integration across the modalities occurs, it may be at a relatively late level of processing, rather than reflecting a peripheral sensory interaction.

To summarize, studies of visual texture perception suggest that roughness is judged from cues that signal the depth and spatial distribution of the surface elements. People find it natural to judge visual textures, and few systematic differences are found between texture judgments based on vision vs. touch. In a context where vision and touch are both used to explore textured surfaces, vision appears to be biased toward encoding pattern or shape descriptions, and touch toward intensive roughness. The relative weights assigned to the senses appear to be controlled, to a large extent, by attentional processes, although there is some evidence that intrusive signals from touched surfaces cannot be ignored in speeded visual texture judgments.

12.5 Auditory Texture Perception

Katz (1925) pointed out that auditory cues that accompany touch are an important contribution to perception. As was noted in the introduction to this chapter, auditory signals for texture are the result of mechanical interactions between an exploring effector and a surface. There is no direct analogue to the textural features encountered in the haptic and visual domain, nor (to our knowledge) have there been efforts to scale auditory texture using MDS.

A relatively small number of studies have explored the extent to which touch-produced sounds convey texture by themselves or in combination with touch. In an early study by Lederman (1979), subjects gave a numerical magnitude to indicate the roughness of metal gratings that they touched with a bare finger, heard with sounds of touching by another person, or both touched and heard. As is typically found for roughness magnitude estimation of surfaces explored with the bare finger, judgments of auditory roughness increased as a power function of the inter-element spacing of the grooves. The power exponent for the unimodal auditory function was smaller than that obtained with touch alone, indicating that differentiation along the stimulus continuum was less when textures were rendered as sounds. In the third, bimodal condition, the magnitude-estimation function was found to be the same as for touch alone. This suggests that the auditory input was simply ignored when touch was available.

Similar findings were obtained by Suzuki et al. (2006). Their magnitude-estimation study included unimodal touch, touch with veridical sound, and touch with frequency-modified sound. The slope of the magnitude-estimation function, a measure of stimulus differentiation, was greatest for the unimodal haptic condition, and, most importantly for present purposes, the bimodal condition with veridical sound produced results very close to those of the touch-only condition. On the whole, the data suggested that there was at best a small effect of sound – veridical or modified – on the touch condition.

Previously we have alluded to studies in which a rigid probe was used to explore textured surfaces, producing a magnitude-estimation function with a pronounced quadratic trend. Under these circumstances, vibratory amplitude has been implicated as a variable underlying the roughness percept (Hollins et al., 2005, 2006; Yoshioka et al., 2007). The auditory counterpart of perceived vibration amplitude is, of course, loudness. This direct link from a parameter governing haptic roughness to an auditory percept suggests that the auditory contribution to perceived roughness might be particularly evident when surfaces were felt with a rigid probe, rather than the bare finger. If rougher surfaces explored with a probe have greater vibratory intensity, and hence loudness, auditory cues to roughness should lead to robust differentiation in magnitude judgments. Further, the roughness of surfaces that are felt with a probe may be affected by auditory cues, indicating integration of the two sources.

These predictions were tested in a study of Lederman, Klatzky, and colleagues (2002), who replicated Lederman's (1979) study using a rigid probe in place of the bare finger. Unimodal auditory, unimodal touch, and bimodal conditions of exploration were compared. The magnitude-estimation functions for all three conditions showed similar quadratic trends. This confirms that auditory cues from surfaces explored with a probe produce roughness signals that vary systematically in magnitude, in the same relation to the structure of the textured surface that is found with haptic cues. The conditions varied in mean magnitude, however, with unimodal haptic exploration yielding the strongest response, unimodal auditory the weakest, and the bimodal condition intermediate between the two. This pattern further suggests that information from touch and audition was integrated in the bimodal conditions; estimated relative weightings for the two modalities derived from the data were 62% for touch and 38% for audition.

Before accepting this as evidence for the integration of auditory cues with haptic cues, however, it is important to note that subsequent attempts by the present authors to replicate this finding failed. Moreover, further tests of the role of auditory cues, using an absolute-identification learning task, found that while stimuli could be discriminated by sounds alone, the addition of sound to haptic roughness had no effect: People under-performed with auditory stimuli relative to the haptic and bimodal conditions, which were equivalent. As with the initial study by Lederman (1979) where surfaces were explored with the bare finger, auditory information appeared to be ignored when haptic cues to roughness were present during exploration with a probe. At least, auditory information appears to be used less consistently than cues produced by touch.

Others have shown, however, that the presence of auditory cues can modulate perceived roughness. Jousmaki and Hari (1998) recorded sounds of participants rubbing their palms together. During roughness judgments these were played back, either identical to the original sounds or modified in frequency or amplitude. Increasing frequency and amplitude of the auditory feedback heightened the perception of smoothness/dryness, making the skin feel more paper-like. The authors named this phenomenon the "parchment-skin illusion."

Guest and colleagues (2002) extended this study to show that manipulating frequency also alters the perceived roughness of abrasive surfaces. The task involved a two-alternative, forced-choice discrimination between two briefly touched surfaces, one relatively rough and one smoother. The data indicated that augmentation of high frequencies increased the perceived roughness of the presented surface, leading to more errors for the smooth sample; conversely, attenuating high frequencies produced a reverse trend. (The authors refer to this effect as a "bias," which suggests a later stage of processing. However, an analysis of the errors reported in Table 1 of their paper indicates a sizeable effect on d', a standard sensitivity [cf. response bias] measure, which dropped from 2.27 in the veridical case to 1.09 and 1.20 for amplified and attenuated sounds, respectively.) The same paper also replicated the parchment-skin illusion and additionally found that it was reduced when the auditory feedback from hand rubbing was delayed. Zampini and Spence (2004) showed similar influences of auditory frequency and amplitude when subjects bit into potato chips and judged their crispness.

The influence of auditory cues on roughness extends beyond touch-produced sounds. Suzuki et al. (2008) showed that white noise, but not pure tones, decreased the slope of the magnitude-estimation function for roughness. In contrast, neither type of sound affected the function for tactile perception of length. This suggests that roughness perception may be tuned to cues from relatively complex sounds.

To summarize, it is clear that people can interpret sounds from surface contact that arise during roughness assessment. Further, sound appears to modulate judgments of roughness based on touch. Evidence is lacking, however, for integration of auditory and haptic cues to roughness, particularly at early levels in perceptual processing.

Further work is needed on many topics related to auditory roughness perception. These include assessment of the features of auditory roughness using techniques like MDS; investigation of visual/auditory roughness interactions; and tests of specific models for inter-sensory integration of roughness cues (see Lederman and Klatzky, 2004, for review) when auditory inputs are present.

12.6 Brain Correlates of Texture Perception

Imaging and lesion studies have been used to investigate the cortical areas that are activated by texture perception within the modalities of vision and touch. Visual textures have been found to activate multiple cortical levels, depending on the particular textural elements that compose the display. Kastner et al. (2000) reported that textures composed of lines activated multiple visual areas, from primary visual cortex (V1) to later regions in the ventral and dorsal streams (V2/VP, V4, TEO, and V3A). In contrast, when the textures were checkerboard shapes, reliable activation was observed only in the relatively later visual areas (excluding V1 and V2/VP), suggesting that the operative areas for texture perception in the visual processing stream depends strongly on scale.

Haptic texture processing has been found to be associated with cortical areas specialized for touch, both primary somatosensory cortex (SI) and the parietal operculum (PO, which contains somatosensory area SII: Burton et al., 1997, 1999; Ledberg et al., 1995; Roland O'Sullivan and Kawashima, 1998; Servos et al., 2001; Stilla and Sathian, 2008). Much of this work compared activation during processing of texture to that when people processed shape.

Another approach is to determine how cortical responses change with gradations in a textured surface. Parietal operculum and insula were activated when people felt textured gratings, whether or not they judged surface roughness, suggesting that these cortical regions are loci for inputs to the percept of roughness magnitude (Kitada et al., 2005). In this same study, right prefrontal cortex (PFC), an area associated with higher level processing, was activated only when roughness magnitude was judged, as opposed to when surfaces were merely explored (see Fig. 12.5). This points to PFC as a component in a neural network that uses the sensory data to generate an intensive response.

Stilla and Sathian (2008) pursued findings by others indicating that shape and texture activated common regions (Ledberg et al., 1995; O'Sullivan et al., 1994; Servos et al., 2001). Their own results suggest that selectivity of neural regions for

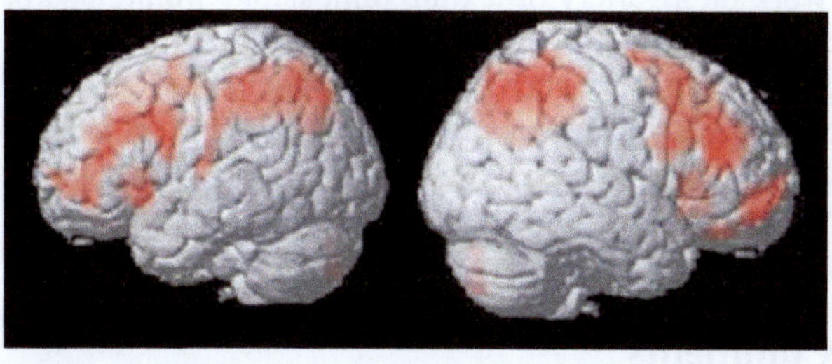

Fig. 12.5 Brain areas selectively activated by magnitude estimation of roughness (cf. no estimation) in the study of Kitada et al. (2005) (adapted from Fig. 3, with permission from Elsevier)

haptic shape and texture is not exclusive, but rather is a matter of relative weighting. Stimuli in the Stilla and Sathian (2008) study were presented for haptic texture processing in the right hand, but the brain areas that were activated more for texture than shape ultimately included bilateral sites, including parietal operculum (particularly somatosensory fields) and contiguous posterior insula. A right medial occipital area that activated preferentially for haptic texture, as opposed to shape, was tentatively localized in visual area V2. This area overlapped with a visual-texture responsive area corresponding primarily to V1; the bisensory overlap was evidenced primarily at the V1/V2 border. However, the lack of correlation between responses to visual and haptic textures in this area suggested that it houses regions that are responsive to one or the other modality, rather than containing neurons that can be driven by either vision or touch.

As Stilla and Sathian (2008) noted, it is critically important in inferring cortical function from fMRI to consider tasks and control conditions. For example, subtracting a shape condition from a texture condition may eliminate spatial processes otherwise associated with roughness. Another consideration is that the processing invoked by a task will change cortical activity, much as instructional set changes the weight of vision vs. touch in texture judgments. For example, imagining how a touched texture will look may invoke visual imagery, whereas imagining how a seen texture would feel could activate areas associated with haptic processing.

In short, measures of brain activation have tended to find that distinct loci for vision and touch predominate, but that some brain regions are responsive to both modalities. Work in this productive area is clearly still at an early stage. In future research, it would be of great interest to evaluate brain responses to auditory texture signals. One relevant fMRI study found that sub-regions of a ventro-medial pathway, which had been associated with the processing of visual surface properties of objects, were activated by the sound of material being crumpled (Arnott et al., 2008). Another question arises from evidence that in the blind, early visual areas take over haptic spatial functions (Merabet et al., 2008; Pascual-Leone and Hamilton, 2001). This gives rise to the possibility that the blind might show quite distinct specialization of cortical areas for texture processing, both in touch and audition, possibly including V1 responses.

Additional work on a variety of texture dimensions would also be valuable, for example, stickiness or friction. Unger (2008) found that the magnitude-estimation function changed dramatically when friction was simulated in textured surfaces, and Hollins and colleagues (2005) found evidence that friction is processed separately, at least to some extent, from other textural properties.

12.7 Final Comments

Our review highlights texture as a multisensory phenomenon. Aspects of texture such as surface roughness can be represented by means of touch, vision, and audition. Variations in surface properties will, within each modality, lead to corresponding variations in the perceived texture. To some extent, the senses interact in arriving at an internal representation of the surface. We should not conclude, however, that surface texture is generally a multisensory percept. The "language" of texture varies across the senses, just as our everyday language for surface properties varies with the input source.

This dynamic research area has already revealed a great deal about human perception of texture and points to exciting areas for further discussion. Moreover, basic research on multisensory texture points to applications in a number of areas, including teleoperational and virtual environments, where simulated textures can enrich the impression of a fully realized physical world.

References

Adelson EH, Bergen JR (1991) The plenoptic function and the elements of early vision. In: Landy MS, Movshon JA (eds) Computational models of visual processing. MIT Press, Cambridge, MA, pp 3–20

Arnott SR, Cant JS, Dutton GN, Goodale MA (2008) Crinkling and crumpling: an auditory fMRI study of material properties. Neuroimage 43:368–378

Bergmann Tiest WM, Kappers A (2006) Haptic and visual perception of roughness. Acta Psychol 124:177–189

Bensmaïa SJ, Hollins M (2003) The vibrations of texture. Somatosens Mot Res 20:33–43

Bensmaïa SJ, Hollins M (2005) Pacinian representations of fine surface texture Percept Psychophys 67:842–854B

Bensmaïa SJ, Hollins M, Yau J (2005) Vibrotactile information in the Pacinian system: a psychophysical model. Percept Psychophys 67:828–841

Binns H (1936) Visual and tactual 'judgement' as illustrated in a practical experiment. Br J Psychol 27: 404–410

Björkman M (1967) Relations between intra-modal and cross-modal matching. Scand J Psychol 8:65–76

Blake DT, Hsiao SS, Johnson KO (1997) Neural coding mechanisms in tactile pattern recognition: the relative contributions of slowly and rapidly adapting mechanoreceptors to perceived roughness. J Neurosci 17:7480–7489

Burton H, MacLeod A-MK, Videen T, Raichle ME (1997) Multiple foci in parietal and frontal cortex activated by rubbing embossed grating patterns across fingerpads: a positron emission tomography study in humans. Cereb Cortex 7:3–17

Burton H, Abend NS, MacLeod AM, Sinclair RJ, Snyder AZ, Raichle ME (1999) Tactile attention tasks enhance activation in somatosensory regions of parietal cortex: a positron emission tomography study. Cereb Cortex 9:662–674

Cascio CJ, Sathian K (2001) Temporal cues contribute to tactile perception of roughness. J Neurosci 21:5289–5296

Chapman CE, Smith AM (2009) Tactile texture. In: Squire L (ed) Encyclopedia of neuroscience. Academic Press, Oxford, pp 857–861

Connor CE, Hsiao SS, Phillips JR, Johnson KO (1990) Tactile rough-ness: neural codes that account for psychophysical magnitude estimates. J Neurosci 10:3823–3836

Connor CE, Johnson KO (1992) Neural coding of tactile texture: comparisons of spatial and temporal mechanisms for roughness perception. J Neurosci 12:3414–3426

Cooke T, Jäkel F, Wallraven C, Bülthoff HH (2007) Multimodal similarity and categorization of novel, three-dimensional objects. Neuropsychologia 45(3):484–495

Cooke T, Kannengiesser S, Wallraven C, Bülthoff HH (2006) Object feature validation using visual and haptic similarity ratings. ACM Trans Appl Percept 3(3):239–261

Drewing K, Ernst MO, Lederman SJ Klatzky RL (2004) Roughness and spatial density judgments on visual and haptic textures using virtual reality. Presented at Euro-Haptics Conference, Munich, Germany
Ernst MO, Banks MS (2002) Humans integrate visual and haptic information in a statistically optimal fashion. Nature 415:429–433
Gamzu E, Ahissar E (2001) Importance of temporal cues for tactile spatial- frequency discrimination. J Neurosci 21(18):7416–7427
Gescheider GA, Bolanowski SJ, Greenfield TC, Brunette KE (2005) Perception of the tactile texture of raised-dot patterns: a multidimensional analysis. Somatosens Mot Res 22(3):127–140
Gibson JJ (1950) The perception of the visual world. Houghton Mifflin, New York
Guest S, Catmur C, Lloyd D, Spence C (2002) Audiotactile interactions in roughness perception. Exp Brain Res 146:161–171
Guest S, Spence C (2003) Tactile dominance in speeded discrimination of pilled fabric samples. Exp Brain Res 150:201–207
Johnson KO, Hsaio SS, Yoshioko T (2002) Neural coding and the basic law of psychophysics. Neuroscientist 8:111–121
Jousmaki V, Hari R (1998) Parchment-skin illusion: sound-biased touch. Curr Biol 8:R190
Harvey LO, Gervais MJ (1981) Internal representation of visual texture as the basis for the judgment of similarity. J Exp Psychol: Human Percept Perform 7(4):741–753
Heller MA (1982) Visual and tactual texture perception: intersensory cooperation. Percept Psychophys 31(4):339–344
Heller MA (1989) Texture perception in sighted and blind observers. Percept Psychophys 45(1):49–54
Ho Y-X, Landy MS, Maloney LT (2006) How direction of illumination affects visually perceived surface roughness. J Vis 6(5):8:634–648, http://journalofvision.org/6/5/8/, doi:10.1167/6.5.8
Hollins M, Bensmaïa S, Karlof K, Young F (2000) Individual differences in perceptual space for tactile textures: evidence from multidimensional scaling. Percept Psychophys 62(8):1534–1544
Hollins M, Bensmaïa S, Risner SR (1998) The duplex theory of texture perception. Proceedings of the 14th annual meeting of the international society for psychophysics, pp 115–120
Hollins M, Bensmaïa SJ, Washburn S (2001) Vibrotactile adaptation impairs discrimination of fine, but not coarse, textures. Somatosens Mot Res 18:253–262
Hollins M, Faldowski R, Rao S, Young F (1993) Perceptual dimensions of tactile surface texture: a multidimensional scaling analysis. Percept Psychophys 54(6):697–705
Hollins M, Lorenz F, Seeger A, Taylor R (2005) Factors contributing to the integration of textural qualities: evidence from virtual surfaces. Somatosens Mot Res 22(3):193–206
Hollins M, Lorenz F, Harper D (2006) Somatosensory coding of roughness: the effect of texture adaptation in direct and indirect touch. J Neurosci 26:5582–5588
Johnson KO, Hsiao SS Yoshioka T (2002) Neural coding and the basic law of psychophysics. Neuroscientist 8:111–121
Julesz B (1984) A brief outline of the texton theory of human vision. Trends Neurosci 7:41–45
Julesz B, Bergen JR (1983) Textons, the fundamental elements in preattentive vision and perception of textures. Bell Syst Tech J 62:1619–1645
Kastner S, De Weerd P, Ungerleider LG (2000) Texture segregation in the human visual cortex: a functional MRI study. J Neurophysiol 83:2453–247
Kirchner E, van den Kieboom G-J, Njo L, Supèr R, Gottenbos R (2007) Observation of visual texture of metallic and pearlescent materials. Color Res Appl 32:256–266
Kitada R, Hashimoto T, Kochiyama T, Kito T, Okada T, Matsumura M, Lederman SJ, Sadata N (2005) Tactile estimation of the roughness of gratings yields a graded response in the human brain: An fMRI study. NeuroImage 25:90–100
Klatzky RL, Lederman SJ (1999) Tactile roughness perception with a rigid link interposed between skin and surface Percept Psychophys 61:591–607
Klatzky RL, Lederman S (2008) Perceiving object properties through a rigid link. In: Lin M, Otaduy M (eds) Haptic rendering: foundations, algorithms, and applications. A K Peters, Ltd, Wellesley, MA, pp 7–19

Klatzky RL, Lederman SJ, Hamilton C, Grindley M, Swendson RH (2003) Feeling textures through a probe: effects of probe and surface geometry and exploratory factors. Percept Psychophys 65:613–631

Klatzky R, Lederman SJ, Reed C (1987) There's more to touch than meets the eye: the salience of object attributes for haptics with and without vision. J Exp Psychol: Gen 116(4):356–369

LaMotte RH, Srinivasan MA (1991) Surface microgeometry: tactile perception and neural encoding. In: Franzen O, Westman J (eds) Information processing in the somatosensory system. Macmillan, London, pp 49–58

Landy MS, Graham N (2004) Visual perception of texture. In: Chalupa LM, Werner JS (eds) The visual neurosciences. MIT Press, Cambridge, MA, pp 1106–1118

Ledberg A, O'Sullivan BT, Kinomura S, Roland PE (1995) Somatosensory activations of the parietal operculum of man. A PET study. Eur J Neurosci 7:1934–1941

Lederman SJ, Klatzky RL (2004) Multisensory texture perception. In: Calvert E, Spence C, Stein B (eds) Handbook of multisensory processes. MIT Press, Cambridge, MA, pp 107–122

Lederman SJ (1974) Tactile roughness of grooved surfaces: the touching process and effects of macro and microsurface structure. Percept Psychophys 16:385–395

Lederman SJ (1979) Auditory texture perception. Perception 8:93–103

Lederman SJ (1983) Tactual roughness perception: spatial and temporal determinants. Can J Psychol 37:498–511

Lederman SJ, Abbott SG (1981) Texture perception: studies of intersensory organization using a discrepancy paradigm, and visual versus tactual psychophysics. J Exp Psychol: Human Percept Perform 7:902–915

Lederman SJ, Klatzky RL, Hamilton C, Grindley M (2000) Perceiving surface roughness through a probe: effects of applied force and probe diameter. Proc ASME Dyn Syst Contr Div DSC-vol. 69–2:1065–1071

Lederman SJ, Klatzky RL, Morgan T, Hamilton C (2002) Integrating multimodal information about surface texture via a probe: relative contributions of haptic and touch-produced sound sources. 10th symposium on haptic interfaces for virtual environment and teleoperator systems. IEEE Computer Society, Los Alamitos, CA, pp 97–104

Lederman SJ, Loomis JM, Williams D (1982) The role of vibration in tactual perception of roughness. Percept Psychophys 32:109–116

Lederman S, Summers C, Klatzky R (1996) Cognitive salience of haptic object properties: role of modality-encoding bias. Perception 25(8):983–998

Lederman SJ, Taylor MM (1972) Fingertip force surface geometry and the perception of roughness by active touch. Percept Psychophys 12:401–408

Lederman SJ, Taylor MM (1972) Fingertip force surface geometry and the perception of roughness by active touch. Percept Psychophys 12:401–408

Lederman SJ, Thorne G, Jones B (1986) Perception of texture by vision and touch: multidimensionality and intersensory integration. J Exp Psychol: Human Percept Perform 12:169–180

Meftah E-M, Belingard L, Chapman CE (2000) Relative effects of the spatial and temporal characteristics of scanned surfaces on human perception of tactile roughness using passive touch. Exp Brain Res 132:351–361

Merabet LB, Hamilton R, Schlaug G, Swisher JD, Kiriakopoulos ET, Pitskel NB, Kauffman T, Pascual-Leone A (2008) Rapid and reversible recruitment of early visual cortex for touch. PLoS ONE 3(8):e3046. doi:10.1371/journal.pone.0003046

O'Sullivan BT, Roland PE, Kawashima R (1994) A PET study of somatosensory discrimination in man. Microgeometry versus macrogeometry. Eur J Neurosci 6:137–148

Pascual-Leone A, Hamilton R (2001) The metamodal organization of the brain. In: Casanova C, Ptito M (eds) Progress in brain research vol. 134, Chapter 27. Amsterdam, Elsevier, pp 1–19

Picard D, Dacremont C, Valentin D, Giboreau A (2003) Perceptual dimensions of tactile textures. Acta Psychol 114(2):165–184

Plomp R, Steeneken HJ (1968) Interference between two simple tones. J Acoust Soc Am 43(4):883–884

Pont SC, Koenderink JJ (2005) Bidirectional texture contrast function. Int J Comp Vis 66:17–34

Rao AR, Lohse GL (1996) Towards a texture naming system: identifying relevant dimensions of texture. Vis Res 36(11):1649–1669

Rasch R, Plomp R (1999) The perception of musical tones. In: Deutsch D (ed) The psychology of music, 2nd edn. Academic Press, San Diego, CA, pp 89–112

Roland PE, O'Sullivan B, Kawashima R (1998) Shape and roughness activate different somatosensory areas in the human brain. Proc Natl Acad Sci 95:3295–3300

Ross HE (1997) On the possible relations between discriminability and apparent magnitude. Br J Math Stat Psychol 50:187–203

Servos P, Lederman S, Wilson D, Gati J (2001) fMRI-derived cortical maps for haptic shape texture and hardness. Cogn Brain Res 12:307–313

Smith AM, Chapman E, Deslandes M, Langlais J-S, Thibodeau M-P (2002) Role of friction and tangential force variation in the subjective scaling of tactile roughness. Exp Brain Res 144: 211–223

Srinivasan MA, Whitehouse JM, LaMotte RH (1990) Tactile detection of slip: surface microgeometry and peripheral neural codes. J Neurophysiol 63:1323–1332

Stilla R, Sathian K (2008) Selective visuo-haptic processing of shape and texture. Human Brain Map 29:1123–1138

Suzuki Y, Gyoba J, Sakamoto S (2008) Selective effects of auditory stimuli on tactile roughness perception. Brain Res 1242:87–94

Suzuki Y, Suzuki M, Gyoba J (2006) Effects of auditory feedback on tactile roughness perception. Tohoku Psychol Folia 65:45–56

Taylor MM, Lederman SJ (1975) Tactile roughness of grooved surfaces: a model and the effect of friction. Percept Psychophys 17:23–36

Treisman A (1982) Perceptual grouping and attention in visual search for features and for objects. J Exp Psychol: Human Percept Perform 8:194–214

Unger BJ (2008) Psychophysics of virtual texture perception. Technical Report CMU-RI-TR-08-45, Robotics Institute, Carnegie Mellon University, Pittsburgh, PA, USA

Unger B, Hollis R, Klatzky R (2007) JND analysis of texture roughness perception using a magnetic levitation haptic device. Proceedings of the second joint eurohaptics conference and symposium on haptic interfaces for virtual environment and teleoperator systems, IEEE Computer Society, Los Alamitos, CA, 22–24 March 2007, pp 9–14

Unger B, Hollis R, Klatzky R (2008) The geometric model for perceived roughness applies to virtual textures. Proceedings of the 2008 symposium on haptic interfaces for virtual environments and teleoperator systems, 13–14 March 2008, IEEE Computer Society, Los Alamitos, CA, pp 3–10

Whitaker TA, Simões-Franklin C, Newell FN (2008) Vision and touch: independent or integrated systems for the perception of texture? Brain Res 1242:59–72

Yoshioka T, Bensmaïa SJ, Craig JC, Hsiao SS (2007) Texture perception through direct and indirect touch: an analysis of perceptual space for tactile textures in two modes of exploration. Somatosens Mot Res 24(1–2):53–70

Zampini M, Spence C (2004) The role of auditory cues in modulating the perceived crispness and staleness of potato chips. J Sens Stud 19:347–363

Chapter 13
Dorsal and Ventral Cortical Pathways for Visuo-haptic Shape Integration Revealed Using fMRI

Thomas W. James and Sunah Kim

13.1 Introduction

Visual object recognition is pervasive and central to many aspects of human functioning. In adults, it seems effortless and nearly automatic. Despite the ease with which we perceive and identify objects, however, computer simulations of object recognition have been largely unsuccessful at mimicking human recognition. Simulations can succeed in constrained environments, but cannot match the flexibility of the human system. One reason machine vision may have had limited success outside of highly constrained contexts is that visual recognition is an extremely difficult computational problem (Lennie, 1998). Another reason, however, may be that computational approaches have largely restricted themselves to modeling the visual system in isolation from other sensory and motor systems, whereas human visual recognition is embedded in interactions between multiple sensory systems (Clark, 1997; de Sa and Ballard, 1998). Although research of multisensory phenomena has a long history (Molyneux, 1688), research into the neural mechanisms of sensory processes in humans and other primates has been dominated in recent years by investigations of unisensory visual function. This has led to a relative paucity of empirical data from – and theoretical treatment of – other sensory systems and, perhaps most importantly, interactions between multiple sensory systems. Our goals in this chapter are twofold. First, we describe an influential theoretical perspective on the organization of the cortical visual system, the two visual streams theory, and apply that perspective to interactions between visual and haptic object shape processing. Second, using a new methodology, we assess neuronal convergence of visual and haptic inputs in regions considered part of those two separable pathways.

T.W. James (✉)
Department of Psychological and Brain Sciences, Cognitive Science Program, Indiana University, 1101 E Tenth St., Bloomington, IN 47405, USA
e-mail: thwjames@indiana.edu

M.J. Naumer, J. Kaiser (eds.), *Multisensory Object Perception in the Primate Brain*,
DOI 10.1007/978-1-4419-5615-6_13,

13.2 Visual Cortical Pathways for Action and Perception

For almost three decades, one of the most dominant theories of visual system organization has been the two visual streams theory (Goodale and Milner, 1992; Ungerleider and Mishkin, 1982). There are two prominent sets of visual projections in the primate cerebral cortex: the ventral stream, which arises in area V1 and projects to the inferotemporal cortex, and the dorsal stream, which also arises in area V1 and projects to the posterior parietal cortex (Baizer et al., 1991; Morel and Bullier 1990; Ungerleider and Mishkin, 1982; Young, 1992). In the early 1990s, Goodale and Milner (Goodale and Milner, 1992; Milner and Goodale, 1995) argued that the ventral stream plays the major role in constructing our perceptual representation of the visual world and the objects within it, while the dorsal stream mediates the visual control of actions that we direct at those objects. This idea was different from the initial conceptualization of "what–where" dual pathways (Ungerleider and Mishkin, 1982). The what–where hypothesis suggested that both streams were involved in perception, but different aspects of perception. The ventral (what) stream was intimately involved in identifying objects, whereas the dorsal (where) stream was involved in locating them in space.

In the Goodale and Milner model, the same information (that is, the same rudimentary features of objects) is processed by both streams, but for different purposes. In other words, the input to both streams is the same, but the outputs are different. Because the inputs are the same and the outputs are different, the calculations and transformations that each stream performs on the input must be different. In the case of the ventral stream, object features and parameters are transformed in such a way that it produces our phenomenological experience of the world, allowing us to deliberate and reason about our choice of actions. In the case of the dorsal stream, the same object information is transformed in such a way that it is useful for *controlling* those actions.

Some of the most compelling evidence for Goodale and Milner's perception–action hypothesis has come from studies of patient DF, a young woman who suffered irreversible brain damage in 1988 as a result of hypoxia from carbon monoxide poisoning (Milner et al., 1991). Studies of DF's visual abilities have shown that DF is unable to report the size, shape, and orientation of an object, either verbally or manually. On the other hand, an analysis of her visuo-motor abilities demonstrates that she shows normal pre-shaping and rotation of her hand when reaching out to grasp the same objects. In other words, DF is able to use visual information about the location, size, shape, and orientation of objects to control her grasping movements (and other visually guided movements), despite the fact that she is unable to perceive and report those same object features. As a concrete example of this dissociation, DF can orient her hand correctly to grasp a rectangular plaque that is placed in front of her at different orientations, as assessed by biomechanical measurements. However, when presented with those same plaques and asked to report the orientation without acting on the object, she cannot do it (Goodale et al., 1991). In fact, she cannot even judge whether a grating stimulus presented on a computer screen is vertical or horizontal (Humphrey et al., 1995).

Structural MRI of DF's brain shows that her lesion is relatively focal and is located bilaterally in an area known as the lateral occipital complex (LOC) (James et al., 2003). The LOC has been studied in healthy individuals for almost 15 years. The LOC is a region on the lateral surface of the cortex at the junction of the occipital and temporal lobes (Malach et al., 1995). The results of many studies have confirmed that the LOC produces more activation with intact objects than with any other class of control stimuli tested. Although the mechanisms of object recognition are unknown and remain the source of intense study, researchers agree that the LOC is intimately involved in visual object recognition, categorization, and naming (Grill-Spector et al., 2001; James et al., 2000; Kourtzi et al., 2003). The location of DF's lesion site and her inability to recognize objects suggests not only that the LOC is involved in visual object recognition but also that an intact LOC is *necessary* for recognition of objects.

Although the lesion to DF's occipito-temporal cortex was relatively focal, there was evidence of smaller scale atrophy throughout the brain, as indicated by enlarged sulci and ventricles (James et al., 2003). Despite the abnormal structural appearance of the atrophied regions, however, BOLD responses from those regions appeared normal. This was in dramatic contrast with the lesion sites, in which the signal resembled those from cerebral spinal fluid, that is, there was no BOLD response. One of the areas that showed striking similarity of functional response between DF and healthy control subjects was the intraparietal sulcus (IPS). In humans and other primates, the IPS has been implicated in planning object-directed actions and spatially representing the environment (Culham and Valyear, 2006; Culham et al., 2006; Frey et al., 2005; Grefkes and Fink, 2005). DF is able to use visual input to guide her object-directed actions, such as reaching and grasping. When she performed grasping actions in the MRI, BOLD activation in her IPS region resembled activation seen in healthy control subjects (James et al., 2003). Thus, DF's case provides clear and converging evidence from behavioral and neuroimaging measures for a separation of function between the dorsal and the ventral visual cortical streams.

The idea of two separate processing pathways, either the what–where hypothesis or the action–perception hypothesis, has been extremely influential in the study of vision, but it has also influenced the study of other sensory systems. A dual-streams approach has been adopted by other researchers to explain the organization of the auditory system (Arnott et al., 2004; Hickok and Poeppel, 2004; Saur et al., 2008) and somatosensory system (Dijkerman and de Haan, 2007; James et al., 2007; Reed et al., 2005).

13.3 Converging Visual and Somatosensory Pathways

Like the visual system, the somatosensory system is organized hierarchically and potentially into two or more separate pathways. There are at least three different two-stream hypotheses that describe the organization of the neural substrates involved in haptic exploration of the environment and, specifically, exploration of

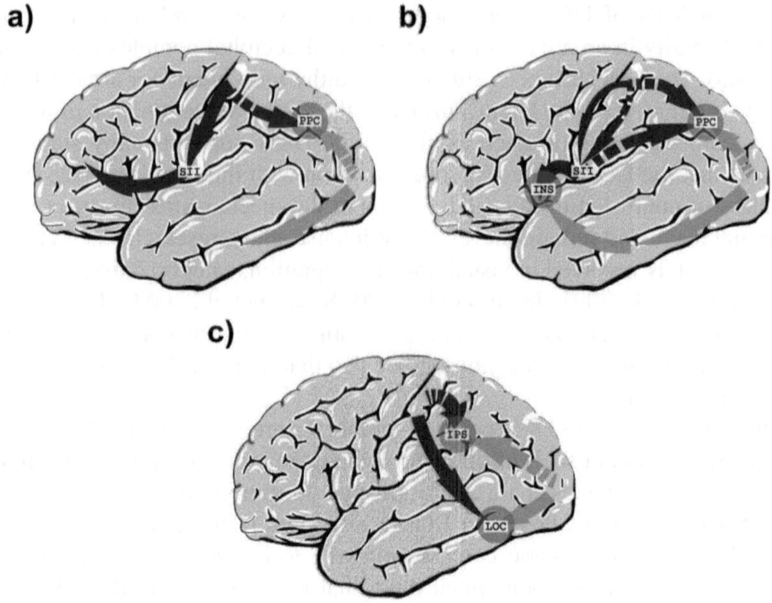

Fig. 13.1 Three two-stream models of haptic object processing. *Light gray arrows* represent visual streams originating in posterior occipital cortex. *Dark gray arrows* represent somatosensory streams originating in post-central cortex. *Solid arrows* denote ventral stream projections, whereas *dashed arrows* denote dorsal stream projections. *Transparent circles* represent zones of multisensory convergence. Abbreviations: LOC, lateral occipital cortex; IPS, intraparietal sulcus; INS, insula; PPC, posterior parietal cortex, SII, secondary somatosensory cortex

physical objects (Dijkerman and de Haan, 2007; James et al., 2007; Reed et al., 2005). The first model (Fig. 13.1a) contends that haptic signals that code object identity are processed in a ventral pathway, which projects from primary somatosensory cortex to the inferior parietal lobe and prefrontal cortex, whereas haptic signals that code the spatial location of objects are processed in a dorsal pathway, which projects from primary somatosensory cortex to the superior parietal lobe (Reed et al., 2005). This theory is similar to the what–where hypothesis in vision, separating the processing of identity from the processing of location (Ungerleider and Mishkin, 1982).

The second model (Fig. 13.1b) contends that processing of haptic signals for object recognition and perception is accomplished by a ventral pathway, which projects from primary and secondary somatosensory cortex to the insula, whereas processing of haptic signals for motor actions is accomplished by a dorsal pathway, which projects from primary and secondary somatosensory cortex to the posterior parietal cortex (Dijkerman and de Haan, 2007). This theory is similar to the visual action–perception hypothesis, separating the processing of objects for perceptual judgments and the processing of objects for actions. An important consideration in this model is the *interaction* between the dorsal and the ventral pathways. Although the visual action–perception hypothesis allows for interactions between the dorsal and the ventral streams, Dijkerman and colleagues suggest that these interactions should be stronger for haptic streams than for vision. A second consideration is the formation of a body representation or body sense. Representations in vision would tend to be of the external environment, even though the receptors are located within the body. Representations of one's own body are completely internal, which may make this type of representation different from visual representations. A third consideration is that the dorsal and ventral streams for vision and haptics may have sites of convergence within the cortex, that is, the pathways may converge at specific sites to integrate information, and these sites may be specific to dorsal and ventral stream function.

The third model (Fig. 13.1c) specifically addresses the convergence of two haptic and two visual streams (James et al., 2007). Haptic object processing is organized into two streams: a ventral stream that projects from the primary somatosensory cortex to the lateral occipito-temporal cortex (LOC) and a dorsal stream that projects from primary somatosensory cortex to the posterior parietal cortex, specifically the intraparietal sulcus (IPS). These two cortical sites (LOC and IPS) mark points of convergence between the visual and the haptic streams of object processing. Convergence of the dorsal visual and haptic pathways is specialized for processing multisensory shape cues to plan object-directed motor actions, whereas convergence of the ventral visual and haptic pathways is specialized for processing multisensory shape cues for object perception, which in turn allows for deliberation and reasoning about our choice of actions directed toward those objects.

Findings to support these models come from a combination of behavioural and neuroimaging studies with patients and healthy subjects, and neurophysiological single-unit recording in nonhuman primates. In the late 1990s, a group of cross-modal haptic–visual priming studies (Easton et al., 1997a, b; Reales and Ballesteros, 1999) changed conceptions in cognitive psychology of how object shape may be represented. In these studies, previous experience with an object facilitated subsequent performance when naming that object. The important finding was that (in most cases) the facilitation occurred whether the sensory modalities of the initial experience and the subsequent test matched or mismatched. The findings of these studies, and subsequent studies (Newell et al., 2001; Norman et al., 2004), suggested that a common representation of shape was used by vision and haptics.

Subsequently, several fMRI studies demonstrated that haptic object recognition recruited areas of putative visual cortex (Amedi et al., 2001; James et al., 2002; Reed et al., 2004; Sathian et al., 1997), suggesting an overlap of visual and haptic *neural* representations for objects. The visual cortical region most consistently recruited in these studies was a sub-region of the LOC, which has been labeled by some as LOtv, for tactile–visual (Amedi et al., 2002). The visual and somatosensory systems process many characteristics of objects. One of the most salient characteristics of objects is their shape (texture is also salient, but shape seems to be the most salient). The results of a number of studies suggest that the key characteristic of objects that leads to recruitment of LOtv is their shape (Amedi et al., 2007; James et al., 2002; Stilla and Sathian, 2008). Some studies also suggest that shape is the most important characteristic for IPS (Culham et al., 2006; Grefkes et al., 2002; Kitada et al., 2006; Peltier et al., 2007). Thus, the processing in LOtv (and perhaps IPS) may not be "visual" or even "visuo-haptic," but instead may be "metamodal." In other words, *shape* information may be processed in LOtv regardless of input modality (Amedi et al., 2007; Pascual-Leone and Hamilton, 2001).

Behavioral data collected from patients with visual agnosia converge with these neuroimaging findings (Feinberg et al., 1986; James et al., 2005; Morin et al., 1984; Ohtake et al., 2001). For instance, patient DF's lesion in the occipito-temporal cortex overlaps with the location of LOC in healthy subjects and has impaired her ability to name or match objects visually, especially when those judgments must be made based on an object's shape (Humphrey et al., 1994; James et al., 2003). However, DF's haptic object recognition ability is also impaired compared to healthy subjects. On three separate tasks, old/new recognition, sequential matching, and paired associates, DF was equally impaired when using vision or haptics (James et al., 2005). In addition to these patient lesion data, transcranial magnetic stimulation (TMS) has been used to produce "transient virtual lesions" in the cortex of healthy individuals. Disrupting neural processing in the occipital cortex caused impairments on a tactile orientation discrimination task (Zangaladze et al., 1999). Taken together, these behavioral and neuroimaging findings from patients and healthy subjects converge to suggest that the visual and haptic systems share overlapping neural substrates for object recognition based on shape analysis (Amedi et al., 2005; James et al., 2007). The most likely candidate for that neural substrate is the LOtv, which resides in what is considered the ventral perceptual stream of visual processing.

Even more intuitive than the convergence of vision and haptics for the recognition of objects is the convergence of vision and haptics for manual interaction with objects. Haptic feedback plays a large role in the calibration of visuo-motor actions (Coats et al., 2008). Neuroimaging studies have shown that at least one area of the intraparietal sulcus (IPS) is involved in bi-modal visuo-haptic processing of shape or geometric properties of objects (Bodegard et al., 2001; Culham and Kanwisher, 2001; Grefkes et al., 2002; Peltier et al., 2007; Roland et al., 1998; Zhang et al., 2004). The IPS is also intimately involved in the planning and execution of sensorimotor actions, including eye movements, pointing, reaching, and grasping (Culham and Valyear, 2006; Grefkes and Fink, 2005). Of particular relevance are the anterior and middle aspects of the IPS. The functional significance of these areas is broad, including the preparation of grasping movements, sensitivity to visual or haptic input, and the processing of object shape and size. Patients with damage to IPS can suffer from tactile apraxia, which is characterized by an inability to recognize objects haptically due to inappropriate use of exploratory movements (Binkofski et al., 1998, 2001; Pause, 1989). These converging lines of evidence have led researchers to conclude that IPS is a site of convergence for several inter-related sensorimotor processes that rely on visual, haptic, and motor information to analyze object shape.

Data from nonhuman primate single-unit recordings strongly support the claim for bi-modal visual and somatosensory processing in IPS (Buneo et al., 2002; Murata et al., 2000; Taira et al., 1990), which is considered homologous with IPS in humans (Grefkes and Fink, 2005; Grefkes et al., 2002). Support for bi-modal processing in LOtv from single-unit recording data, however, is much less sure. One issue is that a nonhuman primate homologue for LOtv has not been established (Tootell et al., 2003). But, despite the issue of homology, there is some evidence that neurons in the ventral visual stream of nonhuman primates do receive both visual and tactile inputs (Maunsell et al., 1991). Thus, the current state of research into visual and haptic shape processing pathways in humans suggests that each sensory system has two functionally specialized cortical pathways, and these pathways converge on at least two separate cortical locations, LOtv and IPS. LOtv is involved in visual and haptic processing of shape information for the purpose of recognition and perception, whereas IPS is responsible for processing visual and haptic shape information for the purposes of guiding object-directed actions. What has not been directly tested, however, is the manner in which signals from the two sensory inputs are combined or integrated in these areas.

13.4 Measuring Neuronal Convergence with BOLD fMRI

When describing the convergence of sensory inputs onto brain regions, researchers in the field of multisensory neurophysiology distinguish between two types of convergence: areal convergence and neuronal convergence (Meredith, 2002). In describing the research on convergence of visual and haptic inputs in the preceding sections, this distinction was not made. *Areal convergence* describes the case when different sensory inputs project to neurons in the same brain region, but do not synapse on the exact same neurons. Because the inputs do not synapse on the same neurons, there is no interaction or integration of the inputs. *Neuronal convergence*, on the other hand, describes the case when inputs from different sensory systems project to the same neurons. By synapsing on the same neurons, the inputs interact at the neural level and can be integrated. Areal convergence and neuronal convergence are relatively simple to dissociate with single-unit recording. If a neuron changes its activity when the animal is simultaneously stimulated through two sensory modalities compared to only one, then the neuron is integrating those inputs (Meredith and Stein, 1983; Stein and Stanford, 2008). Because single-unit recording is difficult or impossible to perform in humans, multisensory integration in the human brain has been investigated using functional neuroimaging techniques. Because BOLD fMRI activation is measured from clusters of voxels that represent large populations of neurons, distinguishing between areal and neuronal convergence with fMRI invokes a different set of criteria than with single units. Because fMRI is a newer methodology than single-unit recording, these criteria are not as well established.

One issue with predicting the strength BOLD activation with multisensory stimuli is that populations of neurons in known multisensory cortical regions contain unisensory as well as multisensory neurons (Barraclough et al., 2005; Benevento et al., 1977; Hikosaka et al., 1988). Under this assumption, the null hypothesis to be rejected is that a multisensory stimulus produces activation equivalent to the sum of that produced by the unisensory components (Calvert et al., 2000; Laurienti et al., 2005). This is because the combination stimulus should excite the unisensory neurons *at least* as effectively as the component stimuli. Only if the combination stimulus produces more activation than this additive null hypothesis ("superadditivity"), do the results imply an interaction between sensory streams. Based on the known neurophysiology, the most likely interpretation of an interaction is that there is a third pool of multisensory neurons in the population, in addition to the two unisensory pools. To be clear, this hypothetical third pool of neurons includes any neuron with a response that is not considered unisensory. Thus, the third pool includes all types of multisensory neurons, including those that show multisensory enhancement and those that show suppression, those with linear or additive responses with multisensory stimuli, and those with nonlinear responses with multisensory stimuli. Known sites of multisensory convergence, such as the superior temporal sulcus (STS) for audio-visual stimuli, have many different types of multisensory neurons combined with unisensory neurons. Thus, a site like STS should produce a pattern of activation that rejects the null hypothesis of only two pools of unisensory neurons. In practice, though, known sites of multisensory convergence like STS rarely show statistically significant evidence of superadditivity with BOLD fMRI signals (Beauchamp, 2005b; Stevenson et al., 2007). Thus, other factors must play a role in determining whether or not BOLD fMRI measurements can detect the presence of a pool of multisensory neurons.

It is generally understood that BOLD fMRI measurements lack a natural zero value or well-defined baseline (Raichle et al., 2001; Stark and Squire, 2001). Because of this constraint, BOLD is considered a "relative" measure of neural activation, rather than an absolute measure. In other words, only *differences* in BOLD response between conditions are meaningful, not the absolute levels. It is possible that the use of absolute BOLD values in the calculation of an additive criterion has led to a lack of consistency across studies assessing multisensory integration of specific brain regions.

The influence of an arbitrary baseline on the superadditive criterion is illustrated graphically in Fig. 13.2. The two top graphs show raw BOLD data collected in two different experiments. Because this is a simulation, we can make the data in the two experiments the same, except for one important factor, which is that experimenters for the different experiments have chosen to use different baselines (or have done so unknowingly). In Experiment 1, the baseline is slightly below the "true baseline" (natural zero), and in Experiment 2, the baseline is slightly above the "true baseline." Raw BOLD activation for the multisensory stimulus condition (VH) is simulated based on a neural population composed of *only unisensory visual and unisensory haptic neurons*. Thus, the result of both experiments should be to *fail to reject* the null hypothesis. The established practice in fMRI is to convert raw BOLD values into percent signal change values by subtracting the value of the baseline condition and then dividing by it. The bottom two graphs in Fig. 13.2 show the data from the top two graphs after undergoing this transformation. Recall that the only difference between experiments was the different baseline activation; therefore, differences between the left and right bottom graphs are due only to a difference in the

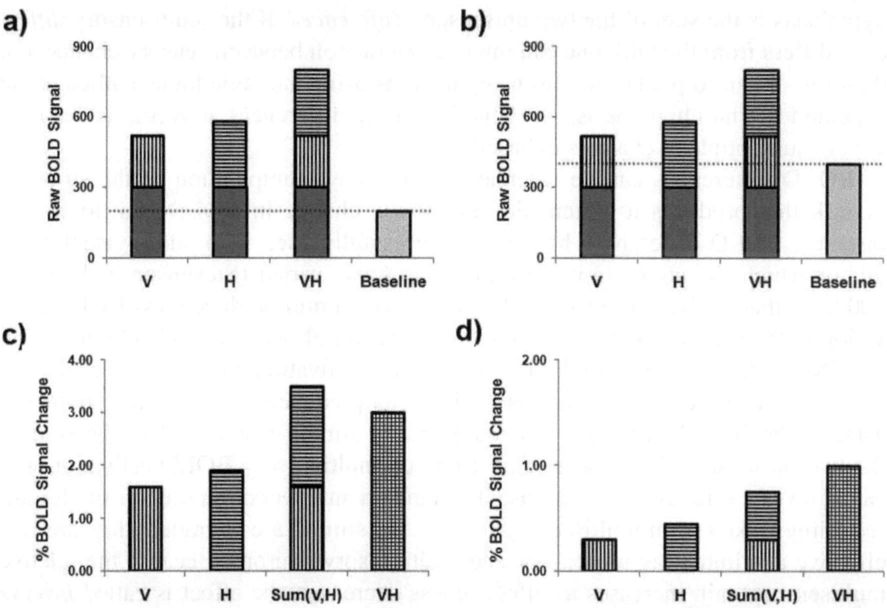

Fig. 13.2 The influence of baseline activation on the absolute additive criterion. (**a**) and (**b**) The height of stacked bars indicates the contribution of different factors to the raw BOLD signal. *Dark gray bars* indicate an arbitrary value added during the reconstruction of MRI images. *Horizontal* and *vertical lines* indicate the contribution of neural activity from visual (V) or haptic (H) sensory channels, respectively. The *light gray bar* is the BOLD activation produced by the "baseline" condition. (**c**) and (**d**) Percent BOLD signal change calculated based on the different baseline activation values shown in (**a**) and (**b**), respectively. Signal change values are proportional to the difference between the total height of the stacked bars and the dotted line indicating the level of baseline activation. The Sum (V,H) is the absolute additive criterion. (**c**) Superadditivity and (**d**) Subadditivity

baseline. The signal change values in the two graphs are clearly different. Both of these experiments would reject the null hypothesis, but the effects are in the opposite direction. More alarming is that rejecting the null hypothesis would reveal nothing about the underlying neural populations, but is completely dependent on the activation of the baseline condition. It is possible that this type of influence may explain the inconsistency in results from different research groups using superadditivity as a criterion (Beauchamp, 2005a, b; Beauchamp et al., 2004a, b; Laurienti et al., 2006; Peiffer et al., 2007; Stevenson and James, 2009; Stevenson et al., 2007).

Because absolute BOLD measurements produce inconsistent results when used as a criterion for the assessment of multisensory integration, we have recently turned to using relative *differences* in BOLD activation. The use of relative differences alleviates the problem of an indeterminate baseline, because the baseline components embedded in the two measurements cancel out when a difference operation is performed. The null hypothesis for these BOLD differences is similar to absolute BOLD measurements and follows a similar hypothesis of additivity, that is, the null hypothesis is the sum of the two unisensory *differences*. If the multisensory *difference* differs from the null, one can infer an interaction between sensory channels in the form of a third pool of multisensory neurons using the same logic applied to the superadditive null hypothesis. The benefit of using differences, however, is that they are not susceptible to changes in baseline.

BOLD differences can be calculated across any manipulation of the stimulus or task that produces a systematic, monotonic change in BOLD activation. For instance, BOLD differences have been successfully used with audio-visual stimuli for which the signal-to-noise ratio (SNR) was varied (Stevenson and James, 2009). In that study, unisensory audio and visual stimuli produced less BOLD activation in the superior temporal sulcus (STS) as stimuli were degraded by lowering the SNR. Multisensory stimuli also showed less activation with reduced SNR, but the decrease in activation was not as large as predicted by the null hypothesis. That is, the BOLD difference for multisensory stimuli was less than the sum of the two unisensory differences. This effect on multisensory BOLD activation was called inverse effectiveness, because it resembled an effect often seen in single-unit recordings taken from multisensory regions. As stimuli are degraded, they are less effective at stimulating unisensory and multisensory neurons. Because the relative multisensory gain increases as effectiveness decreases, the effect is called *inverse effectiveness* (Meredith and Stein, 1986). Although a pattern of inverse effectiveness in BOLD activation does not necessarily imply that neurons in that area are inversely effective, it does imply a difference from the null hypothesis, and thus an interaction between sensory channels.

13.5 Sites of Visuo-haptic Neuronal Convergence

Although there is considerable evidence for bi-modal visual and haptic processing of object shape in the primate brain in at least two cortical sites, LOtv and IPS, until very recently (Tal and Amedi, 2009) a test for neuronal convergence in humans using fMRI had not been reported. Using the BOLD differences method described above, we designed an experiment to test for neuronal convergence of visual and haptic inputs in the LOC and IPS (Kim and James in press). Our stimulus manipulation was to vary the level of stimulus quality. Finding a significant "difference of differences" across levels of stimulus quality would confirm the presence of multisensory integration, even in the absence of superadditivity. Based on previous results with audio-visual integration using this method, we hypothesized that the direction of that difference would be in the direction predicted by inverse effectiveness, that is, as stimulus quality was reduced, there should be a smaller drop in multisensory activation than predicted based on the drop in unisensory activation.

We localized ROIs using a standard method that contrasts visual objects (VO) with visual textures (VT) and haptic objects (HO) with haptic textures (HT) (Amedi et al., 2001). The visual contrast is the same standard functional localizer used to isolate the LOC visually (Malach et al., 1995). The haptic contrast typically activates a large cluster along the entire IPS. A conjunction of the two contrasts isolates overlapping regions that are object selective for both sensory modalities. In the ventral stream, the overlapping region is labeled LOtv and is found consistently with the conjunction contrast. An overlapping region in the dorsal stream is less consistently found with the conjunction contrast (Amedi et al., 2005; Lacey et al., 2009), perhaps because researchers do not control the parameters of active exploration. Nevertheless, our use of the conjunction contrast did find statistically reliable clusters of voxels in the IPS. Figure 13.3 shows the two functional ROIs, LOtv and IPS, localized on the group-average functional data ($N = 7$) and superimposed on group-average anatomical images. The analysis used to produce the maps was a conjunction of four contrasts, HO–HT and VO–VT and HO–VT and VO–HT. This analysis localizes regions that are bi-modal and object selective and that also have equal contribution from visual and haptic conditions. Images on the left and right of Fig. 13.3 are shown at different statistical thresholds. At the more conservative threshold (Fig. 13.3a, c), the right-hemisphere cluster does not survive. At the more liberal threshold (Fig. 13.3b, d), the left-hemisphere cluster is much larger than the right. The same left-hemisphere bias for haptic or visuo-haptic object processing has been shown in previous studies; however, the significance of the possible lateralization of functional is unknown. We wanted to analyze both the left and the right hemispheres and wanted to equate the size (i.e., number of voxels) of the left- and right-hemisphere clusters. Due to the reliable left-hemisphere bias, this

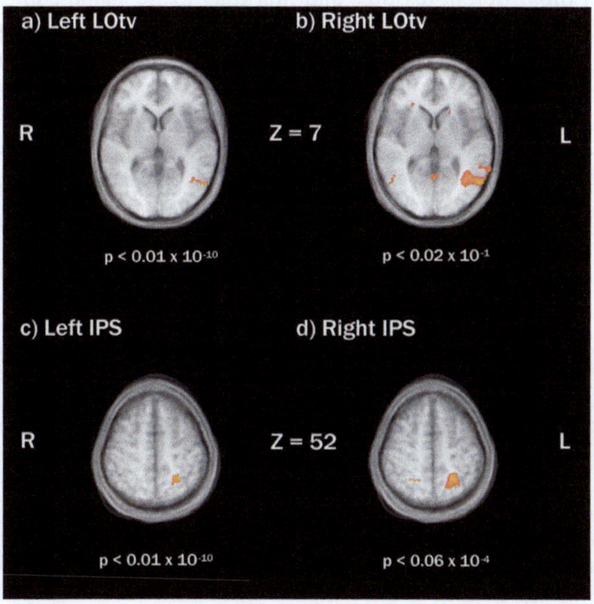

Fig. 13.3 Bi-modal visuo-haptic object-selective regions of interest. The boundaries of the regions of interest used in the analyses are outlined in *bright yellow*. The heights of the axial slices in Talairach space are indicated by the Z = labels

meant using a different statistical threshold for determining the left- and right-hemisphere ROIs. Thus, the left-hemisphere ROIs were the yellow-outlined clusters in Panels A and C and the right-hemisphere ROIs were the yellow-outlined clusters in Panels B and D. The large, non-outlined clusters in the left hemisphere in Panels B and D were not analyzed. Importantly, the threshold for the left-hemisphere ROIs was set at a typically conservative value. A more liberal threshold was used only for the right-hemisphere ROI. The results of these contrasts were consistent with previous research suggesting that LOtv and IPS are instrumental in processing object shape information (Amedi et al., 2005). They are also consistent with previous work suggesting that LOtv and IPS are sites of convergence for visual and haptic sensory inputs (James et al., 2007).

To assess inverse effectiveness and superadditivity in these ROIs, we presented subjects with novel objects from two categories and instructed them to perform a two-alternative forced-choice decision. Sixteen objects were used. All objects were created by attaching four wooden geon-like geometric components in a standard configuration to provide differences in shape information (Biederman, 1987). In this experiment, only one of the components was diagnostic of category membership. Eight objects in Category 1 had a half-circle-shaped diagnostic component, and the other eight objects in Category 2 had a triangle-shaped diagnostic component (Fig. 13.4a). Each stimulus was 14 cm wide and 9.5 cm long. Texture on the stimuli was determined by the size of nylon beads that were glued to the surface. Textures and non-diagnostic features could not be used to perform the 2AFC task. The purpose of the different textures and non-diagnostic shape components was to add complexity and variability to the psychological object similarity space and keep subjects more interested in the task, but subjects were explicitly instructed to use only the diagnostic shape feature to perform the task. The distribution of textures and non-diagnostic components was the same across the two object categories.

For visual presentation, a grayscale picture of each stimulus was presented using an LCD projector and rear-projection screen. Subjects viewed the images using a rear-face mirror attached to the head coil. Visual stimuli were presented at $12° \times 8°$

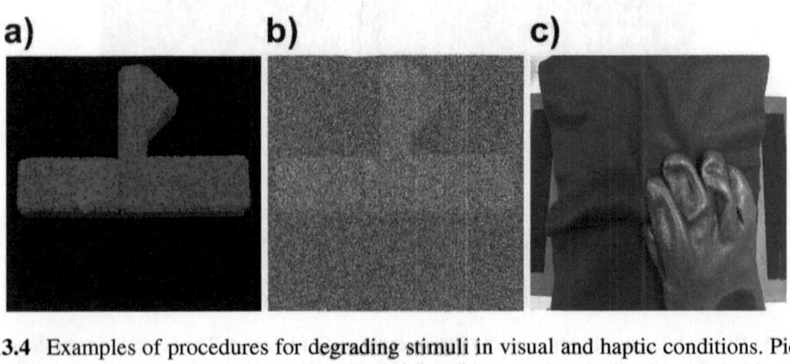

Fig. 13.4 Examples of procedures for degrading stimuli in visual and haptic conditions. Pictures of an undegraded object (**a**), a visually degraded object (**b**), and a haptically degraded object being palpated (**c**)

of visual angle. To establish different levels of salience, a fixed level of external Gaussian noise was added to the stimuli (Fig. 13.4b) and the stimulus signal contrast was reduced to two different levels. These levels were calibrated independently for each individual and reflected their 71 and 89% performance thresholds.

For haptic presentation, an angled "table" was placed on the subject's midsection. An experimenter standing in the MRI room delivered tangible stimuli to the subject by placing them on a designated spot on the table. Subjects palpated the objects with eyes closed with both hands and were instructed not to lift the objects from the table. To establish different levels of salience, subjects wore a pair of PVC gloves, which reduced their tactile sensitivity. Individual 71 and 89% performance thresholds were measured by covering objects with a different number of layers of thick felt fabric (Fig. 13.4c). The layered fabric further reduced tactile sensitivity, but unlike the gloves, allowed the experimenter to rapidly change between performance (or sensitivity) levels.

Subjects and the experimenter both wore headphones and listened for a sequence of auditory cues that indicated when subjects should start and stop hand movements and when the experimenter should switch out the stimulus. The final design had two factors: stimulus quality and stimulus modality, with two levels of quality (high and low) and three modalities (visual [V], haptic [H], and visuo-haptic [VH]). Seven subjects participated in the experiment. Imaging parameters and data pre-processing steps were standard and are described elsewhere (Kim and James in press). Accuracy and reaction time measures showed strong effects of stimulus quality. In the unisensory conditions, accuracy levels were close to 71% for low quality and 89% for high quality, the performance levels for which the conditions were calibrated. Reaction times were also slower for the low quality than high quality condition.

BOLD activations in left and right LOtv and IPS are shown in Fig. 13.5 for all conditions in the 2×3 design. The null hypothesis for the additive model is also presented for each level of stimulus quality and labeled S(V,H). It is apparent from comparing the VH stimulus condition to the null hypothesis that there is no evidence of superadditivity in either LOtv or IPS. The VH stimulus condition produced subadditive activation in all cases except for the high quality condition in right LOtv, which was additive. Thus, if superadditivity were the only criterion, we would infer from these data that LOtv and IPS do not represent sites of neuronal convergence for visual and haptic sensory inputs.

Figure 13.6 shows the results of the new BOLD differences analysis, performed on the same data shown in Fig. 13.5. Instead of comparing absolute levels of BOLD activation, recall that this analysis compares *differences* in BOLD activation. Thus, the height of the bars in Fig. 13.6 represents the difference in BOLD activation between high- and low-signal quality conditions. This difference is represented for visual (ΔV), haptic (ΔH), and visuo-haptic (ΔVH) stimulus conditions. The null hypothesis to be rejected is represented by the sum of unisensory differences (i.e., S(ΔV, ΔH) bar). There is a clear difference between the ΔVH and S(ΔV, ΔH) for both LOtv and IPS in the left hemisphere. The difference in the right hemisphere is in the same direction, but is less robust. Based on this rejection of the

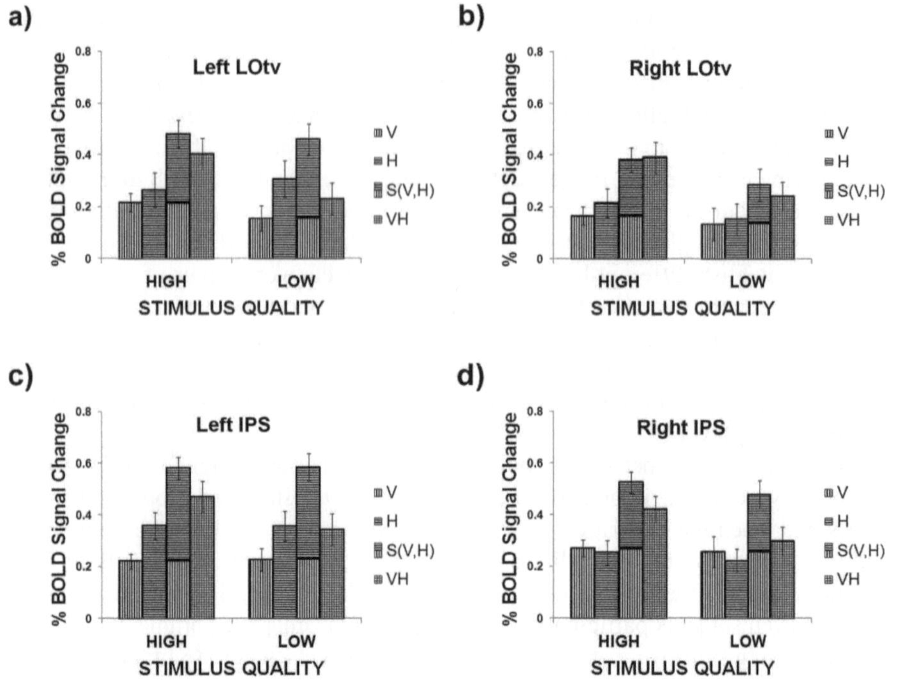

Fig. 13.5 BOLD percent signal change as a function of sensory condition, stimulus quality, region of interest, and hemisphere shown for the following regions: a) left LOtv, b) right LOtv, c) left IPS and d) right IPS. S(V, H) represented the absolute additive criterion

null hypothesis, we can infer that the underlying neuronal population is not composed of two pools of unisensory neurons: one visual and one haptic. Based on the neurophysiology of known multisensory areas, a likely inference is that these two areas contain a mixture of unisensory and multisensory neurons (Meredith and Stein, 1983, 1986).

Rejection of our null hypothesis, though, simply means that there was a difference between ΔVH and the sum of ΔV and ΔH. A difference in either direction implies an interaction between sensory modalities, possibly due to a third pool of multisensory neurons. But, the direction shown in Fig. 13.6 was unexpected. One of the general principles of single-unit recordings from multisensory neurons is that multisensory enhancement increases as the stimuli are degraded in quality. That is, the gain in activity with a multisensory stimulus over and above the activity with a unisensory stimulus increases with decreasing quality. Low-quality stimuli are also less effective at stimulating unisensory and multisensory neurons. Thus, inverse effectiveness describes the phenomenon that as the effectiveness of a stimulus decreases, the multisensory gain increases. Given that inverse effectiveness is a known principle of neural activity in multisensory neurons, we predicted that activity in LOtv and IPS, if the null hypothesis was rejected, would also show evidence of inverse effectiveness. However, the pattern of change shown in Figs. 13.5 and 13.6

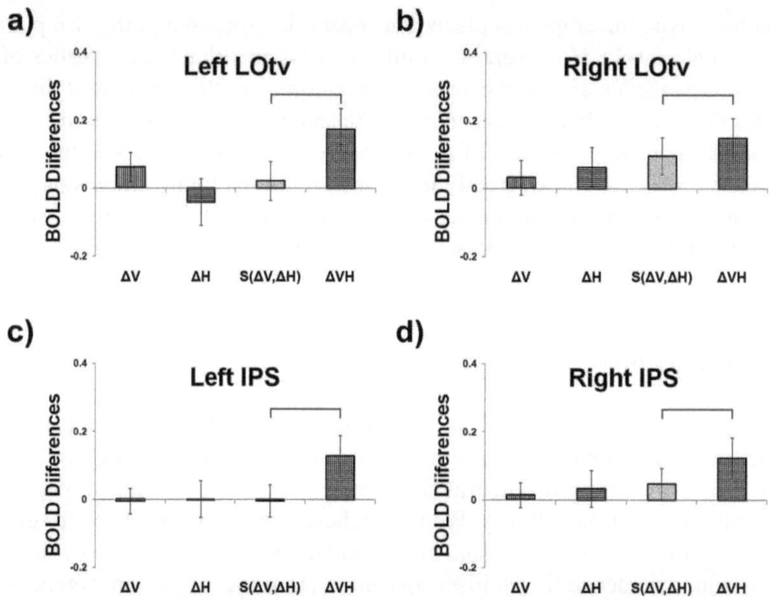

Fig. 13.6 BOLD differences as a function of sensory condition, region of interest, and hemisphere shown for the following regions: a) left LOtv, b) right LOtv, c) left IPS and d) right IPS. Differences were calculated as high quality minus low quality. S(ΔV,ΔH) represents the additive-differences criterion. ΔVH represents the multisensory difference, which is compared to S(ΔV,ΔH) to establish the presence of enhanced or inverse effectiveness

is the opposite. The multisensory gain is stronger for high-quality stimuli than for low-quality stimuli. It is important to note that the opposite direction of the effect is not due to an indirect relation between stimulus quality and brain activation (effectiveness). That is, if high-quality stimuli produced less activation in these regions than low-quality stimuli, then that alone could make the change in gain appear to go in the opposite direction.

The multisensory gains shown in Figs. 13.5 and 13.6 are clearly stronger with the high-quality stimuli. This effect is the opposite of that seen with inverse effectiveness (Meredith and Stein, 1986). Thus, we suggest that this effect should be called "enhanced effectiveness" (Kim and James in press), because as the effectiveness of the stimuli is enhanced, the multisensory gain is also disproportionately enhanced.

Both LOtv and IPS showed evidence for integration of visual and haptic sensory inputs and both LOtv and IPS showed enhanced effectiveness. The effect was in the same direction in both the left and the right hemispheres, but was much stronger in the left. The results provide further evidence that the visual and haptic systems process object shape through two pathways and that LOtv and IPS represent points of convergence for those two pathways (James et al., 2007). LOtv forms part of the ventral or perceptual pathway for vision and haptics, and IPS forms part of the dorsal or action pathway for vision and haptics. The results suggest that integration of visual and haptic sensory inputs is *similar* in LOtv and IPS. This result may be unexpected, given the emphasis placed on separable processes in the two pathways (Goodale et al., 1991). However, our results suggest only that the properties of multisensory convergence are similar in the two regions, not that the underlying neural processes that rely on that convergence are the same. Although speculative, our data may suggest that the properties of multisensory convergence are similar in different areas of cortex. To more directly test this hypothesis, future studies of neuronal convergence of visual and haptic sensory channels in ventral and dorsal pathways should investigate both object recognition and object-directed actions.

13.6 Conclusions

We have reviewed the evidence for separable action and perception pathways in both the visual and the haptic systems for the analysis of object shape. These systems converge on at least two neural sites: one in the dorsal action pathway and one in the ventral perception pathway. We tested these sites of convergence for evidence of multisensory integration of visual and haptic inputs. In the absence of superadditivity, we found evidence for multisensory integration (neuronal convergence) using a new method that employs relative BOLD differences instead of absolute BOLD values. Dorsal and ventral sites showed the same general pattern of multisensory integration. This suggests that integration of object shape information in the dorsal and ventral streams may occur by the same general mechanisms.

Acknowledgments This research was supported in part by the Faculty Research Support Program, administered by the Indiana University Office of the Vice President of Research, and in part by the Indiana METACyt Initiative of Indiana University, funded in part through a major grant from the Lilly Endowment, Inc.

References

Amedi A, Jacobson G, Hendler T, Malach R, Zohary E (2002) Convergence of visual and tactile shape processing in the human lateral occipital complex. Cereb Cortex 12:1202–1212

Amedi A, Malach R, Hendler T, Peled S, Zohary E (2001) Visuo-haptic object-related activation in the ventral visual pathway. Nat Neurosci 4:324–330

Amedi A, Stern WM, Camprodon JA, Bermpohl F, Merabet L, Rotman S, Hemond C, Meijer P, Pascual-Leone A (2007) Shape conveyed by visual-to-auditory sensory substitution activates the lateral occipital complex. Nat Neurosci 10:687–689

Amedi A, von Kriegstein K, van Atteveldt NM, Beauchamp MS, Naumer MJ (2005) Functional imaging of human crossmodal indentification and object recognition. Exp Brain Res 166: 559–571

Arnott SR, Binns MA, Grady CL, Alain C (2004) Assessing the auditory dual-pathway model in humans. Neuroimage 22:401–408

Baizer JS, Ungerleider LG, Desimone R (1991) Organization of visual inputs to the inferior temporal and posterior parietal cortex in macaques. J Neurosci 11:168–190

Barraclough NE, Xiao D, Baker CI, Oram MW, Perrett DI (2005) Integration of visual and auditory information by superior temporal sulcus neurons responsive to the sight of actions. J Cogn Neurosci 17:377–391

Beauchamp MS (2005a) See me, hear me, touch me: multisensory integration in lateral occipital-temporal cortex. Curr Opin Neurobiol 15:145–153

Beauchamp MS (2005b) Statistical criteria in FMRI studies of multisensory integration. Neuroinformatics 3:93–113

Beauchamp MS, Argall BD, Bodurka J, Duyn JH, Martin A (2004a) Unraveling multisensory integration: patchy organization within human STS multisensory cortex. Nat Neurosci 7: 1190–1192

Beauchamp MS, Lee KE, Argall BD, Martin A (2004b) Integration of auditory and visual information about objects in superior temporal sulcus. Neuron 41:809–823

Benevento LA, Fallon J, Davis BJ, Rezak M (1977) Auditory--visual interaction in single cells in the cortex of the superior temporal sulcus and the orbital frontal cortex of the macaque monkey. Exp Neurol 57:849–872

Biederman I (1987) Recognition-by-components: a theory of human image understanding. Psychol Rev 94:115–147

Binkofski F, Dohle C, Posse S, Stephan KM, Hefter H, Seitz RJ, Freund HJ (1998) Human anterior intraparietal area subserves prehension: a combined lesion and functional MRI activation study. Neurology 50:1253–1259

Binkofski F, Kunesch E, Classen J, Seitz RJ, Freund HJ (2001) Tactile apraxia: unimodal apractic disorder of tactile object exploration associated with parietal lobe lesions. Brain 124: 132–144

Bodegard A, Geyer S, Grefkes C, Zilles K, Roland PE (2001) Hierarchical processing of tactile shape in the human brain. Neuron 31:317–328

Buneo CA, Jarvis MR, Batista AP, Andersen RA (2002) Direct visuomotor transformations for reaching. Nature 416:632–636

Calvert GA, Campbell R, Brammer MJ (2000) Evidence from functional magnetic resonance imaging of crossmodal binding in the human heteromodal cortex. Curr Biol 10:649–657

Clark A (1997) Being there. MIT Press, Cambridge

Coats R, Bingham GP, Mon-Williams M (2008) Calibrating grasp size and reach distance: interactions reveal integral organization of reaching-to-grasp movements. Exp Brain Res 189:211–220

Culham JC, Cavina-Pratesi C, Singhal A (2006) The role of parietal cortex in visuomotor control: what have we learned from neuroimaging? Neuropsychologia 44:2668–2684

Culham JC, Kanwisher NG (2001) Neuroimaging of cognitive functions in human parietal cortex. Curr Opin Neurobiol 11:157–163

Culham JC, Valyear KF (2006) Human parietal cortex in action. Curr Opin Neurobiol 16:205–212

de Sa VR, Ballard DH (1998) Perceptual learning from cross-modal feedback. In: Goldstone RL, Schyns PG, Medin DL (eds) Psychology of learning and motivation, vol 36. Adademic Press, San Diego, CA

Dijkerman HC, de Haan EH (2007) Somatosensory processes subserving perception and action. Behav Brain Sci 30:189–201; discussion 201–139

Easton RD, Greene AJ, Srinivas K (1997a) Transfer between vision and haptics: memory for 2-D patterns and 3-D objects. Psychonomic Bulletin and Review 4:403–410

Easton RD, Srinivas K, Greene AJ (1997b) Do vision and haptics share common representations? Implicit and explicit memory within and between modalities. J Exp Psychol: Learn, Mem, Cogn 23:153–163

Feinberg TE, Gonzalez Rothi LJ, Heilman KM (1986) Multimodal agnosia after unilateral left hemisphere lesion. Neurology 36:864–867

Frey SH, Vinton D, Norlund R, Grafton ST (2005) Cortical topography of human anterior intraparietal cortex active during visually guided grasping. Brain Res Cogn Brain Res 23:397–405

Goodale MA, Milner AD (1992) Separate visual pathways for perception and action. Trends Neurosci 15:20–25

Goodale MA, Milner AD, Jakobson LS, Carey DP (1991) A neurological dissociation between perceiving objects and grasping them. Nature 349:154–156

Grefkes C, Fink GR (2005) The functional organization of the intraparietal sulcus in humans and monkeys. J Anat 207:3–17
Grefkes C, Weiss PH, Zilles K, Fink GR (2002) Crossmodal processing of object features in human anterior intraparietal cortex: an fMRI study implies equivalencies between humans and monkeys. Neuron 35:173–184
Grill-Spector K, Kourtzi Z, Kanwisher N (2001) The lateral occipital complex and its role in object recognition. Vision Res 41:1409–1422
Hickok G, Poeppel D (2004) Dorsal and ventral streams: a framework for understanding aspects of the functional anatomy of language. Cognition 92:67–99
Hikosaka K, Iwai E, Saito H, Tanaka K (1988) Polysensory properties of neurons in the anterior bank of the caudal superior temporal sulcus of the macaque monkey. J Neurophysiol 60: 1615–1637
Humphrey GK, Goodale MA, Corbetta M, Aglioti S (1995) The McCollough effect reveals orientation discrimination in a case of cortical blindness. Curr Biol 5:545–551
Humphrey GK, Goodale MA, Jakobson LS, Servos P (1994) The role of surface information in object recognition: studies of a visual form agnosic and normal subjects. Perception 23: 1457–1481
James TW, Culham JC, Humphrey GK, Milner AD, Goodale MA (2003) Ventral occipital lesions impair object recognition but not object-directed grasping: an fMRI study. Brain 126: 2463–2475
James TW, Humphrey GK, Gati JS, Menon RS, Goodale MA (2000) The effects of visual object priming on brain activation before and after recognition. Curr Biol 10:1017–1024
James TW, Humphrey GK, Gati JS, Servos P, Menon RS, Goodale MA (2002) Haptic study of three-dimensional objects activates extrastriate visual areas. Neuropsychologia 40:1706–1714
James TW, James KH, Humphrey GK, Goodale MA (2005) Do visual and tactile object representations share the same neural substrate? In: Heller MA, Ballesteros S (eds) Touch and blindness: psychology and neuroscience. Lawrence Erlbaum, Mahwah, NJ
James TW, Kim S, Fisher JS (2007) The neural basis of haptic object processing. Can J Exp Psychol 61:219–229
Kim S, James TW (in press) Enhanced effectiveness in visuo-haptic object-selective brain regions with increasing stimulus salience. Hum Brain Mapp
Kitada R, Kito T, Saito DN, Kochiyama T, Matsumura M, Sadato N, Lederman SJ (2006) Multisensory activation of the intraparietal area when classifying grating orientation: a functional magnetic resonance imaging study. J Neurosci 26:7491–7501
Kourtzi Z, Tolias AS, Altmann CF, Augath M, Logothetis NK (2003) Integration of local features into global shapes: monkey and human fMRI studies. Neuron 37:333–346
Lacey S, Tal N, Amedi A, Sathian K (2009) A putative model of multisensory object representation. Brain Topogr 21:269–274
Laurienti PJ, Burdette JH, Maldjian JA, Wallace MT (2006) Enhanced multisensory integration in older adults. Neurobiol Aging 27:1155–1163
Laurienti PJ, Perrault TJ, Stanford TR, Wallace MT, Stein BE (2005) On the use of superadditivity as a metric for characterizing multisensory integration in functional neuroimaging studies. Exp Brain Res 166:289–297
Lennie P (1998) Single units and visual cortical organization. Perception 27:889–935
Malach R, Reppas JB, Benson RR, Kwong KK, Jiang H, Kennedy WA, Ledden PJ, Brady TJ, Rosen BR, Tootell RB (1995) Object-related activity revealed by functional magnetic resonance imaging in human occipital cortex. Proc Natl Acad Sci U S A 92:8135–8139
Maunsell JH, Sclar G, Nealey TA, DePriest DD (1991) Extraretinal representations in area V4 in the macaque monkey. Vis Neurosci 7:561–573
Meredith MA (2002) On the neuronal basis for multisensory convergence: a brief overview. Brain Res Cogn Brain Res 14:31–40
Meredith MA, Stein BE (1983) Interactions among converging sensory inputs in the superior colliculus. Science 221:389–391

Meredith MA, Stein BE (1986) Visual, auditory, and somatosensory convergence on cells in superior colliculus results in multisensory integration. J Neurophysiol 56:640–662
Milner AD, Goodale MA (1995) The visual brain in action. Oxford University Press, Oxford, UK
Milner AD, Perrett DI, Johnston RS, Benson RS, Jordan PJ, Heeley TR, Bettucci DW, Mortara D, Mutani F, Terazzi R, Davidson DLW (1991) Perception and action in visual form agnosia. Brain 114:405–428
Molyneux W (1688) Letter to John Locke. In: de Beer ES (ed) The correspondence of John Locke, vol 3. Clarendon Press, Oxford
Morel A, Bullier J (1990) Anatomical segregation of two cortical visual pathways in the macaque monkey. Vis Neurosci 4:555–578
Morin P, Rivrain Y, Eustache F, Lambert J, Courtheoux P (1984) Visual and tactile agnosia. Rev Neurol 140:271–277
Murata A, Gallese V, Luppino G, Kaseda M, Sakata H (2000) Selectivity for the shape, size, and orientation of objects for grasping in neurons of monkey parietal area AIP. J Neurophysiol 83:2580–2601
Newell FN, Ernst MO, Tjan BS, Bulthoff HH (2001) Viewpoint dependence in visual and haptic object recognition. Psychol Sci 12:37–42
Norman JF, Norman HF, Clayton AM, Lianekhammy J, Zielke G (2004) The visual and haptic perception of natural object shape. Percept Psychophys 66:342–351
Ohtake H, Fujii T, Yamadori A, Fujimori M, Hayakawa Y, Suzuki K (2001) The influence of misnaming on object recognition: a case of multimodal agnosia. Cortex 37:175–186
Pascual-Leone A, Hamilton R (2001) The metamodal organization of the brain. Prog Brain Res 134:427–445
Pause M, Kunesch E, Binkofsku F, Freund HJ (1989) Sensorimotor disturbances in patients with lesions of the parietal corex. Brain 112:1599–1625
Peiffer AM, Mozolic JL, Hugenschmidt CE, Laurienti PJ (2007) Age-related multisensory enhancement in a simple audiovisual detection task. Neuroreport 18:1077–1081
Peltier S, Stilla R, Mariola E, LaConte S, Hu X, Sathian K (2007) Activity and effective connectivity of parietal and occipital cortical regions during haptic shape perception. Neuropsychologia 45:476–483
Raichle ME, MacLeod AM, Snyder AZ, Powers WJ, Gusnard DA, Shulman GL (2001) A default mode of brain function. Proc Natl Acad Sci U S A 98:676–682
Reales JM, Ballesteros S (1999) Implicit and explicit memory for visual and haptic objects: corss-modal priming depends on structural descriptions. J Exp Psychol: Learn, Mem, Cogn 25: 644–663
Reed CL, Klatzky RL, Halgren E (2005) What vs. where in touch: an fMRI study. Neuroimage 25:718–726
Reed CL, Shoham S, Halgren E (2004) Neural substrates of tactile object recognition: an fMRI study. Hum Brain Mapp 21:236–246
Roland PE, O'Sullivan B, Kawashima R (1998) Shape and roughness activate different somatosensory areas in the human brain. Proc Natl Acad Sci 95:3295–3300
Sathian K, Zangaladze A, Hoffman JM, Grafton ST (1997) Feeling with the mind's eye. Neuroreport 8:3877–3881
Saur D, Kreher BW, Schnell S, Kummerer D, Kellmeyer P, Vry MS, Umarova R, Musso M, Glauche V, Abel S, Huber W, Rijntjes M, Hennig J, Weiller C (2008) Ventral and dorsal pathways for language. Proc Natl Acad Sci U S A 105:18035–18040
Stark CE, Squire LR (2001) When zero is not zero: the problem of ambiguous baseline conditions in fMRI. Proc Natl Acad Sci U S A 98:12760–12766
Stein BE, Stanford TR (2008) Multisensory integration: current issues from the perspective of the single neuron. Nat Rev Neurosci 9:255–266
Stevenson RA, Geoghegan ML, James TW (2007) Superadditive BOLD activation in superior temporal sulcus with threshold non-speech objects. Exp Brain Res 179:85–95

Stevenson RA, James TW (2009) Audiovisual integration in human superior temporal sulcus: Inverse effectiveness and the neural processing of speech and object recognition. Neuroimage 44:1210–1223

Stilla R, Sathian K (2008) Selective visuo-haptic processing of shape and texture. Hum Brain Mapp 29:1123–1138

Taira M, Mine S, Georgopoulos AP, Murata A, Sakata H (1990) Parietal cortex neurons of the monkey related to the visual guidance of hand movement. Exp Brain Res 83:29–36

Tal N, Amedi A (2009) Multisensory visual-tactile object related network in humans: insights gained using a novel crossmodal adaptation approach. Exp Brain Res 198:165–182

Tootell RB, Tsao D, Vanduffel W (2003) Neuroimaging weighs in: Humans meet macaques in "primate" visual cortex. J Neurosci 23:3981–3989

Ungerleider LG, Mishkin M (1982) Two cortical visual systems. In: Ingle DJ, Goodale MA, Mansfield RJ (eds) The analysis of visual behavior. MIT Press, Cambridge, MA, pp 549–586

Young MP (1992) Objective analysis of the topological organization of the primate cortical visual system. Nature 358:152–155

Zangaladze A, Epstein CM, Grafton ST, Sathian K (1999) Involvement of visual cortex in tactile discrimination of orientation. Nature 401:587–590

Zhang M, Weisser VD, Stilla R, Prather SC, Sathian K (2004) Multisensory cortical processing of object shape and its relation to mental imagery. Cogn Affect Behav Neurosci 4:251–259

Chapter 14
Visuo-haptic Perception of Objects and Scenes

Fiona N. Newell

14.1 Introduction

Although both the visual and the tactile modalities can extract, encode, and process spatial information from objects for the purpose of recognition and localisation, relatively little is understood about how such information is shared across these modalities. The aim of this chapter is to provide an overview of our current understanding of multisensory integration for the purpose of high-level perception. This chapter is structured around evidence in support of the idea that vision and touch contribute to two basic perceptual tasks, namely object recognition and the perception of the spatial location of objects, and that these sensory modalities process information for the purpose of recognition or localisation in a very similar manner. Moreover, visual and tactile processing of spatial object information is underpinned by shared neural substrates, although the extent to which these substrates are shared seems to be task dependent. I will argue that it is the common way in which information is processed by vision and touch, underpinned by shared neural substrates, that allows for efficient sharing of information across these modalities for recognition and localisation.

The reader may be interested to note the historical context to the topic of this chapter. The year 2009 marks the tercentenary of the publication of George Berkeley's (1709) *An Essay Towards a New Theory of Vision* [first edition] which he penned whilst he was a scholar in Trinity College Dublin. In his essay he declares that he will "consider the Difference there is betwixt the Ideas of Sight and Touch, and whether there be any Idea common to both Senses". The proposals outlined in his essay offered such a unique insight into the issue of how vision and touch contribute to perception that they still resonate today. Indeed, 300 years following the publication of this essay, many researchers around the world are grappling with the very same issues laid out in Berkeley's essay. On the face of it, this may sound rather

F.N. Newell (✉)
School of Psychology and Institute of Neuroscience, Lloyd Building, Trinity College, Dublin 2, Ireland
e-mail: fiona.newell@tcd.ie

M.J. Naumer, J. Kaiser (eds.), *Multisensory Object Perception in the Primate Brain*,
DOI 10.1007/978-1-4419-5615-6_14,

pessimistic as is suggests that not much progress has been achieved in the science of visuo-tactile integration in 300 years. On the contrary, the current state of knowledge is a significant achievement but before we could provide empirical evidence for the philosophical questions raised by Berkeley it was first necessary for several other significant scientific discoveries to be made. These include (but are most certainly not limited to) Darwin's theory of evolution and the consequent development of comparative studies to provide insight into neuronal processes in multisensory integration; advances in neuroscience such as Cajal's discovery of the synapse and Hubel and Wiesel's discovery of the structure of the visual cortex; the emergence of new scientific disciplines such as experimental psychology and cognitive neuroscience; and the advent of technology such as computers and neuroimaging. These and many other advances have afforded us a multidisciplinary account of how the senses contribute to and result in a coherent perception of the world around us.

To highlight the advances in our understanding particularly of how vision and touch share information about objects and the layout of objects that surround us, research studies have generally focused on either the perception of object information for recognition or the perception of space for localisation. Over the following sections of this chapter, I will review to what extent information is shared across these modalities for the recognition of objects and spatial arrangements of objects in scenes and provide evidence for common functional organisation of these modalities for the purpose of each task.

14.2 Evidence for Common Principles of Functional Organisation Across Vision and Touch

Over the past few decades, research into visual processing has provided evidence that this system is structurally and functionally separated into two streams, namely the occipitotemporal (i.e. "what") and occipitoparietal (i.e. "where") stream. Accordingly, each stream is involved in the processing of visual information for different, goal-directed, purposes. The occipitotemporal pathway projects from primary visual cortex to the ventral areas of the brain in the temporal cortex and the occipitoparietal pathway projects to the dorsal areas of the brain to the parietal cortex and there is thought to be limited crosstalk between these streams (Ungerleider and Haxby, 1994; Young, 1992). Functionally, these areas are specified as being involved in either the recognition of objects (i.e. "what") or the perception of space for localisation ("where") or action ("how"). Evidence for this structural and functional dichotomy has been provided through lesion studies in animals (Mishkin et al., 1983) and from neuropsychological patient studies (Goodale and Milner, 1992). Furthermore, it is thought that this dual processing in the visual system is optimal to allow for the efficient processing of visual information for the purposes of recognition or action (e.g. Young, 1992).

If the visual system has an underlying structure that facilitates the efficient processing of information for perception, then we might ask whether the other senses are also similarly organised. Recent neuroimaging studies have suggested that this dual processing may also apply to both the human auditory system (Romanski et al., 1999) and the tactile system (Reed et al., 2005; Van Boven et al., 2005). In particular, Reed et al. conducted a neuroimaging study of tactile processing in which they asked participants to conduct either an object recognition or an object localisation task using the same, everyday familiar objects as stimuli and the same motor movements across conditions. They reported that these tasks selectively activated different brain regions. In particular brain regions involved in feature integration, such as inferior parietal areas, were activated during object recognition whereas brain regions involved in spatial processing, such as superior parietal regions, were activated during the object location task. Interestingly, a brain area typically involved in the recognition of familiar objects through vision (Grill-Spector et al., 1999; Malach et al., 1995) and touch (Amedi et al., 2001), that is, the lateral occipital complex, was not selectively activated during the object recognition task, suggesting that cortical activation of brain regions involved in object recognition may also depend on the nature of the exploration (i.e. recognition based on global shape or collection of local features) or on the possible role of imagery especially in the recognition of more familiar objects (see Lacey et al., 2009).

In a related study, we investigated brain activation during a tactile shape-matching and feature location task using the same unfamiliar objects interchangeably across these tasks (Newell et al., 2010b). See Fig. 14.1a for an example of some of our stimuli. Our results corroborate those reported by Reed et al. and suggest that, similar to the visual system, the tactile system is organised around distinct functional regions of the brain, each selectively involved in the processing of object information for recognition and for spatial localisation. Specifically, we found activation in ventral regions of the cortex, corresponding to the areas within and around the right lateral occipital complex, which were activated during the shape-matching task. In contrast, brain regions activated during the feature location task included more dorsal areas such as the left supramarginal and angular gyrus and also more ventral areas such as areas in and around the middle to lateral temporal region. Figure 14.1b provides an illustration of these activations.

Taken together, both the Reed et al. study and our study provide evidence for distinct processing of object shape and spatial location in the tactile system. In particular, clear associated areas of activation, particularly in the lateral occipital complex, are observed for object recognition tasks across both vision and touch (see also Amedi et al. 2001; Lacey et al., 2009). That the same cortical substrates underpin both visual and tactile object recognition suggests that similar object information is extracted by both modalities and consequently accessible to each of these senses. For the spatial tasks, although there is no evidence that a single cortical area was consistently activated by tactile and visual spatial perception across these studies, areas which were activated tended to lie within dorsal regions. Thus, the results from these studies suggest that the functional distinction between the object recognition and spatial tasks is preserved in the tactile system. The fact that specific cortical areas were not commonly activated across the studies may reflect task- and/or stimulus-specific spatial information processing. Nevertheless, the lack of

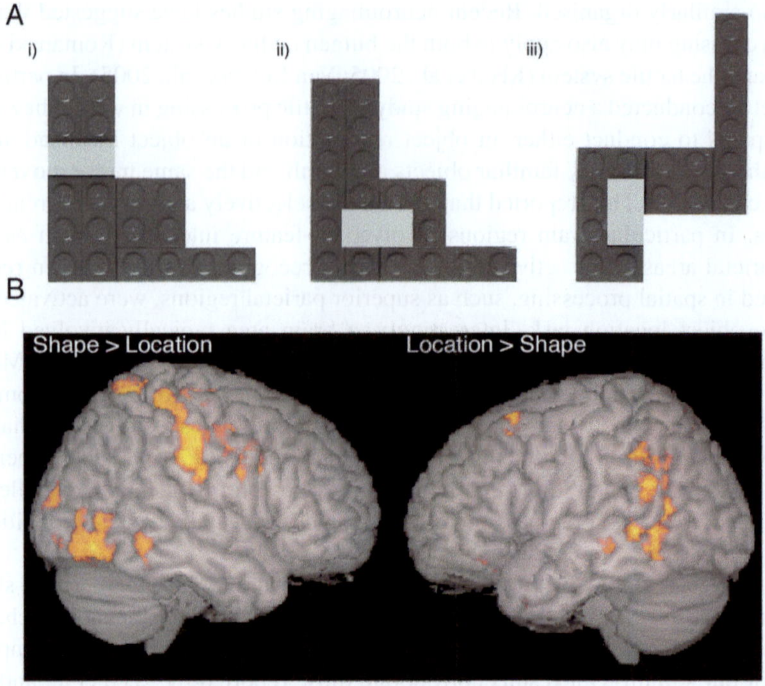

Fig. 14.1 (**a**) An example of the unfamiliar objects used in the Newell et al. (2010a, b) and Chan and Newell (2008) studies. Objects (i) and (ii) represent identically shaped objects, but the location of a feature on the objects changed across object shapes (i.e. same shape but different feature location). Objects (i) and (iii) represent differently shaped objects. (**b**) An illustration of the differential activations for each of the tasks ($z > 2.3$, $p < 0.05$ corrected)

common areas involved in visual and tactile spatial tasks raises the question of how efficiently spatial information is integrated across vision and touch since it suggests that common spatial information may not be extracted by these modalities. This question will be addressed in the following sections.

Recent behavioural evidence has corroborated the findings from our neuroimaging studies into the dual processing of object recognition and spatial localisation within and across the visual and tactile senses. We investigated the extent to which object information is processed separately and independently for the purpose of recognition and spatial location within vision and touch and also across these modalities using a dual-interference task (Chan and Newell, 2008). Using the same set of novel object stimuli interchangeably across tasks (see Fig. 14.1a), participants were required to perform an object shape or a feature location delayed match-to-sample task. In the shape task, for example, participants were presented with an unfamiliar object shape which they learned either through vision or touch. Following an inter-stimulus interval (ISI) of about 20 s, they were then presented with a second object shape which they had to judge as either the same or different to the shape of the first stimulus. The feature localisation task followed the same protocol but participants had to judge whether an object feature remained in the same relative location across the object surfaces or not. We then embedded a dual-interference, or secondary, task into these primary tasks by presenting a second matching task during the ISI of the primary task. This secondary task was either the same as the primary task (but with different stimuli) or the opposite task. In other words, if participants conducted a shape-based primary task then the embedded secondary task was either another shape-based task or a feature localisation task. Our aim was to investigate whether visual and haptic memory performance during a shape-matching task was interfered by another shape-related task or by a spatial task or by both, and whether visual and haptic performance on an object localisation task was affected by either an interfering shape task or a spatial task. We found that performance on a within-modal visual and haptic object shape-matching task was affected by a secondary shape-related task only and not a spatial task. Furthermore, our results suggested a double dissociation of task function since performance on a spatial task was affected by a secondary spatial task only and not a shape-matching task. Thus, our results suggest that there is a functional independence within both the visual system and the tactile system for the processing of object information for the purpose of recognition or spatial location. In our final experiment in that series, we investigated whether task-related interference was modality specific or independent of modality by embedding a secondary task that was conducted in a different modality to the primary task. Our results suggested that performance on spatial location tasks was affected by a crossmodal spatial task but not by a crossmodal shape-related task. In contrast, however, performance on a shape-matching task was affected by both shape and feature localisation secondary tasks. Thus, whereas a primary spatial task was not affected by a secondary, crossmodal shape-related task, a primary shape task was affected by both a crossmodal shape and a crossmodal spatial task.

Interestingly, this latter finding is consistent with our neuroimaging study discussed earlier in which we found that an object localisation task not only activates more dorsal areas in the parietal lobe but also areas in the temporal lobe which have been previously associated with shape perception, particularly the middle to lateral occipital areas (see Location > Shape activation image in Fig. 14.1b). In sum, evidence from both neuroimaging and behavioural studies converge to support the idea for functional independence of task-based information processing related to recognition and localisation not only within the visual and tactile modalities but also across these modalities, with the caveat that there may be some sharing of resources for object-based, feature localisation tasks across modalities. Since it has previously been argued that the functional distinction between "what" and "where" streams facilitates efficient information processing for recognition or action within the visual system (Young, 1992), we can probably assume that the distinction in the somatosensory system between cortical areas involved in object recognition or spatial localisation similarly benefits tactile perception. However, in order to maintain a coherent, multisensory perception of our world the brain must also allow for efficient cross-sensory information processing for the purpose of the task at hand. This may either be achieved by allowing the most appropriate sensory modality to dominate the perceptual outcome (Welch and Warren, 1986) or by merging the sensory information into a robust representation for perception or action (e.g. Ernst and Bülthoff, 2004). The following sections will examine the evidence supporting the idea that information is shared across vision and touch for the purpose of object recognition and spatial perception. Based on this evidence, I will argue that efficient crossmodal interactions seem to be determined by the extent to which principles of information processing are shared across vision and touch. Moreover, evidence is also emerging from neuroimaging studies that the cortical areas subserving object recognition and object localisation, although distinct within each modality, are largely overlapping across these modalities.

14.3 Evidence for Common Principles of Information Processing Across Vision and Touch for Object Recognition

Up to relatively recently, very little was understood about how we seem to effortlessly recognise objects under different ambient conditions such as changes in illumination, colour, viewpoint, and non-rigid changes due to movement. Although many researchers have investigated how object recognition is achieved by the visual system, a particular focus of this research has been on how vision solves the problem of object constancy, that is, how objects are recognised independent of viewpoint. This research effort was particularly driven by a need for computer scientists and engineers to develop systems of object recognition that could be adaptable for the purpose of automated recognition for manufacturing, security, and medical purposes. Much effort was conducted in the 1980s and 1990s on understanding how the human brain solves such an intractable problem as object constancy, with the idea that once this is understood in humans then these principles can be adopted into the design and development of computer systems and robots that could recognise objects at least well as humans, if not better. Needless to say, technology still lags far behind the capabilities of the human perceptual system although significant advances have been made through experimental psychology, cognitive neuroscience, and physiology into our understanding of how object constancy is achieved in the visual system (e.g. Biederman, 1987; Tarr and Bülthoff, 1998; Ullman, 1998).

14.3.1 Shape Constancy and the Visual Recognition of Objects Across Changes in Viewpoint

To account for view-invariant object recognition, many visual theorists initially proposed that objects were represented as structural descriptions in memory (e.g. Biederman, 1987; Marr, 1982). According to this proposal, the image of an object is deconstructed into its component 3D parts and it is the unique structural specification of these parts and their relative positions which allows the object to

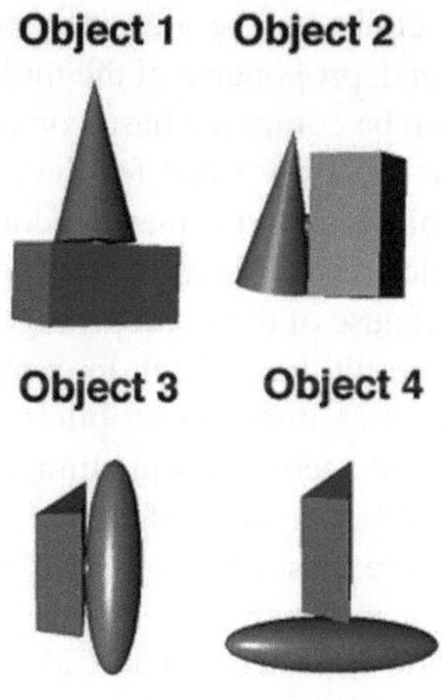

Fig. 14.2 An illustration of four distinct objects defined by either unique parts or unique arrangement of their parts. Objects 1 and 2 differ in the structural arrangements of their parts, as do objects 3 and 4. Objects 1 and 3, for example, differ both on their component parts and the structural arrangements of these parts

be recognised. Object constancy is effectively achieved provided the object parts, and the relations between them, can be resolved from the image of the object (e.g. Biederman and Gerhardstein, 1993). For example, Fig. 14.2 illustrates different objects consisting of same or different parts and arrangements of parts. All objects are recognised as different objects even though some share part shapes (e.g. Objects 1 and 2 comprise different arrangements of the same parts as do Objects 3 and 4). By reducing shapes to their basic component parts and specifying the arrangement of these parts, then object recognition should be efficient and, moreover, independent of viewpoint.

Biederman and colleagues provided evidence in support of this structural descriptions approach. They reported, for example, that recognition performance of images of objects from which their parts can be resolved is largely independent of viewpoint (Biederman and Gerhardstein, 1993) but when part information cannot be resolved then recognition is impaired (Biederman and Ju, 1998). However, other studies found evidence that object recognition is not independent of viewpoint and that even slight changes in the view of an object can disrupt performance (Bülthoff and Edelman, 1992; Newell and Findlay, 1997; Tarr, 1995; Tarr and Bülthoff, 1998). These findings led many researchers to propose that objects are represented in memory as a collection of select views and that recognition is consequently most efficient to views that match these stored views or the nearest stored view. This model was referred to as the "multiple views" model (e.g. Tarr and Bülthoff, 1998). Studies showing view dependency in object recognition tended to involve highly similar and unfamiliar object shapes as stimuli suggesting that the task demands may affect view-dependent performance (see, e.g. Newell, 1998; Hayward & Williams, 2000). Nevertheless, the recognition of highly familiar objects also seemed to be more efficient from some but not all views, known as canonical views (e.g. Palmer et al., 1981).

Thus it appeared that whereas the structural descriptions approach could account for the recognition of objects from different basic level categories, it could not account for the view-dependent recognition of novel, similar objects. On the other hand, proponents of the multiple views approach largely ignored the fact that objects can be compared based on their structure, suggesting that structural descriptions can be readily formed for the purpose of object perception. For example, in Fig. 14.2, objects 2 and 3 may be considered similar because of their basic part structure of one part adjacent to another. Similarly, objects 1 and 4 may be considered similar because of their part arrangements. If objects were stored as collection of views, then it would be difficult to explain how such structural comparisons could be achieved from an image description.

To account for the limitations of these approaches, Newell et al. (2005) proposed a hybrid model of object recognition. In a series of experiments, we found evidence in support of the idea that objects are represented as image-based parts but where the relative spatial locations of these parts are specified. Thus, our model took into account the idea that many objects can be deconstructed into component shapes and, moreover, that recognition can often be achieved when some of these components are obscured (such as a handle of a mug positioned at the back of the mug or a handbag positioned on its side). Since this model assumes that object parts are stored as images then it also predicts that recognition would be view dependent when these image-based parts are not presented in familiar or canonical orientations.

In sum, there is strong evidence from behavioural studies to suggest that the visual system does not solve the object constancy problem completely and that recognition is not completely invariant to changes in viewpoint. This evidence is also supported by studies investigating the neural correlates of object recognition which report view-dependent activations at the level of the single neuron in ventral areas of macaque brain (Logothetis and Pauls, 1995) and in the BOLD response in ventral cortical areas of the human brain (Andresen et al., 2009; Grill-Spector et al. 1998). However, other studies reported view invariance in the ventral stream with changes in rotation specifically affecting activation in more dorsal areas, particularly the IPS, more related to perception for action (e.g. James et al., 2002) or view invariance in neurons within later visual areas of the medial temporal lobe (Quiroga et al., 2005) possibly involving a consolidation of view-dependent activity from earlier visual areas. Given that the results from behavioural and neurophysiological studies converge to support the idea that recognition may not be invariant to viewpoint, it may thus seem like a conundrum why our everyday perception of objects is seemingly so robust and efficient. One potential solution may be that recognition occurs using a network of brain areas that act in unison to overcome the viewpoint problem either via a distributed coding of object information (e.g. Haxby et al., 2001) or with sparse, image-based representations (e.g. Reddy and Kanwisher, 2006) coupled with spatial processes such as mental rotation (e.g. Schendan and Stern, 2007; 2008). However, the fact that object recognition in the real world is not confined to one sensory modality may offer a clue as to how object constancy is achieved in the brain in that it may be via a combination of sensory information that objects are recognised most efficiently.

14.3.2 *Shape Constancy and the Recognition of Static Objects in Vision and Touch*

Although efforts in understanding how objects are recognised have mainly been concentrated on visual processing, object recognition is clearly not confined to the visual sense. Although objects can be identified through, for example, their characteristic sounds (e.g. the roar of a lion) or smells (e.g. a freshly peeled orange), it is only through the visual and tactile systems that object shape can be determined. Haptic perception of object shape can be very efficient and many studies have shown that familiar objects can be easily recognised using touch only (Gibson, 1962; Klatzky et al., 1985), even with very limited exposure to the object (Klatzky and Lederman, 1995). Since object information can be perceived through both vision and touch then it is possible that redundant shape information encoded across the senses may offer in a more robust representation of the object in memory (Ernst and Bülthoff, 2004).

If this is indeed the case, then it begs the question as to whether or not object recognition through the tactile system can provide the key to solving the object constancy problem. With this in mind, we investigated whether view-dependent object recognition was specific to visual processing and whether tactile object recognition is invariant to changes in object position (Newell et al., 2001). In agreement with Berkeley's statement that a blind man "By the Motion of his Hand he might discern the Situation of any Tangible Object placed within his Reach" we reasoned that since the hand is free to explore all surfaces of a 3D object (within certain size constraints), unlike vision which is constrained by optics, then haptic recognition should not necessarily be dependent on the object's position in the hand (i.e. its "view"). Surprisingly, we found that the haptic recognition of unfamiliar objects was dependent on the view of the object presented and that this cost in haptic recognition performance to a change in object view was similar to that found in visual object performance. Similar to the visual recognition of familiar objects, we also reported that the haptic system recognises some views of familiar objects more efficiently than other views (Woods et al., 2009).

However, in the Newell et al. (2001) study we found that the object views which promoted the most efficient recognition performance differed across modalities: whereas visual recognition was best for familiar frontal views of objects, the haptic system seemed to recognise better the back views of objects (with reference to the direction the observer is facing). Since each sensory system recognises different views most efficiently, then it seemed vision and haptics do not provide redundant information about object shape for the purpose of recognition. How then would object constancy be maintained if each system processes object information in a qualitatively different way? Although this seems impossible, the results of our study suggest a means by which object constancy is achieved across the senses: since vision and haptics encode different aspects of the object's shape then the combination of this non-redundant but complementary object information should result in a rich description of an object in memory that would help maintain object constancy over changes in object view. Such a rich representation should, therefore, mean that subsequent recognition of the object would be very efficient and indeed would likely not be dependent on viewpoint. In a recent study, Lacey et al. (2007) provided evidence to suggest that this is indeed the case. They investigated visual, haptic, and crossmodal object discrimination using a set of novel objects presented at different views rotated across one of three axes (*X*, *Y*, or *Z*). Their results corroborated previous findings that both visual and haptic object recognition is dependent on the view of the object but when combined they found that crossmodal object recognition performance is independent of viewpoint.

Both the Newell et al. (2001) and the Lacey et al. (2007) studies investigated visual and haptic object recognition across views of objects which were, by necessity, constrained in order to control the view information presented during training and test. In the real world, however, hand-sized objects are often picked up and palpated under free exploration. Indeed, in some previous studies on visual, haptic, and crossmodal recognition of familiar objects, haptic exploration was unconstrained whereas the object was presented in a fixed position for visual testing. These studies consistently reported that the sharing of object information across modalities is efficient (e.g. Easton et al. 1997; Reales and Ballesteros, 1999), suggesting that the information encoded by both modalities can be combined to allow for a rich, multisensory representation in memory. However, when familiar objects are used in a task, it is unclear the extent to which verbal labelling or other semantic information mediates crossmodal performance, especially when objects are presented in different views or positions across modalities. Indeed Berkeley hinted at the role which verbal labelling may play in crossmodal object recognition by stating that “Every Combination of Ideas is considered as one thing by the Mind, and in token thereof, is marked by one Name”.

The use of unfamiliar objects that are not readily associated with distinct names gets around this issue. For example, Norman et al. (2004) used shapes based on natural objects from the same category (i.e. pepper or capsicum shapes) and found that crossmodal recognition performance was as good as within-modal performance. Norman et al. concluded that there are important similarities between vision and touch that allow for the same information to be represented in object memory (see also Gibson, 1979 for a similar conclusion). We also investigated whether the recognition of freely explored objects is efficient in vision and touch, and across these modalities, using the same set of novel objects as in our previous study (Ernst et al., 2007). Specifically, we tested both unimodal and bimodal (multisensory) recognition performance and found that unimodal performance was more efficient than bimodal performance. This finding seems to contradict the suggestions from earlier studies that vision and touch provide complementary information about an object which ultimately leads to a rich representation of the object in memory. However, close video analysis of the exploration procedures adopted during haptic exploration suggested that information encoding was optimised for efficient within-modal performance but not for crossmodal or bimodal performance.

The results from studies based on unconstrained exploration of objects suggest that object familiarity has no effect on the degree to which object information can be shared across the senses. Moreover, investigations into the recognition of static objects suggest that object constancy can be achieved by combining the inputs from both vision and touch into a rich representation of the object in memory.

14.3.3 Shape Constancy and the Recognition of Dynamic Objects in Vision and Touch

Whilst most investigations into object constancy have centred on the issue of viewpoint dependency, object shape can also be deformed by motion and any recognition system would need to account for how object constancy is maintained despite such changes to the shape. For example, in the animal world, shape can dramatically change whilst the animal is in motion and, moreover, this information may differ from when the animal is stationary and at rest. Furthermore, many small artefact objects can also change shape as a result of object motion: e.g. the opening of a book, mobile phone, or Swiss Army knife; or the rotation of a scissors blade; or the flipping of a stapler, can result in overall shape changes or reveal shape features not otherwise present in the object's image when it is closed. The shape information of these types of objects can change dramatically from one moment to the next but, as with changes in object view, we nevertheless seem to maintain object constancy with seemingly little effort.

Previous investigations on the visual recognition of dynamic objects have suggested that rigid motion information when combined with shape information can offer a unique cue to the identity of the object (e.g. Stone, 1998, 1999; Vuong and Tarr, 2004). For example, we found that object shapes associated with a particular movement pattern during learning were subsequently recognised more efficiently when shown with the same motion pattern than when shown moving in a different way (Newell et al., 2004). More recently, Setti and Newell (2009) reported that the visual recognition of unfamiliar objects in which non-rigid shape changes occur during motion is also affected by a change in the motion pattern of the parts of the objects. Thus, the findings of recent studies on the recognition of dynamic objects suggest that motion information can play an important role and that, moreover, objects may be stored in memory as spatiotemporal representations rather than static images of objects.

The tactile system obviously can play an important role in the perception of moving objects since many artefacts move as a consequence of haptic interactions. For example, a scissors changes shape as a consequence of movement of the hand. This begs the question, therefore, whether or not object motion affects the recognition of objects encoded through touch in the same way as those encoded through vision. Since motion is a cue for recognition in the visual domain then we can also ask whether the manner in which an object moves is a useful cue for recognition in the haptic domain. To that end we recently conducted a series of behavioural studies in which we investigated the role of motion information on the visual, haptic, and crossmodal recognition of object shapes (Whitaker et al., in prep.). We first created a set of unfamiliar objects, each with a moveable part which could rotate, flip, or slide on the object. At the beginning of the experiment, participants first learned a set of target objects either through vision only (by observing the object being moved by the experimenter) or through touch only (by actively palpating the object and moving the object part). Our results suggest that object motion is an important cue for recognition through touch: a moving target object was recognised better than its static counterpart. Furthermore, we found that both within-modal visual and haptic recognition benefited from the presence of the motion cue and the movement of the target objects also facilitated crossmodal recognition.

In order to assess whether motion is indeed an important cue for object recognition that is shared across modalities, we conducted a neuroimaging experiment to elucidate the neural correlates of crossmodal recognition of dynamic objects (Chan et al., 2010). We were specifically interested in investigating whether cortical areas known to be involved in visual motion and the visual recognition of dynamic objects were also involved in the haptic recognition of moving objects. Previous studies have found that area MT/MST is activated by dynamic information (e.g. Tootell et al., 1995) and also that motion implied in a static image is sufficient to activate this area (Kourtzi and Kanwisher, 2000). We first trained a group of participants to recognise a set of unfamiliar, moveable, and static objects using either vision or touch. We then presented static visual images of these objects to the participants whilst we recorded brain activations using fMRI. We found that area MT/MST was activated to images of objects previously learned as moving but not to objects learned as static. Surprisingly, this activation occurred to images of objects which were previously learned using either vision or touch (see Fig. 14.3). In other words, area MT/MST was active to both within modality and crossmodal presentations of previously learned dynamic objects.

These findings, together with the behavioural results reported earlier, suggest that both vision and touch contribute to the perception of moving objects and that, as such, both modalities may combine and share information in order to maintain object constancy not just in situations which involve changes in viewpoint but also those in which movement changes the shape information. Moreover, these findings are in contrast with Berkeley's conclusions on whether motion information is shared across modalities: he asserted that "... it clearly follows, that Motion perceivable by Sight is of a Sort distinct from Motion perceivable by Touch". However, although Berkeley did concede that for visuo-tactile perception "The Consideration of Motion, may furnish a new Field for Inquiry" it is perhaps surprising to note that this particular field remains relatively new three centuries later!

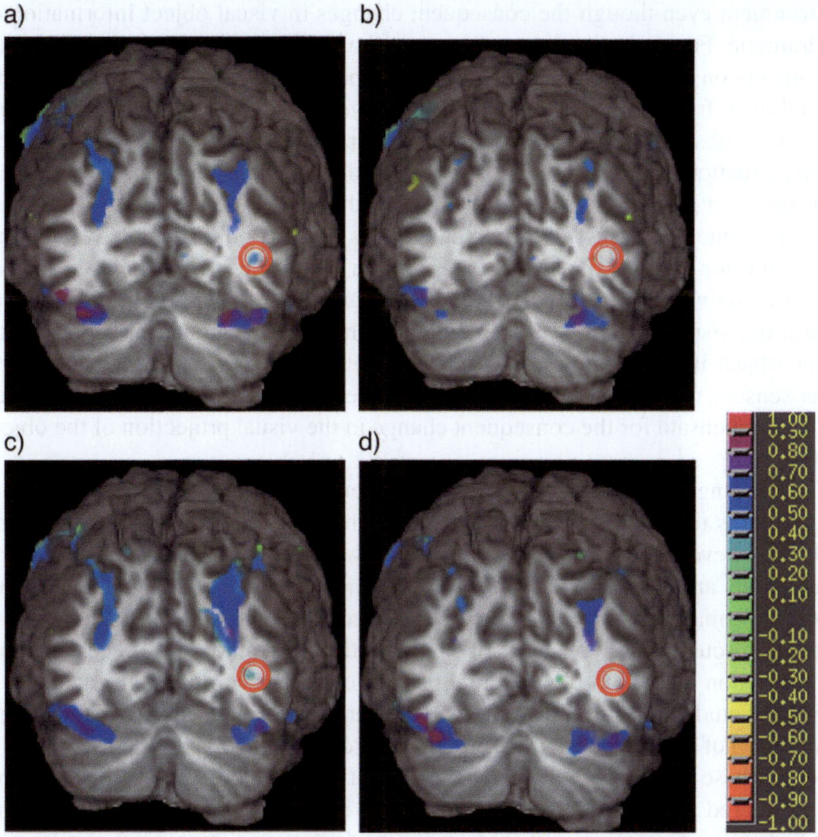

Fig. 14.3 Plot showing activation patterns across the four different learning conditions: **(a)** haptic motion; **(b)** haptic static; **(c)** visual motion and **(d)** visual static. Colours used represent positive and negative percentage change in the BOLD response (see key for colour coding). Area MT/MST (MT+) is highlighted on each map by a *red circle*

14.4 Evidence for Common Information Processing of "Where" Information Across Vision and Touch

14.4.1 Visual and Haptic Spatial Updating of Scenes

As previously mentioned, one of the greatest achievements of the human brain is the ability to perceive objects as constant despite dramatic changes in the projected retinal image, or in the tactile impressions, with changes in object position or shape deformities due to motion. An example of object constancy in the real world is that object recognition appears to remain invariant whilst we move around our environment even though the consequent changes in visual object information can be dramatic. For example, as we walk around a desk, the projected retinal image of the objects on that table can differ greatly depending on whether we have walked behind or in front of the desk. Although changes in object viewpoint consequently occur with observer motion, recognition performance does not seem to be affected in this situation. On the contrary, the recognition of views of an object that occur with observer motion is more efficient than the recognition of those same views when presented to a passive observer. Simons and his colleagues have attempted to account for this invariant object recognition with observer motion by proposing that extra-retinal information, such as vestibular or proprioception information, can inform the visual system of movement and consequently update the representation of the object in visual memory (e.g. Simons et al., 2002). Thus, information from other sensory modalities can update the representation of the object in visual memory to compensate for the consequent change in the visual projection of the object's image.

The finding that extra-retinal cues can update object representations in memory also pertains to the recognition of arrays of objects or scenes (Wang and Simons, 1999). However, up until recently, very little was known as to whether haptic representations are also updated with observer motion. Indeed it would seem that if object information is shared or accessible across modalities, as is suggested by research discussed previously, then spatial updating should be a process common to both vision and touch if a coherent perception of our world is to be achieved. In a series of studies, we investigated first whether the processes involved in the haptic recognition of object scenes is similar to the processes involved in the visual recognition of these scenes such that spatial information about object locations in a scene can be shared across modalities. Our previous research suggested that both "what" and "where" information is shared across modalities and, moreover, similar cortical substrates underpin these processes across modalities (Chan and Newell, 2008; Newell et al., 2010b). Similar to previous reports on the view-dependent recognition of scenes of objects in visual perception (Diwadkar and McNamara, 1997), we established that scene perception is also view dependent in haptic recognition in that the recognition of rotated scenes is more error prone than the recognition of scenes from a familiar view (Newell et al., 2005). Indeed, we recently found similar effects of scene rotation on haptic recognition using novel objects as we previously found using familiar shapes (see Fig. 14.4). However, in our study involving scenes of familiar shapes, we found that crossmodal recognition was less efficient than within-modal performance, although it was nevertheless better than chance. This cost in performance when crossing modalities did not seem to be due to differences in encoding across vision and touch (i.e. that vision can encode an object array in parallel, or from a single glance (Biederman et al., 1974; Rousselet et al., 2002) whereas haptics requires serial encoding of object positions). Instead, we argued that whilst some information can be shared across modalities, other spatial information is more modality specific and does not readily transfer across the senses.

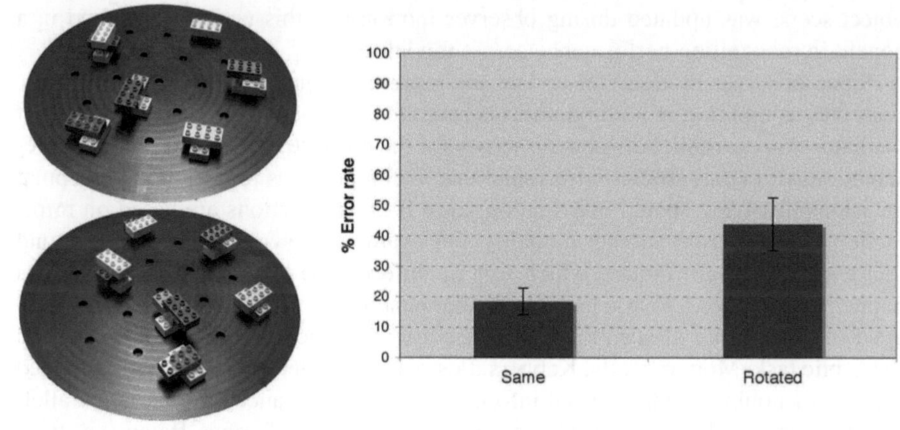

Fig. 14.4 An example of a scene of novel objects which participants were required to learn using haptics only (*top left* of image). An example of a test scene is shown in *bottom left* scene where an object has been displaced and the entire scene rotated by 60°. The plot shown on the *right* is the mean percentage errors made in novel scene recognition across non-rotated and rotated scenes

14.4.2 Crossmodal Updating in Scene Perception

The finding that scene recognition is better within than across modalities then led us to ask whether or not both visual and haptic scene perception benefit from spatial updating with observer movement. We found evidence to suggest that this is the case: the cost in both the visual and the haptic scene recognition performance as a consequence of passive rotation of the scene was prevented when the change in viewpoint was induced by observer motion (Pasqualotto et al., 2005). In particular, we replicated the results found in previous studies that the visual recognition of an array of familiar shapes is impaired when the scene is rotated relative to the observer and that this cost in recognition performance is removed when the rotated view is induced by the observer moving to a new position. We also extended this finding to the haptic domain and found that haptic scene representations are also updated with observer movement. The next question we asked was whether or not the representations of objects scenes are shared across modalities. In our most recent experiments, participants were required to learn the position of objects in a scene using either touch or vision only and we then tested their recognition in the other modality (Newell et al., 2010a). Prior to recognition the experimenter displaced one of the objects in the scene and the participant had to indicate which object had changed position. Furthermore, between learning and test the scene was either rotated relative to the passive observer, or the participant changed position. We found a cost in crossmodal performance when the scene was passively rotated. However, this cost was significantly reduced when the change in scene view was caused by a change in the observer's position. In other words, the visual or haptic representation of the object scene was updated during observer motion and this updating resulted in a benefit in recognition performance across modalities.

Since observer motion can update the representation of an object in memory, such that recognition in all modalities is benefited, this begs the question as to what mediates this crosstalk between the senses for the purpose of updating spatial representations. Previous studies have found that vision provides precision in perceptual decisions involving spatial information, even if those decisions are based on information encoded from another modality. For example, Newport et al. (2002) found that even visual information which was noninformative to a haptic task improves haptic performance.[1] Vision had a particular benefit when participants were encouraged to use a more allocentric than egocentric reference frame when performing the haptic task. More recently, Kappers, Postma, and colleagues further investigated the role of noninformative visual information on performance in a haptic parallel-matching task and found that vision affects the type of reference frame (i.e. from egocentric to allocentric) used to encode the haptic stimuli (Postma et al., 2008; Volcic et al., 2008). Moreover, they found that visual interfering information presented during the haptic task resulted in a cost in the haptic performance. Kaas et al. (2007) also found that noninformative visual information that is incongruent to haptic information can affect haptic performance on a parallelity task but only if the haptic information is encoded relative to an allocentric rather than an egocentric reference frame. These studies suggest that vision has a direct effect on haptic processing of spatial information by providing an allocentric reference frame to which haptic information is encoded.

14.4.3 The Role of Noninformative Visual Information on Haptic Scene Perception

We recently investigated the role of noninformative visual information on memory for object scenes encoded through touch. In these experiments, participants learned and were tested on their recognition of a scene of familiar objects using touch only. In separate conditions, participants could either view their surroundings (the test scene was never seen) or they were blindfolded during the task. We found evidence that noninformative vision can improve the haptic recognition of a scene of familiar object shapes (Pasqualotto et al., in prep.). However, the availability of visual information (albeit noninformative to the task) did not reduce the cost of recognising these haptic scenes when rotated, suggesting that scenes of familiar objects are stored as egocentric representations in memory, irrespective of the availability of noninformative visual information or of the encoding modality (see also Diwadkar and McNamara, 1997; Newell et al., 2005). Using virtual scenes, we manipulated

[1] Non-informative visual information is information that would not, on its own, be sufficient to solve the task. For example, seeing the surrounding room but without seeing the test stimuli would be considered ‘non-informative’ visual information.

the type of ambient visual information available during the haptic task to investigate the precise nature of the visual information that improves haptic performance. Specifically, participants could either see a furnished room, an empty room, or an image of the furniture without the room context. We found that spatial information (i.e. the presence of the room), not object landmarks (i.e. furniture only), was necessary in the visual image to benefit haptic performance. In conclusion, these studies suggest that vision can provide the optimum reference frame for encoding and retrieving spatial information through other senses although this benefit seems to be context dependent and does not necessarily affect the reference frame relative to which haptic spatial information is represented in memory.

If vision affects haptic spatial perception and memory then we might ask whether spatial perception is compromised in persons without visual experience. Some recent studies (e.g. Pasqualotto and Newell, 2007; Postma et al., 2008) tested haptic recognition of scenes of objects or haptic orientation perception in congenitally blind, late blind, and sighted people. There seems to be a consistent finding that tactile spatial perception is compromised in individuals who have impaired visual abilities, particularly those who were blind from early on in the course of development. Indeed it is well known from neurodevelopmental studies that early visual experience is required for normal development of the visual system (see, e.g. Lewis and Maurer, 2005 for a review) and that late intervention in repairing visual abnormalities can have long-term detrimental effects on the development of visual processing. The behavioural findings from haptic spatial perception suggest that the absence of visual experience can also affect the development of efficient spatial processing in another modality, namely touch. This finding has led some researchers to suggest that vision is the spatial sense which calibrates or modulates spatial perception in other, less spatially precise modalities (Thinus-Blanc and Gaunet, 1997).

14.5 Conclusions and Future Directions

The results of studies discussed above largely contradict Berkeley's assumption that "The Extension, Figures, and Motions perceived by Sight are specifically Distinct from the Ideas of Touch ... nor is there any such thing as an Idea, or kind of Idea common to both Senses". On the contrary, evidence from the literature suggests that the manner in which object information is processed does not depend on the encoding modality. Moreover, both the visual and the tactile processing of object information seem to be underpinned by shared neural substrates. Since principles of information processing and neural resources are, to a large extent, shared across modalities for object recognition and localisation, this suggests that unisensory information is pooled together at some stage in perceptual processing. Although the time course of visuo-tactile interactions in the brain for the purpose of object perception has yet to be elucidated these interactions could occur either later on, according to the purpose of the task, or earlier such that all information is encoded into a multisensory representation to which each modality has access. Research on audio-visual processing for object recognition suggests that these interactions occur earlier on in perceptual processing than previously thought (e.g. Molholm et al., 2004). In any case, there seems to be little evidence that the sharing of information across modalities requires a distinct and separate recoding process that allows for vision and touch to share information for object recognition and localisation.

Although Berkeley concluded in his essay that visual and tactile processing are independent, he also asserted that where associations do exist between vision and touch these associations are not innate but are, instead, arbitrary and built from experience with the world. Specifically he proposed that "this Naming and Combining together of Ideas is perfectly Arbitrary, and done by the Mind in such sort, as Experience shows it to be most convenient". Although many studies have now provided evidence that vision and touch can efficiently share "ideas" for the purpose of object recognition and spatial perception, the extent to which these ideas are innate or hard wired versus the extent to which the associations are acquired through experience is, as yet, undetermined. However, evidence from developmental studies suggests that experience may be required for efficient crossmodal interactions to occur. For example, although some studies have provided evidence for crossmodal shape perception in neonates (e.g. Meltzoff and Borton, 1979) others have found that this crossmodal performance is not very efficient (Sann and Streri, 2007). Furthermore, some studies have found that whereas adult perception of spatial characteristics of object shape is based on a statistically optimal integration of information across vision and touch (Ernst and Banks, 2002) there is no evidence for this optimal integration in young children and indeed it does not seem to emerge until later on in development (Gori et al., 2008). As such, these studies suggest that although the sensory systems seem to be hard wired from birth to share information, the precision and efficiency with which information is integrated across the senses in adult perception seems to be dependent on experience. As Berkeley himself stated, ". . . this Connexion with Tangible Ideas has been learnt, at our first Entrance into the World, and ever since, almost every Moment of our Lives, it has been occurring to our Thoughts, and fastening and striking deeper on our Minds". However, very little is known about how this developmental process occurs and, moreover, what factors influence the normal development of multisensory integration. Future research into these areas would be very useful not only in elucidating the neurodevelopmental processes of perception but also in offering potential rehabilitative procedures to restore sensory function in a damaged brain or to counteract sensory decline due to either trauma or normal ageing.

In sum, an essay penned by the philosopher, George Berkeley, 300 years ago tapped into issues that are still relevant in the field of perception today. Although recent research has addressed many of the questions raised in that essay, some important issues on the nature and development of integration across vision and touch remain outstanding. It is encouraging to note, though, how much progress has been made on elucidating the behavioural and neural correlates of multisensory recognition in the last couple of decades and we can look forward to providing further empirical evidence in response to Berkeley's musings on the nature and development of visuo-tactile interactions in the near future.

References

Amedi A, Malach R, Hendler T, Peled S, Zohary E (2001) Visuo-haptic, object-related activation in the ventral visual pathway. Nat Neurosci 4:324–330

Andresen DR, Vinberg J, Grill-Spector K (2009) The representation of object viewpoint in human visual cortex. Neuroimage 45(2):522–536

Berkeley G (1709) An essay towards a new theory of vision. Jeremy Pepyat Booksellers, Dublin, Ireland

Biederman I (1987) Recognition-by-components: a theory of human image understanding. Psychol Rev 94:115–147

Biederman I, Gerhardstein PC (1993) Recognizing depth-view, rotated objects: evidence and conditions for three-dimensional viewpoint invariance. J Exp Psychol Hum Percept Perform 19:1162–1182

Biederman I, Ju G (1988) Surface versus edge-based determinants of visual recognition. Cogn Psychol 20(1):38–64

Biederman I, Rabinowitz JC, Glass AL, Stacy EW Jr (1974) On the information extracted from a glance at a scene. J Exp Psychol 103(3):597–600

Bülthoff HH, Edelman S (1992) Psychophysical support for a two-dimensional view interpolation theory of object recognition. Proc Natl Acad Sci USA 89:60–64

Chan JS, Newell FN (2008) Behavioral evidence for task-dependent "what" versus "where" processing within and across modalities. Percept Psychophys 70(1):36–49

Chan JS, Whitaker TA, Simões-Franklin C, Garavan H, Newell FN (2010) Implied haptic object motion activates visual area MT/MST. Neuroimage, 49(2):1708–1716

Diwadkar VA, McNamara TP (1997) Viewpoint dependence in scene recognition. Psycholog Sci 8:302–307

Easton RD, Srinivas K, Greene AJ (1997) Do vision and haptics share common representations? Implicit and explicit memory within and between modalities. J Exp Psychol Learn Mem Cogn 23(1):153–163

Ernst MO, Banks MS (2002) Humans integrate visual and haptic information in a statistically optimal fashion. Nature 415(6870):429–433

Ernst MO, Bülthoff HH (2004) Merging the senses into a robust percept. Trends Cogn Sci 8: 162–169

Ernst MO, Lange C, Newell FN (2007) Multisensory recognition of actively explored objects. Can J Exp Psychol 61(3):242–253

Gibson JJ (1962) Observations on active touch. Psycholog Rev 69:477–491

Goodale MA, Milner AD (1992) Separate visual pathways for perception and action. Trends Neurosci 15:20–25

Gori M, Del Viva M, Sandini G, Burr DC (2008) Young children do not integrate visual and haptic form information. Curr Biol 18(9):694–698

Grill-Spector K, Kushnir T, Edelman S, Avidan G, Itzchak Y, Malach R (1999) Differential processing of objects under various viewing conditions in the human lateral occipital complex. Neuron 24:187–203

Haxby JV, Gobbini MI, Furey ML, Ishai A, Schouten JL, Pietrini P (2001) Distributed and overlapping representations of faces and objects in ventral temporal cortex. Science 293: 2425–2430

Hayward WG, Williams P (2000) Viewpoint dependence and object discriminability. Psychol Sci 11:7–12

James TW Humphrey GK Gati JS Menon RS, Goodale MA (2002) Differential effects of viewpoint on object-driven activation in dorsal and ventral streams. Neuron 35:793–801

Kaas AL, van Mier HI, Lataster J, Fingal M, Sack AT (2007) The effect of visuo-haptic congruency on haptic spatial matching. Exp Brain Res 183(1):75–85

Klatzky RL, Lederman SJ (1995) Identifying objects from a haptic glance. Percept Psychophys 57(8):1111–1123

Klatzky RL, Lederman SJ, Metzger V (1985) Identifying objects by touch: an expert system. Percept Psychophys 37:299–302
Kourtzi Z, Kanwisher N (2000) Activation in human MT/MST by static images with implied motion. J Cogn Neurosci 12(1):48–55
Lacey S, Peters A, Sathian K (2007) Cross-modal object recognition is viewpoint-independent. PLoS ONE 2(9):e890
Lacey S, Tal N, Amedi A, Sathian K (2009) A putative model of multisensory object representation. Brain Topogr 21(3–4):269–274
Lewis TL, Maurer D (2005) Multiple sensitive periods in human visual development: evidence from visually deprived children. Dev Psychobiol 46(3):163–183
Logothetis NK, Pauls J (1995) Psychophysical and physiological evidence for viewer-centered object representations in the primate. Cereb Cortex 5(3):270–288
Malach R, Reppas JB, Benson RR, Kwong KK, Jiang H, Kennedy WA, Ledden PJ, Brady TJ, Rosen BR, Tootell RBH (1995) Object-related activity revealed by functional magnetic resonance imaging in human occipital cortex. Proc Natl Acad Sci USA 92:8135–8139
Marr D (1982) Vision. Freeman Publishers, San Francisco
Meltzoff AN, Borton RW (1979) Intermodal matching by human neonates. Nature 282:403–404
Mishkin M, Ungerleider LG, Macko KA (1983) Object vision and spatial vision: two cortical pathways. Trends Neurosci 6:414–417
Molholm S, Ritter W, Javitt DC, Foxe JJ (2004) Multisensory visual-auditory object recognition in humans: a high-density electrical mapping study. Cereb Cortex 4:452–465
Newell FN (1998) Stimulus context and view dependence in object recognition. Perception 27(1):47–68
Newell FN, Ernst MO, Tjan BS, Bülthoff HH (2001) Viewpoint dependence in visual and haptic object recognition. Psychol Sci 12(1):37–42
Newell FN, Findlay JM (1997) The effect of depth rotation on object identification. Perception 26(10):1231–1257
Newell FN, Finucane CM, Lisiecka D, Pasqualotto A, Vendrell I (2010a) Active perception allows for spatial updating of object locations across modalities. Submitted
Newell, Hansen, Steven, Calvert (2010b) Tactile discrimination and localisation of novel object features activates both ventral and dorsal streams. Manuscript in preparation
Newell FN, Sheppard DM, Edelman S, Shapiro KL (2005) The interaction of shape- and location-based priming in object categorisation: evidence for a hybrid "what + where" representation stage. Vision Res 45(16):2065–2080
Newell FN, Wallraven C, Huber S (2004) The role of characteristic motion in object categorization. J Vis 4(2):118–129
Newell FN, Woods AT, Mernagh M, Bülthoff HH (2005) Visual, haptic and crossmodal recognition of scenes. Exp Brain Res 161(2):233–242
Newport R, Rabb B, Jackson SR (2002) Noninformative vision improves haptic spatial perception. Curr Biol 12(19):1661–1664
Norman JF, Norman HF, Clayton AM, Lianekhammy J, Zielke G (2004) The visual and haptic perception of natural object shape. Percept Psychophys 66(2):342–351
Palmer S, Rosch E, Chase P (1981) Canonical perspective and the perception of objects. In: Long JB, Baddeley AD (eds) Attention and performance, vol. IX. Erlbaum, Hillsdale, NJ, pp 135–151
Pasqualotto A, Finucane CM, Newell FN (2005) Visual and haptic representations of scenes are updated with observer movement. Exp Brain Res 166(3–4):481–488
Pasqualotto A, Newell FN (2007) The role of visual experience on the representation and updating of novel haptic scenes. Brain Cogn 65(2):184–194
Postma A, Zuidhoek S, Noordzij ML, Kappers AM (2008) Keep an eye on your hands: on the role of visual mechanisms in processing of haptic space. Cogn Proc 9(1):63–68
Postma A, Zuidhoek S, Noordzij ML, Kappers AM (2008) Haptic orientation perception benefits from visual experience: evidence from early-blind, late-blind, and sighted people. Percept Psychophys 70(7):1197–1206
Quiroga RQ, Reddy L, Kreiman G, Koch C, Fried I (2005) Invariant visual representation by single neurons in the human brain. Nature 435:1102–1107

Romanski LM, Tian B, Fritz J, Mishkin M, Goldman-Rakic PS, Rauschecker JP (1999) Dual streams of auditory afferents target multiple domains in the primate prefrontal cortex. Nat Neurosci 2:1131–1136

Reales JM, Ballestersos S (1999) Implicit and explicit memory for visual and haptic objects: cross-modal priming depends on structural descriptions. J Exp Psychol Learn Mem Cogn 25_(3) 644–663

Reddy L, Kanwisher N (2006) Coding of visual objects in the ventral stream. Curr Opin Neurobiol 16:408–414

Reed CL, Klatzky RL, Halgren E (2005) What vs. where in touch: an fMRI study. Neuroimage 25:718–726

Rousselet GA, Fabre-Thorpe M, Thorpe SJ (2002) Parallel processing in high-level categorization of natural images. Nat Neurosci 5(7):629–630

Sann C, Streri A (2007) Perception of object shape and texture in human newborns: evidence from cross-modal transfer tasks. Dev Sci 10(3):399–410

Schendan HE, Stern CE (2007) Mental rotation and object categorization share a common network of prefrontal and dorsal and ventral regions of posterior cortex. Neuroimage 35(3):1264–1277

Schendan HE, Stern CE (2008) Where vision meets memory: prefrontal-posterior networks for visual object constancy during categorization and recognition. Cereb Cortex 18(7):1695–1711

Setti A, Newell FN (2009) The effect of body and part-based motion on the recognition of unfamiliar objects. Vis Cogn 18(3):456–480

Simons DJ, Wang RF, Roddenberry D (2002) Object recognition is mediated by extraretinal information. Percept Psychophys 64(4):521–530

Stone JV (1999) Object recognition: view-specificity and motion-specificity. Vis Res 39(24): 4032–4044

Stone JV (1998) Object recognition using spatiotemporal signatures. Vis Res 38(7):947–951

Tarr MJ (1995) Rotating objects to recognize them: a case study of the role of viewpoint dependency in the recognition of three-dimensional objects. Psychon Bull Rev 2:55–82

Tarr MJ, Bülthoff HH (1998) Image-based object recognition in man, monkey and machine. Cognition 67:1–20

Thinus-Blanc C, Gaunet F (1997) Representation of space in blind persons: vision as a spatial sense? Psychol Bull 121(1):20–42

Tootell RB, Reppas JB, Kwong KK, Malach R, Born RT, Brady TJ, Rosen BR, Belliveau JW (1995) Functional analysis of human MT and related visual cortical areas using magnetic resonance imaging. J Neurosci 15(4):3215–3230

Ungerleider LG, Haxby JV (1994) 'What' and 'where' in the human brain. Curr Opin Neurobiol 4, 157–165

Ullman S (1998) Three-dimensional object recognition based on the combination of views. Cogn 67:21–44

Van Boven RW, Ingeholm JE, Beauchamp MS, Bikle PC, Ungerleider LG (2005) Tactile form and location processing in the human brain. Proc Natl Acad Sci U S A 102(35):12601–12605

Volcic R, van Rheede JJ, Postma A, Kappers AM (2008) Differential effects of non-informative vision and visual interference on haptic spatial processing. Exp Brain Res 190(1):31–41

Vuong QC, Tarr MJ (2004) Rotation direction affects object recognition. Vis Res 44(14): 1717–1730

Welch RB, Warren DH (1986) Intersensory interactions. In: Boff KR, Kaufman L, Thomas JP (eds) Handbook of perception and performance, vol. 1. Sensory processes and perception. John Wiley and Sons, New York, pp 25–1–25–36

Whitaker TA, Chan J, Newell FN (2010) The role of characteristic motion in haptic and visuo-haptic object recognition. Manuscript in preparation

Wang RF, Simons DJ (1999) Active and passive scene recognition across views. Cognition 70(2):191–210

Woods AT, Moore A, Newell FN (2009) Canonical views in haptic object perception. Perception 37(12):1867–1878

Young MP (1992) Objective analysis of the topological organization of the primate cortical visual system. Nature 358(6382):152–155

Chapter 15
Haptic Face Processing and Its Relation to Vision

Susan J. Lederman, Roberta L. Klatzky, and Ryo Kitada

15.1 Overview

Visual face processing has strong evolutionary significance for many biological species because the face conveys information that is critically important to biological survival: predator or prey? friend or enemy? potential mate? A substantial research literature in cognitive psychology, cognitive science, and cognitive neuroscience has established that face processing is an essential function of visual perception, to such an extent that a subset of visual abilities, sometimes referred to as a "face module," may be dedicated to it. Criteria for such face skills are often derived from arguments that face processing is not only universal in humans but also observed in other species, developmentally early to emerge, and performed by highly specialized cortical areas.

Numerous detailed published reviews and books have been published on visual face processing (e.g., Adolphs, 2002; Bruce and Young, 1998; Peterson and Rhodes, 2003). Compared to other object categories, the general hallmarks of visual processing for facial identity are that it is (a) highly practiced (Gauthier et al., 2003; see also McKone and Kanwisher, 2004); (b) predominantly based on overall configuration (de Gelder and Rouw, 2000; Farah et al., 1995; Maurer et al. 2002); (c) orientation specific (e.g., Diamond and Carey, 1986; Farah et al., 1995; Freire et al., 2000; Leder and Bruce, 2000; Maurer et al., 2002; Sergent, 1984); and (d) identity specific (Bruce and Young, 1998).

Not surprisingly, almost all of the face research from which these general principles derive involves the visual system. However, recent research reveals that humans are also capable of haptically recognizing both facial identity and facial expressions of emotion in live faces, 3-D face masks (rigid molds taken from live faces) and 2-D raised-line depictions. Such results clearly confirm that face processing is not unique to vision.

S.J. Lederman (✉)
Department of Psychology, Queen's University, Kingston, ON K7L 3N6, Canada
e-mail: susan.lederman@queensu.ca

M.J. Naumer, J. Kaiser (eds.), *Multisensory Object Perception in the Primate Brain*, DOI 10.1007/978-1-4419-5615-6_15, © Springer Science+Business Media, LLC 2010

The current chapter will focus primarily on the research literature pertaining to the haptic extraction of facial identity and facial expressions of emotion, two aspects of face perception that have received considerable attention from vision scientists. In particular, we will address a number of questions concerning functional aspects that pertain to how humans *process* and *represent* facial identity and emotional expressions, together with the *neural mechanisms* that underlie these functions. We note that information-processing theorists make a fundamental distinction between representation and process: whereas "representation" refers to the data on which computational operations are performed, "process" refers to the operations themselves.

Theoretically, we conceptualize face processing as involving haptic or visual object-recognition systems that are likely to show both commonalities *and* differences in facial processes and representations, whether expressed in functional or neural terms. Hence, our discussions of facial identity and emotion perception will each begin with a brief survey of the relevant vision literature, followed by a more extensive consideration of haptic face processing, on its own and as it relates to vision.

15.2 Facial Identity

15.2.1 Visual Perception of Facial Identity

Humans are clearly highly effective at perceiving, recognizing, and identifying individuals on the basis of information that is visually extracted from faces.

15.2.1.1 How Does the Visual System *Process* Facial Identity?

Of critical concern to vision researchers is whether face processing is based primarily on the facial configuration (i.e., features *and* their spatial interrelations) or more on the features themselves. Much of the visual face research has unequivocally emphasized the primacy of configural processes (Maurer et al., 2002).

Early studies used evidence of the "face inversion effect" (i.e., upright faces are better recognized than inverted faces) to confirm the importance of configural processing, arguing that inverting faces impair the viewer's ability to process faces configurally (e.g., Diamond and Carey, 1986; Sergent, 1984; Yin, 1969). However, Maurer et al. (2002) have since noted that on its own, the face inversion paradigm does not unambiguously evaluate the contribution of configural processing. After all, face inversion may impair performance for other reasons, such as disrupting information about the features *per se*, or because of inexperience in identifying inverted faces.

A number of subsequent studies have included other experimental manipulations in conjunction with the face inversion paradigm to provide stronger support for the claim that inverting the face interferes with configural processing. For example,

Farah et al. (1995) first trained participants to identify a series of upright faces by name, based on either the whole face or only one part (e.g., nose). In a subsequent identification test using new upright or inverted versions of the training faces, "whole-face" training produced an inversion effect, but "part-face" training failed to do so (see Fig. 15.1). The latter result suggests that the initial part-face presentation impeded configural processing.

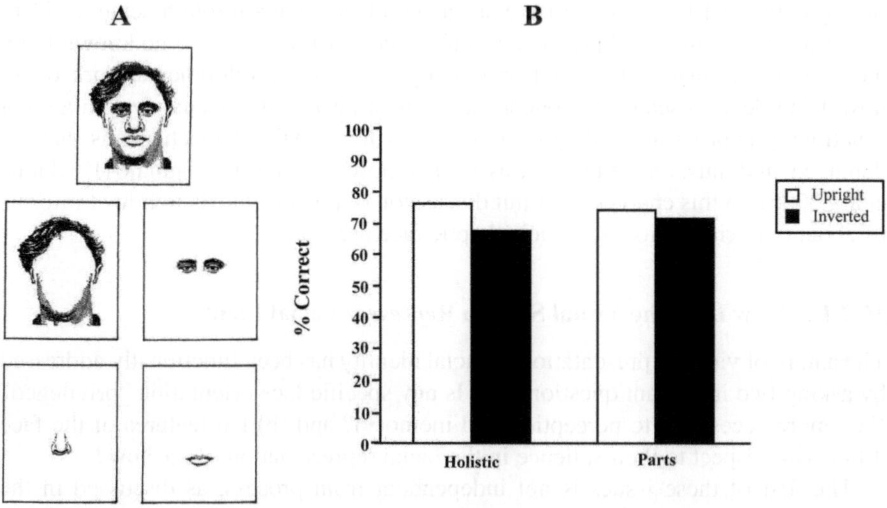

Fig. 15.1 One of the stimulus faces employed by Farah et al. (1995) Experiment 2. (**a**) The top is a sample "holistic" face; below are four "part" versions of the same face. Note that during the subsequent test phase, only the holistic versions were presented. (**b**) Results from Farah et al. (Experiment 2) indicate that the face inversion effect was eliminated when participants were trained to encode faces by their parts. Reprinted with permission of the American Psychological Association

Freire et al. (2000) asked participants to discriminate upright or inverted faces that differed in configural information, that is, the eyes and mouths were slightly shifted in their locations although the features remained unchanged. Using the figural discrepancies, they could easily discriminate the upright faces but not the inverted ones. In contrast, participants could discriminate faces that differed in their features equally well in upright and inverted conditions. This pattern of results further suggests that face inversion disrupts configural, but not featural, processes.

Researchers have also used several other experimental paradigms that strongly support the primacy of the face configuration in processing upright faces. For example, "scrambling" facial features (e.g., Collishaw and Hole, 2000) and combining or "morphing" the halves of different faces (e.g., Hole, 1994) both alter the normal facial configuration while leaving the features unchanged. Inverting features within a face (the "Thatcher effect") also appears to disrupt configural processing (Boutsen and Humphreys, 2003). Conversely, "blurring" facial features alter the features themselves while leaving the face configuration unchanged. Collectively, the results of such studies confirm that the accuracy with which facial identity is recognized is disrupted more when the configuration of the face is altered, as opposed to the features themselves.

In addition to studying those who are neurologically intact, visual face researchers have focused on a clinical population of persons diagnosed with "prosopagnosia." While such people can identify faces at the "basic" category level (Rosch et al., 1976), they demonstrate limited, if any, success in visually differentiating individual faces. Some have sustained clear trauma through acute accident or disease ("adventitious" prosopagnosia); in contrast, others have no known brain damage and normal early visual processing systems ("developmental" prosopagnosia). While all researchers concur that some aspects of the visual system are not functioning appropriately, they disagree as to the domain of mechanisms that are damaged, and thus, on the object classes that those mechanisms regulate (Duchaine et al., 2006). In this chapter, we limit discussion of prosopagnosia to how it informs us about our primary focus, namely, haptic face perception.

15.2.1.2 How Does the Visual System R*epresent* Facial Identity?

The nature of visual representation of facial identity has been functionally addressed by asking two important questions: (a) Is any specific face orientation "privileged" (i.e., more accessible to perception and memory)? and (b) Do features of the face differ with respect to their salience in the facial representation; if so, how?

The first of these issues is not independent from process, as discussed in the previous section. In addition to indicating configuration-based processing of facial identity, a strong face inversion effect speaks to the role of orientation in the facial representation. That the identity of upright visually presented faces is commonly recognized more accurately than inverted faces suggests that the upright orientation is canonical or "privileged."

With respect to the second question, a variety of techniques have been employed to assess the relative salience of facial features. These include, for example, the study of eye movements, psychophysical experiments with spatially filtered stimuli, multidimensional scaling, and the use of subjective questionnaires. In general, the region around the eyes appears to be most important for visual face recognition (e.g., Keating and Keating, 1982; Leder et al., 2001; Mangini and Biederman, 2004; Schyns et al., 2002; Sekuler et al., 2004); more precisely, people visually attend foremost to the eyebrows (Schyns et al., 2002), followed in descending order of importance by the eyes, mouth, and, finally, nose (Fraser et al., 1990; Haig, 1986; Janik et al., 1978).

15.2.1.3 What Are the *Neural Mechan*isms that Underlie Visual Perception of Facial Identity?

Visual neuroscience has further contributed to our understanding of human face perception by investigating the neural mechanisms that underlie a variety of important functions, including but not restricted to the perception of facial identity. This topic has received much attention, and the interested reader may consult Haxby et al. (2000) and Posamentier and Abdi, (2003).

Haxby et al. (2000) have proposed that human face perception is mediated by a hierarchically organized, distributed neural network that involves multiple bilateral regions (Fig. 15.2), including but not limited to the FFA. This model functionally distinguishes between the representation of invariant facial characteristics, such as identity, and variable aspects such as expression, eye gaze, and lip movement that all contribute to social communication. Collectively, fMRI studies have revealed the significance of three regions in the occipitotemporal extrastriate area: (a) bilateral regions in the lateral fusiform gyrus (i.e., fusiform face area, or FFA: Clark et al., 1996; Hadjikhani and de Gelder, 2002; Halgren et al., 1999; Haxby et al., 1994; Kanwisher et al., 1997; McCarthy et al., 1997), (b) lateral inferior occipital gyri (Hadjikhani and de Gelder, 2002; Halgren et al., 1999; Hoffman and Haxby, 2000; Levy et al., 2001; Puce et al., 1996), and (c) posterior superior temporal sulcus (Halgren et al., 1999; Haxby et al., 1999; Hoffman and Haxby, 2000; McCarthy et al., 1997; Puce et al., 1998). Researchers have long argued that the FFA uniquely contributes to face recognition, although others have recast its role in terms of a module for the perception of faces *and* non-face objects for which the observer possesses a high degree of expertise (Gauthier et al., 1999).

Ultimately, the ever-increasing sophistication in technologies (e.g., combination of different neuroimaging techniques, multi-electrode stimulation/recording techniques, and computational modeling) will enhance our understanding of the distributed neural networks and computations that underlie the multiple functions of face perception.

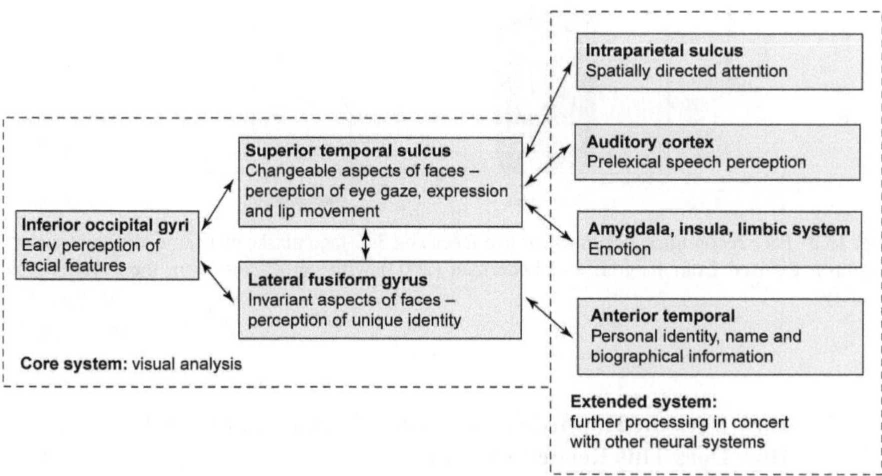

Fig. 15.2 Haxby et al.'s model of a distributed neural system for perceiving human faces. The system is composed of a core system used for visual face analysis and an extended system used for additional processing of the meaning of the facial information. Reprinted from Haxby et al. (2000) with permission of Elsevier Ltd

15.2.2 *Haptic Perception of Facial Identity*

Whether sighted or blind, individuals rarely choose to recognize a person by manually exploring their face. While this may be true, are they *capable* of doing so; alternatively, is face perception strictly a visual phenomenon? We now know that people can haptically discriminate and identify both unfamiliar and familiar live faces and corresponding clay face masks at levels considerably above chance (Kilgour and Lederman, 2002; see also Casey and Newell, 2003; Pietrini et al., 2004). In the Kilgour and Lederman (2002) study, blindfolded sighted college students haptically matched the facial identity of live actors with a success rate of 80% (chance = 33%), as shown in Fig. 15.3. When rigid face masks were used, accuracy declined to about 60%; nevertheless, performance remained well above chance. Kilgour et al. (2005) subsequently showed that with considerable training, people could learn to identify face masks by name perfectly. More recently, McGregor et al. (2010) have shown that such individuals are also capable of learning to name individual 2-D raised-line drawings of individual faces, with accuracy improving by $\sim$ 60% after only five blocks of training with feedback. Finally, Kilgour et al. (2004) have confirmed the first known case of haptic prosopagnosia, a condition in which the individual was unable to haptically differentiate faces.

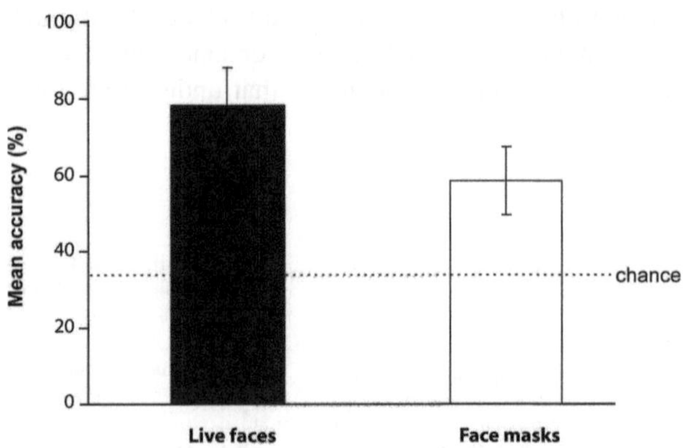

Fig. 15.3 Face recognition accuracy for live faces and 3-D face masks in a same–different matching task. Revised from Kilgour and Lederman (2002) with permission from the Psychonomic Society

15.2.2.1 How Does the Haptic System *Process* Facial Identity and How Does This Relate to Vision?

Two noteworthy questions about haptic face processing have been considered to date: (a) what is the relative importance of configural, as opposed to feature-based, processes in the haptic perception of facial identity and (b) to what extent is visual mediation used to process facial information haptically?

In Section 15.2.1.1, we noted that vision scientists have used evidence of a "face inversion effect" – upright faces are better recognized than inverted faces – to argue that the recognition of upright faces strongly involves configural processing. Recent studies indicate a parallel in haptic face processing for neurologically intact individuals. People haptically differentiate the identity of 3-D face masks better when they are presented upright, as opposed to inverted (Kilgour and Lederman, 2006). It is further noteworthy that in addition to being unable to haptically differentiate upright faces at levels above chance, the prosopagnosic individual (LH) in the Kilgour et al. (2004) study demonstrated a paradoxical inversion effect (i.e., *better* performance for inverted than upright faces) haptically (Fig. 15.4), as well as visually (for possible neural explanations of the paradoxical inversion effect, see Farah et al., 1998; de Gelder and Rouw, 2000). To this extent, then, haptic processing and visual processing of facial identity are similarly influenced by orientation.

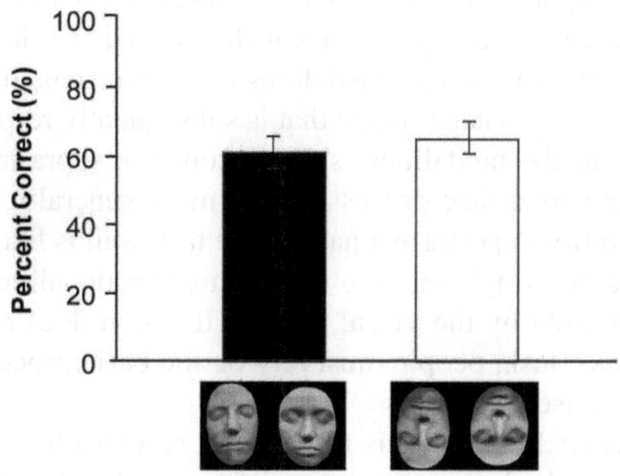

Fig. 15.4 Recognition accuracy for a prosopagnosic individual (LH) for upright and inverted face masks in a same/different matching task. Performance was at chance level for upright faces and higher than chance for inverted faces, reflecting a paradoxical face-inversion effect. Revised and reprinted from Kilgour et al. (2004) with permission from Elsevier Ltd

Acknowledging the caveat raised by Maurer et al. (2002) with respect to using the face inversion paradigm on its own to assess the role of configural processing, it would be desirable to employ one of the more direct methodologies (morphing, scrambling, blurring, etc.). Accordingly, McGregor et al. (2010) used 2-D raised-line drawings in a face-identity learning task involving scrambled, as well as upright and inverted faces. The upright and scrambled displays produced equivalent performance. Because scrambling faces alters the global facial configuration, McGregor et al. concluded that it was not used to haptically process facial identity portrayed in 2-D raised-line drawings. Scrambled faces also produced higher accuracy than inverted faces. Because face inversion alters the local configural information about the features, McGregor et al. further concluded that participants haptically processed only *local* configural information about the 2-D features, the features themselves being treated as oriented objects within a body-centered frame of reference. Because this study focused on the haptic system, a visual control was not included; however, a parallel visual study would clearly be informative.

Casey and Newell (2007) also used more direct methods to assess cross-modal transfer and the contribution of configural processing to such transfer. Participants in a haptic–visual cross-modal transfer task matched a haptically presented unfamiliar 3-D face mask to one of three subsequently presented colored 2-D visual displays that were normal, blurred, or scrambled. As mentioned earlier, blurring the image leaves the global configural arrangement of the features unchanged, while altering details about the features *per se*. Conversely, scrambling the features alters the facial configuration while leaving the features unaffected. Although only limited haptic–visual cross-modal transfer was observed, performance in the normal and blurred conditions was equivalent. The authors concluded that to the limited extent that cross-modal transfer did occur in processing facial identity, both modalities used global configuration – as opposed to feature-based processing.

Whether or not haptic processing of facial identity involves the use of visual mediation is highly controversial among touch and vision scientists. Any performance similarities between the two modalities may be attributed to the transformation of haptic inputs into a visual image that is subsequently re-processed by visual mechanisms and/or to the modalities' sharing common supra-modal processes. If vision does mediate haptic face processing (or, more generally, haptic object processing), then the ability to perform a haptic face task well is likely the result of the haptic system "piggybacking" onto the underlying functionalities, representations, and brain structures used by the visual system. If vision does not mediate performance in a haptic face task, people must rely on the basic processing mechanisms associated with the sense of touch.

What little behavioral data on this topic exists reveals little support for the use of visual mediation in the haptic perception of facial identity. Minimal correlation was observed between VVIQ test results (Visual Vividness Imagery Questionnaire: Marks, 1973) and performance on a haptic face-identity task involving 3-D face masks (e.g., Kilgour and Lederman, 2002). Moreover, Pietrini et al. (2004) demonstrated that totally blind subjects (two congenital; two early blind with no memory of any visual experiences) achieved >90% accuracy in a one-back haptic discrimination task also involving lifelike face masks. We will present additional converging neuroimaging evidence from Kitada et al. (2009) in Section 15.2.2.3.

15.2.2.2 How Does the Haptic System *Represent* Facial Identity and How Does This Relate to Vision?

When considering the nature of haptically derived representations of facial identity, it is important to note the greater efficiency with which the haptic system can process material, as opposed to geometric, properties. This is likely due to several factors: the haptic system's low spatial acuity, the relative costs and benefits of different manual exploratory procedures (Lederman and Klatzky, 1987), and the high demands on spatiotemporal integration and memory given that haptic exploration is typically sequential. The converse is generally true for vision, that is, vision's excellent spatial acuity and its ability to process edges in an object or display simultaneously render this modality particularly efficient when processing geometric, as opposed to material, properties. Such differences in efficiency in turn affect the relative salience of material and geometric features for the haptic and visual systems, respectively (see, e.g., Klatzky et al., 1987; Lederman et al., 1996).

In keeping with this material/geometry distinction, recall that haptic face matching was 20% more accurate with live faces than with face masks (Kilgour and Lederman, 2002). This finding implicates material-specific properties of the face as important sources of haptic information about facial identity. It is important to note as well that haptic performance with the 3-D masks remained well above chance, confirming the importance of 3-D structural information. Evidence of bi-directional cross-modal transfer (whether partial or complete) in priming studies using homogeneous non-face objects and face masks (Reales and Ballesteros, 1999; Easton et al., 1997a, b; Hadjikhani and Roland, 1998; Kilgour and Lederman, 2002; Casey and Newell, 2003, 2007; Norman et al., 2004) further confirms that vision and touch processes have access to at least some common structural representations. However, to the extent that the transfer is incomplete (particularly for faces), the two modalities may well represent different aspects of the object in light of the material–geometry distinction above, that is, a relatively stronger emphasis on structure for vision and material for touch.

We turn now to the role of orientation in representations of facial identity derived from haptic inputs and comparisons with vision (see Section 15.2.1.2). Collective evidence of a haptic face inversion effect for facial identity (McGregor et al., 2010; Kilgour and Lederman, 2006; Kilgour et al., 2004) suggests that with respect to facial orientation, vision and haptics share a common constraint in representing facial identity: Representations derived from exploring 3-D face masks and 2-D drawings are orientation dependent within the x–y fronto-parallel plane; thus, like vision, the upright face is "preferred" or "canonical" for haptics (but see Newell et al., 2001, which has shown different orientation preferences in the sagittal plane for 3-D nonsense objects).

15.2.2.3 What Are the *Neural Mechanisms* that Underlie Haptic Perception of Facial Identity and How Does This Relate to Vision?

Several studies have now begun to employ neuroimaging techniques (e.g., fMRI) to determine the underlying components of the distributed neural network recruited by haptic identification of faces (vs. non-face control objects) and by faces presented haptically vs. visually. Thus, in this section we address (a) studies that have focused on brain activity specifically induced by haptically presented faces and (b) studies that have compared brain activity elicited by haptic vs. visual face presentation.

We begin by noting that a number of fMRI studies have collectively shown that haptic processing of common non-face objects activates extrastriate areas (e.g., lateral occipital complex) traditionally believed to serve only visual functions (Amedi et al., 2001; Deibert et al., 1999; James et al., 2002; Reed et al., 2004). Researchers have now extended this work by examining haptic processing of face masks by both blindfolded neurologically intact and congenitally blind observers (Kilgour et al., 2004; Pietrini et al., 2004; Kilgour et al., 2005).

In one study (Kilgour et al., 2005), after learning to successfully identify a set of 3-D face masks by name via unimanual exploration with the left hand, neurologically intact individuals performed the same task in the scanner with a subset of those masks. Among other findings, the fusiform gyrus (left) was activated more strongly when faces, as opposed to non-face nonsense objects of similar shape and size, were explored haptically. The study by Kilgour et al. (2004) (see Section 15.2.2.2), which required a prosopagnosic individual (LH) and neurologically intact controls to haptically match 3-D face masks, complements the fMRI findings obtained by Kilgour et al. (2005) with neurologically intact individuals. Kilgour et al. (2004) proposed that LH's inability to haptically match faces was due to damage in the occipitotemporal cortex. Together, these two studies suggest that the occipitotemporal region plays an important role in haptic processing of faces, as well as other objects.

An additional fMRI study that further extends our inchoate understanding of the neural basis of haptic processing of facial identity specifically investigated the influence of familiarity on haptic face identification (James et al., 2006). Subjects were carefully trained to unimanually identify a subset of 3-D plaster face masks ("familiar" subset) using their left hand. In the scanner, they were then haptically presented with old and new objects to judge for familiarity. The left fusiform gyrus was activated more strongly by the haptic presentation of familiar (cf. unfamiliar) objects, suggesting that this area specifically differentiates between haptically familiar and unfamiliar faces.

We now compare brain organization and neural substrates in face identification under haptic vs. visual face presentations. The studies by Kilgour et al. (2005) and James et al. (2006) suggest that both haptic and visual identification of faces activate the left fusiform gyrus, although the sub-regions that are activated within that area by each modality may be different. In contrast, it is well known that visually presented faces recruit the right hemisphere more strongly than the left (Gauthier et al., 1999; Kanwisher et al., 1997). Because a corresponding visual face-presentation condition was not included in the initial exploratory studies on haptic face processing (James et al., 2006; Kilgour et al., 2005), it is possible that regions activated by haptic have been activated by visual face presentations. The occurrence of strong activation in the left hemisphere coupled with no significant right hemisphere activation suggests, rather, that the neural systems that mediate haptic and visual face recognition diverge. Because manual exploration is so often sequential, perhaps activation in the left fusiform gyrus was greater than in the right because haptically derived inputs about the facial configuration must be integrated over time. Conversely, visually derived inputs about the face may be simultaneously integrated over space. Dissociation of temporal- and spatial-integration processes in left and right hemispheres, respectively, has received support from theories of brain lateralization (Kolb and Whishaw, 2003) (for alterative explanations, see James et al., 2006).

Two additional fMRI studies offer complementary evidence for the suggestion that there is some overlap between vision and touch in their neural representations of facial identity, but that at least some information is preserved in separate modality-specific channels. Pietrini et al. (2004) used pattern classification methods (drawn from the fields of statistics and machine learning) in conjunction with fMRI data to determine if the specific class of object (bottles, shoes, faces) could be predicted from patterns of activity in ventral temporal extrastriate cortex that were derived during match-to-sample and simple exploration tasks involving visual vs. haptic presentations. While there was strong overlap and correlation between visual and haptic activation patterns for the non-biological categories (bottles, shoes), this was not the case for faces. Thus, the representation of biological common objects (i.e., the face) may not be fully cross-modally shared by the widely distributed neural network proposed by Haxby et al. (2002).

In a related study, Kitada et al. (2009) examined brain organization for haptic and visual identification of human body parts (faces, hands and feet) vs. a non-biological category of control objects (bottles). In accord with Pietrini et al. (2004), haptic and visual object identification activated largely disjoint networks (Fig. 15.5a). However, it is possible that face sensitivity may be shared across sensory modalities in small regions, of which locations are spatially varied across subjects. The authors examined two regions which produced the strongest activation in haptic and visual face identification. These two discrete areas, HFR ("haptic face region") and FFA ("fusiform face area"), were sensitive to 3-D face masks (cf. controls) whether presented haptically or visually. Nevertheless, the corresponding activation patterns across object categories (faces, feet, hands, and bottles) were different for FFA and HFR regions (Fig. 15.5). Kitada et al. concluded that although both regions within the fusiform gyrus are sensitive to faces, independent of sensory modality, the sub-region that is most sensitive to haptically presented faces (HFR) is functionally distinct from that which is most sensitive to visually presented faces.

A number of tactile/haptic functional neuroimaging studies with non-face patterns and objects have now confirmed that the visual cortex is generally involved in normal tactual perception by sighted and blind observers (for further details, see review by Sathian and Lacey, 2007). What remains unclear for both non-face and face objects, however, is whether this visual involvement consists of knowledge-directed processes (e.g., anticipatory visual imagery or visual memory) that may assist or mediate tactual performance, stimulus-directed activation of visual cortical areas by tactual inputs, which in turn implies that these putative "visual" areas are in fact "multisensory," or both stimulus-driven and knowledge-driven processes (Lacey, Campbell and Sathian, 2007; Sathian and Lacey, 2008). Further research on this issue is clearly required.

Recently, Kitada et al. (2009) addressed the use of visual imaging vs. multisensory processing of faces and other body parts by including a third condition in which subjects were required to visually image targeted exemplars of face masks (as well as other body parts). Several measures of visual imagery were obtained, involving both behavioral (i.e., VVIQ: Marks, 1973; subjective reports regarding the extent to which subjects used visual imagery) and neuroimaging measures (i.e., neural activation in visual imagery vs. haptic conditions). Various correlational analyses converged in showing that at best, visual mediation could account for only a

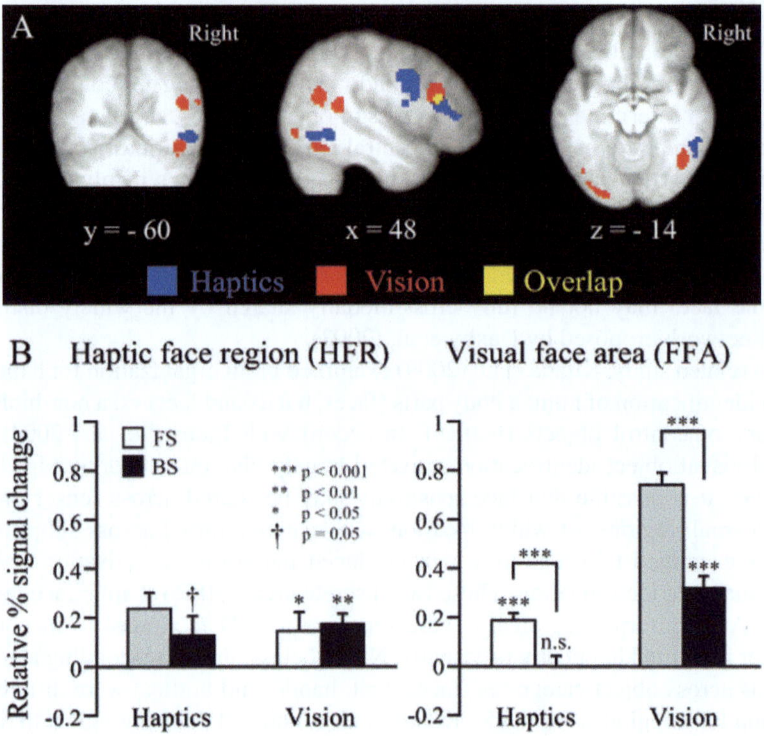

Fig. 15.5 (**a**) Group analysis. Activation patterns during identification of faces compared to control objects were superimposed on the coronal, sagittal, and transverse planes of the anatomical image averaged across the subjects. (**b**) Signal change of faces and other body parts relative to the control objects in HFR and FFA. The *gray bar* indicates the condition used to define the region. Data are presented as the mean ± SEM. *Asterisks and n.s.* above each bar indicate the result of a one-sample *t*-test on the sensitivity score for faces (FS) and other body parts (BS). *Asterisks* above a pair of bars show the result of a post hoc pair-wise comparison. Reprinted from Kitada et al. (2009) with permission of the MIT Press and the Cognitive Neuroscience Society

relatively minor portion of the increase in category-specific signal observed with haptically presented faces (and other body parts) (for further details, see Kitada et al., 2009). The authors concluded that visual imagery is not *necessary* to achieve good haptic perception of facial identity (or other body parts).

15.2.3 Summary

In Section 15.2 we considered how facial identity is visually vs. haptically processed, focusing specifically on the relative importance of configural vs. feature-based processes and the extent to which visual mediation is used to process haptically derived inputs about facial identity and the extent and nature of cross-modal transfer between visual and haptic processing of facial identity. We then considered the nature of visual and haptic face representations, addressing primary issues pertaining to whether a specific face orientation is "privileged," and whether and how facial features differ with respect to their salience in facial representations. Finally, we examined the unisensory and multisensory neural substrates that underlie facial identity. In Section 15.3, we address many of the same questions as they pertain to facial expressions of emotion.

15.3 Facial Expressions of Emotion

15.3.1 Visual Perception of Emotion from Facial Expressions

A second critical component of visual face processing that has attracted much attention by vision researchers concerns how people communicate their emotions nonverbally by means of varying their facial expressions (Darwin, 1972/1955). Facial expressions exhibit invariant features with respect to both the static musculoskeletal pattern when the expression is fully formed and from brief changes in these patterns over time. A small set of facial expressions of emotion that include anger, disgust, fear, happiness, sadness, and surprise are universally recognized (Ekman et al., 1987). Details of these expressions have been expressed in terms of specific "facial action patterns" (Ekman and Friesen, 1975), with visually detectable consequences that are used to process facial expressions of emotion in static photographs, line drawings, and artificial dynamic displays. In keeping with the organization of Section 15.2.1, we now address significant issues pertaining to facial processing and representations of facial expressions of emotion and to their underlying neural substrates.

15.3.1.1 How Does the Visual System *Process* Facial Expressions of Emotion?

As previously noted, one of the hallmarks of face processing is the influence of face orientation on face perception and recognition, with important implications for how faces are processed and for the manner in which the visual inputs are represented. With respect to process, several studies have demonstrated clear face inversion effects relating to the visual perception of facial expressions of emotion by both neurologically intact and almost all prosopagnosic observers (Calder et al., 2000; de Gelder et al., 2003; Fallshore and Bartholow, 2003; Prkachin, 2003). Such studies suggest that global face configuration plays a primary role in processing facial expressions of emotion. Thus, neurologically intact individuals consistently show standard inversion effects in face identity and emotion tasks. Prosopagnosic participants, however, do not. Although unable to identify faces, they are still capable of perceiving and identifying facial expressions of emotion and do show inversion effects. Such performance differences in the prosopagnosic group support traditional claims of functional dissociation between identity and emotion (Bruce and

Young, 1986). However, such putative independence has been challenged by de Gelder and colleagues (2003), who showed that facial expressions of emotion modulate facial identification by neurologically intact and prosopagnosic participants (although in opposite directions).

15.3.1.2 How Does the Visual System *Represent* Facial Expressions of Emotion?

We now address three important issues with regard to how facial expressions of emotion are represented. Paralleling Section 15.2.2.2 on facial identity, we begin by considering two issues (a) Are face representations orientation independent; if the answer to this question is affirmative, is a specific orientation "privileged"? (b) What is the relative salience of facial features? A third significant issue (c) pertains to theoretical approaches used to address the visual representation of facial expressions of emotion. With respect to the first issue, we briefly note that evidence of inversion effects for emotions implies that orientation is important in the representation of emotional expressions, and more particularly, that the upright orientation is privileged. This point will be discussed further below.

As for the second issue, studies show that people use a combination of features to recognize emotions, including the local shape of face components (e.g., eyes wide open/narrowed, brows raised/lowered, corners raised/lowered) and pigmentation or texture differences (e.g., mouth open/closed; teeth showing/behind the lips) (for summaries, see e.g., Ekman and Friesen, 1975; Bruce and Young, 1986). While shape plays a very important role in visual recognition of facial expressions of emotion, the contributions of surface features (e.g., skin pigmentation) are more limited (Bruce and Young, 1998).

Turning to theoretical approaches, researchers in the area of social psychology have proposed both category-based and dimension-based models of the visual recognition of facial expressions of emotion. Category-based models generally argue for a limited set of cross-culturally recognizable expressions that most frequently include happiness, sadness, anger, disgust, fear, and surprise (Ekman et al., 1987). People are very good at classifying these primary categories of facial emotions when visually presented in static photographs (e.g., Ekman, 1972; Wehrle et al., 2000), line drawings (Etcoff and Magee, 1992) and simulated dynamic presentations (e.g., Calder et al., 1996; Ekman, 1972; Etcoff and Magee, 1992).

In contrast, dimension-based models have argued that visual recognition of facial emotions is based on the location of faces within an *n*-dimensional continuous psychological space, typically described by a 2-D solution (Galati et al., 1997; Katsikitis, 1997; Russell, 1980). For example, Katsikitis (1997) conducted multidimensional scaling of the similarity structure of facial-emotion space using photos of six primary emotions plus a neutral expression. In a 2-D solution, the expressions were approximately distributed in a circle with "neutral" close to the centre, as shown in Fig. 15.6. One dimension was identified as pleasant/unpleasant (i.e., going from happiness and surprise to disgust, anger, and sadness). Katsikitis further proposed that participants used landmark features on a second dimension (upper

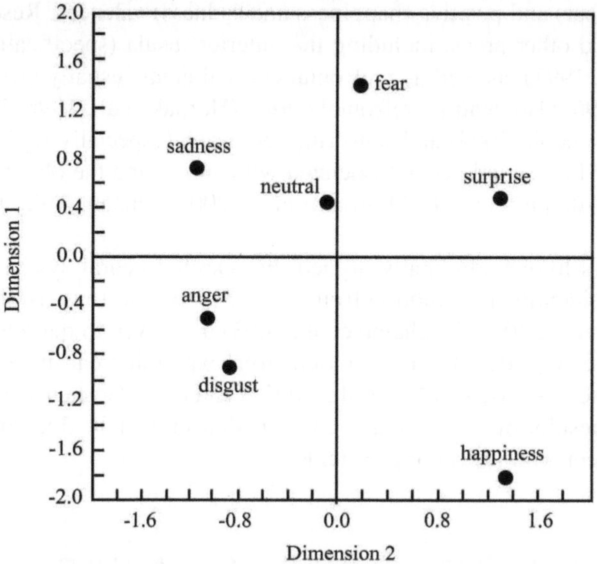

Fig. 15.6 Visual face space for facial expressions of emotion in 2-D line drawings based on a 2-D MDS solution. Reprinted from Katsikitis (1997) with permission from Pion Ltd

face/lower face) as clues for differentiating emotions, with surprise, fear, and sadness tending to involve the upper face, and happiness, disgust, and anger the lower face (see also Galati et al., 1997). In Russell's (1980) "circumplex" model, the two dimensions are pleasure–displeasure and arousal–sleepiness.

15.3.1.3 What Are the *Neural* Mechanisms that Underlie Visual Perception of Facial Expressions of Emotion?

In this section, we briefly address two significant aspects: (a) localized neural regions vs. spatially distributed neural networks for visually perceiving emotional expressions and (b) dissociation vs. interaction in the visual processing of facial identity and emotional expressions.

In their model of visual face perception, Haxby et al. (2000) propose an extended neural network that is used in conjunction with the core system underlying facial identity (see Fig. 15.2) in order to extract social relevance. The extension is thought to be phylogenetically older and faster and to involve the amygdala, which receives inputs (especially those related to threat) from sub-cortical mechanisms via a retinal-collicular-pulvinar pathway (Morris et al., 1998). Much of the relevant fMRI research has implicated a strong role for the amygdala, particularly in processing negative facial expressions, namely fear and anger (e.g., Adams et al., 2003; Adolphs et al., 1999). However, Winston et al. (2003) used an event-related design to show that the amygdala (together with extrastriate and fusiform cortex and posterior STS) actually responds similarly across basic emotions with negative (disgust and fear) and positive (happiness and sadness) valences. Research has further implicated other areas, including the anterior insula (specifically for disgust, Phillips et al., 1997), as well as prefrontal cortical areas (usually orbitofrontal cortex: Rolls, 1996; also ventral prefrontal cortex (Hornak et al., 1996; Winston et al., 2003; Nomura et al., 2004) and sensorimotor cortex (especially right hemisphere), perhaps related to somatic cues associated with simulating the observed emotional expressions (Adolphs et al., 1999; Adams et al., 2003; Ohman, 2002; Winston et al., 2003).

Researchers have traditionally argued that parallel neural systems are used to process facial identity and emotion from facial expressions (e.g., Bruce and Young, 1986; Calder et al., 2001; Duchaine et al., 2003). However, in neurologically intact viewers, processing identity and emotion from expression can interact (Rotshtein et al., 2001; see also de Gelder et al., 2003; Ganel and Goshen-Gottstein, 2004). Overall, the results of such studies confirm that cross talk does occur between identity- and emotion-processing systems.

15.3.2 Haptic Processing of Emotion from Facial Expressions

Haptic researchers have shown that people are also capable of haptically perceiving and recognizing the six culturally universal facial expressions of emotion. Lederman et al. (2007) showed that with only 5–10 minutes training, young adults were able to use touch alone to classify these expressions at levels usually well above chance (17%). In one experiment, blindfolded subjects actively and bimanually explored the six emotional expressions portrayed statically by live actors. Classification accuracy was 51% and increased substantially to 74% in a second experiment when the live expressions were dynamically formed under the subjects' stationary hands (Fig. 15.7).

Clearly the perception of universal facial expressions of emotion is also bimodal. This finding seems reasonable inasmuch as the invariant features of the musculoskeletal facial displays of each emotion are accessible to the hands, as well as to the eyes. Subsequent studies have confirmed that people can haptically classify this same primary set of emotional expressions above chance levels when displayed on rigid 3-D face masks, which retain 3-D structure but not material information (Baron, 2008), and in 2-D raised-line drawings, which simplify the remaining 2-D information but eliminate both 3-D structural and material information normally available in static live faces (Lederman et al., 2008).

15.3.2.1 How Does the Haptic System *Process* Facial Expressions of Emotion and How Does This Relate to Vision?

Paralleling haptic research on facial identity, researchers have also investigated whether face inversion effects occur with respect to haptic processing of facial expressions of emotion. Sizable (~15%) inversion effects have been documented in

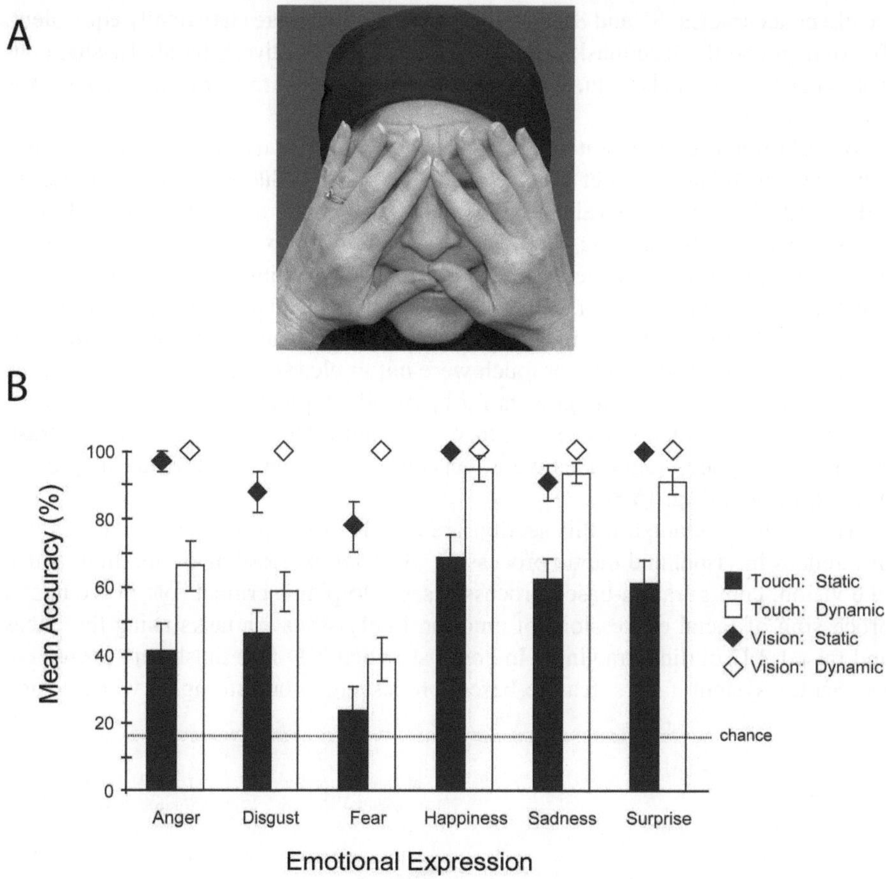

Fig. 15.7 (**a**) Initial or continuing hand positions used to haptically process happiness depicted statically and dynamically, respectively, by a live model; (**b**) haptic emotion-classification accuracy as function of emotional expression for static and dynamic displays. Reprinted from Lederman et al. (2007) with permission of Wiley and Sons Canada

studies involving seven-alternative-forced-choice classification (six primary emotional expressions + "neutral") tasks with both live faces (Direnfeld, 2007) and 2-D raised-line drawings, for which it was possible to include a third scrambled-face condition (Lederman et al., 2008). In both studies, emotions were classified more accurately when faces were presented upright, as opposed to inverted. In the latter study, classification accuracy was higher for upright faces than for either scrambled or inverted faces, for which accuracy was equivalent. Collectively, these two studies suggest that as with vision, configural processing of facial expressions of emotion plays a very important role haptically. A notable exception to this statement is a study by Baron (2008), which presented expressions on 3-D face masks in upright or inverted orientations. For the masks, upright and inverted faces both produced excellent accuracies (81 and 84%, respectively), which were statistically equivalent. To confirm that the face mask displays were indeed effective, a parallel visual control experiment was also run. Unlike the haptic results, a face inversion effect was now observed.

An additional experiment in the Lederman et al. study (2008) focused on a different, but related, aspect of face inversion effects. Subjects were required to judge both the emotional valence (positive vs. negative) and the associated emotional magnitude (on a scale of 1–5) of each of seven expressions (six culturally universal expressions and neutral) in both upright and inverted 2-D orientations using either touch or vision. When faces were presented visually, an inversion effect (lower magnitude for inverted faces) reliably occurred across all emotions except for happy. The results for touch were not as clear-cut. Relative to vision, the signed emotional valence judgments for haptically explored faces were quite variable, with no reliable evidence of a face inversion effect (Fig. 15.8a). In contrast, when the unsigned magnitudes were considered, the haptic and visual judgments were equivalent (Fig. 15.8b).

The studies included in this section are also interesting for the similarities and differences in visual and haptic processing of facial expressions of emotion. Much like vision, configuration-based processes seem to play a critical role in the haptic processing of facial expressions of emotion in classification tasks using live faces and raised 2-D outline drawings. In contrast, when 3-D face masks are presented, the haptic system favors feature-based processing more strongly, while vision

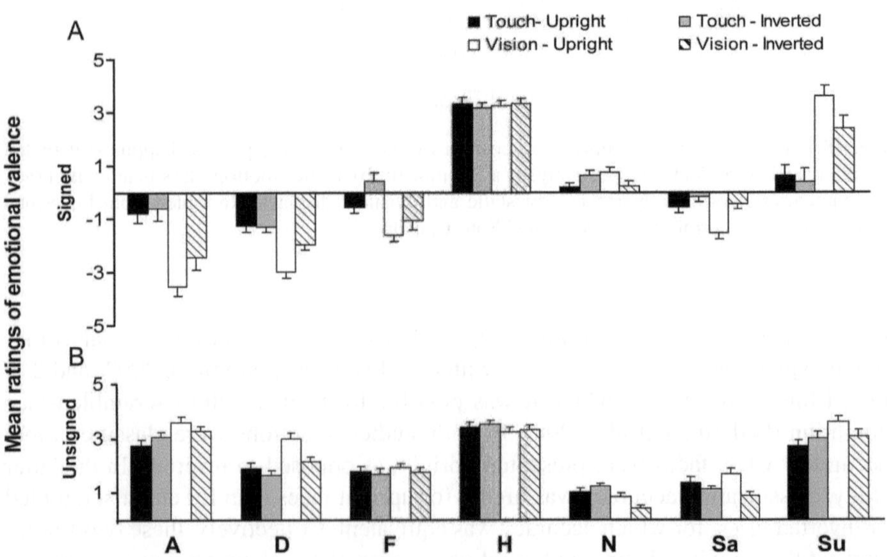

Fig. 15.8 Mean haptic and visual ratings of the emotional valence of facial expressions of emotion presented in upright vs. inverted orientations. (**a**) Signed ratings (+ 1 SEM); (**b**) unsigned ratings (+1 SEM). Reprinted from Lederman et al. (2008) with permission of the IEEE Computer Society

emphasizes configuration-based processes. Visual processing and haptic processing of facial expressions of emotion also differ with respect to judgments of emotional valence. Unlike vision, the haptic system appears unable to judge emotional valence consistently, likely because of the subtle differences in the spatial information available to a perceptual system with poor spatial resolving capacity (cf. vision). The two modalities are fairly similar in their ability to scale emotional magnitude of the relatively intense emotions portrayed in these studies, possibly because the magnitude of the differences along this dimension is more perceptually accessible to touch, as well as to vision.

15.3.2.2 How Does the Haptic System *Represent* Facial Expressions of Emotion and How Does This Relate to Vision?

In this section, we return to several issues raised in Section 15.3.1.2 with respect to vision. We now ask (a) Is haptic perception of facial expressions of emotion dependent on orientation, and if so, is there a canonical orientation? (b) Which facial features are primary for the haptic system, and what is their relative salience? (c) Finally, we consider the relevance of two major theoretical approaches to visual representation of emotional expressions for touch.

Based on the above-mentioned orientation effects, we conclude that the upright orientation is generally "privileged" in the representations of haptically encoded facial expressions of emotion. However, this conclusion applies only to live faces and 2-D raised-line drawings, inasmuch as face orientation had no observable effect with rigid 3-D face masks (Baron, 2008). The excellent performance obtained with 3-D masks may be attributed to the availability of a variety of 2-D and 3-D features, including teeth within the mouth, a feature that was either available but not explored (live faces) or absent (2-D drawings).

Which features/regions of static live face displays are most important for haptic processing of facially expressed emotions? Abramowicz (2006) compared performance when exploration of live faces was restricted either to the eyes/brows/forehead/nose (upper two-thirds) or to the nose/mouth/jaw/chin region (lower two-thirds). She found that while neither region was *necessary* for above chance classification accuracy, both were *sufficient*, with the lower two-thirds of the face producing more accurate performance. Using raised-line drawings, Ho (2006) similarly found that participants tended to be more accurate (as well as faster and more confident) when the eye/brow/forehead contours were deleted, as opposed to the mouth/chin/jaw contours. Vision, which was also assessed in this study, showed a similar response pattern across emotional expressions (see also Sullivan and Kirkpatrick, 1996, for related results, but cf. Gouta and Miyamoto, 2000).

Execution of facial action patterns that underlie the facial communication of human emotions produces transient changes in the musculoskeletal structure of the face and in the associated material properties of the skin and underlying tissue. Thus, dynamic facial displays of emotion may offer additional valuable haptic information to the perceiver, particularly given the temporal acuity of the hand in comparison to the eye (Jones and Lederman, 2006). Lederman et al. (2008) directly compared static and dynamic facial displays of emotion and found a marked improvement in haptic accuracy with dynamic information (51 vs. 74%, respectively). Visual studies tend to confirm the importance of dynamic cues in the representations of basic facial expressions of emotion (see, e.g., Ambadar et al., 2005; Atkinson et al., 2004; Cunningham et al., 2005; Kamachi et al., 2001; but see Bassili, 1978).

Although it is too early in the investigation of haptic face processing to produce a detailed model of how facial expressions derived from haptic inputs are represented, we may still address this issue from the perspectives of dimensional and category models previously proposed with respect to vision. In terms of a dimensional approach (e.g., Calder et al., 2001; Russell, 1980; Woodworth and Schlossberg, 1959), we highlight the two dimensions along which Lederman et al. (2008) required participants to haptically judge emotional valence – emotional "mood" (scale sign) and emotional intensity (scale magnitude). These two dimensions would appear tangentially related to Russell's visual pleasure–displeasure and arousal–sleepiness dimensions, respectively.

In keeping with a categorical approach (Ekman et al., 1987), one may ask whether certain feature regions of the face are more salient than others when haptically judging facial expressions of emotion. Subjective reports of the relative importance of different regions of the face and videotapes of hand movements during manual face exploration in many of our earlier studies suggested that both the lower-level mouth and upper-level eye regions (cf. mid-level nose region) of live faces and 2-D facial depictions may prove to be of particular importance both haptically and visually. The experimental studies by Abramowicz (2006) and Ho (2006) provide empirical confirmation of these subjective observations.

15.3.2.3 What Are the *Neural Mechanisms* that Underlie Haptic Perception of Facial Expressions of Emotion and How Does This Relate to Vision?

We are aware of only one study that has addressed the underlying neural substrates of haptic identification of facial expressions of emotion. To date, visual studies have shown that the inferior frontal gyrus, inferior parietal lobe, and cortical areas within and near the superior temporal sulcus are components of a cortical network – possibly the human analogue of the "mirror-neuron system in animals" (e.g., Rizzolatti et al., 1996) – used to process visual information about human actions, including facial expressions of emotion (Carr et al., 2003; Montgomery and Haxby, 2008; see also Kilts et al., 2003). Using fMRI techniques, Kitada et al. (2010) hypothesized that these regions are also involved in the haptic identification of facial expressions of emotion portrayed on 3-D rigid face masks. Subjects identified three different emotional expressions (disgust, neutral, and happiness) and three different types of control objects (shoes) haptically vs. visually. In brief, this study found that haptic and visual identification of facial expressions of emotion activated overlapping, yet separate, neural mechanisms including the three hypothesized regions that form part of a cortical neural network for understanding human actions. On the basis of these results, the authors suggested that this action network may partly underlie the perception of facial expressions of emotion that are presented haptically, as well as visually.

15.3.3 Summary

In Section 15.3, we addressed how humans – both neurologically intact and prosopagnosic individuals – visually and haptically process and represent facial expressions of emotion and the nature of the underlying neural substrates that support these functions. With respect to process, we focused primarily on the debate regarding configural vs. feature-based processing. With respect to representation, we addressed three issues: (a) Are face representations orientation independent and if so, is any specific orientation "privileged"? (b) What is the relative salience of facial features? and (c) Finally, what theoretical approaches have been used to address visual representation of facial expressions of emotion? We then considered the nature of the underlying neural mechanisms (uni- and multisensory) that are involved in the visual and haptic perception of facial expressions of emotion.

15.4 General Summary, Conclusions, and Future Directions

We have considered functional issues pertaining to how humans *process* and *represent* facial identity and emotional expressions, together with the *neural mechanisms* that underlie these important functions. In this section, we review those issues as they pertain to haptic face processing and its relation to vision, and we suggest directions for future research.

Is face processing solely a visual phenomenon? Research described here highlights the fact that face processing is a bimodal perceptual phenomenon that is accessible through manual, as well as visual, exploration, as confirmed with live faces, rigid 3-D face masks, and even 2-D raised-line drawings.

Is face processing unique? Interpreting the results obtained in visual studies has proved highly controversial and is beyond the scope of the current chapter. At one end of the controversy, a number of studies with neurologically intact and prosopagnosic individuals have contributed empirical support for the uniqueness of face perception in haptic and visual modalities. At the other side of the controversy, some researchers have argued that any object class for which individuals have special expertise may be processed differently than other object classes. The validity of this alternate interpretation with respect to haptic face processing certainly deserves consideration.

15.4.1 How Are Faces Processed?

Configural vs. feature-based processing. The results of many studies collectively confirm that visual processing of facial identity and facial expressions of emotion is highly configural. Global configural processing appears to be a prominent aspect of haptic face processing as well, but haptic studies considered in this chapter have also implicated roles for haptic feature-based processing as befits an information-processing system that extracts information more sequentially than the visual system. Considering wholistic vs. feature-based processing dichotomously tends to mask the complexity of the issues and the answers. Further research should consider constraints on configural processing and effects of task and context.

Visual-mediation vs. multisensory processing. Current data suggest that even if visual imagery is sometimes used, it is not necessary to achieve excellent performance on haptic facial identity tasks. To expand our current understanding, the nature of visual imagery in face processing tasks must be further clarified; in this regard, a more extensive battery of evaluation tasks would prove very helpful.

Intersensory transfer. Only limited intersensory transfer between vision and haptics takes place with respect to facial identity. To the extent that such transfer occurs, it appears to be globally configural, as opposed to feature based. The amount and nature of intersensory transfer has yet to be addressed with respect to facial expressions of emotion.

15.4.2 How Are Faces Represented?

Role of orientation. Research reported in this chapter has obtained face inversion effects with both visually and haptically presented displays (with some exceptions: identity in 2-D face displays and emotion in 3-D face masks). To the extent that face perception is orientation dependent, it implies that the upright position is "canonical" in face representation.

Relative importance of different facial regions to visual vs. haptic face processing. Whereas the eye and brow regions of the face are emphasized relatively more in visual facial identity tasks, the mouth region appears to be favored somewhat more overall when haptics is used. In terms of its potential application, this issue could be more systematically compared across the three major types of haptic face display (compliant live faces, 3-D face masks and 2-D drawings), inasmuch as the type and the amount of information that can be extracted haptically will vary. Despite comparable evaluations of the intensity of facial emotions in 2-D drawings, only vision reliably judges the extent to which the valence is positive or negative. The performance with haptic displays is not surprising, given the subtlety of 2-D facial cues to emotional valence or mood.

Theoretical approaches to the study of human facial emotions. Although both category-based and dimensional models have been proposed for visual representations of facial expressions of emotion, we know of no such comparable investigations with respect to haptics to date.

15.4.3 What Are the Underlying Neural Mechanisms and How Does this Relate to Vision?

Although neuroimaging studies have shown that visual and haptic systems share similarities in face processing, it is not clear to what extent haptic and visual systems share neural substrates for face perception; whether other brain regions, such as the inferior occipital gyrus, are involved in processing in both vision and haptics; and how multiple brain regions communicate during face processing. Other neuroimaging techniques such as the electroencephalogram (EEG) or magnetoencephalogram (MEG) may be able to elucidate temporal processing of haptic, as well as visual, face recognition. Models of effective connectivity based on functional neuroimaging data (e.g., Friston et al., 2003) are also needed to understand how multiple areas interact.

Finally, since neural mechanisms underlying visual and haptic face recognition are similar, one may ask whether neural mechanisms dedicated to face perception still exist without visual experience. An fMRI study on congenitally blind individuals may be able to answer this question.

Acknowledgments This chapter was prepared with the financial assistance of a grant to SL from the Canadian Institutes for Health (CIHR). We thank Cheryl Hamilton for her help in preparing the manuscript for publication.

References

Abramowicz A (2006) Haptic identification of emotional expressions portrayed statically vs. dynamically on live faces: the effect of eliminating eye/eyebrow/forehead vs. mouth/jaw/chin regions. Undergraduate thesis, Queen's University, Kingston, Ontario

Adolphs R (2002) Neural systems for recognizing emotion. Curr Opin Neurobiol 12:169–177

Adolphs R, Tranel D, Hamann S, Young AW, Calder AJ, Phelps EA, Anderson A, Lee GP, Damasio AR (1999) Recognition of facial emotion in nine individuals with bilateral amygdala damage. Neuropsychologia 37(10):1111–1117

Adams RB Jr, Gordon HL, Baird AA, Ambady N, Kleck RE (2003) Effects of gaze on amygdala sensitivity to anger and fear faces. Science 300(5625):1536

Ambadar Z, Schooler JW, Cohn JF (2005) Deciphering the enigmatic face: the importance of facial dynamics in interpreting subtle facial expressions. Psychol Sci 16(5):403–410

Amedi A, Malach R, Kendler T, Peled S, Zohary E (2001) Visuo-haptic object-related activation in the ventral visual pathway. Nat Neurosci 4:324–330

Atkinson AP, Dittrich WH, Gemmell AJ, Young AW (2004) Emotion perception from dynamic and static body expressions in point-light and full-light displays. Perception 33(6):717–746

Baron M (2008) Haptic classification of facial expressions of emotion on 3-D facemasks and the absence of a haptic face-inversion effect. Undergraduate honours thesis, Queen's University, Kingston, Ontario

Bassili JN (1978) Facial motion in the perception of faces and of emotional expression. J Exp Psychol Human Percept Perform 4(3):373–379

Bousten L, Humphreys GW (2003) The effect of inversion on the encoding of normal and "thatcherized" faces. Q J Expt Psychol 56A(6):955–975

Bruce V, Young A (1986) Understanding face recognition. Br J Psychol 77(3):305–327

Bruce V, Young A (1998) In the eye of the beholder: the science of face perception. Oxford University Press, New York, NY

Calder AJ, Burton AM, Miller P, Young AW, Akamatsu S (2001) A principal component analysis of facial expressions. Vision Res 41(9):1179–1208

Calder AJ, Young AW, Keane J, Dean M (2000) Configural information in facial expression perception. J Exp Psychol Hum Percept 26(20):527–551

Calder AJ, Young AW, Perrett DI, Etcoff NL, Rowland D (1996) Categorical perception of morphed facial expressions. Vis Cognit 3(2):81–117

Carr L, Iacobon M, Dubeau MC, Mazziotta JC, Lenzi GL (2003) Neural mechanisms of empathy in humans: a relay from neural systems for imitation to limbic areas. Proc Natl Acad Sci U S A 100:5497–5502

Casey S, Newell F (2003) Haptic own-face recognition. Proc 3rd Int Conf Eurohapt 2003:424–429

Casey SJ, Newell FN (2007) Are representations of unfamiliar faces independent of encoding modality? Neuropsychologia 45(3):506–513

Clark VP, Keil K, Maisog J Ma, Courtney S, Ungerleider LG, Haxby JV (1996) Functional magnetic resonance imaging of human visual cortex during face matching: a comparison with positron emission tomography. Neuroimage 4:1–15

Collishaw SM, Hole GJ (2000) Featural and configurational processes in the recognition of faces of different familiarity. Perception 29(8):893–909

Cunningham DW, Kleiner M, Wallraven C, Bulthoff HH (2005) Manipulating video sequences to determine the components of conversational facial expressions. ACM Trans Appl Percept 2(3):251–269

Darwin C (1872/1955). The expression of emotions in man and animals. Philosophical Library, New York

de Gelder B, Frissen I, Barton J, Hadjikhani N (2003) A modulatory role for facial expressions in prosopagnosia. Proc Natl Acad Sci U S A 100(22):13105–13110

de Gelder B, Rouw R (2000) Paradoxical configuration effects for faces and objects in prosopagnosia. Neuropsychologia 38:1271–1279

Deibert E, Kraut M, Kremen S, Hart J Jr (1999) Neural pathways in tactile object recognition. Neurology 52:1413–1417

Diamond R, Carey S (1986) Why faces are and are not special: An effect of expertise. J Exp Psychol Gen 115:107–117

Direnfeld E (2007) Haptic classification of live facial expressions of emotion: a face-inversion effect. Undergraduate honours thesis, Queen's University, Kingston, Ontario

Duchaine BC, Parer H, Nakayama K (2003) Normal recognition of emotion in a prosopagnosic. Perception 32:827–838

Duchaine BC, Yovel G, Butterworth EJ, Nakayama K (2006) Prosopagnosia as an impairment to face-specific mechanisms: elimination of the alternative hypotheses in a developmental case. Cogn Neuropsychol 23(5):714–747

Easton RD, Greene AJ, Srinivas K (1997a) Transfer between vision and haptics: memory for 2-D patterns and 3-D objects. Psychon Bull Rev 4:403–410

Easton RD, Greene AJ, Srinivas K (1997b) Do vision and haptics share common representations? Implicit and explicit memory within and between modalities. J Exp Psychol Learn Mem Cognit 23:153–163

Ekman P (1972) Universals and cultural differences in facial expressions of emotion. In: Cole J (ed) Nebraska symposium on motivation. University of Nebraska Press, Lincoln, NE, pp 207–238

Ekman P, Friesen WV (1975) Unmasking the face: a guide to recognising emotions from facial clues. Prentice Hall, Engelwood Cliffs, NJ

Ekman P, Friesen WV, O'Sullivan M, Chan A, Diacoyanni-Tarlatzis I, Heider K et al. (1987) Universals and cultural differences in the judgment of facial expression of emotions. J Pers Soc Psychol 53:712–717

Etcoff NL, Magee JJ (1992) Categorical perception of facial expressions. Cognition 44(3):227–240

Fallshore M, Bartholow J (2003) Recognition of emotion from inverted schematic drawings of faces. Percept Mot Skills 96(1):236–244

Farah MJ, Tanaka JW, Drain HM (1995) What causes the face inversion effect? J Exp Psychol Hum Percept Perform 21:628–634

Farah MJ, Wilson KD, Drain HM, Tanaka JR (1995) The inverted face inversion effect in prosopagnosia: evidence for mandatory, face-specific perceptual mechanisms. Vis Res 35(14):2089–2093

Farah MJ Wilson KD, Drain M, Tanaka JN (1998) What is "special" about face perception? Psychol Rev 105:482–498

Fraser IH, Craig GL, Parker DM (1990) Reaction time measures of feature saliency in schematic faces. Perception 19(5):661–673

Freire A, Lee K, Symons LA (2000) The face-inversion effect as a deficit in the encoding of configural information: direct evidence. Perception 29:159–170

Galati D, Scherer KR, Ricci-Bitti PE (1997) Voluntary facial expression of emotion: comparing congenitally blind with normally sighted encoders. J Pers Soc Psychol 73(6):1363–1379

Ganel T, Goshen-Gottstein Y (2004) Effects of familiarity on the perceptual integrality of the identity and expression of faces: the parallel-route hypothesis revisited. J Exp Psychol Hum Percept Perform 30(3):583–597

Gauthier I, Tarr MJ, Anderson AW, Skudlarski P, Gore JC (1999) Activation of the middle fusiform "face area" increases with expertise in recognizing novel objects. Nat Neurosci 2:568–573

Gauthier I, Curran T, Curby KM, Collins D (2003) Perceptual interference supports a non-modular account of face processing. Nat Neurosci 6(4):428–432

Gouta K, Miyamoto M (2000) Emotion recognition: facial components associated with various emotions. Jpn J Psychol 71(3):211–218

Hadjikhani N, de Gelder B (2002) Neural basis of prosopagnosia: an fMRI study. Hum Brain Mapp 16(3):176–182

Hadjikhani N, Roland PE (1998) Crossmodal transfer of information between the tactile and visual representations in the human brain: a positron emission tomographic study. J Neurosci 18:1072–1084

Haig ND (1986) Exploring recognition with interchanged facial features. Perception 5(3):235–247

Halgren E, Dale AM, Sereno MI, Tootell RBH, Marinkovic K, Rosen BR (1999) Location of human face-selective cortex with respect to retinotopic areas. Hum Brain Mapp 7(1):29–37

Haxby JV, Hoffman EA, Gobbini MI (2000) The distributed human neural system for face perception. Trends Cogn Sci 4:223–233

Haxby JV, Hoffman EA, Gobbini MI (2002). Human neural systems for face recognition and social communication. Biol Psychiat 51(1):59–67

Haxby JV, Horwitz B, Ungerleider LG, Maisog J Ma, Pietrini P, Grady CL (1994) The functional organization of human extrastriate cortex: a PET-rCBF study of selective attention to faces and locations. J Neurosci 14:6336–6353

Haxby JV, Ungerleider JG, Clark VP, Schouten JL, Hoffman EA, Martin A (1999) The effect of face inversion on activity in human neural systems for face and object perception. Neuron 22:189–199

Ho E (2006) Haptic and visual identification of facial expressions of emotion in 2-D raised-line depictions: relative importance of mouth versus eyes + eyebrow regions. Undergraduate honours thesis, Queen's University, Kingston, Ontario

Hoffman EA, Haxby JV (2000) Distinct representations of eye gaze and identity in the distributed human neural system for face perception. Nat Neurosci 3(1):80–84

Hole GJ (1994) Configurational factors in the perception of unfamiliar faces. Perception 23(1): 65–74

Hornak J, Rolls ET, Wade D (1996) Face and voice expression identification in patients with emotional and behavioural changes following ventral frontal lobe damage. Neuropsychologia 34(4):247–261

James TW, Humphrey GK, Gati JS, Servos P, Menon RS, Goodale MA (2002) Haptic study of three-dimensional objects activates extrastriate visual areas. Neuropsychologia 40:1706–1714

James TW, Servos P, Huh E, Kilgour A, Lederman SJ (2006) The influence of familiarity on brain activation during haptic exploration of 3-D facemasks. Neurosci Lett 397:269–273

Janik SW, Wellens AR, Goldberg ML, Dell'Osso LF (1978) Eyes as the center of focus in the visual examination of human faces. Percept Mot Skills 47(3):857–858

Jones LA, Lederman SJ (2006) Human hand function. Oxford University Press, New York

Kamachi M, Bruce V, Mukaida S, Gyoba J, Yoshikawa S, Adamatsu S (2001) Dynamic properties influence the perception of facial expressions. Perception 30(7):875–887

Kanwisher N, McDermott J, Chun MM (1997) The fusiform face area: a module in human extrastriate cortex specialized for face perception. J Neurosci 17:4302–4311

Katsikitis M (1997) The classification of facial expressions of emotion: a multidimensional-scaling approach. Perception 26(5):613–626

Keating CF, Keating EG (1982) Visual scan patterns of rhesus monkeys viewing faces. Perception 11(2):211–219

Kilgour A, Lederman S (2002) Face recognition by hand. Percept Psychophys 64(3):339–352

Kilgour A, Lederman SJ (2006) A haptic face-inversion effect. Perception 35:921–931

Kilgour AK, de Gelder B, Lederman SJ (2004) Haptic face recognition and prosopagnosia. Neuropsychologia 42:707–712

Kilgour A, Kitada R, Servos P, James T, Lederman SJ (2005) Haptic face identification activates ventral occipital and temporal areas: an fMRI study. Brain Cogn 59:246–257

Kilts CD, Egan G, Gideon DA, Ely TD, Hoffman JM (2003) Dissociable neural pathways are involved in the recognition of emotion in static and dynamic facial expressions. Neuroimage 18(1):156–68

Kitada R, Johnsrude I, Kochiyama T, Lederman SJ (2009) Functional specialization and convergence in the occipitotemporal cortex supporting haptic and visual identification of human faces and body parts: an fMRI study. J Cogn Neurosci 21(10):2027–2045

Kitada R, Johnsrude I, Kochiyama T, Lederman SJ (2010). Brain networks involved in haptic and visual identification of facial expressions of emotion: an fMRI study. Neuroimage 49(2): 1677–1689

Klatzky RL, Lederman SJ, Reed C (1987) There's more to touch than meets the eye: relative salience of object dimensions for touch with and without vision. J Exp Psychol Gen 116(4):356–369

Kolb B, Whishaw IQ (2003) Fundamentals of human neuropsychology, 5th edn.: Worth Publishers, New York, NY

Lacey S, Campbell C, Sathian, K (2007) Vision and touch: multiple or multisensory representations of objects? Perception 36(10):1513–1521

Leder H, Bruce V (2000) When inverted faces are recognized: the role of configural information in face recognition. Quart J Exp Psychol A Hum Exp Psychol 53A(2):513–536

Leder H, Candrian G, Huber O, Bruce V (2001) Configural features in the context of upright and inverted faces. Perception 30(1):73–83

Lederman SJ, Klatzky RL (1987) Hand movements: a window into haptic object recognition. Cogn Psychol 19(3):342–368

Lederman SJ, Klatzky RL, Abramowicz A, Salsman K, Kitada R, Hamilton C (2007) Haptic recognition of static and dynamic expressions of emotion in the live face. Psyc Sci 1 18(2):158–164

Lederman SJ, Klatzky RL, Rennert-May E, Lee JH, Ng K, Hamilton C (2008) Haptic processing of facial expressions of emotion in 2-D raised-line drawings. IEEE Trans Haptics 1:27–38

Lederman S, Summers C, Klatzky R (1996) Cognitive salience of haptic object properties: role of modality-encoding bias. Perception 25(8):983–998

Levy I, Hasson U, Avidan G, Hendler T, Malach R (2001) Center-periphery organization of human object areas. Nat Neurosci 4:533–539

Mangini MC, Biederman I (2004) Making the ineffable explicit: estimating the information employed for face classifications. Cogn Sci 28:209–226

Maurer D, Le Grand R, Mondloch CJ (2002) The many faces of configural processing. Trends Cogn Sci 6:255–226

Marks D (2003) Visual imagery differences in the recall of pictures. Brit J Psychol 64:17–24

McCarthy G, Puce A, Gore JC, Allison T (1997) Face specific processing in human fusiform gyrus. J Cogn Neurosci 9:605–610

McGregor TA, Klatzky RL, Hamilton C, Lederman SJ (2010). Haptic classification of facial identity in 2-D displays: learning and inversion effects. IEEE: Transactions on Haptics 1(2):48–55

McKone E, Kanwisher N (2004) Does the human brain process objects of expertise like faces? A review of the evidence. In: Dehaene S, Duhamel JR, Hauser M, Rizzolatti (eds) From monkey brain to human brain MIT Press, Boston

Montgomery K, Haxby JV (2008) Mirror neuron system differentially activated by facial expressions and social hand gestures: a functional magnetic resonance imaging study. J Cogn Neurosci 20(10):1866–1877
Morris JS, Öhman A, Dolan RJ (1998) Conscious and unconscious emotional learning in the human amygdala. Nature 393:467–470
Newell FN, Ernst MO, Tjan BJ, Bülthoff HH (2001) View-dependence in visual and haptic object recognition. Psyc Sci 12:37–42
Nomura M, Ohira H, Haneda K, Iidaka T, Sadato N, Okada T et al. (2004) Functional association of the amygdala and ventral prefrontal cortex during cognitive evaluation of facial expressions primed by masked angry faces: an event-related fMRI study. Neuroimage 21(1):352–363
Norman JF, Norman HF, Clayton AM, Lianekhammy J, Zielke G (2004) The visual and haptic perception of natural object shape. Percept Psychophys 66(2):342–351
Ohman A (2002) Automaticity and the amygdala: Nonconscious responses to emotional faces. Curr Dir Psychol Sci 11(2):62–66
Peterson MA, Rhodes G (2003) Perception of faces, objects, and scenes: analytic and holistic processes. Advances in visual cognition. Oxford University Press, New York
Phillips ML, Young AW, Senior C, Brammer M., Andrews C, Calder AJ et al. (1997) A specific neural substrate for perceiving facial expressions of disgust. Nature 389(6650):495–498
Pietrini P, Furey ML, Ricciardi E, Gobbini MI, Wu WH, Cohen L et al. (2004) Beyond sensory images: object-based representation in the human ventral pathway. Proc Natl Acad Sci U S A 101:5658–5663
Posamentier MT, Abdi H (2003) Processing faces and facial expressions. Neuropsychol Rev 13(3):113–143
Prkachin GC (2003) The effects of orientation on detection and identification of facial expressions of emotion. Br J Psychol 94(1):45–62
Puce A, Allison T, Asgari M, Gore JC, McCarthy G (1996) Differential sensitivity of human visual cortex to faces, letterstrings, and textures: a functional magnetic resonance imaging study. J Neurosci 16(16):5205–5215
Puce A, Allison T, Bentin S, Gore JC, McCarthy G (1998) Temporal cortex activation in humans viewing eye and mouth movements. J Neurosci 18(6):2188–2199
Reales JM, Ballesteros S (1999) Implicit and explicit memory for visual and haptic objects: cross-modal priming depends on structural descriptions. J Exp Psychol Learn Mem Cogn 25(3): 644–663
Reed CL, Shoham S, Halgren E (2004) Neural substrates of tactile object recognition: an fMRI study. Hum Brain Mapp 21:236–246
Rizzolatti G, Fadiga L, Gallese V, Fogassi L (1996) Premotor cortex and the recognition of motor actions. Brain Res Cogn Brain Res 3(2):131–141
Rolls E (1996) The orbitofrontal cortex. Philos Trans R Soc Lond B Biol Sci B351:1433–1444
Rosch E, Mervis C, Gray W, Johnson D, Boyes-Braem P (1976) Basic objects in natural categories. Cogn Psychol 8:382–439
Rotshtein P, Malach R, Hadar U, Graif M, Hendler T (2001) Feeling or features: different sensitivity to emotion in high-order visual cortex and amygdala. Neuron 32(4):747–757
Russell J (1980) A circumplex model of affect. J Pers Soc Psychol 39(6):1161–1178
Sathian K, Lacey S (2007). Journeying beyond classical somatosensory cortex. Can J Exp Psychol 61(3):254–264
Sathian K, Lacey S (2008) Visual cortical involvement during tactile perception in blind and sighted individuals. In: Rieser JJ, Ashmead DH, Ebner FF, Corn AL (eds) Blindness and brain plasticity in navigation and object perception. Erlbaum, Mahwah, NJ, pp 113–125
Schyns PG, Bonnar L, Gosselin F (2002) Show me the features! Understanding recognition from the use of visual information. Psychol Sci 13(5):402–409
Sekuler AB, Gaspar CM, Gold JM, Bennett PJ (2004) Inversion leads to quantitative, not qualitative changes in face processing. Curr Biol 14(5):391–396
Sergent J (1984) An investigation into component and configural processes underlying face perception. Br J Psychol 75(2):221–242

Sullivan LA, Kirkpatrick SW (1996) Facial interpretation and component consistency. Genet Soc Gen Psychol Monogr 122(4):389–404
Wehrle T, Kaiser S, Schmidt S, Scherer KR (2000) Studying the dynamics of emotional expression using synthesized facial muscle movements. J Pers Soc Psychol 78(1):105–119
Winston RS, O'Doherty J, Dolan RJ (2003) Common and distinct neural responses during direct and incidental processing of multiple facial emotions. Neuroimage 20(1):84–97
Woodworth RS, Schlosberg H (1959) Experimental psychology. Slovenska Akademia Vied, Oxford, England
Yin RK (1969) Looking at upside-down faces. J Exp Psychol 81(1):141–145

Part IV
Plasticity

Chapter 16
The Ontogeny of Human Multisensory Object Perception: A Constructivist Account

David J. Lewkowicz

16.1 Introduction

Our perceptual world is multisensory in nature (J. J. Gibson, 1966; Marks, 1978; Stein and Meredith, 1993; Werner, 1973). This means that the objects in our natural world are usually specified by some combination of visual, auditory, tactile, olfactory, and gustatory attributes. From a theoretical perspective, it is possible that the mélange of multisensory object attributes might be confusing. Fortunately, however, the human perceptual system and its underlying neural mechanisms have evolved to enable us to integrate the various multisensory attributes that are usually available. As a result, we are able to perceive multisensory attributes as part-and-parcel of coherent, spatiotemporally continuous, and bounded physical entities.

The fact that adults easily perceive coherent multisensory objects raises the obvious and central question regarding the developmental origins of this ability. Specifically, the question is when and how does the ability to perceive the coherent nature of multisensory objects emerge? The answer is that this ability takes time to emerge. Furthermore, the answer is decidedly contrary to recent nativist claims that infants come into this world with "core knowledge" that endows them with a predetermined set of principles about objects and their properties (Spelke and Kinzler, 2007). The empirical findings that will be reviewed here will show that the developmental emergence of the ability to perceive specific types of intersensory relations is a heterochronous affair (Lewkowicz, 2002) and it will be argued that this reflects the outcome of a complex and dynamic interaction between constantly changing neural, behavioral, and perceptual processes. As a result, the theoretical conclusion that will be reached here will be opposite to the nativist one. It will be argued that the current empirical evidence is consistent with constructivist and developmental systems approaches to perceptual and cognitive development (Cohen et al., 2002; Piaget, 1954; Spencer et al., 2009). According to these approaches, perceptual skills

D.J. Lewkowicz (✉)
Department of Psychology, Florida Atlantic University, 777 Glades Rd., Boca Raton, FL, USA
e-mail: lewkowic@fau.edu

M.J. Naumer, J. Kaiser (eds.), *Multisensory Object Perception in the Primate Brain*,
DOI 10.1007/978-1-4419-5615-6_16,

at particular points in the life of an organism are the product of a complex developmental process that involves the co-action of multiple factors at different levels of organization (cellular, neural, behavioral, and extraorganismic).

16.2 Setting the Theoretical Problem

It might be argued that multisensory objects present a difficult challenge for developing infants. Their nervous system is highly immature at birth and they are perceptually and cognitively inexperienced. As a result, it is reasonable to expect that infants come into the world with no prior knowledge of their physical world and, thus, that they would not know a priori whether certain multisensory attributes belong together. For example, they would not be expected to know whether a particular melodious and high-pitched voice belongs to one or another person's face. Indeed, the relation between various modality-specific object attributes (e.g., color, pitch, taste, or temperature) is arbitrary and infants must learn to bind them on a case-by-case basis in order to perceive multisensory objects as coherent entities. This is complicated by the fact that the sensory systems develop and mature over the first few months of life and, as a result, may not always be able to detect relevant modality-specific attributes with sufficient precision to bind them. Interestingly, however, as will be shown here, infants have a way of getting around this problem by essentially ignoring many multisensory object features (e.g., the various attributes that specify a person's identity) and by simply binding them on the basis of their spatial and temporal coincidence. In other words, even though young infants posses rather poor auditory and visual sensory skills, and even though they may have difficulty resolving the detailed nature of auditory and visual inputs, they possess basic mechanisms that enable them to bind modality-specific multisensory attributes and that permit them to take advantage of the redundancy inherent in multisensory objects and events. As a result, despite their limitations, infants possess one basic but powerful mechanism that enables them to begin the construction of a multisensory object concept.

A second way for infants to construct a multisensory object concept is to rely on the various amodal invariant attributes that are normally inherent in multisensory inputs. These types of attributes provide equivalent information about objects across different modalities. For example, objects can be specified in audition and vision by their common intensity, duration, tempo, and rhythm. Likewise, in vision and touch objects can be specified by their common shape and texture and in audition and touch objects can be specified by their common duration, tempo, and rhythm. In general, amodal information is inherently relational and, as a result, intersensory binding is not necessary. What is necessary, however, is that the perceiver be able to detect the amodal invariance and here, again, evidence indicates that this mechanism emerges gradually during infancy (Lewkowicz, 2000a, 2002).

The behavioral literature has clearly demonstrated that human adults are very good at perceiving coherent multisensory objects and events (Calvert et al., 2004;

Marks, 1978; Welch and Warren, 1986) and, thus, that they can bind modality-specific attributes and detect amodal invariant attributes. In addition, a broader comparative literature has demonstrated that the redundancy inherent in multisensory objects and events is highly advantageous because it facilitates detection, discrimination, and learning in human adults as well as those of many other species (Partan and Marler, 1999; Rowe, 1999; Stein and Stanford, 2008; Summerfield, 1979). Underlying this ability is a nervous system that has evolved mechanisms specifically devoted to the integration of multisensory inputs and to the perception of amodal invariance (Calvert et al., 2004; Stein and Meredith, 1993; Stein and Stanford, 2008). Indeed, intersensory integration mechanisms are so pervasive throughout the primate brain that some have gone so far as to suggest that the primate brain is essentially a multisensory organ (Ghazanfar and Schroeder, 2006).

The behavioral developmental literature also has provided impressive evidence of intersensory perception in infancy. It has shown that infants can bind multisensory inputs and perceive various types of intersensory relations (Lewkowicz, 2000a, 2002; Lewkowicz and Lickliter, 1994; Lickliter and Bahrick, 2000; Walker-Andrews, 1997), and that they can take advantage of multisensory redundancy (Bahrick et al., 2004; Lewkowicz, 2004; Lewkowicz and Kraebel, 2004). What is particularly interesting about this evidence is that it is broadly consistent with the dominant theory of intersensory perceptual development put forth by E. Gibson (1969). Gibson's theory is based on the concept of perceptual differentiation and holds that infants are equipped with the ability to detect amodal invariance at birth and that, as development progresses, they learn to differentiate increasingly finer and more complex forms of amodal invariance. Although at first blush the extant empirical evidence appears to be consistent with Gibson's theory, a more detailed look at the evidence indicates that the developmental picture is considerably more complex. In essence, a good bit of the empirical evidence on infant intersensory perception amassed since Gibson proposed her theory has indicated that certain intersensory perceptual skills are absent early in life, that specific ones emerge at specific times during infancy, and that following their emergence they continue to improve and change as infants grow and acquire perceptual experience (Bremner et al., 2008; Lewkowicz, 2000a, 2002; Lewkowicz and Lickliter, 1994; Lickliter and Bahrick, 2000; Walker-Andrews, 1997). This evidence is actually consistent with the theory that posits the opposite process to that of differentiation, namely developmental integration (Birch and Lefford, 1963; Piaget, 1952). According to this theory, intersensory perceptual abilities are initially absent and only emerge gradually as infants acquire experience with multisensory inputs and discover the relations among them.

Given that evidence can be found to support the developmental differentiation and the developmental integration views, one reasonable theoretical stance might be that both developmental differentiation and developmental integration processes contribute to the emergence of the multisensory object concept. It turns out, however, that some recent empirical evidence suggests that a third process, that involves perceptual narrowing, also contributes in an important way to the development of intersensory perceptual functions. The importance of this process in the development of intersensory perception has recently been uncovered by Lewkowicz and

Ghazanfar (2006) and is described later in this chapter. As a result, the most reasonable theoretical framework that best describes the processes involved in the development of intersensory perception is one that incorporates the processes of developmental differentiation and integration as well as developmental narrowing.

The new theoretical framework raises several distinct but highly inter-related questions that must be answered if we are to better understand how infants come to acquire a stable and coherent conception of their multisensory world. First, do infants possess the necessary perceptual mechanisms for perceiving various multisensory object attributes as part-and-parcel of coherent object representations? Second, if they do, when do these mechanisms first begin to emerge? Third, what is the nature of these early mechanisms and how do they change over development? Finally, what underlying processes govern the emergence of these early mechanisms? These four questions will be addressed by reviewing research from my laboratory as well as related research from other laboratories, with a specific focus on infant perception of audio-visual (A-V) relations.

16.3 Response to A-V Intensity Relations

There is little doubt that infants' ability to perceive multisensory coherence is hampered by their neurobehavioral immaturity, relative perceptual inexperience, and the multisensory and diverse character of the external world. As noted earlier, however, the relational and invariant nature of the typical multisensory perceptual array provides many ready-made sources of coherence (J. J. Gibson, 1966). Indeed, it is precisely for this reason that some theorists have asserted that infants come into the world prepared to pick up multisensory coherence (E. J. Gibson, 1969; Thelen and Smith, 1994). Although this view is a reasonable one, given their neurobehavioral and perceptual limitations, young infants are not likely to perceive multisensory coherence in the same way that adults do. That is, young infants probably come into the world armed with some basic perceptual skills that enable them to discover relatively simple types of multisensory coherence and only later, as they gradually acquire the ability to perceive higher-level types of intersensory relations, discover more complex forms of multisensory coherence. Thus, initially, infants are likely to be sensitive to relatively low-level kinds of intersensory perceptual relations such as intensity and temporal synchrony.

Consistent with the above developmental scenario, Lewkowicz and Turkewitz (1980) found that newborn infants are, in fact, sensitive to A-V intensity relations. This study was prompted by Schneirla's (1965) observation that young organisms of many different species are sensitive to the quantitative nature of stimulation, defined as a combination of the physical intensity of stimulation and organismic factors (i.e., level of arousal). Lewkowicz and Turkewitz (1980) reasoned that if young infants are primarily responsive to the effective intensity of stimulation, then young infants should be able to perceive intensity-based A-V relations. To test this prediction, Lewkowicz and Turkewitz (1980) habituated 3-week-old infants to a constant-intensity visual stimulus (a white disc) and then tested their response to different-intensity white-noise stimuli. Importantly, the white-noise stimuli spanned an intensity range that included one stimulus that adults judged to be most equivalent in terms of its intensity to the visual stimulus presented to the infants during the habituation phase. As a result, Lewkowicz and Turkewitz expected that if infants spontaneously attempted to equate the auditory and visual stimuli, they would exhibit differential response to the auditory stimuli. Specifically, it was expected that infants would exhibit the smallest response recovery to the auditory stimulus whose intensity was judged by adults to be equivalent to the visual stimulus and increasingly greater response recovery to those auditory stimuli whose intensities were increasingly more discrepant from the visual stimulus. The results were consistent with this prediction and, thus, indicated that neonates do, indeed, have the capacity to perceive A-V equivalence on the basis of intensity.

In a follow-up study, Lewkowicz and Turkewitz (1981) tested the generality of intensity-based intersensory responsiveness by testing newborn infants' response to different intensities of visual stimulation following exposure to auditory stimulation of a constant intensity. If intensity-based intersensory responsiveness is a general characteristic of neonatal perceptual responsiveness then newborns' response to visual stimulation varying in intensity should be affected in a systematic way by prior auditory stimulation (presumably mediated by the induction of arousal changes). To test this possibility, Lewkowicz and Turkewitz (1981) presented pairs of visual stimuli varying in intensity to two groups of newborns. One group first heard a moderate-intensity white-noise stimulus whereas the other group did not. Consistent with expectation, results indicated that the group that heard the white-noise stimulus preferred to look at a low-intensity visual stimulus whereas the group that did not hear the white-noise stimulus preferred to look at a higher-intensity visual stimulus. This shift in the looking preference was interpreted to reflect the effects of increased arousal caused by exposure to the white-noise stimulus. Presumably, when the increased internal arousal was combined with the physical intensity of the moderate-intensity stimulus, the effective intensity of stimulation exceeded the infants' optimal level of preferred stimulation and caused them to shift their attention to the lower-intensity stimulus in order to re-establish the optimal level of stimulation.

16.4 Response to A-V Temporal Synchrony Relations

The natural multisensory world is characterized by patterns of correlated and amodally invariant information (J. J. Gibson, 1966). Part of this invariance is due to the fact that, under normal conditions, the auditory and visual sensory attributes that specify multisensory objects and events are usually temporally coincident. That is, the temporal patterns of auditory and visual information that specify everyday objects and events always have concurrent onsets and offsets and, thus, are always temporally synchronous (e.g., a talker's vocalizations always start and stop whenever the talker's lips start and stop moving). Audio-visual synchrony can provide infants with an initial opportunity to perceive audible and visible object and event attributes as part-and-parcel of coherent entities (E. J. Gibson, 1969; Lewkowicz, 2000a; Thelen and Smith, 1994). For infants, this is especially important for two reasons. First, it provides them with an initial opportunity to detect the multisensory coherence of their world. Second, once infants discover that auditory and visual inputs are synchronous, this sets the stage for the discovery of other types of A-V relations including those based on equivalent durations, tempos, and rhythmical structure, gender, affect, and identity. In other words, an initial perceptual sensitivity to A-V temporal synchrony relations (as well as A-V intensity relations) provides infants with the initial scaffold that enables them to subsequently discover more complex forms of intersensory correspondence.

16.4.1 Infant Perception of A-V Synchrony Relations

Detection of temporal A-V synchrony is relatively easy because it only requires the perception of synchronous energy onsets and offsets and because it is mediated by relatively low-level, subcortical tecto-thalamo-insular pathways that make the detection of A-V temporal correspondence at an early level of cortical processing possible (Bushara et al., 2001). This suggests that despite the fact that the infant nervous system is immature, infants should be able to detect A-V synchrony relations quite early in postnatal life. Indeed, studies at the behavioral level have indicated that, starting early in life, infants are sensitive and responsive to A-V temporal synchrony relations. For example, Lewkowicz (1996) conducted systematic studies to investigate the detection of A-V temporal synchrony relations across the first year of life by studying 2-, 4-, 6-, and 8-month-old infants and compared their performance to that of adults tested in a similar manner. The infants were habituated to a two-dimensional object that bounced up and down on a computer monitor and made an impact sound each time it changed direction at the bottom of the monitor. Following habituation, infants were given a set of test trials to determine the magnitude of their A-V asynchrony detection threshold. This was done either by presenting the impact sound prior to the object's visible bounce (sound-first condition) or by presenting it following the visible bounce (sound-second condition).

Results yielded no age differences and indicated that the lowest A-V asynchrony that infants detected in the sound-first condition was 350 ms and that the lowest asynchrony that they detected in the sound-second condition was 450 ms. In contrast, adults who were tested in a similar task and with the same stimuli detected the asynchrony at 80 ms in the sound-first condition and at 112 ms in the sound-second condition. Based on these results, Lewkowicz concluded that the intersensory temporal contiguity window (ITCW) is considerably wider in infants than in adults and that, as a result, infants tend to perceive temporally more disparate multisensory inputs as coherent.

16.4.2 Infant Perception of A-V Speech Synchrony and Effects of Experience

In subsequent studies, Lewkowicz has found that the ITCW is substantially larger for multisensory speech than for simple abstract events. In the first of these studies, Lewkowicz (2000b) habituated 4-, 6- and 8-month-old infants to audio-visually synchronous syllables (/ba/ or /sha/) and then tested their response to audio-visually asynchronous versions of these syllables. Results indicated that infants successfully detected a 666 ms asynchrony. In a subsequent study, Lewkowicz (2003) found that 4- to 8-month-old infants detected an asynchrony of 633 ms.

In the most recent study, Lewkowicz (2010) investigated whether detection thresholds for audio-visual speech asynchrony are affected by initial short-term experience. The purpose of this study was, in part, to determine whether the effects of short-term exposure to synchronous vs. asynchronous audio-visual events have effects similar to those observed in adults. Studies with adults have shown that when they are first tested with audio-visually asynchronous events, they perceive them as such, but that after they are given short-term exposure to such events they respond to them as if they are synchronous (Fujisaki et al., 2004; Navarra et al., 2005; Vroomen et al., 2004). In other words, short-term adaptation to audio-visually asynchronous events appears to broaden the ITCW in adults. If this adaptation effect is partly due to an experience-dependent synchrony bias that is due to a lifetime of nearly exclusive exposure to synchronous events – resulting in what Welch and Warren (1980) call the "unity assumption" then infants may not exhibit such a bias because of their relative lack of experience with synchronous multisensory events. Lewkowicz (2010) conducted a series of experiments to test this prediction. In the first experiment, 4- to 10-month-old infants were habituated to an audio-visually synchronous audio-visual syllable and then were tested to see if they could detect three increasing levels of asynchrony (i.e., 366, 500, and 666 ms). Figure 16.1a shows the data from the habituation trials – indicating that infants habituated to the synchronous audio-visual syllable – and Fig. 16.1b shows the results from the test trials. Planned contrast analyses of the test trial data, comparing the duration of looking in each novel test trial, respectively, with the duration of looking in the FAM 0 test trial indicated that response recovery was not significant in the NOV 366 and NOV 500 test trials but that it was significant in the NOV 666 test trial. These results are consistent with previous findings of infant detection of asynchrony in showing that infants do not detect A-V speech asynchrony below 633 ms. In the second experiment, 4- to 10-month-old infants were habituated to an asynchronous syllable (666 ms) and were then tested for their ability to detect decreasing levels of asynchrony (i.e., 500, 366, and 0 ms). Figure 16.2a shows the data from the habituation trials – indicating that infants habituated to the asynchronous audio-visual syllable – and Fig. 16.2b shows the results from the test trials. Planned contrast analyses of the test trial data, comparing the duration of looking in each novel test trial, respectively, with the duration of looking in the FAM 666 test trial indicated that infants did not exhibit significant

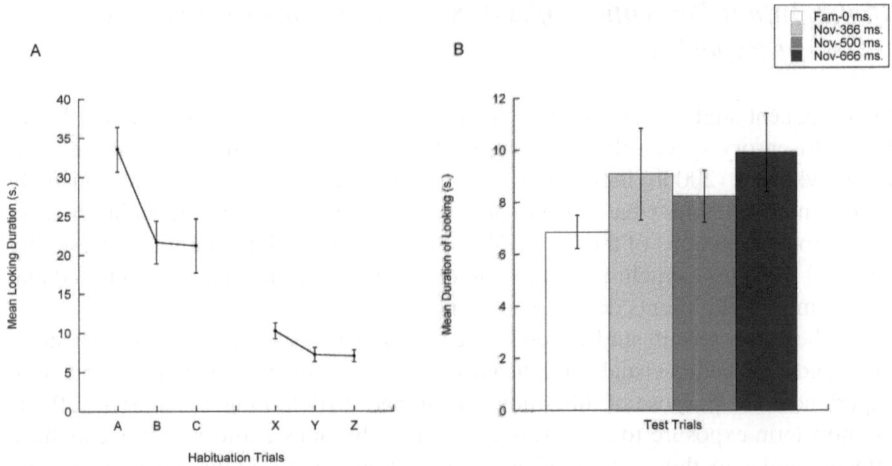

Fig. 16.1 Infant detection of A-V synchrony relations following habituation to a synchronous audio-visual syllable. (**a**) The mean duration of looking during the first three (A, B, and C) and the last three (X, Y, and Z) habituation trials is shown. (**b**) The mean duration of looking in response to the various levels of asynchrony during the test trials is shown. *Error bars* indicate the standard error of the mean

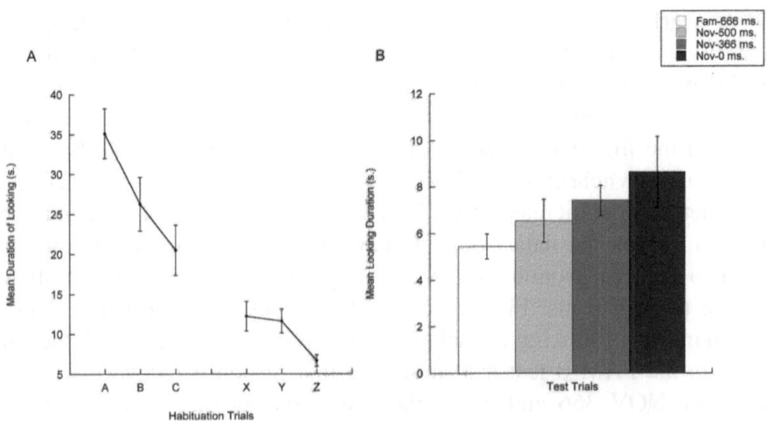

Fig. 16.2 Infant detection of A-V synchrony relations following habituation to an asynchronous audio-visual syllable. (**a**) The mean duration of looking during the habituation phase is shown and (**b**) the mean duration of looking in response to the various levels of asynchrony during the test trials is shown. *Error bars* indicate the standard error of the mean

response recovery in the NOV 500 test trial but that they did in the NOV 366 and the NOV 0 test trials. The pattern of responsiveness in this experiment is opposite to what might be expected on the basis of the adult adaptation findings. Rather than exhibit broadening of the ITCW following short-term adaptation to a discriminable A-V asynchrony, the infants in the second experiment exhibited narrowing of the ITCW in that they discriminated not only between the 666 ms asynchrony and synchrony (0 ms) but also between the 666 ms asynchrony and an asynchrony of 366 ms.

Together, the findings from studies of infant response to audio-visual speech indicate that it is more difficult for infants to detect a desynchronization of the auditory and visual streams of information when that information is a speech syllable. That is, following initial learning of a synchronous audio-visual speech token infants only detect an asynchrony of 666 ms, whereas following initial learning of a synchronous bouncing/sounding object infants detect an asynchrony of 350 ms, a difference of more than 300 ms. The most likely reason for this difference is that the audio-visual speech signal consists of continuous changes in facial gestures and vocal information, making it more difficult to identify the precise point where the desynchronization occurs. The findings also indicate that an initial short-term experience with an asynchronous speech event facilitates subsequent detection of asynchrony in infants, an effect opposite to that found in adults. This seemingly unexpected adaptation effect actually makes sense when the infant's relative lack of perceptual experience and the presumed lack of a unity assumption are taken into account. In the absence of a unity assumption, short-term exposure to an asynchronous multisensory event does not cause infants to treat it as synchronous but rather focuses their attention on the event's temporal attributes and, in the process, appears to sharpen their detection of the A-V temporal relationship.

16.4.3 Binding of Multisensory Attributes

A number of studies have shown that from birth on infants can bind spatiotemporally congruent abstract objects and the sounds that accompany them and that they can bind human faces and the vocalizations they make (Bahrick, 1983, 1988, 1992; Brookes et al., 2001; Lewkowicz, 1992a, b, 2000b, 2003; Reardon and Bushnell, 1988; Slater et al., 1997). What makes these findings particularly interesting is that despite the fact that infants are able to detect A-V temporal synchrony relations from a very young age on, and the fact that their asynchrony detection thresholds do not change during the first year of life (Lewkowicz, 1996, 2010), infants' ability to associate different types of multisensory attributes varies with age and depends on the familiarity of the attributes. Thus, when the attributes are human faces and voices, infants as young as 3 months of age can associate them (Brookes et al., 2001), but when the attributes are various object attributes such as color, shape, taste, or temperature, only older infants are able to associate them. For example, whereas neither 3.5- nor 5-month-old infants can associate the color/shape of an object and the pitch of an accompanying sound, 7-month-old infants can (Bahrick, 1992, 1994). Likewise, only 6-month-old infants can bind the color or pattern of an object with its particular shape (Hernandez-Reif and Bahrick, 2001). Finally, it is not until 7 months of age that infants can bind the color of an object and its taste (Reardon and Bushnell, 1988) and even at this age they do not associate color and temperature (Bushnell, 1986). Together, this set of findings indicates that the ability to bind familiar modality-specific object properties emerges relatively early in infancy but that the ability to form more arbitrary associations of less familiar object properties emerges later. This suggests that sensitivity to A-V temporal synchrony relations can facilitate intersensory binding but that the facilitation is greatest for familiar modality-specific object properties.

16.4.4 Binding of Nonnative Faces and Vocalizations

Although the familiarity of modality-specific attributes seems to contribute to successful intersensory binding, recent evidence from our studies indicates that sensitivity to synchronous intersensory relations is much broader in younger than in older infants. This evidence comes from one of our recent studies (Lewkowicz and Ghazanfar, 2006) in which we found that young infants bind nonnative faces and vocalizations but that older infants do not. The developmental pattern of initial broad perceptual tuning followed by narrowing of that tuning a few months later was not consistent with the conventional view that development is progressive in nature and that it usually leads to a broadening of perceptual skills. It was consistent, however, with a body of work on infant unisensory perceptual development showing that some forms of unisensory perceptual processing also narrow during the first year of life. For example, it has been found that young infants can detect nonnative speech contrasts but that older infants do not (Werker and Tees, 1984) and that young infants can discriminate nonnative faces (i.e., of different monkeys) as well as the faces of other races, but that older infants do not (Kelly et al., 2007; Pascalis et al., 2002). Based on these unisensory findings, we (Lewkowicz and Ghazanfar, 2006) asked whether the perceptual narrowing that has been observed in the unisensory perceptual domain might reflect a general, pan-sensory process. If so, we hypothesized that young infants should be able to match nonnative faces and vocalizations but that older infants should not. We put this hypothesis to test by showing side-by-side faces of a monkey producing two different visible calls (see Fig. 16.3) to groups of 4-, 6-, 8-, and 10-month-old infants. During the initial two preference trials we showed the faces in silence while during the second two trials we showed the faces together with the audible call that matched one of the two visible calls (Fig. 16.3). The different calls (a coo and a grunt) differed in their durations and, as a result, the matching visible and audible calls corresponded in terms of their onsets and offsets as well as their durations. In contrast, the non-matching ones only corresponded in terms of their onsets.

We expected that infants would look longer at the visible call that matched the audible call if they perceived the correspondence between them. As predicted, and consistent with the operation of a perceptual narrowing process, we found that the two younger groups of infants matched the corresponding faces and vocalizations but that the two older groups did not. These findings confirmed our prediction that intersensory perceptual tuning is broad early in infancy and that it narrows over time. We interpreted the narrowing effects as a reflection of increasing specialization for

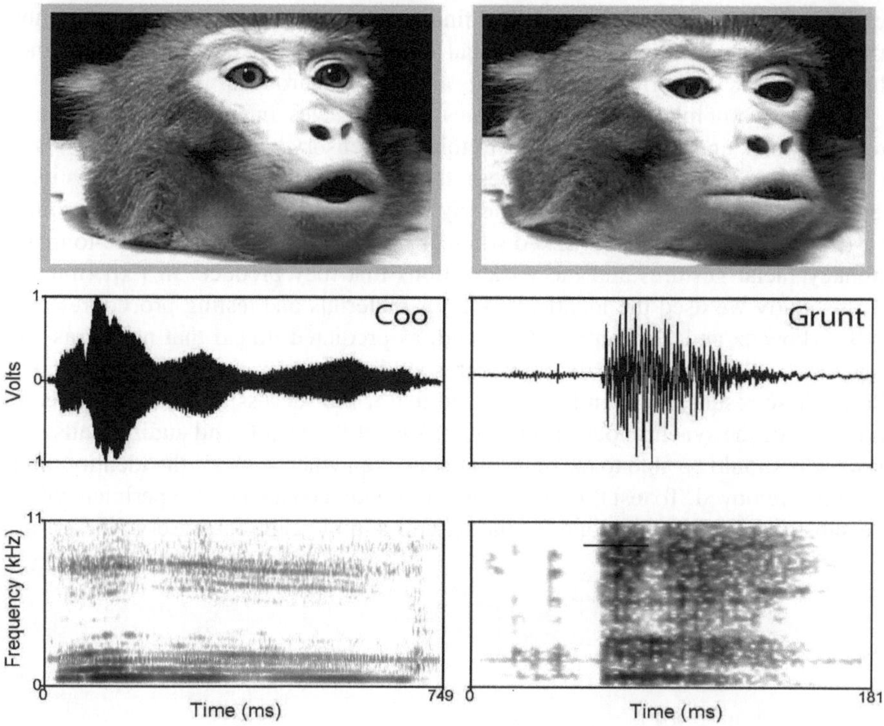

Fig. 16.3 Single video frames showing the facial gestures made by one of the monkeys when producing the coo and the grunt. The gestures depicted are at the point of maximum mouth opening. Below the facial gestures are the corresponding sonograms and spectrograms of the audible call

human faces and vocalizations that is the direct result of selective experience with native faces and vocalizations and a concurrent lack of experience with nonnative ones.

Because the matching faces and vocalizations corresponded in terms of both their onset and offset synchrony and their durations, the obvious question was whether the successful matching was based on one or both of these perceptual cues. Thus, in a subsequent study (Lewkowicz et al., 2008), we tested this possibility by repeating the Lewkowicz and Ghazanfar (2006) study except that this time we presented the monkey audible calls out of synchrony with respect to both visible calls. This meant that the corresponding visible and audible calls were only related in terms of their durations. Results indicated that A-V temporal synchrony mediated the successful matching in the younger infants because this time neither 4- to 6-month-old nor 8- to 10-month-old infants exhibited intersensory matching. The fact that the younger infants did not match despite the fact that the corresponding faces and vocalizations still corresponded in terms of their duration shows that duration alone was not sufficient to enable infants to make intersensory matches. Indeed, this latter finding is consistent with previous findings (Lewkowicz, 1986) indicating that infants do not match auditory and visual inputs that are equated in terms of their duration unless the corresponding inputs also are synchronous.

If A-V temporal synchrony mediates cross-species intersensory matching in young infants, and if responsiveness to this intersensory perceptual cue depends on a basic and relatively low-level process, then it is possible that cross-species intersensory matching may emerge very early in development. To test this possibility, we (Lewkowicz et al., 2010) asked whether newborns also might be able to match monkey facial gestures and the vocalizations that they produce. In Experiment 1 of this study we used the identical stimulus materials and testing procedures used by Lewkowicz and Ghazanfar (2006) and, as predicted, found that newborns also matched visible monkey calls and corresponding vocalizations (see Fig. 16.4a). Given these results, we then hypothesized that if the successful matching reflected matching of the synchronous onsets and offsets of the visible and audible calls then newborns should be able to make the matches even when some of the identity information is removed. To test this possibility, we conducted a second experiment where we substituted a complex tone for the natural audible call. To preserve the critical

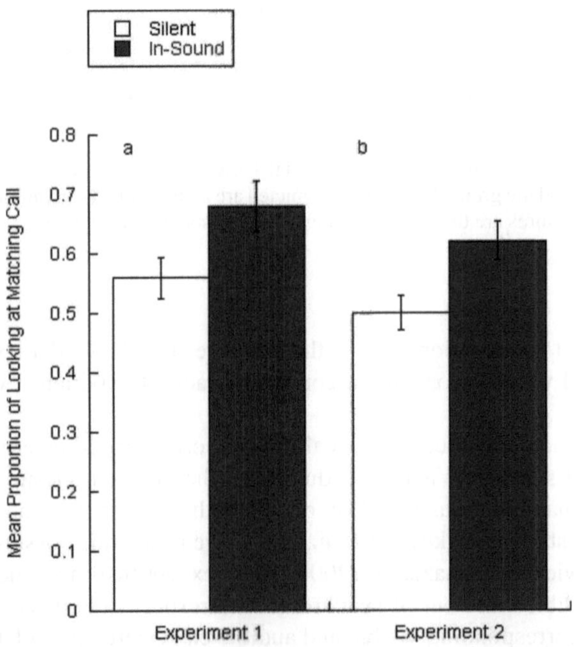

Fig. 16.4 Newborns' visual preference for matching visible monkey calls in the absence and presence of the matching natural audible call or a tone. (**a**) The mean proportion of looking at the matching visible call when it was presented during silent and in-sound trials is shown, respectively, when the natural audible call was presented. (**b**) The mean proportion of looking at the matching visible call in the silent and in-sound test trials is shown, respectively, when the complex tone was presented. *Error bars* represent the standard errors of the mean

temporal features of the audible call, we ensured that the tone had the same duration as the natural call and that, once again, its onsets and offsets were synchronous with the matching visible call. Results indicated that despite the absence of identity information and despite the consequent absence of the correlation between the dynamic variations in facial gesture information and the amplitude and formant structure inherent in the natural audible call, newborns still performed successful intersensory matching (see Fig. 16.4b). These results suggest that newborns' ability to make cross-species matches in Experiment 1 was based on their sensitivity to the temporally synchronous onsets and offsets of the matching faces and vocalizations and that it was not based on their identity information nor on the dynamic correlation between the visible and audible call features.

Together, the results from the two experiments with newborns demonstrate that they are sensitive to a basic feature of their perceptual world, namely stimulus energy onsets and offsets. As suggested earlier, it is likely that this basic sensitivity bootstraps newborns' entry into the world of multisensory objects and enables them to detect the coherence of multisensory objects despite the fact that this process ignores the specific identity information inherent in the faces and vocalizations. Moreover, it is interesting to note that our other recent studies (Lewkowicz, 2010) have shown that the low-level sensitivity to stimulus onsets and offsets continues into the later months of life; this is indicated by the finding that infants between 4 and 10 months of age detect an A-V desynchronization even when the stimulus consists of a human talking face and a tone stimulus. Thus, when the results from the newborn study are considered together with the results from this latter study with older infants it appears that sensitivity to a relatively low-level kind of intersensory relation makes it possible for infants to detect the temporal coherence of multisensory inputs well into the first year of life.

Needless to say, although an early sensitivity to energy onsets and offsets provides infants with a very powerful and useful cue for discovering the coherence of multisensory objects and events, it has its obvious limitations. In particular, if young infants are primarily responsive to energy onsets and offsets then the finding that it is not until the latter half of the first year of life that infants become capable of making arbitrary intersensory associations (e.g., color/shape and pitch or color and taste) makes sense. That is, making an association between a static perceptual attribute such as shape, color, or temperature is far more difficult than between attributes representing dynamic naturalistic events where energy transitions are available and clear. For example, when we watch and hear a person talking, we can see when the person's lips begin and stop moving and can hear the beginning and end of the accompanying vocalizations. In contrast, when we watch and hear a person talking and need to detect the correlation between the color and shape of the person's face and the pitch of the person's voice, it is more difficult to detect the correlation because the color and shape of the person's face do not vary over time. Likewise, because the shape and color of an object are features that do not vary over time, their correlation with an object's taste (e.g., an apple) is arbitrary and, thus, less salient to an infant who is primarily responsive to energy onsets and offsets as well as the dynamic properties of objects.

The pervasive and fundamental role that A-V temporal synchrony plays in infant perceptual response to multisensory attributes suggests that sensitivity to this intersensory perceptual cue reflects the operation of a fundamental early perceptual mechanism. That is, even though sensitivity to A-V temporal synchrony is mediated by relatively basic and low-level processing mechanisms, as indicated earlier, this intersensory relational cue provides infants with a powerful initial perceptual tool for gradually discovering, via the process of perceptual learning and differentiation (E. J. Gibson, 1969), that multisensory objects are characterized by other forms of intersensory invariance. In essence, the initial developmental pattern consists of the discovery of basic synchrony-based intersensory relations enabling infants to perceive multisensory objects and events as coherent entities. This, however, does not permit them to perceive higher-level, more complex features. For example, at this point infants essentially ignore the rich information that is available in-between the energy onsets and offsets of auditory and visual stimulation. The situation changes, however, once infants discover synchrony-based multisensory coherence because now they are in a position to proceed to the discovery of the more complex information that is located "inside" the stimulus. For example, once infants start to bind the audible and visible attributes of talking faces, they are in a position to discover that faces and the vocalizations that accompany them also can be specified by common duration, tempo, and rhythm, as well as by higher-level amodal and invariant attributes such as affect, gender, and identity.

Newborns' sensitivity to A-V temporal synchrony relations is particularly interesting when considered in the context of the previously discussed sensitivity to the effective intensity of multisensory stimulation. One possibility is that newborn's ability to detect stimulus energy onsets and offsets in A-V synchrony detection tasks is actually directly related to their sensitivity to multisensory intensity. In other words, it may be that A-V intensity and temporal synchrony detection mechanisms work together to bootstrap newborns' entry into the multisensory world and that, together, they enable newborns to discover a coherent, albeit relatively simple, multisensory world. As they then grow and learn to differentiate increasingly finer aspects of their perceptual world, they gradually construct an increasingly more complex picture of their multisensory world.

16.4.5 The Importance of Spatiotemporal Coherence in Object Perception

In addition to permitting infants to bind and match auditory and visual object attributes, evidence from my laboratory and that of my colleagues (Scheier et al., 2003) has shown that A-V spatiotemporal synchrony cues can help infants disambiguate what are otherwise ambiguous visual events. Specifically, when two identical visual objects move toward one another, pass through each other, and then continue to move away, the majority of adults watching such a display report seeing the two objects streaming through one another. When, however, a simple sound is presented at precisely the point when the objects coincide, a majority of adults report that the objects now bounce against each other (Sekuler et al., 1997). The specific spatiotemporal relationship between the sound and the motion of the two objects is responsible for this illusion and helps resolve the visual ambiguity (Watanabe and Shimojo, 2001). We (Scheier et al., 2003) asked whether infants might also profit from such A-V spatiotemporal relations to resolve visual ambiguity. Given the already reviewed evidence of the power of A-V temporal synchrony to organize the infant's multisensory world, it would be reasonable to expect that infants might also exhibit the bounce illusion. To test this possibility, we habituated 4-, 6-, and 8-month-old infants either to the streaming display with the sound occurring at the point of coincidence or to the same display with the sound presented either prior to coincidence or after it. Then, we tested infants in the two groups with the opposite condition. Results indicated that both groups exhibited response recovery at 6 and 8 months of age but not at 4 months of age, indicating that the two older age groups detected the specific spatiotemporal relationship between the sound and the spatial position of the moving objects. Thus, the specific temporal relationship between a sound and an ambiguous visual event can disambiguate the event for infants starting at 6 months of age. Given that this phenomenon is dependent on attentional factors (Watanabe and Shimojo, 2001), we interpreted these findings to mean that the emergence of this multisensory object perception system reflects the emergence of more advanced attentional mechanisms located in the parietal cortex that can quickly and flexibly switch attention. In essence, one perceives the bounce illusion when attention to the motion of the objects is briefly interrupted by the sound. This, in turn, requires the operation of parietal attentional mechanisms that emerge by around 6 months of age (Ruff and Rothbart, 1996).

16.4.6 Summary of Effects of A-V Temporal Synchrony on Infant Perception

Overall, the findings to date on infant response to A-V temporal synchrony and their reliance on it for the perception of coherent multisensory objects yield the following interim conclusions. First, infants are sensitive to A-V temporal synchrony relations from birth on and this sensitivity does not appear to change during early human development but does between infancy and adulthood. Second, response to A-V temporal synchrony relations appears to be based mainly on sensitivity to stimulus energy onsets and offsets. Third, despite the absence of age-related changes in the A-V asynchrony detection threshold during infancy (a) the threshold can be modified by short-term experience and (b) the effects of such experience are opposite to those found in adults. Fourth, early infant sensitivity to A-V temporal synchrony relations is so broad that it permits younger but not older infants to even bind non-native facial gestures and accompanying vocalizations. Fifth, infant ability to bind modality-specific multisensory attributes on the basis of A-V temporal synchrony changes over the first months of life in that young infants can bind the audible and visible attributes of familiar and naturalistic types of objects and events but only older infants can bind the multisensory attributes that specify more abstract, less common, types of relations. Finally, response to A-V temporal synchrony cues can be overridden by competing temporal pattern cues during the early months of life and only older infants can respond to synchrony cues when they compete with rhythmic pattern cues.

16.5 Response to A-V Colocation

Findings from studies of infant response to multisensory colocation cues, like those from studies of infant response to A-V temporal synchrony cues, also have yielded evidence of developmental changes. In the aggregate, these findings indicate that even though infants are sensitive to multisensory colocation cues, they also exhibit major changes in the way they respond to such cues (Morrongiello, 1994). Thus, even though starting at birth infants exhibit coordinated auditory and visual responses to lateralized sounds, these kinds of responses are reflexive in nature and, as a result, do not constitute adequate evidence of true intersensory perception. Nonetheless, despite their being reflexive, such responses provide infants with an initial opportunity to experience collocated auditory and visual stimulus attributes. As a result, from birth on infants have coordinated multisensory experiences and have an initial basis for gradually learning to expect to see objects where they hear sounds and, in the process, can construct a coordinated map of audio-visual space.

The developmental improvement in the ability to perceive spatial colocation is likely due to a combination of maturational and experiential factors working in concert with one another. In terms of sheer sensory processing capacity, the auditory and visual systems undergo major changes. In the auditory system, accuracy of sound localization changes dramatically during infancy as indicated by the fact that the minimum audible angle is 27° at 8 weeks, 23.5° at 12 weeks, 21.5° at 16 weeks, 20° at 20 weeks, 18° at 24 weeks, and 13.5° at 28 weeks of age (Morrongiello et al., 1990) and then declines more slowly reaching a minimum audible angle of around 4° by 18 months of age (Morrongiello, 1988). In the visual modality, visual field extent is relatively small early in infancy – its magnitude depends on the specific perimetry method used but is around 20° at 3.5 months – and then roughly doubles in the nasal visual field and roughly triples in the lateral visual field, reaching adult-like levels by 6–7 months of age (Delaney et al., 2000). Although data on infants younger than 3.5 months of age are not available, it is safe to assume that the extent of the visual field is smaller in younger infants. In addition to the dramatic changes in the minimum audible angle and visual field extent during infancy, major changes occur in spatial resolution, saccadic localization, smooth pursuit abilities, and localization of moving objects (Colombo, 2001). The fact that all of these sensory/perceptual skills are rather poor at first and then gradually improve over the first months of life means that localization and identification of objects is rather crude early in infancy and only improves gradually. Doubtless, as the various sensory/perceptual skills improve, concurrent experiences with collocated auditory and visual sources of stimulation help refine multisensory localization abilities.

My colleagues and I conducted a comprehensive study of infant localization of auditory, visual, and audio-visual stimulation across the first year of life (Neil et al., 2006). In this study, we investigated responsiveness to lateralized unisensory and bisensory stimuli in 2-, 4-, 6-, 8-, and 10-month-old infants as well as adults using a perimetry device that allows the presentation of auditory, visual, or audio-visual targets at different eccentricities. We presented targets at 25° or 45° to the right and left of the subjects. At the start of each trial, the infant's attention was centered by presenting a central stimulus consisting of a set of flashing light-emitting diodes (LEDs) and bursts of white noise. As soon as the infant's attention was centered, the lateralized targets – a vertical line of flashing LEDs, a burst of white noise, or both – were presented at one of the two eccentricities and either to the subject's right or left in a random fashion. To determine whether and how quickly infants localized the targets, we measured latency of eye movements and/or head turns to the target.

As expected, response latency decreased as a function of age regardless of type of stimulus and side on which it was presented. Second, even the 8- to 10-month-old infants' response latencies were longer than those found in adults, indicating that the orienting systems continue to improve beyond this age. Third, response latency was slowest to the auditory targets, faster to visual targets, and fastest to the audio-visual targets, although the difference in latency to visual as opposed to audio-visual was relatively small. Of particular interest from the standpoint of the development of a multisensory object concept, we found the greatest age-related decline in response latency to audio-visual targets during the first 6 months of life, suggesting that it is during the first 6 months of life that infants are acquiring the ability to localize the auditory and visual attributes of objects in a coordinated fashion. Consistent with this pattern, it was only by 8–10 months of age that infants first exhibited some evidence of an adult-like non-linear summation of response latency to audio-visual as opposed to auditory or visual targets. In other words, prior to 8 months of age responsiveness to audio-visual stimuli reflected the faster of the two unisensory responses whereas by 8 months of age it reflected non-linear summation for the first time. Importantly, however, this kind of multisensory enhancement was only found in the 8- to 10-month-old infants' response to targets presented at 25°. The absence of this effect at 45° indicates that integration of auditory and visual localization signals is still not fully mature by 8–10 months of age and, therefore, that it undergoes further improvement past that age. In sum, these results are consistent with a constructivist account by showing that adult-like localization responses do not emerge until the end of the first year of life and that even then they are not nearly as good as they are in adults.

When the kinds of developmental changes in infant localization responses described above are considered together with the previously cited evidence of long-term changes in A-V temporal synchrony thresholds, they suggest that the multisensory object concept develops slowly during infancy and probably continues to develop well past the first year of life. In brief, the evidence shows that the ability to integrate various multisensory object attributes is quite poor at birth and that it emerges gradually over the first months of postnatal life. This gradual emergence is likely to reflect the concurrent development of a myriad of underlying and interdependent factors such as neural growth and differentiation, improvement in sensory thresholds and sensory processing, improvement in various motor response systems, and accumulating sensory, perceptual, and motor experience. Given that so many underlying processes are changing and interacting with one another during early development, and given that the nature of the information to be integrated and the specific task requirements affect integration, it is not surprising that the various and heterogeneous intersensory integration skills emerge in a heterochronous fashion (Lewkowicz, 2002).

Regardless of the heterochronous emergence of heterogeneous intersensory integration skills and regardless of the fact that the mature multisensory object concept emerges relatively slowly, there is little doubt that young infants have some rudimentary integration skills that enable them to embark on the path to adult-like integration. Nonetheless, the relatively slow emergence of the multisensory object concept means that nativist accounts that claim that the object concept is present at birth are incorrect and that they mis-characterize the complex developmental processes underlying its emergence. This conclusion receives additional support from studies of the development of the unisensory (i.e., visual) object concept. For example, it is well-known that 4-month-old infants can take advantage of the spatiotemporal coherence of the parts of a moving but seemingly incomplete object to perceive the parts as belonging to a single object (Kellman and Spelke, 1983). Consistent with the development of multisensory object coherence, studies have shown that the ability to perceive coherent visual objects is not present at birth and that it only emerges gradually during the first months of life (S. P. Johnson, 2004). Furthermore, its gradual emergence is partly due to the development of a voluntary, cortical, and attention-driven saccadic eye movement system that slowly comes to dominate an initially reflexive, subcortically controlled, saccadic system (M. H. Johnson, 2005).

16.6 Perception of Multisensory Sequences in Infancy

Multisensory objects often participate in complex actions that are sequentially organized. For example, when people speak they produce sequences of vocalizations along with highly correlated facial gestures. Similarly, when a drummer plays the drum he produces a patterned series of sounds that are correlated with the pattern of his hand and arm motions. In both cases, the patterns produced are perceived as unitary events that carry specific meanings. In the case of an audio-visual speech utterance, it is the syntactically prescribed order of the syllables and words that imbue the utterance with a specific meaning. In the case of the drummer, it is the order of different drum beats that imbues the musical passage with specific meaning (e.g., a particular rhythm and mood). Infants must at some point become capable of perceiving, learning, and producing sequences to function adaptively. This fact raises the obvious question of when this ability might emerge in development. Prior theoretical accounts have claimed that sequence perception and learning abilities are innate (Greenfield, 1991; Nelson, 1986). More recent research on infant pattern and sequence perception has clearly demonstrated, however, that this is simply not the case.

Although pattern and sequence perception skills are related, they also are distinct. They are related because both enable perceivers to detect the global organization of a set of distinct stimulus elements and they are distinct because only sequence perception skills enable the extraction of specific ordinal relations among the distinct elements making up specific sequences. Bearing the similarities and differences in mind, studies of pattern perception have shown that infants are sensitive to unisensory as well as bisensory rhythmic patterns from birth on (Lewkowicz, 2003; Lewkowicz and Marcovitch, 2006; Mendelson, 1986; Nazzi et al., 1998; Pickens and Bahrick, 1997) and, as shown below, studies also have shown that infants exhibit rudimentary sequence perception and learning abilities. Although no studies to date have investigated sequence perception and learning at birth, the extant evidence indicates that infants can learn adjacent and distant statistical relations, simple sequential rules, and ordinal position information. Of particular relevance to the current argument that the multisensory object concept is constructed during early life is the fact that these various sequence perception and learning abilities emerge at different points in infancy.

The earliest sequence learning ability to emerge is the ability to perceive and learn adjacent statistical relations. Thus, beginning as early as 2 months of age infants can learn the adjacent statistical relations that link a series of looming visual shapes (Kirkham et al., 2002; Marcovitch and Lewkowicz, 2009), by 8 months of age they can learn the statistical relations that link adjacent static object features (Fiser and Aslin, 2002) as well as adjacent nonsense words in a stream of sounds (Saffran et al., 1996), and by 15 months of age infants begin to exhibit the ability to learn distant statistical relations (Gómez and Maye, 2005). The most likely reason why the ability to perceive and learn adjacent statistical relations emerges earliest is because it only requires the perception and learning of the conditional probability relations between adjacent sequence elements and, thus, only requires the formation of paired-associates. The more complex ability to learn abstract sequential rules (e.g., AAB vs. ABB) emerges by 5 months of age when they are specified by abstract objects and accompanying speech sounds (Frank et al., 2009), by 7.5 months of age when the rules are instantiated by nonsense syllables (Marcus et al., 2007; Marcus et al., 1999), and by 11 months of age when the rules are instantiated by looming objects (S. P. Johnson et al., 2009). Finally, it is not until 9 months of age that infants can track the ordinal position of a particular syllable in a string of syllables (Gerken, 2006).

It is interesting to note that most of the studies of infant sequence learning to date have investigated this skill by presenting unisensory stimuli. As already noted, however, our world is largely multisensory in nature and multisensory redundancy facilitates learning and discrimination (Bahrick et al., 2004; Lewkowicz and Kraebel, 2004). Consequently, it is important to determine how and when infants are able to perceive and learn multisensory sequences. With this goal in mind, we have conducted several studies of sequence perception and learning in infancy. In some of these studies, we provided infants with an opportunity to learn a single audio visual sequence consisting of distinct moving objects and their impact sounds whereas in others we allowed infants to learn several different sequences with each composed of different objects and their distinct impact sounds. In either case, during the habituation phase, the different objects could be seen appearing one after another at the top of a computer monitor and then moving down toward a ramp at the bottom of the stimulus display monitor. When each object reached the ramp, it produced its distinct impact sound, turned to the right and moved off to the side and disappeared. This cycle was repeated for the duration of each habituation trial. Following habituation, infants were given test trials during which the order of one or more of the various sequence elements was changed and the question was whether infants detected it.

In an initial study (Lewkowicz, 2004), we asked whether infants can learn a sequence composed of three moving/impacting objects and, if so, what aspects of that sequence they encoded. Results indicated that 4-month-old infants detected serial order changes only when the changes were specified concurrently by audible and visible attributes during both the learning and the test phase, and only when the impact part of the event – a local event feature that by itself was not informative about the overall sequential order – was blocked from view. In contrast, 8-month-old infants detected order changes regardless of whether they were specified by unisensory or bisensory attributes and regardless of whether they could see the impact that the objects made or not. In sum, younger infants required multisensory redundancy to learn the sequence and to detect the serial order changes whereas the older infants did not.

In a follow-up study (Lewkowicz, 2008), we replicated the earlier findings, ruled out primacy effects, extended the earlier findings by showing that even 3-month-old infants can perceive, learn, and discriminate three-element dynamic audio visual sequences, and that 3-month-olds also require multisensory redundancy to successfully learn the sequences. In addition, we found that object motion plays an important role because infants exhibited less robust responsiveness to audio visual sequences consisting of looming/sounding objects than of explicitly moving/sounding objects. Finally, we found that 4-month-old infants can perceive and discriminate longer (i.e., 4-element) sequences, that they exhibit more robust responsiveness to the longer sequences, and, somewhat surprisingly, that they do so even when they can see the impact part of the event. At first blush, this last result appeared to be paradoxical in relation to the earlier results with 4-month-olds but when the fact that a sequence composed of more objects/sounds makes it harder to focus on the individual objects is taken into account, this finding makes sense. That is, when there are more objects/sounds, infants' attention is shifted to the more global aspects of the sequence because there is less time to attend to each individual object and its impact sound.

The two previous studies established firmly that young infants are able to perceive and learn audio visual sequences and that they can detect changes in the order of the sequence elements. What is not clear from these findings, however, is what specific sequence property underlies infants' ability to detect order changes. As indicated earlier, infants are sensitive to statistical relations from an early age. The changes in sequential order presented in our initial two studies involved changes not only in the order of a particular object and its sound but also in its statistical relations to other sequence elements. As a result, it was necessary to investigate the separate contribution of each of these sequential attributes to infant sequence learning and discrimination. We did so in our most recent study (Lewkowicz and Berent, 2009). Here, we investigated directly whether 4-month-old infants can track the statistical relations among specific sequence elements (e.g., AB, BC) and/or whether they also can encode abstract ordinal position information (e.g., that B is the second element in a sequence of ABC). Thus, across three experiments we habituated infants to sequences of four moving/sounding objects. In these sequences, three of the objects and their sounds varied in their ordinal positions whereas one target object/sound maintained its invariant ordinal position (e.g., A**B**CD and C**B**DA). Following habituation to such sequences, we then presented sequences where the target sequence element's ordinal position was changed and the question was whether infants detected this change. Figure 16.5 shows that when the ordinal change disrupted the statistical relations between adjacent sequence elements, infants exhibited significant response recovery (i.e., discrimination). If, however, the statistical relations were controlled for when the ordinal change was made (i.e.,

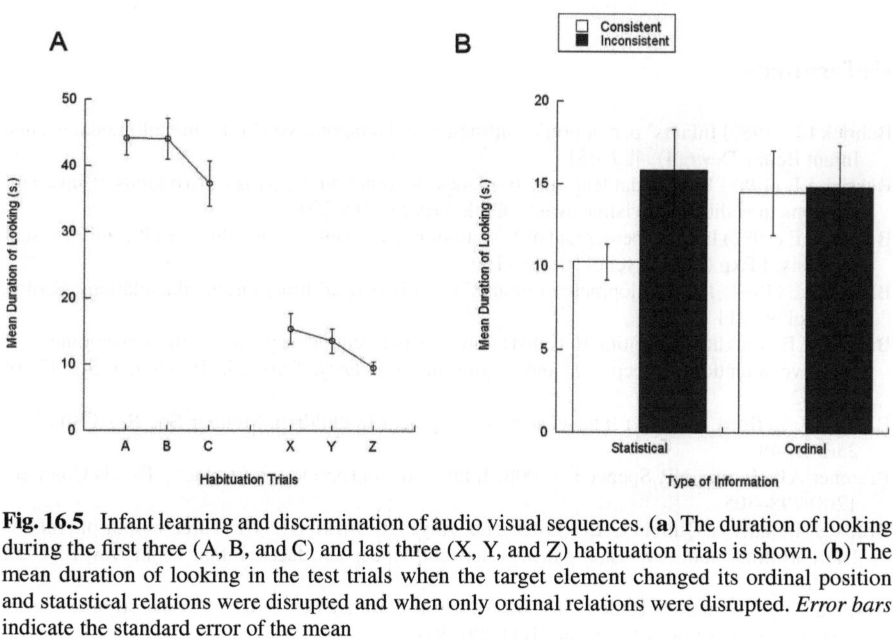

Fig. 16.5 Infant learning and discrimination of audio visual sequences. (**a**) The duration of looking during the first three (A, B, and C) and last three (X, Y, and Z) habituation trials is shown. (**b**) The mean duration of looking in the test trials when the target element changed its ordinal position and statistical relations were disrupted and when only ordinal relations were disrupted. *Error bars* indicate the standard error of the mean

when no statistical relations were disrupted), infants did not exhibit evidence of successful learning and discrimination. These results indicate that 4-month-old infants can learn the order of sequence elements and that they do so by tracking their statistical relations but that they do not track their invariant ordinal position. When these findings are combined with the previously reviewed findings on sequence learning in infancy, they clearly show that different sequence learning abilities emerge at different times, with more sophisticated ones emerging later than less sophisticated ones. In addition, it is reasonable to assume that the developmental emergence of these different sequence learning abilities is largely a function of experience and that, as a result, these abilities are constructed early in life by infants through their interactions with their everyday spatiotemporally and sequentially structured world.

16.7 Conclusion

Infants' perception and conception of multisensory objects is a developmental product of a complex set of dynamic and ever-changing processes. As is obvious from the foregoing, major changes in how infants respond to multisensory objects occur during infancy. The challenge of explicating the full set of interactions that contribute to this critical skill is still ahead of us. The hope is that the current review of some of the extant empirical evidence on infants' response to multisensory objects will spur further inquiry into this fundamental topic and will shed additional light on the exquisite complexity of the developmental processes underlying the emergence of the multisensory object concept.

References

Bahrick LE (1983) Infants' perception of substance and temporal synchrony in multimodal events. Infant Behav Dev 6(4): 429–451

Bahrick LE (1988) Intermodal learning in infancy: learning on the basis of two kinds of invariant relations in audible and visible events. Child Dev 59:197–209

Bahrick LE (1992) Infants' perceptual differentiation of amodal and modality-specific audio-visual relations. J Exp Child Psychol 53:180–199

Bahrick LE (1994) The development of infants' sensitivity to arbitrary intermodal relations. Ecolog Psychol 6(2):111–123

Bahrick LE, Lickliter R, Flom R (2004) Intersensory redundancy guides the development of selective attention, perception, and cognition in infancy. Curr Dir Psycholog Sci 13(3): 99–102

Birch HG, Lefford A (1963) Intersensory development in children. Monogr Soc Res Child Dev 25(5):1–48

Bremner AJ, Holmes NP, Spence C (2008) Infants lost in (peripersonal) space? Trends Cogn Sci 12(8):298–305

Brookes H, Slater A, Quinn PC, Lewkowicz DJ, Hayes R, Brown E (2001) Three-month-old infants learn arbitrary auditory-visual pairings between voices and faces. Infant Child Dev 10(1–2): 75–82

Bushara KO, Grafman J, Hallett M (2001) Neural correlates of auditory-visual stimulus onset asynchrony detection. J Neurosci 21(1):300–304

Bushnell EW (1986) The basis of infant visual-tactual functioning: Amodal dimensions or multimodal compounds? Adv Infan Res 4:182–194

Calvert G, Spence C, Stein B (eds) (2004) The handbook of multisensory processes. MIT Press, Cambridge, MA

Cohen LB, Chaput HH, Cashon CH (2002) A constructivist model of infant cognition. Cogn Dev. Special Issue: Constructivism Today 17(3–4):1323–1343

Colombo J (2001) The development of visual attention in infancy. Annu Rev Psychol 52:337–367

Delaney SM, Dobson V, Harvey EM, Mohan KM, Weidenbacher HJ, Leber NR (2000) Stimulus motion increases measured visual field extent in children 3.5 to 30 months of age. Optom Vis Sci 77(2):82–89

Fiser J, Aslin RN (2002) Statistical learning of new visual feature combinations by infants. Proc Natl Acad Sci 99(24):15822–15826

Frank MC, Slemmer JA, Marcus GF, Johnson SP (2009) Information from multiple modalities helps 5-month-olds learn abstract rules. Dev Sci 12:504–509

Fujisaki W, Shimojo S, Kashino M, Nishida S (2004) Recalibration of audiovisual simultaneity. Nat Neurosci 7(7):773–778

Gerken L (2006) Decisions, decisions: infant language learning when multiple generalizations are possible. Cogn 98(3):B67–B74

Ghazanfar AA, Schroeder CE (2006) Is neocortex essentially multisensory? Trends Cogn Sci 10(6):278–285. Epub 2006 May 2018

Gibson EJ (1969) Principles of perceptual learning and development. Appleton, New York

Gibson JJ (1966) The senses considered as perceptual systems. Houghton-Mifflin, Boston

Gómez RL, Maye J (2005) The developmental trajectory of nonadjacent dependency learning. Infancy 7(2):183–206

Greenfield PM (1991) Language, tools and brain: the ontogeny and phylogeny of hierarchically organized sequential behavior. Behav Brain Sci 14(4):531–595

Hernandez-Reif M, Bahrick LE (2001) The development of visual-tactual perception of objects: modal relations provide the basis for learning arbitrary relations. Infancy 2(1):51–72

Johnson MH (2005) Developmental cognitive neuroscience, 2nd edn. Blackwell, London

Johnson SP (2004) Development of perceptual completion in infancy. Psychol Sci 15(11):769–775

Johnson SP, Fernandes KJ, Frank MC, Kirkham N, Marcus GF, Rabagliati H et al. (2009) Abstract rule learning for visual sequences in 8- and 11-month-olds. Infancy 14:2–18

Kellman PJ, Spelke ES (1983) Perception of partly occluded objects in infancy. Cogn Psychol 15(4):483–524

Kelly DJ, Quinn PC, Slater AM, Lee K, Ge L, Pascalis O (2007) The other-race effect develops during infancy: evidence of perceptual narrowing. Psycholog Sci 18(12):1084–1089

Kirkham NZ, Slemmer JA, Johnson SP (2002) Visual statistical learning in infancy: evidence for a domain general learning mechanism. Cognition 83(2):B35–B42

Lewkowicz DJ (1986) Developmental changes in infants' bisensory response to synchronous durations. Infant Behav Dev 9(3):335–353

Lewkowicz DJ (1992a). Infants' response to temporally based intersensory equivalence: the effect of synchronous sounds on visual preferences for moving stimuli. Infant Behav Dev 15(3): 297–324

Lewkowicz DJ (1992b). Infants' responsiveness to the auditory and visual attributes of a sounding/moving stimulus. Percept Psychophys 52(5):519–528

Lewkowicz DJ (1996) Perception of auditory-visual temporal synchrony in human infants. J Exp Psychol: Hum Percept Perform 22(5):1094–1106

Lewkowicz DJ (2000a). The development of intersensory temporal perception: an epigenetic systems/limitations view. Psycholog Bull 126(2):281–308

Lewkowicz DJ (2000b). Infants' perception of the audible, visible and bimodal attributes of multimodal syllables. Child Dev 71(5):1241–1257

Lewkowicz DJ (2002) Heterogeneity and heterochrony in the development of intersensory perception. Cogn Brain Res 14:41–63

Lewkowicz DJ (2003) Learning and discrimination of audiovisual events in human infants: the hierarchical relation between intersensory temporal synchrony and rhythmic pattern cues. Dev Psychol 39(5): 795–804
Lewkowicz DJ (2004) Perception of serial order in infants. Dev Sci 7(2):175–184
Lewkowicz DJ (2008) Perception of dynamic and static audiovisual sequences in 3- and 4-month-old infants. Child Dev 79(5):1538–1554
Lewkowicz DJ (2010) Infant perception of audio-visual speech synchrony. Dev Psycholog 46(1):66–77
Lewkowicz DJ, Berent I (2009) Sequence learning in 4 month-old infants: do infants represent ordinal information? Child Dev 80(6):1811–1823
Lewkowicz DJ, Ghazanfar AA (2006) The decline of cross-species intersensory perception in human infants. Proc Natl Acad Sci U S A 103(17):6771–6774
Lewkowicz DJ, Kraebel K (2004) The value of multimodal redundancy in the development of intersensory perception. In: Calvert G, Spence C, Stein B (eds) Handbook of multisensory processing. MIT Press, Cambridge, pp 655–678
Lewkowicz DJ, Leo I, Simion F (2010) Intersensory perception at birth: newborns match non-human primate faces & voices. Infancy 15(1):46–60
Lewkowicz DJ, Lickliter R (eds) (1994) The development of intersensory perception: comparative perspectives. : Lawrence Erlbaum Associates, Inc., Hillsdale, NJ
Lewkowicz DJ, Marcovitch S (2006) Perception of audiovisual rhythm and its invariance in 4- to 10-month-old infants. Dev Psychobiol 48:288–300
Lewkowicz DJ, Sowinski R, Place S (2008) The decline of cross-species intersensory perception in human infants: Underlying mechanisms and its developmental persistence. Brain Res 1242:291–302. [Epub (ahead of print)]
Lewkowicz DJ, Turkewitz G (1980) Cross-modal equivalence in early infancy: Auditory-visual intensity matching. Dev Psychol 16:597–607
Lewkowicz DJ, Turkewitz G (1981) Intersensory interaction in newborns: modification of visual preferences following exposure to sound. Child Dev 52(3):827–832
Lickliter R, Bahrick LE (2000) The development of infant intersensory perception: advantages of a comparative convergent-operations approach. Psycholog Bull 126(2):260–280
Marcovitch S, Lewkowicz DJ (2009) Sequence learning in infancy: the independent contributions of conditional probability and pair frequency information. Dev Sci 12(6):1020–1025
Marcus GF, Fernandes KJ, Johnson SP (2007) Infant rule learning facilitated by speech. Psycholog Sci 18(5):387–391
Marcus GF, Vijayan S, Rao S, Vishton P (1999) Rule learning by seven-month-old infants. Sci 283(5398):77–80
Marks L (1978) The unity of the senses. Academic Press, New York
Mendelson MJ (1986) Perception of the temporal pattern of motion in infancy. Infant Behav Dev 9(2):231–243
Morrongiello BA (1988) Infants' localization of sounds in the horizontal plane: estimates of minimum audible angle. Dev Psychol 24:8–13
Morrongiello BA (1994) Effects of colocation on auditory-visual interactions and cross-modal perception in infants. In: Lewkowicz DJ, Lickliter R (eds) The development of intersensory perception: comparative perspectives. Lawrence Erlbaum, Hillsdale, NJ, pp 235–263
Morrongiello BA, Fenwick KD, Chance G (1990) Sound localization acuity in very young infants: an observer-based testing procedure. Dev Psychol 26:75–84
Navarra J, Vatakis A, Zampini M, Soto-Faraco S, Humphreys W, Spence C (2005) Exposure to asynchronous audiovisual speech extends the temporal window for audiovisual integration. Cogn Brain Res, 25(2):499–507
Nazzi T, Bertoncini J, Mehler J (1998) Language discrimination by newborns: toward an understanding of the role of rhythm. J Exp Psychol: Hum Percept Perform 24(3):756–766
Neil PA, Chee-Ruiter C, Scheier C, Lewkowicz DJ, Shimojo S (2006) Development of multisensory spatial integration and perception in humans. Dev Sci 9(5):454–464

Nelson K (1986) Event knowledge: structure and function in development. Erlbaum, Hillsdale, NJ
Partan S, Marler P (1999) Communication goes multimodal. Science 283(5406):1272–1273
Pascalis O, Haan M de, Nelson CA (2002) Is face processing species-specific during the first year of life? Science 296(5571):1321–1323
Piaget J (1952) The origins of intelligence in children. International Universities Press, New York
Piaget J (1954) The construction of reality in the child. Routledge & Kegan, London
Pickens J, Bahrick LE (1997) Do infants perceive invariant tempo and rhythm in auditory-visual events? Infant Behav Dev 20:349–357
Reardon P, Bushnell EW (1988) Infants' sensitivity to arbitrary pairings of color and taste. Infant Behav Dev 11(2):245–250
Rowe C (1999) Receiver psychology and the evolution of multicomponent signals. Animal Behav 58:921–931
Ruff HA, Rothbart MK (1996) Attention in early development: themes and variations. Oxford University Press, New York, NY
Saffran JR, Aslin RN, Newport EL (1996) Statistical learning by 8-month-old infants. Science 274(5294):1926–1928
Scheier C, Lewkowicz DJ, Shimojo S (2003) Sound induces perceptual reorganization of an ambiguous motion display in human infants. Dev Sci 6:233–244
Schneirla TC (1965) Aspects of stimulation and organization in approach/withdrawal processes underlying vertebrate behavioral development. In: Lehrman DS, Hinde RA, Shaw E (eds) Advances in the study of behavior. Academic Press, New York, pp 1–71
Sekuler R, Sekuler AB, Lau R (1997) Sound alters visual motion perception. Nature 385:308
Slater A, Brown E, Badenoch M (1997) Intermodal perception at birth: newborn infants' memory for arbitrary auditory-visual pairings. Early Dev Parent 6(3–4):99–104
Spelke ES, Kinzler KD (2007) Core knowledge. Dev Sci 10(1):89–96
Spencer JP, Blumberg MS, McMurray B, Robinson SR, Samuelson LK, Tomblin JB (2009) Short arms and talking eggs: why we should no longer abide the nativist-empiricist debate. Child Dev Perspect 3(2):79–87
Stein BE, Meredith MA (1993) The merging of the senses. The MIT Press, Cambridge, MA
Stein BE, Stanford TR (2008) Multisensory integration: current issues from the perspective of the single neuron. Nat Rev Neurosci 9(4):255–266
Summerfield AQ (1979) Use of visual information in phonetic perception. Phonetica 36:314–331
Thelen E, Smith LB (1994) A dynamic systems approach to the development of cognition and action. MIT Press, Cambridge, MA
Vroomen J, Keetels M, de Gelder B, Bertelson P (2004) Recalibration of temporal order perception by exposure to audio-visual asynchrony. Cogn Brain Res 22(1):32–35
Walker-Andrews AS (1997) Infants' perception of expressive behaviors: differentiation of multimodal information. Psycholog Bull 121(3):437–456
Watanabe K, Shimojo S (2001) When sound affects vision: effects of auditory grouping on visual motion perception. Psychol Sci 12(2):109–116
Welch RB, Warren DH (1980) Immediate perceptual response to intersensory discrepancy. Psycholog Bull 88:638–667
Welch RB, Warren DH (1986) Intersensory interactions. In: Boff KR, Kaufman L, Thomas JP (eds) Handbook of perception and human performance: Sensory processes and perception, vol. 1. J. Wiley & Sons, New York, pp 1–36
Werker JF, Tees RC (1984) Cross-language speech perception: evidence for perceptual reorganization during the first year of life. Infant Behav Dev 7(1):49–63
Werner H (1973) Comparative psychology of mental development. International Universities Press, New York

Chapter 17
Neural Development and Plasticity of Multisensory Representations

Mark T. Wallace, Juliane Krueger, and David W. Royal

17.1 A Brief Introduction to Multisensory Processes

As highlighted throughout this volume, we live enmeshed within a world rich in sensory information. Some of this information is derived from a common source (think of a bouncing ball and the associated visual and auditory cues) and must be "bound" together in order to create a unified perceptual representation. In contrast, much of the sensory information that is continually bombarding us is unrelated and must remain segregated in order to have a coherent perceptual representation. In an effort to efficiently sort this wealth of (multi)sensory information, the brain has evolved an architecture that not only segregates and processes information on a sensory-specific basis but that also combines and synthesizes sensory information from different modalities.

The numerous examples of multisensory influences on behavioral and perceptual reports serve as illustrations of the power and prevalence of multisensory "integration" in shaping our everyday interactions with and perceptions of the world around us. These examples range from a speeding of simple reaction times to improvements in target detection and localization and gains in speech intelligibility under noisy conditions (for a review of these and other examples, see Calvert et al., 2004; Stein and Meredith, 1993). In addition, a number of perceptual illusions serve to further highlight the continual interplay between the senses. These include the ventriloquism effect (Jack and Thurlow, 1973; Thurlow and Jack, 1973), in which the source of an auditory signal (i.e., the ventriloquist's voice) is shifted by conflicting visual cues (i.e., the dummy's head and lip movements), and the McGurk effect (McGurk and MacDonald, 1976), in which the concurrent presentation of conflicting visual and auditory speech cues results in a novel and fused percept (e.g., auditory /ba/ + visual /ga/ frequently gives rise to reports of /da/). Nonetheless, despite our seemingly intuitive understanding of multisensory phenomena, it is only recently that our knowledge of its neural substrates has been the subject of intensive investigation.

M.T. Wallace (✉)
Vanderbilt Brain Institute, Vanderbilt University, Nashville, TN 37232, USA
e-mail: mark.wallace@vanderbilt.edu

M.J. Naumer, J. Kaiser (eds.), *Multisensory Object Perception in the Primate Brain*,
DOI 10.1007/978-1-4419-5615-6_17,

17.2 Neural Processing in Adult Multisensory Circuits

Since the mid-1980s, neurophysiological and behavioral research has focused on furthering our understanding of the basic organizational features of multisensory brain circuits. Initially, the vast majority of this work focused on a midbrain structure, the superior colliculus (SC; for reviews of the SC, see King, 2004; Stein and Meredith, 1991; Sparks, 1986; Sparks and Groh, 1995). The SC has been an exceptional model for these studies for several reasons. First, its intermediate and deep layers contain a large population of multisensory neurons, defined as those that are responsive to or influenced by stimuli from two or more sensory modalities. In the two principal model species for multisensory research, the rhesus monkey and the cat, the proportion of multisensory neurons in the deeper layers of the SC ranges from approximately 28% (monkey – Wallace et al., 1996) to more than 60% (cat – Meredith and Stein, 1986b; Wallace and Stein, 1997; Wallace et al., 1993). The second strength of the SC as a model is its well-ordered spatiotopic structure, in which the visual, auditory, and somatosensory representations are aligned with one another, in essence creating a unified multisensory representation of space (Stein and Meredith, 1993 for a review). Finally, the SC has a well-defined behavioral role, playing an important role in the control of saccades and coordinated eye and head movements (Sparks, 1999). Indeed, in addition to the (multi)sensory responses that characterize its deep layer neuronal populations, many of these same neurons also have premotor responses that are linked to the generation of saccades and head movements (Sparks, 1986; Wurtz and Munoz, 1995).

Work in the SC has defined the basic operational principles of multisensory neurons when confronted with stimuli from multiple sensory modalities. Most notable in this regard has been the striking non-linearities that can be seen in the response profiles of these neurons when challenged with stimuli from two or more modalities. Depending on the referent, these changes can be represented along a continuum from enhancement (a multisensory response that is significantly larger than the best unisensory response) to depression (a multisensory response that is significantly smaller than the best unisensory response), or when both unisensory responses are considered, from superadditivity to subadditivity (Meredith and Stein, 1986b; Perrault et al., 2005; Stanford and Stein, 2007; Stanford et al., 2005; Wallace et al., 1996). The direction (i.e., enhancement vs. depression, super- vs. subadditivity) and magnitude of these response changes have been shown to be dependent on the physical characteristics of the stimuli that are combined and their relationships to one another. These stimulus-dependent determinants of the multisensory product have been largely captured in the three major principles of multisensory integration first put forth by Meredith and Stein (Meredith and Stein, 1986a, b; Meredith et al., 1987). The first two of these principles have highlighted the critical importance of spatial and temporal proximity of the paired stimuli in generating multisensory interactions. Such an operational principle makes intuitive sense, in that multisensory stimuli that occur close together in space and time are likely to be derived from a common event. The third principle, inverse effectiveness, also has intuitive appeal, in that it posits that stimuli that are weakly effective when presented alone give rise to the largest changes when paired (i.e., the largest multisensory interactions). Such a principle makes great adaptive sense, since stimuli that are highly effective on their own need little additional information in order to be salient. Conversely, the major benefits of multisensory stimulation come when each of the component stimuli provides weak or ambiguous information.

Although first described at a neural level, these principles have also been shown to be applicable when describing behavioral or perceptual changes associated with multisensory circumstances in both animal models and humans (Bushara et al., 2001; Cappe et al., 2009; Colonius and Diederich, 2004; Stein et al., 1989, 1988; Frens et al., 1995; Kayser et al., 2005; Macaluso et al., 2004; Senkowski et al., 2007; Stevenson and James, 2009; Teder-Salejarvi et al., 2005). Thus, in the cat, spatially and temporally coincident pairings of weak visual and auditory stimuli result in a significant facilitation of orientation behaviors (Stein et al., 1988, 1989). Conversely, when these pairings are spatially disparate orientation behaviors are actively inhibited. A similar set of results is seen in both non-human primate and human responses to saccadic targets. For example, spatially and temporally coincident visual and auditory stimulus pairs result in speeded saccadic eye movements, and this facilitation of response disappears as the visual and auditory cues are separated in space and/or time (Corneil et al., 2002a; Frens and van Opstal, 1995, 1998; Harrington and Peck, 1998; Hughes et al., 1998).

Furthermore, even though first detailed in the SC, multisensory neurons and their interactive profiles have now been described in numerous brain structures, including classic areas of "association" cortex (e.g., cat anterior ectosylvian sulcus – (Wallace et al., 1992); monkey and human superior temporal sulcus – (Barraclough et al., 2005; Beauchamp et al., 2004; Benevento et al., 1977; Bruce et al., 1981; Calvert et al., 2000; Stevenson et al., 2007; Stevenson and James, 2009); monkey and human intraparietal areas – (Avillac et al., 2005, 2007; Schlack et al., 2005; Kitada et al., 2006; Makin et al., 2007) as well as areas traditionally considered the exclusive domain of a single sensory modality (Ghazanfar and Schroeder, 2006). However, despite this explosive growth in our understanding of how adult multisensory circuits operate and their relationship to behavioral and/or perceptual processes, our understanding of the developmental antecedents that result in these mature multisensory processing architectures has lagged behind until relatively recently.

17.3 The Development of Subcortical Multisensory Representations

Neurophysiological studies have now begun to characterize the maturation of multisensory neurons and the circuits with which they are associated. These initial studies focused again on the SC, largely because of the rich foundation of knowledge about the processing capabilities of its multisensory population in adult animals. These studies have mainly been carried out in two species, the cat and monkey, and reveal

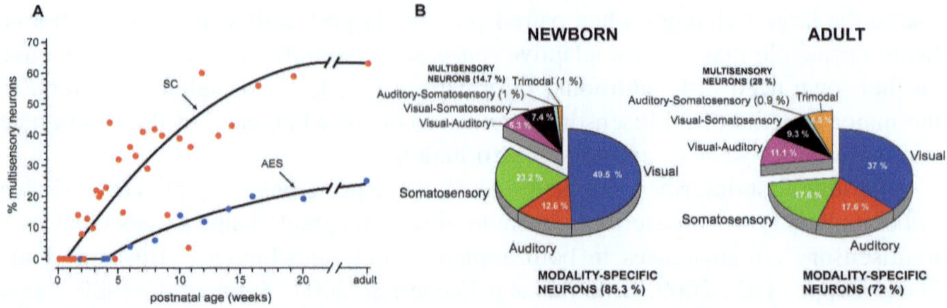

Fig. 17.1 The developmental chronology of multisensory neurons in the SC of the cat (*left*) and monkey (*right*). In cat, note that multisensory neurons are absent at birth. Beginning between the first and second postnatal week the first multisensory neurons appear, and then this population gradually increases in incidence over the ensuing 4 months. In the rhesus monkey, pie charts show the percentages of sensory-responsive neurons in the newborn and adult (inset). Note here that multisensory neurons are indeed present at birth in this species, but are far from adult-like in their incidence. Adapted from Wallace and Stein (1997) and Wallace et al. (1996)

striking parallels (and some key differences) in their multisensory developmental chronology. In the cat, multisensory neurons are absent in the SC at birth, and the only sensory-responsive neuronal population evident at this time is responsive to somatosensory stimuli (Wallace et al., 1997; Fig. 17.1). As development progresses, auditory neurons appear at approximately 5 days postnatal, at around the time of opening of the ear canals. With the presence of both somatosensory and auditory responsiveness in the SC, the first multisensory neurons emerge at about this same time frame. Several weeks subsequent to this, visual responses first appear in the deep layers of the SC (they are present in the superficial layers much earlier; see Kao et al., 1994), and the first visually responsive multisensory neurons are seen at this time. Over the ensuing several months of postnatal life, the proportion of multisensory neurons in the deep SC grows until it reaches adult levels (i.e., approximately 60%) by about 4 months of age (Fig. 17.1).

In the non-human primate (i.e., rhesus monkey) SC, although the developmental progression has yet to be fully elucidated, the timeline for multisensory development appears to be advanced relative to cat. Thus, recordings immediately after birth have revealed the presence of somatosensory, auditory, and visual responses in the deep SC, along with the presence of a small but significant multisensory population (Wallace and Stein, 2001; Fig. 17.1). Such a difference between species is not surprising given the relatively altricial developmental state of the cat and the relatively precocial state of the newborn monkey. In addition to this developmental difference, the final distribution of multisensory neurons in the deep SC differs substantially between these two species. As previously stated, whereas multisensory neurons become the majority population in the adult cat SC, in the rhesus monkey this final value is only between 25 and 30%. This difference likely reflects the strong visual dominance of the monkey, resulting in a larger proportion of unisensory visual neurons in this important oculomotor structure.

Despite the presence of multisensory neurons in the newborn monkey SC, these neurons (and those in the newborn cat SC) are strikingly immature in their response properties when compared with neurons recorded at later developmental time points. Although significant differences are seen in the responses of these neurons in a variety of domains (Wallace and Stein, 2001; Wallace et al., 1997), the most germane in the current context is the lack of any integrative capabilities (Fig. 17.2).

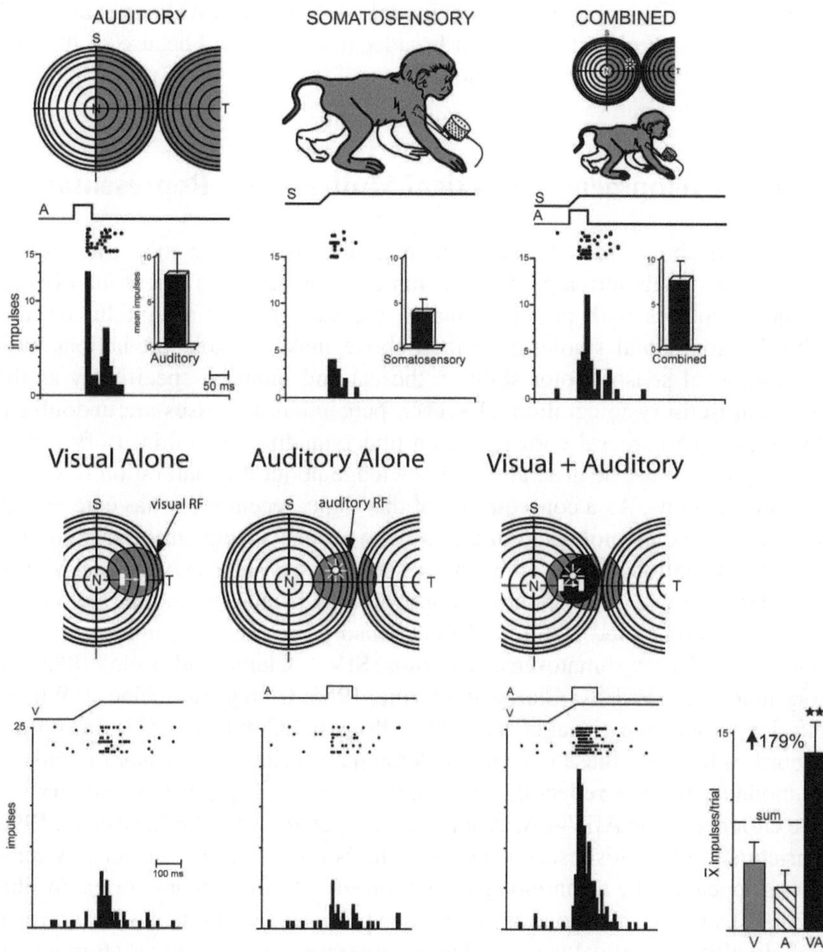

Fig. 17.2 Multisensory integration is absent in the earliest population of multisensory neurons. Shown here is data from an auditory–somatosensory neuron in the newborn monkey SC (*top*) and data from a visual–auditory neuron in the adult monkey SC (*bottom*). In both, receptive fields are represented by the shading, and the location of stimuli used for sensory tests is shown on the receptive field plots. Rasters and histograms below this show each of the neuron's responses to the individual unisensory stimuli and to the multisensory combination, and bar graphs show the mean responses for each condition. Note that whereas the neuron from the newborn animal responds to each of the stimuli in a very similar manner, the neuron from the adult responds to the stimulus combination with a significant and superadditive response enhancement. Adapted from Wallace and Stein (2001) and Wallace et al. (1996)

Thus, although early multisensory neurons have the capacity to respond to inputs from multiple sensory modalities (and are thus characterized as multisensory), their responses to combined stimulation fail to exhibit the significant enhancements seen at later ages. Even though a detailed accounting of the developmental changes taking place in the primate SC has yet to be done, evidence from cat would suggest that just as for the maturation of multisensory neurons themselves, the maturation of integrative capacity takes place over an extended period of postnatal life. While in cat this process is complete by approximately 4–5 months (Wallace et al., 1997), in primate it is likely to extend over a broader time window because of the more extended time for neurologic maturation.

17.4 The Development of Cortical Multisensory Representations

As highlighted above, the SC plays an integral sensorimotor role, transforming (multi)sensory signals into a premotor command that serves to orient the eyes and head (and in species with mobile pinnae – the ears) toward a stimulus of interest. The developmental studies described above make testable predictions about the maturation of sensorimotor skills in the cat and monkey, specifically as they relate to multisensory integration. However, perceptual processes are undoubtedly the domain of the cerebral cortex, and an understanding of multisensory perceptual development must be grounded in knowledge about the maturation of cortical multisensory circuits. As a consequence of this logic, recent work has extended the developmental studies pioneered in the SC into cortex, specifically focusing on an area of cat association cortex – the anterior ectosylvian sulcus (AES; see, Wallace et al., 2006). The cat AES lies at the apposition of frontal, temporal, and parietal cortices and is comprised of three distinct sensory-specific (i.e., unisensory) subdivisions – the fourth somatosensory cortex (SIV – Clemo and Stein, 1983), the auditory field AES (FAES – Clarey and Irvine, 1986; Clarey and Irvine, 1990), and the anterior ectosylvian visual area (AEV – Olson and Graybiel, 1987). In addition, at the borders between these unisensory domains is a rich multisensory population whose modality profiles reflect the neighboring cortices (e.g., visual–auditory neurons are enriched at the AEV–FAES border) (Jiang et al., 1994; Wallace et al., 1992). The enrichment of multisensory neurons at the borders between unisensory representations appears to be a common parcellation scheme for sensory cortex (Wallace et al., 2004b), and examinations of non-human primates (and humans) have revealed this to hold also for cortical areas within the superior temporal sulcus (Barraclough et al., 2005; Beauchamp et al., 2004; Calvert, 2001; Maier et al., 2008; Noesselt et al., 2007), the intraparietal sulcus (Avillac et al., 2005, 2007; Calvert, 2001; Duhamel et al., 1998; Miller and D'Esposito, 2005; Schlack et al., 2005), and the lateral occipital temporal area (Beauchamp, 2005). Although to date the contributions of AES toward multisensory perception have not been identified, converging evidence suggests that it is likely to play a role(s) in motion perception/binding, coordinate transformations, and eye movements (Benedek et al., 2004; Nagy et al., 2003; Scannell et al., 1995; Tamai et al., 1989).

Developmental studies conducted in cat AES outlined a progression in multisensory maturation much like that seen in subcortical structures, with a slight delay in cortex relative to the SC (Wallace et al., 2006). Thus, at the earliest postnatal ages, only somatosensory-responsive neurons are present in what is the presumptive SIV. Slightly later (i.e., at approximately 2 weeks after birth), auditory neurons are first seen in what will become FAES, and still later (i.e., at about 4–6 weeks) visual responses are seen in AEV. As in the SC, with the appearance of two sensory-responsive populations in AES come the first multisensory neurons, in this case somatosensory–auditory neurons at the border between SIV–FAES. Not surprisingly, as the individual unisensory representations mature, there is a concomitant growth in the multisensory border representations. A comparison between SC and AES developmental trajectories in cat reveals that the maturation of cortical multisensory circuits is delayed relative to subcortex by several weeks, a result in keeping with what is typically seen in the development of unisensory systems.

Again paralleling the subcortical results, the earliest multisensory cortical neurons lack integrative capacity (Wallace et al., 2006). As postnatal maturation proceeds, these neurons acquire adult-like integrative features, most notably the ability to generate large changes in their responses when presented with multisensory cues. One intriguing commonality between the multisensory populations in the SC and AES is that the transition to adult-like integration is a rapid and seemingly binary process (Wallace et al., 1998, 2006). Neurons either lack the ability to integrate their different sensory inputs or do so in a manner very similar to adults; little evidence exists for an intermediary state in which integrative capacity is only partially adult-like. Such a result is suggestive of the presence of a developmental "switch" that gates the transition to adult multisensory abilities. Good evidence that this switch is cortically derived for the SC has come from studies in which cortical deactivation or neonatal ablation compromises multisensory integration in SC neurons (Jiang et al., 2006; Wallace and Stein, 2000). The presence of such critical descending influences has yet to be established in the primate model, as have the mechanisms gating the maturation of cortical multisensory integration.

17.5 The Maturation of the Multisensory Integrative Principles

In addition to detailing the functional chronology of multisensory subcortical and cortical representations, the studies cited above have also provided important insights into developmentally gated changes in the spatial, temporal, and inverse effectiveness principles of multisensory integration. One of the most dramatic changes that takes place in multisensory neurons with development is a striking reduction in the size of the individual receptive fields. In both SC and AES these receptive fields are exceedingly large in the earliest multisensory neurons, often encompassing all of sensory space represented by the peripheral organ (Wallace and Stein, 1997; Wallace et al., 2006). As development progresses, these fields grow smaller, revealing the strong correspondence that characterizes adult neurons. Intriguingly, when individual neurons are examined, this consolidation of receptive fields appears to be a rapid process and is tightly linked to the appearance of integrative capacity. Thus, if receptive fields are immature (i.e., large) in area, there is a high probability that the neuron will lack integrative capacity. Conversely, as soon as receptive fields are close to their adult size, there is a high likelihood that the neuron will be able to integrate its different sensory inputs. As soon as this receptive field contraction and integrative maturation takes place, neurons appear to abide by the spatial principle. In contrast, the development of temporal multisensory processing appears to be a more gradual process. Thus, whereas adult multisensory neurons have a fairly broad temporal "window" within which multisensory interactions can be generated, integrating multisensory neurons in young animals frequently have a much narrower temporal "tuning curve." For this property, there indeed appears to be a developmental sequence in which the size of this window gradually broadens during postnatal life. Finally, the principle of inverse effectiveness appears to be in place as soon as neurons are capable of multisensory interactions.

17.6 Sensory Experience as a Key Determinant in Multisensory Development

The immaturity of multisensory processes in the developing brain, and the protracted period of postnatal life during which multisensory neurons and their integrative capacity mature, strongly suggests that sensory experience plays an important role in the development of multisensory representations. In order to evaluate the importance of this experience in these processes, recent work has eliminated sensory information in one sensory system (i.e., vision) during postnatal life and examined the impact of this rearing condition on both subcortical and cortical multisensory representations (Carriere et al., 2007; Wallace et al., 2004a). These studies have been conducted in cat because of the wealth of data on the organization of its multisensory circuits and because of the strong foundation of knowledge on the impact of deprivation on the organization and function of the visual system (Hubel and Wiesel, 1998).

Raising animals in an environment devoid of visual cues (i.e., dark-rearing) was found to have a profound impact on multisensory development. In these animals, recordings were done as adults after a minimum of 6 months of visual deprivation and thus in the adult animal. Somewhat surprisingly, in both the SC and the AES, a substantial (but still less than in normally reared animals) number of visually responsive multisensory neurons were found following dark-rearing (Carriere et al., 2007; Wallace et al., 2004a). However, when the responses of these neurons were examined under multisensory (i.e., combined visual–auditory) conditions, they were found to be strikingly different from normally reared animals. Interestingly, the results were somewhat different between SC and AES. In the SC, dark-rearing appears to completely abolish multisensory integration, with neurons responding to

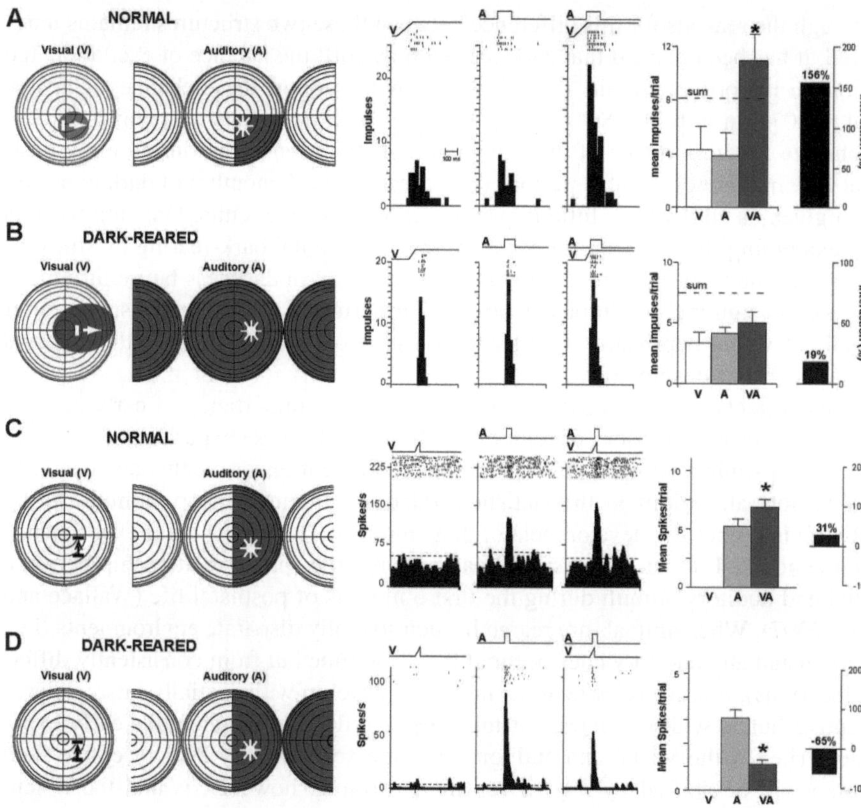

Fig. 17.3 Dark-rearing compromises the development of multisensory integration. Shown in the top two panels are data from a visual–auditory multisensory neuron recorded in the SC of a normally reared cat (**a**) and from a dark-reared cat (**b**). On the left the receptive fields (shading) and stimulus locations are shown. On the right, rasters, histograms and bar graphs show the responses of these neurons to the visual, auditory, and combined (visual–auditory) stimulation. Note the significant enhancement in the normally reared animal, and the lack of this enhancement in the dark-reared animal. Shown on the bottom is a parallel data set from the AES cortex (**c** and **d**). In this case, the effects of dark-rearing manifest as a response depression when stimulus combinations that typically result in enhancement are paired. Adapted from Wallace et al. (2004) and Carriere et al. (2007)

stimulus combinations in much the same manner as they responded to the individual unisensory stimuli (very much resembling the young animal) (Wallace et al., 2004a) (Fig. 17.3a). In contrast, in AES the multisensory population responded to this same stimulus combination not with no interactions, but rather with an active response depression (Carriere et al., 2007) (Fig. 17.3b). It is important to emphasize here that the stimulus combinations used in both circumstances were identical – the pairing of spatially and temporally coincident and weakly effective stimuli – those that typically give rise to large response enhancements under normal conditions.

Although the reasons for this difference between these two structures remains unresolved, it has been posited that dark-rearing may shift the balance of excitation and inhibition in cortical circuits in such a way as to support these changes (Carriere et al., 2007). In both the SC and the AES, recent experiments (unpublished data) emphasize the importance of the experiential deprivation being during early postnatal life, in that comparable periods (i.e., greater than 6 months) of dark exposure when given as adults have little impact on multisensory circuits. Ongoing work is now examining the reversibility of the effects of neonatal dark-rearing by studying changes in multisensory processing in animals reared in darkness but reintroduced into normal lighting conditions as adults. Furthermore, future work using shorter periods of visual deprivation will try to converge on delimiting the critical period for multisensory development.

These effects of dark-rearing clearly illustrate the importance of early sensory experience for multisensory development. To examine these experience-dependent effects in a slightly more sophisticated manner, recent experiments have begun to look at how alterations in the statistical relations of multisensory stimuli during early life influence the development of these circuits. The first of these experiments, again conducted in cat, has systematically altered the spatial relationship between visual and auditory stimuli during the first 6 months of postnatal life (Wallace and Stein, 2007). When animals are reared in such spatially disparate environments (i.e., where visual and auditory cues occur at the same time but from consistently different locations), multisensory neurons in the SC develop with spatially misregistered receptive fields, with the degree of misregistry reflecting the animal's early experiences (i.e., if the visual and auditory cues are separated by 30°, receptive field overlap will be shifted by 30°). Even more striking, however, is that these neurons now "prefer" spatially disparate visual and auditory stimulus combinations in a manner that is in keeping with the receptive field misalignments (Fig. 17.4a). Response enhancements are now seen for combinations that are most frequently encountered during postnatal life, whereas combinations that are more reflective of "normal" development (spatially coincident pairings) now typically result in little change.

In a parallel manner, other experiments probed the importance of the temporal relationship of the paired stimuli for the development of multisensory representations. In these experiments, visual and auditory stimuli were now always presented in spatial coincidence but at a fixed set of temporal disparities. In one group, this disparity was set at 0 ms (simultaneity – a control group), in another group at 100 ms (visual leading auditory), and in a final group by 250 ms (again visual leading auditory). As predicted, visual–auditory SC multisensory neurons in animals raised at the 0 ms disparity had temporal tuning functions much like what is seen under normal rearing circumstances, with peak multisensory enhancement being observed at or near temporal coincidence (Fig. 17.4b). In contrast, neurons from animals reared in the 100 ms temporal-disparity group had rightward shifted tuning functions, being peaked much closer to the experienced temporal disparity (i.e., 100 ms). However, animals in the final group (i.e., 250 ms) showed an absence of multisensory integration. This final result suggests that the large temporal gap between the visual and

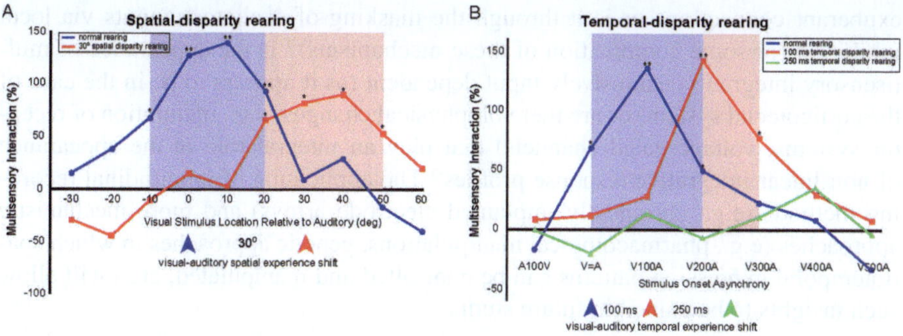

Fig. 17.4 The development of multisensory integration is highly malleable and can be strongly shaped by manipulating early postnatal sensory experience. (**a**) The magnitude of multisensory interactions is plotted as a function of the spatial relationship between paired visual and auditory stimuli for two neurons from cat SC. In one (*blue*), the animal was reared under normal conditions. In the other (*red*), the animal was reared in an environment in which the visual and auditory cues were presented in temporal coincidence but at a fixed spatial disparity of 30°. Note that for the neuron from the normally reared animal, peak multisensory interactions are seen when the visual and auditory stimuli are presented in close spatial coincidence (i.e., 0° and 10°). In contrast, for the neuron from the spatial-disparity-reared animal, peak interactions are now seen at disparities close to those presented during early life. Blue and red shading depict the spatial "window" within which multisensory interactions are elicited. (**b**) A similar plot to that shown in **a**, except that the important experiential variable is the temporal relationship of the visual and auditory stimuli. This panel contrasts the results from animals raised under three different conditions: normal rearing (*blue*), 100 ms temporal-disparity rearing (*red*), and 250 ms temporal-disparity rearing (*green*) (see text for additional detail on these rearing conditions). Note that for the normally reared animal the peak interactions are seen when the visual stimulus precedes the auditory stimulus by 100 ms. This peak is shifted by 100 ms in the 100 ms temporal-disparity condition. In comparison, in the 250 ms temporal-disparity condition multisensory integration is abolished

auditory stimuli lies beyond the plastic capacity of these neurons to reflect the statistical relations of the altered sensory world. Such a result has important implications for the biophysical mechanisms capable of supporting plasticity in multisensory circuits and provides an important temporal "window" for the manifestation of such plastic changes.

17.7 The Mechanistic Bases for Multisensory Development?

An open and fascinating set of questions in regard to the development of both subcortical and cortical representations is the mechanistic underpinnings for the dramatic changes that take place during early postnatal life. For example, as the multisensory representation grows in each of these structures, is it through the addition of inputs onto unisensory neurons (e.g., does an early somatosensory neuron become a visual–somatosensory neuron) or is it through the unique addition of multisensory elements? Is receptive field consolidation the result of the pruning of exuberant connections or is it through the masking of excitatory inputs via local inhibition (or some combination of these mechanisms)? Is the appearance of multisensory integration exclusively input dependent (as it appears to be in the case of the corticotectal system) or are there biophysical changes (e.g., maturation of receptor systems, voltage-gated channels) that play an integral role in the appearance of non-linear integrative response profiles? The application of longitudinal recording methods (e.g., chronically implanted electrode arrays) and more mechanistic approaches (e.g., pharmacological manipulations, genetic approaches in which spatiotemporal expression patterns can be controlled and manipulated, etc.) will allow such insights to be gained in future studies.

17.8 Spatial Receptive Fields and Spatiotemporal Receptive Fields: New Tools to Evaluate Multisensory Representations and Their Development

One common characteristic of virtually all multisensory systems examined to date is the striking overlap between the individual spatial receptive fields of individual neurons and the large size of these fields (Avillac et al., 2005; Benedek et al., 1996, 2004; Carriere et al., 2008; Dong et al., 1994; Kadunce et al., 2001; Meredith and Stein, 1986a; Schlack et al., 2005; Stein et al., 1976; Royal et al., 2009; Wallace and Stein, 1996; Wallace et al., 1992). These receptive fields are classically depicted as large bordered areas within which a stimulus evokes an excitatory response. This is even more dramatic in a developmental context, where receptive fields tend to be substantially larger when compared with the adult. Early studies looking at how spatial location modulates multisensory interactions hinted at dramatic differences in responsiveness dependent on stimulus location within these large receptive fields (Meredith and Stein, 1986a, b). In order to begin to quantify this, recent studies have now focused on the creation of spatial receptive fields (SRFs), constructs that provide a view into this response heterogeneity (Carriere et al., 2008; Royal et al., 2009). To date, this work has focused on cat cortical multisensory representations (i.e., AES), but will soon be used to also evaluate subcortical receptive field architecture in the SC of both cat and monkey. The results for AES have illustrated marked response heterogeneity within the individual unisensory receptive fields of multisensory neurons (Fig. 17.5). More important perhaps than this complexity in SRF structure is the consequent impact that it has on multisensory processing. Thus, spatially coincident pairings at different locations within the SRF revealed a very interesting pattern of multisensory interactions. Those in which stimuli were paired at weakly responsive regions of the SRF typically gave rise to large, superadditive response enhancements (Fig. 17.5). In contrast, pairings at highly effective locations within the SRF resulted in subadditive interactions. Taken together, these results are strongly suggestive that space is only important in modulating multisensory interactions through its impact on response strength – arguing for a preeminence of inverse effectiveness over space per se in these interactions.

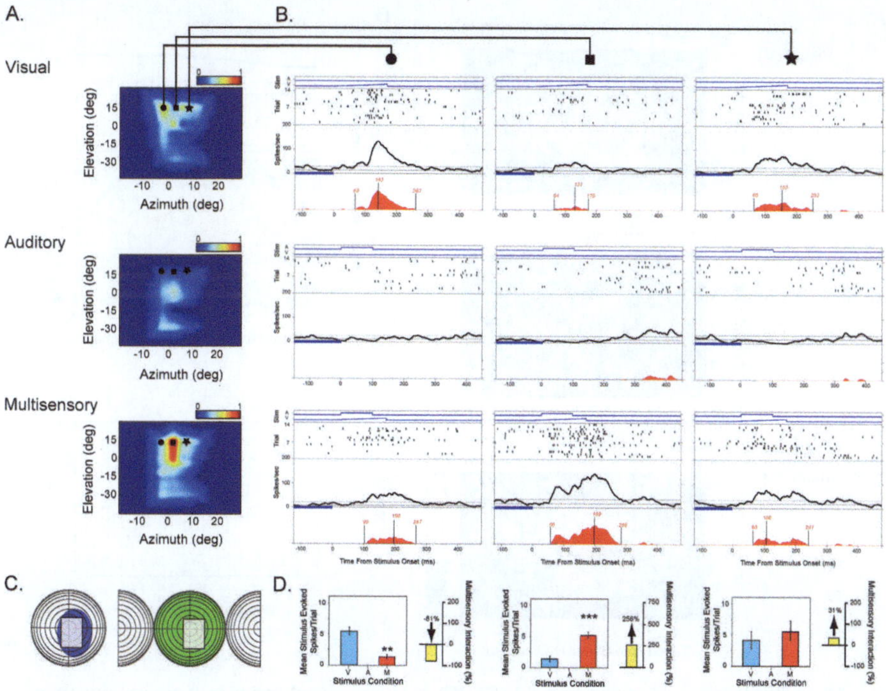

Fig. 17.5 Receptive field heterogeneity is a major determinant of multisensory interactions in cortical multisensory neurons. (**a**) Spatial receptive field (SRF) plots for a visual–auditory AES neuron and in which the pseudocolored representation reflects relative firing rate for the visual, auditory, and combined visual–auditory conditions. (**b**) Visual (*top*), auditory (*middle*), and multisensory (*bottom*) responses for stimuli positioned at three different locations within the SRFs (represented with a *circle*, *square,* and *star*). Note that when the visual stimulus is positioned at a location in which the visual response is strong (*left panel*), the multisensory pairing results in a response that is less than the visual response. In contrast, when the visual response is weak, a large superadditive interaction is seen to the multisensory pairing (*middle panel*). Locations at which intermediate levels of visual response are elicited result in very little interaction (*right panel*). (**c**) Locations of SRF analyses (*gray shading*) are shown on a standard receptive field representations. (**d**) Bar graphs summarize the responses depicted in panel **b**. From Carriere et al. (2008)

Although initially restricted to the spatial dimensions of azimuth and elevation (i.e., SRF plots are typically rendered in two dimensions), these analyses have now been expanded to a third dimension – time. Evaluation of the multisensory response profile as a function of both spatial location and time relative to stimulus onset results in the construction of spatiotemporal receptive fields (STRFs). The evaluation of the spatial (or in the case of the auditory system – spectral) and temporal features of a sensory response is not unique to multisensory systems, and similar analyses have been employed within the visual (Cai et al., 1997; DeAngelis et al., 1993; Reid et al., 1991), auditory (Fritz et al., 2003; Miller et al., 2002; Shamma, 2008) and somatosensory (Ghazanfar et al., 2001) systems to characterize the response dynamics of unisensory neuronal populations. In multisensory cortical

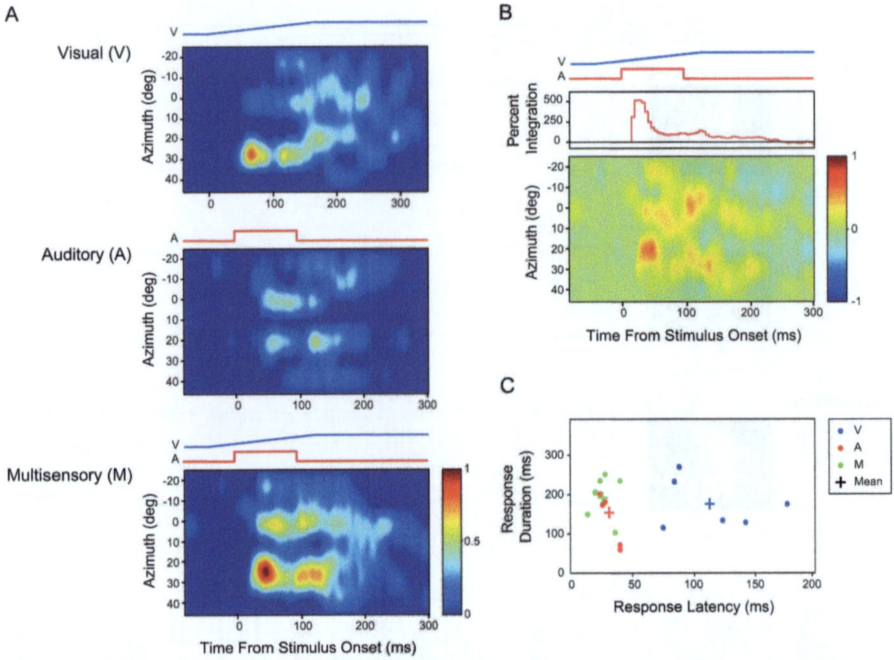

Fig. 17.6 Spatiotemporal receptive field (STRF) analyses reveal complex temporal dynamics in the responses of cortical multisensory neurons. (**a**) Pseudocolor STRF plots for the visual, auditory, and multisensory conditions represent neuronal firing rate as a function of azimuthal location (*y*-axis) and time post-stimulus (*x*-axis). (**b**) When the actual multisensory response is compared with that predicted from a simple summative model, the contrast plot reveals regions of superadditivity (warm colors) that vary as a function of spatial location and time. (**c**) When the latency and duration of the multisensory response is compared relative to the visual and auditory responses, the multisensory response is both faster (i.e., of shorter latency) and has a longer duration than either of the unisensory responses. From Royal et al. (2009)

circuits, STRF analyses show not only a pattern of spatial heterogeneity to these responses (as have been derived for the SRF analyses described above) but also a response that has complex temporal dynamics (Royal et al., 2009) (Fig. 17.6). These STRF analyses have revealed that the integrated multisensory response of these neurons can be divided into several important phases. The first of these is an early superadditive response that is in keeping with recent studies that have shown there to be a latency shift under multisensory conditions and in which response onset is slightly accelerated relative to the shortest latency unisensory response (Bell et al., 2005; Rowland et al., 2007). In addition, a second phase of superadditivity is seen late in the sensory response, and reflects an increase in the duration of the discharge elicited under multisensory conditions. Whereas the initial speeding is the likely mechanism behind the facilitated saccade dynamics observed under multisensory (i.e., visual–auditory) conditions (Bell et al., 2005; Corneil et al., 2002b; Frens and van Opstal, 1998; van Wanrooij et al., 2009), the prolongation of the response has yet to be linked to a behavioral or perceptual role. The longer discharge duration may simply be a way of encoding increased stimulus salience and may represent a mechanism to disambiguate stimuli when they are close to perceptual thresholds.

Together these new SRF and STRF findings reveal a previously unappreciated complexity to multisensory receptive field architecture, a complexity that has important deterministic consequences for the evoked multisensory discharges. The functional utility of this organization has yet to be established, but one suggestion is that it may play an important role in coding the relative motion of a multisensory stimulus (Carriere et al., 2008). In addition, the application of these new methods of analysis to questions concerning development and plasticity in multisensory systems will undoubtedly reveal important functional and mechanistic properties to these processes. Preliminary data suggest that SRF heterogeneity is dramatically reduced in early neonatal multisensory neurons and point to this complexity being a product of early sensory experience.

17.9 Multisensory Studies in Awake and Behaving Preparations

Much of the previously described work looking at development and plasticity in multisensory subcortical and cortical circuits has been derived from studies conducted in anesthetized animals. These studies have allowed the time-intensive analyses of receptive field microstructure in individual neurons, and in which the location and temporal relationships of stimuli can be systematically and parametrically varied. In addition, this preparation is ideally suited to developmental studies in which behavioral control is often impossible. However, these studies are limited in their ability to provide insights into behavioral and perceptual processes, and parallel work is now beginning to examine multisensory encoding under awake and behaving circumstances. Indeed, early work done in the awake cat SC has both established the presence of multisensory interactions and shown that these interactions are qualitatively very similar to those detailed in anesthetized animals (Peck, 1987; Wallace et al., 1998). More recently these studies have been extended to the non-human primate SC, where again the basic integrative features of multisensory neurons appear to be fundamentally similar between the two preparations (Bell et al., 2001, 2005). Although initially focused on the SC, these studies are now beginning to move into cortical domains more likely to play important roles in multisensory perceptual and cognitive processes (Avillac et al., 2005; Barraclough et al., 2005; Schlack et al., 2005).

As alluded to above, one limitation of these experiments in awake and behaving animals is the much shorter periods of isolation and consequent characterization of individual neurons. In an effort to overcome this hurdle, we have recently embarked on the development of a new and more dynamic stimulus array that will allow the definition of SRF and STRF structure in a time frame much reduced from current analyses. This method is a derivative of the reverse correlation methods (e.g., spike triggered averaging) used in unisensory systems to define receptive field architecture and dynamism (Cai et al., 1997; deCharms et al., 1998). In addition to the time benefits, such a dynamic stimulus array can be argued to be more in keeping with "ethological" stimuli, in which the spatial and temporal relationships of ongoing stimuli are continually changing.

An additional and exciting opportunity that is allowed by the shift toward these more dynamic preparations is an ability to relate between the perceptual development literature, the development of multisensory behavioral and perceptual processes, and the neural circuits that subserve these processes. For example, the fundamental observations concerning perceptual narrowing in both human and animal models (see Lewkowicz and Ghazanfar, 2006; Zangenehpour et al., 2009) make explicit predictions about the abilities of multisensory neurons and multisensory networks to process amodal vs. multisensory stimulus sets – predictions that can be tested in developing animals in which correlations between neuronal activity and behavior/perception can be made. For example, as perceptual narrowing progresses during maturation, do we see a comparable narrowing in the filter sets that comprise multisensory neurons and that may form the basis for these changes in perceptual selectivity? Addressing such questions will provide an integrated neural/perceptual view into multisensory development that has been lacking to date.

17.10 Moving Toward the Development and Plasticity of Multisensory Object Representations

To date, the vast majority of efforts in describing multisensory development and plasticity has come from looking at spatial, rather than object, representations. This is largely a result of the historical emphasis of multisensory research, which being grounded initially in the SC, has been almost exclusively focused on spatial processes. However, as covered in other chapters of this volume, there is growing interest in the contributions of multisensory processes to the formation of object representations. As with spatial representations, the importance of multisensory integration in object processing is intuitive, as many of the objects in our dynamic world are specified by cues from multiple sensory modalities. Indeed, studies have shown significant benefits from the combination of multisensory cues in the building of an object representation, often reflected in processes such as speeded identification times (Helbig and Ernst, 2007; Giard and Peronnet, 1999; Molholm et al., 2004; Schneider et al., 2008). Nonetheless, despite its obvious importance in building our perceptual gestalt and despite the progress we have made of late in detailing the maturation of multisensory spatial representations, little is known about how multisensory object representations develop. Work in the inferior temporal (IT) cortex of developing rhesus monkeys may provide some insights, although this work has largely been restricted to studies of visual development (Rodman, 1994; Rodman et al., 1993). Here, the functional maturation of neuronal responses in IT, a key area of the ventral visual stream thought to be central in object processing, appears to occur over a protracted period of postnatal life lasting at least until 1 year of age (and likely later). The parallels of these findings to those seen in developing spatial representations, coupled with the connectivity of IT to areas integral in multisensory processing (e.g., the superior temporal polysensory area – STP; see Felleman and van Essen, 1991), strongly suggests that the maturation of the neural substrates for multisensory object processing take place during a prolonged period of postnatal life and during which experiences gained with multisensory stimuli are a key driving force in this maturational process. Evidence for this has come from a recent study conducted in human infants, and in which multisensory (i.e., visual-tactile) exploration of objects increased infants' sensitivity to color information (Wilcox et al., 2007). Most importantly, the selective benefits of multisensory exploration were not seen until the infants were 10.5 months old. Taken together, these results suggest that a fruitful line of future research will be dedicated to unraveling the brain sites and developmental events leading up to the creation of a mature multisensory object representation.

References

Avillac M, Ben Hamed S et al. (2007) Multisensory integration in the ventral intraparietal area of the macaque monkey. J Neurosci 27(8):1922–1932

Avillac M, Deneve S et al. (2005) Reference frames for representing visual and tactile locations in parietal cortex. Nat Neurosci 8(7):941–949

Barraclough NE, Xiao D et al. (2005) Integration of visual and auditory information by superior temporal sulcus neurons responsive to the sight of actions. J Cogn Neurosci 17(3):377–391

Beauchamp MS (2005) Statistical criteria in FMRI studies of multisensory integration. Neuroinformatics 3(2):93–113

Beauchamp MS, Argall BD et al. (2004) Unraveling multisensory integration: patchy organization within human STS multisensory cortex. Nat Neurosci 7(11):1190–1192

Bell AH, Corneil BD et al. (2001) The influence of stimulus properties on multisensory processing in the awake primate superior colliculus. Can J Exp Psychol 55(2):123–132

Bell AH, Meredith MA et al. (2005) Crossmodal integration in the primate superior colliculus underlying the preparation and initiation of saccadic eye movements. J Neurophysiol 93(6):3659–3673

Benedek G, Eordegh G et al. (2004) Distributed population coding of multisensory spatial information in the associative cortex. Eur J Neurosci 20(2):525–529

Benedek G, Fischer-Szatmari L et al. (1996) Visual, somatosensory and auditory modality properties along the feline suprageniculate-anterior ectosylvian sulcus/insular pathway. Prog Brain Res 112:325–334

Benevento LA, Fallon J et al. (1977) Auditory–visual interaction in single cells in the cortex of the superior temporal sulcus and the orbital frontal cortex of the macaque monkey. Exp Neurol 57(3):849–872

Bruce C, Desimone R et al. (1981) Visual properties of neurons in a polysensory area in superior temporal sulcus of the macaque. J Neurophysiol 46(2):369–384

Bushara KO, Grafman J et al. (2001) Neural correlates of auditory-visual stimulus onset asynchrony detection. J Neurosci 21(1):300–304

Cai D, DeAngelis GC et al. (1997) Spatiotemporal receptive field organization in the lateral geniculate nucleus of cats and kittens. J Neurophysiol 78(2):1045–1061

Calvert GA (2001) Crossmodal processing in the human brain: insights from functional neuroimaging studies. Cereb Cortex 11(12):1110–1123

Calvert GA, Campbell R et al. (2000) Evidence from functional magnetic resonance imaging of crossmodal binding in the human heteromodal cortex. Curr Biol 10(11):649–657

Calvert GA, Spence C et al. (eds) (2004) The handbook of multisensory processes. Cambridge, MA, The MIT Press

Cappe C, Thut G et al. (2009) Selective integration of auditory-visual looming cues by humans. Neuropsychologia 47(4):1045–1052

Carriere BN, Royal DW et al. (2007) Visual deprivation alters the development of cortical multisensory integration. J Neurophysiol 98(5):2858–2867

Carriere BN, Royal DW et al. (2008) Spatial heterogeneity of cortical receptive fields and its impact on multisensory interactions. J Neurophysiol 99(5):2357–2368

Clarey JC, Irvine DR (1986) Auditory response properties of neurons in the anterior ectosylvian sulcus of the cat. Brain Res 386(1–2):12–19

Clarey JC, Irvine DR (1990) The anterior ectosylvian sulcal auditory field in the cat: I. An electrophysiological study of its relationship to surrounding auditory cortical fields. J Comp Neurol 301(2):289–303

Clemo HR, Stein BE (1983) Organization of a fourth somatosensory area of cortex in cat. J Neurophysiol 50(4):910–925

Colonius, H, Diederich A (2004) Multisensory interaction in saccadic reaction time: a time-window-of-integration model. J Cogn Neurosci 16(6):1000–1009

Corneil BD, Olivier E et al. (2002a). Neck muscle responses to stimulation of monkey superior colliculus. I. Topography and manipulation of stimulation parameters. J Neurophysiol 88(4):1980–1999

Corneil BD, van Wanrooij M et al. (2002b). Auditory-visual interactions subserving goal-directed saccades in a complex scene. J Neurophysiol 88(1):438–454

DeAngelis GC, Ohzawa I et al. (1993) Spatiotemporal organization of simple-cell receptive fields in the cat's striate cortex. I. General characteristics and postnatal development. J Neurophysiol 69(4):1091–1117

deCharms RC, Blake DT et al. (1998) Optimizing sound features for cortical neurons. Science 280(5368):1439–1443

Dong WK, Chudler EH et al. (1994) Somatosensory, multisensory, and task-related neurons in cortical area 7b (PF) of unanesthetized monkeys. J Neurophysiol 72(2):542–564

Duhamel JR, Colby CL et al. (1998) Ventral intraparietal area of the macaque: congruent visual and somatic response properties. J Neurophysiol 79(1):126–136

Felleman DJ, van Essen DC (1991) Distributed hierarchical processing in the primate cerebral cortex. Cereb Cortex 1(1):1–47

Frens MA, van Opstal AJ (1998) Visual-auditory interactions modulate saccade-related activity in monkey superior colliculus. Brain Res Bull 46(3):211–224

Frens MA, van Opstal AJ et al. (1995) Spatial and temporal factors determine auditory-visual interactions in human saccadic eye movements. Percept Psychophys 57(6):802–816

Frens MA, van Opstal AJ (1995) A quantitative study of auditory-evoked saccadic eye movements in two dimensions. Exp Brain Res 107(1):103–117

Fritz J, Shamma S et al. (2003) Rapid task-related plasticity of spectrotemporal receptive fields in primary auditory cortex. Nat Neurosci 6(11):1216–1223

Ghazanfar AA, Krupa DJ et al. (2001) Role of cortical feedback in the receptive field structure and nonlinear response properties of somatosensory thalamic neurons. Exp Brain Res 141(1): 88–100

Ghazanfar AA, Schroeder CE (2006) Is neocortex essentially multisensory? Trends Cogn Sci 10(6):278–285

Giard MH, Peronnet F (1999) Auditory-visual integration during multimodal object recognition in humans: a behavioral and electrophysiological study. J Cogn Neurosci 11(5): 473–490

Harrington LK, Peck CK (1998) Spatial disparity affects visual-auditory interactions in human sensorimotor processing. Exp Brain Res 122(2):247–252

Helbig HB, Ernst MO (2007) Optimal integration of shape information from vision and touch. Exp Brain Res 179(4):595–606
Hubel DH, Wiesel TN (1998) Early exploration of the visual cortex. Neuron 20(3):401–412
Hughes HC, Nelson MD et al. (1998) Spatial characteristics of visual-auditory summation in human saccades. Vision Res 38(24):3955–3963
Jack CE, Thurlow WR (1973) Effects of degree of visual association and angle of displacement on the "ventriloquism" effect. Percept Mot Skills 37(3):967–979
Jiang H, Lepore F et al. (1994) Sensory modality distribution in the anterior ectosylvian cortex (AEC) of cats. Exp Brain Res 97(3):404–414
Jiang W, Jiang H et al. (2006) Neonatal cortical ablation disrupts multisensory development in superior colliculus. J Neurophysiol 95(3):1380–1396
Kadunce DC, Vaughan JW et al. (2001) The influence of visual and auditory receptive field organization on multisensory integration in the superior colliculus. Exp Brain Res 139(3):303–310
Kao CQ, McHaffie JG et al. (1994) Functional development of a central visual map in cat. J Neurophysiol 72(1):266–272
Kayser C, Petkov CI et al. (2005) Integration of touch and sound in auditory cortex. Neuron 48(2):373–384
King AJ (2004) The superior colliculus. Curr Biol 14(9): R335–338
Kitada R, Kito T et al. (2006) Multisensory activation of the intraparietal area when classifying grating orientation: a functional magnetic resonance imaging study. J Neurosci 26(28): 7491–7501
Lewkowicz DJ, Ghazanfar AA (2006) The decline of cross-species intersensory perception in human infants. Proc Natl Acad Sci U S A 103:6771–6774
Macaluso E, George N et al. (2004) Spatial and temporal factors during processing of audiovisual speech: a PET study. Neuroimage 21(2):725–732
Maier JX, Chandrasekaran C et al. (2008) Integration of bimodal looming signals through neuronal coherence in the temporal lobe. Curr Biol 18(13):963–968
Makin TR, Holmes NP et al. (2007) Is that near my hand? Multisensory representation of peripersonal space in human intraparietal sulcus. J Neurosci 27(4):731–740
McGurk H, MacDonald J (1976) Hearing lips and seeing voices. Nature 264(5588):746–748
Meredith MA, Nemitz JW et al. (1987) Determinants of multisensory integration in superior colliculus neurons. 1. Temporal factors. J Neuroscience 7:3215–3229
Meredith MA, Stein BE (1986a). Spatial factors determine the activity of multisensory neurons in cat superior colliculus. Brain Res 365:350–354
Meredith MA, Stein BE (1986b). Visual, auditory, and somatosensory convergence on cells in superior colliculus results in multisensory integration. J Neurophysiol 56:640–662
Miller LM, D'Esposito M (2005) Perceptual fusion and stimulus coincidence in the cross-modal integration of speech. J Neurosci 25(25):5884–5893
Miller LM, Escabi MA et al. (2002) Spectrotemporal receptive fields in the lemniscal auditory thalamus and cortex. J Neurophysiol 87(1):516–527
Molholm S, Ritter W et al. (2004) Multisensory visual-auditory object recognition in humans: a high-density electrical mapping study. Cereb Cortex 14(4):452–465
Nagy A, Eordegh G et al. (2003) Spatial and temporal visual properties of single neurons in the feline anterior ectosylvian visual area. Exp Brain Res 151(1):108–114
Noesselt T, Rieger JW et al. (2007) Audiovisual temporal correspondence modulates human multisensory superior temporal sulcus plus primary sensory cortices. J Neurosci 27(42): 11431–11441
Olson CR, Graybiel AM (1987) Ectosylvian visual area of the cat: location, retinotopic organization, and connections. J Comp Neurol 261(2):277–294
Peck CK (1987) Visual-auditory interactions in cat superior colliculus: their role in the control of gaze. Brain Res 420:162–166
Perrault TJ Jr, Vaughan JW et al. (2005) Superior colliculus neurons use distinct operational modes in the integration of multisensory stimuli. J Neurophysiol 93(5):2575–2586

Reid RC, Soodak RE et al. (1991) Directional selectivity and spatiotemporal structure of receptive fields of simple cells in cat striate cortex. J Neurophysiol 66(2):505–529
Rodman HR (1994) Development of inferior temporal cortex in the monkey. Cereb Cortex 5: 484–498
Rodman HR, Scalaidhe SP et al. (1993) Response properties of neurons in temporal cortical visual areas of infant monkeys. J Neurophysiol 70(3):1115–1136
Rowland BA, Quessy S et al. (2007) Multisensory integration shortens physiological response latencies. J Neurosci 27(22):5879–5884
Royal DW, Carriere BN et al. (2009) Spatiotemporal architecture of cortical receptive fields and its impact on multisensory interactions. Exp Brain Res
Scannell JW, Blakemore C et al. (1995) Analysis of connectivity in the cat cerebral cortex. J Neurosci 15(2):1463–1483
Schlack A, Sterbing-D'Angelo SJ et al. (2005) Multisensory space representations in the macaque ventral intraparietal area. J Neurosci 25(18):4616–4625
Schneider TR, Engel AK et al. (2008) Multisensory identification of natural objects in a two-way crossmodal priming paradigm. Exp Psychol 55(2):121–132
Senkowski D, Talsma D et al. (2007) Good times for multisensory integration: effects of the precision of temporal synchrony as revealed by gamma-band oscillations. Neuropsychologia 45(3):561–571
Shamma S (2008) Characterizing auditory receptive fields. Neuron 58(6):829–831
Sparks DL (1986) Translation of sensory signals into commands for control of saccadic eye movements: role of primate superior colliculus. Physiol Rev 66(1):118–171
Sparks DL (1999) Conceptual issues related to the role of the superior colliculus in the control of gaze. Curr Opin Neurobiol 9(6):698–707
Sparks DL, JM Groh (1995) The superior colliculus: a window for viewing issues in integrative neuroscience. In: Gazzaniga MS (ed) The cognitive neurosciences. Cambridge, MA, The MIT Press: 565–584
Stanford TR, Quessy S et al. (2005) Evaluating the operations underlying multisensory integration in the cat superior colliculus. J Neurosci 25(28):6499–6508
Stanford TR, BE Stein (2007) Superadditivity in multisensory integration: putting the computation in context. Neuroreport 18(8):787–792
Stein B, Meredith M et al. (1989) Behavioral indices of multisensory integration: orientation to visual cues is affected by auditory stimuli. J Cogn Neurosci 1:12–24
Stein BE, Huneycutt WS et al. (1988) Neurons and behavior: the same rules of multisensory integration apply. Brain Res 448:355–358
Stein BE, Magalhaes-Castro B et al. (1976) Relationship between visual and tactile representations in cat superior colliculus. J Neurophysiol 39(2):401–419
Stein BE, Meredith MA (1991) Functional organization of the superior colliculus. In: Leventha AG (ed) The neural basis of visual function. l. Macmillan, Hampshire, UK, pp 85–110
Stein BE, Meredith MA (1993) The Merging of the Senses. MIT Press, Cambridge, MA
Stevenson RA, Geoghegan ML et al. (2007) Superadditive BOLD activation in superior temporal sulcus with threshold non-speech objects. Exp Brain Res 179(1):85–95
Stevenson RA, James TW (2009) Audiovisual integration in human superior temporal sulcus: inverse effectiveness and the neural processing of speech and object recognition. Neuroimage 44(3):1210–1223
Tamai Y, Miyashita E et al. (1989) Eye movements following cortical stimulation in the ventral bank of the anterior ectosylvian sulcus of the cat. Neurosci Res 7(2):159–163
Teder-Salejarvi WA, Di Russo F et al. (2005) Effects of spatial congruity on audio-visual multimodal integration. J Cogn Neurosci 17(9):1396–1409
Thurlow WR, Jack CE (1973) Certain determinants of the "ventriloquism effect". Percept Mot Skills 36(3):1171–1184
van Wanrooij MM, Bell AH et al. (2009) The effect of spatial-temporal audiovisual disparities on saccades in a complex scene. Exp Brain Res 198(2–3):425–437

Wallace MT, Carriere BN et al. (2006) The development of cortical multisensory integration. J Neurosci 26(46):11844–11849
Wallace MT, McHaffie JG et al. (1997) Visual response properties and visuotopic representation in the newborn monkey superior colliculus. J Neurophysiol 78(5):2732–2741
Wallace MT, Meredith MA et al. (1992) Integration of multiple sensory modalities in cat cortex. Exp Brain Res 91(3):484–488
Wallace MT, Meredith MA et al. (1993) Converging influences from visual, auditory, and somatosensory cortices onto output neurons of the superior colliculus. J Neurophysiol 69(6):1797–1809
Wallace MT, Meredith MA et al. (1998) Multisensory integration in the superior colliculus of the alert cat. J Neurophysiol 80(2):1006–1010
Wallace MT, Perrault TJ Jr et al. (2004a) Visual experience is necessary for the development of multisensory integration. J Neurosci 24(43):9580–9584
Wallace MT, Ramachandran R et al. (2004b) A revised view of sensory cortical parcellation. Proc Natl Acad Sci U S A 101(7):2167–2172
Wallace MT, Stein BE (1996) Sensory organization of the superior colliculus in cat and monkey. Prog Brain Res 112:301–311
Wallace MT, Stein BE (1997) Development of multisensory neurons and multisensory integration in cat superior colliculus. J Neurosci 17(7):2429–2444
Wallace MT, Stein BE (2000) Onset of cross-modal synthesis in the neonatal superior colliculus is gated by the development of cortical influences. J Neurophysiol 83(6):3578–3582
Wallace MT, Stein BE (2001) Sensory and multisensory responses in the newborn monkey superior colliculus. J Neurosci 21(22):8886–8894
Wallace MT, Stein BE (2007) Early experience determines how the senses will interact. J Neurophysiol 97(1):921–926
Wallace MT, Wilkinson LK et al. (1996) Representation and integration of multiple sensory inputs in primate superior colliculus. J Neurophysiol 76(2):1246–1266
Wilcox T, Woods R et al. (2007) Multisensory exploration and object individuation in infancy. Dev Psychol 43(2):479–495
Wurtz RH, Munoz DP (1995) Role of monkey superior colliculus in control of saccades and fixation.In: Gazzaniga MS (ed) The cognitive neurosciences. The MIT Press, Cambridge, MA, pp 533–548
Zangenehpour S, Ghazanfar AA et al. (2009) Heterochrony and cross-species intersensory matching by infant vervet monkeys. PLoS ONE 4:e4302

Chapter 18
Large-Scale Brain Plasticity Following Blindness and the Use of Sensory Substitution Devices

Andreja Bubic, Ella Striem-Amit, and Amir Amedi

18.1 Introduction

Living with a sensory impairment is challenging and those who have lost the use of one sensory modality need to find ways to deal with numerous problems encountered in daily life. When vision is lost, this may include navigating through space, finding objects, recognizing people or surroundings, reading or even communicating without much access to nonverbal signs provided by others such as eye gaze or facial expressions. Nevertheless, the blind manage to function efficiently in their environment, often so to a surprisingly high degree. The same is true for the deaf. How is this level of functioning achieved? What sort of cognitive restructuring is needed to allow the blind to, for instance, develop spatial representations using only auditory or tactile information, recognize and navigate through familiar environments, or build a representation of a novel space? How is such restructuring implemented in the brain? Finally, how does the nervous system deal with an initially large silent cortical surface: does this area simply remain silent or does it become integrated with the rest of the brain in an atypical, but nevertheless functional manner?

Answering these questions is not just of great theoretical interest but also has important implications for improving current and developing new rehabilitation approaches for blindness, deafness, and other clinical situations such as stroke. This includes "classical" approaches such as educational programs that teach blind children how to recognize and efficiently exploit tactile or auditory cues for spatial processing or initiatives aimed at providing wider and earlier access to the most efficient rehabilitation techniques. However, understanding the links between the brain's ability to remodel itself and behavior as well as the factors which influence

A. Amedi (✉)
Department of Medical Neurobiology, Institute for Medical Research Israel-Canada (IMRIC), Hebrew University-Hadassah Medical School, Jerusalem 91220, Israel
e-mail: amir.amedi@ekmd.huji.ac.il
website: http://brain.huji.ac.il

Andreja Bubic and Ella Striem-Amit contributed equally to the chapter.

M.J. Naumer, J. Kaiser (eds.), *Multisensory Object Perception in the Primate Brain*,
DOI 10.1007/978-1-4419-5615-6_18, © Springer Science+Business Media, LLC 2010

this linkage is crucial for developing novel rehabilitation techniques aimed not just at compensating, but also restoring parts of the lost sensory functions. These primarily include neuroprostheses which attempt to restore visual function to the impaired visual system of the blind and sensory substitution devices (SSDs) which use a human–machine interface to provide visual information to the blind via a non-visual modality. Since these techniques depend crucially on the possibility of teaching the blind brain new complex perceptual skills involved in vision, they can be developed more efficiently if enough is known about the plasticity of our neural system as well as the neural foundations of information processing, especially sensory processing within and across individual modalities. Efficient use of these techniques is based on implicit assumptions that we are able to exploit and channel the brain's ability to reorganize itself and to link or translate information from individual senses to multisensory and back to the unimodal (visual) percepts with the goal of restoring some features of the lost modality.

However, it is difficult to control something that one does not adequately understand. This may explain why restoration of truly functional sensory modalities using neuroprostheses is still not possible and why, despite significant recent progress, SSDs have yet to reach their full potential. Solving these problems depends not only on increasing our knowledge of the general principles of brain plasticity, but also acknowledging the impact of individual differences on sensory rehabilitation. Although many factors might be important in this respect, the onset of sensory loss is the most prominent source of individual variance. Thus, although surgically restoring sight at an early age might result in almost full sensory recovery, attempting the same in an adult who has never had visual experience poses enormous challenges because vision is, in many ways, a learned skill. Giving the occipital cortex access to visual signals will therefore not automatically guarantee the emergence of normal vision and, if done later in life, may even hamper a reorganized functional system by interfering with tactile, auditory, language, or memory processing that has been rerouted to this cortical surface.

Thus, sight restoration and sensory substitution offer a unique and highly successful key to understanding brain plasticity, perception, sensory integration, and the binding problem – the link between brain activity and conscious perception, as well as other fundamental issues in neuroscience, psychology, philosophy, and other disciplines. Clearly, developing new approaches and improving existing rehabilitation techniques aimed at compensating for sensory loss depends to a great extent on how well current knowledge about normal sensory processing and the brain's potential for change can be integrated and applied. This chapter will attempt to present and integrate some of this knowledge, mainly concentrating on blindness and rehabilitation techniques available for the blind. Before introducing these, we will explore the minds and brains of those who need such devices in more detail, depict their cognitive adjustments to sensory impairments, and describe different types of neuroplastic changes which support such cognitive restructuring. We will then look at the importance of individual, especially developmental, differences in experimental and practical rehabilitation settings. Following this, we will review the currently available rehabilitation techniques, primarily sensory substitution devices, and, to a somewhat lesser degree, clinical ophthalmologic and neuroprosthetic approaches. These techniques are designed to exploit one of the brain's fundamental intrinsic properties – its plasticity. This property manifests itself not only following sensory loss but in all normal or pathological contexts. The brain constantly changes and yet, despite the fact that it can (especially in some circumstances) undergo extensive modifications in basic morphology, connectivity, physiology, or neurochemistry, manages to preserve stability and continuity. Some of these general features of neuroplasticity will be discussed in detail as they can help understand specific changes that occur in cases of sensory impairment.

18.2 The Plastic Brain

Plasticity in the brain, i.e., neuroplasticity, reflects the brain's ability to change its structure and function throughout the course of a lifetime. This intrinsic property of the nervous system is visible across different levels of brain functioning which include genetic, neuronal, and synaptic, as well as the level of brain networks and the nervous system as a whole. Consequently, plasticity is also manifested in the dynamics of emergent cognitive processes and overt behavior. Each one of these levels can incorporate different types of changes, such as structural changes in axon terminals, dendritic arborization, and spine density in neurons, as well as changes in glial cells in case of synaptic plasticity (Kolb, 1995). Although the importance of these types of plastic changes has been recognized for many years (Malenka and Bear, 2004), Merabet et al. (2008) have recently argued that the capacity for plastic modulation at the level of single neurons is likely to be somewhat limited due to the high complexity of individual neurons and other constrains in higher vertebrates. As a result, this level must be complemented by higher potential for change at the neural network level. This is thought to be mediated through an architecture of distributed neural networks composed of nodes that perform computations somewhat independently of the properties of their inputs, thus enabling their integration into different networks. In this way, neural networks can remain highly dynamic and adaptable to changing environmental demands without endangering the stability of individual nodes. Some of these nodes or brain regions may be more or less susceptible to change. For example, although plasticity has mainly been investigated in regions such as the hippocampus, which is even characterized by a certain degree of adult neurogenesis (Eriksson et al., 1998), substantial plastic changes can also occur in neocortical regions (Kolb, 1995). In particular, a much higher degree of plasticity has been reported in associative unimodal or multisensory regions than in the primary sensory cortices, which may partly be due to high sensitivity of higher level areas to crossmodal inputs (Fine, 2008). Although useful, this separation into different levels of plastic changes is somewhat artificial because individual levels of brain organization are not mutually independent, but directly or indirectly influenced by all other levels (Shaw and McEachern, 2000). In addition, other non-developmental factors such as the previous history of synapse's activity may also influence this potential for future plasticity (Abraham and Bear, 1996). Finally, it needs to be emphasized that the potential for change is itself not static, as it varies dramatically throughout the course of life. This potential is at its highest in early childhood, whereas it is typically assumed that large-scale reorganization in adulthood primarily occurs in response to pathological states. However, the adult brain also changes in non-pathological states, and such use-induced plasticity may differ from lesion-induced change in terms of its extent and the underlying mechanisms (Dinse and Merzenich, 2000). Finally, although constantly changing, the brain needs to remain stable to a certain degree. Therefore, neuronal plasticity must be balanced by neuronal stability through homeostatic control within and between different levels of neural functioning (Shaw and McEachern, 2000).

18.2.1 Plasticity Across the Lifespan

It is generally believed that the nervous system is the most plastic during its development, both in the case of normal development and following brain injury. Although the brain is thought to retain the ability to change throughout life, especially in pathological cases, this assumption is mostly corroborated by experimental findings. The developing brain is a highly dynamic system which undergoes several distinct phases from cell formation to the rapid growth and subsequent elimination of unused synapses before finally entering into a more stable phase following puberty (Chechik et al., 1998). The functional assignment of individual brain regions that occurs during this time is crucially dependent on synaptic development which undergoes drastic changes that often take place in spurts. In the visual cortex, during the first year after birth, the number of synapses grows tremendously and is subsequently scaled down to the adult level around the age of 11 through extensive decreases in synaptic and spine density, dendritic length, or even the number of neurons (Kolb, 1995). This process is primarily determined by experience and neural activity: synapses which are used are strengthened while others are not either reinforced or actively eliminated. Synaptic development is highly dependent on competition between incoming inputs, the lack of which can result in a decreased level of synaptic revision and persistence of redundant connections in adulthood (De Volder et al., 1997). This process of synaptic pruning represents fairly continuous and extended tuning of neural circuits and can be contrasted with other types of changes which occur at very short timescales. During such periods of intensified development (i.e., critical or, more broadly, sensitive periods; Knudsen, 2004; Michel and Tyler, 2005), the system is the most sensitive to abnormal environmental inputs or injuries (Wiesel and Hubel, 1963). Thus, injuries affecting different stages of development, even when they occur at a roughly similar age, may trigger distinct patterns of compensatory neuroplastic changes and lead to different levels of recovery. Specifically, early studies of recovery after visual loss (Wiesel and Hubel, 1963, 1965) suggested that visual deprivation of even short duration, but occurring at an early developmental stage when vision is particularly sensitive to receiving natural input, may irreversibly damage the ability to normally perceive visual input at older ages. Conversely, recent sparse evidence of visual recovery after early-onset blindness (Fine et al., 2003; Gregory and Wallace, 1963), which will be discussed in more length in the next sections addressing visual restoration, demonstrates that this may not necessarily apply in all cases.

The potential for neuroplasticity after puberty is considered to be either much lower than in childhood, or possible only in cases of pathological states and neural overstimulation (Shaw and McEachern, 2000). However, recovery following different types of pathological states occurring in adulthood (Brown, 2006; Chen et al., 2002), changes in neural count and compensatory increase in the number of synapses in aging (Kolb, 1995), and the profound changes revealed by functional neuroimaging following short periods of blindfolding (Amedi et al., 2006; Pascual-Leone et al., 2005; Pitskel et al., 2007) suggest otherwise. In reconciling these seemingly contradictory conclusions, it is useful to take into account the multi-faceted nature of plasticity which includes different forms of changes occurring at different timescales and at different levels of neural functioning. For example, synaptic changes occurring in aging develop over an extended period of time and in synergy with altered experiences and needs characteristic for the later periods in life. The robust, short-term plasticity witnessed in blindfolding may arise from the recruitment of already existing, but commonly unused, inhibited, or masked pathways which become available once the source or reason for such masking (e.g., availability of visual input in those who have been blindfolded) is removed. Therefore, some forms of adult plasticity do not reflect "plasticity de novo" which is characterized by the creation of new connectivity patterns (Burton, 2003). In contrast, in pathological states, injuries, or late sensory loss, both of these types of changes can co-occur and mutually interact. Rapid changes reflecting the unmasking of existing connections occurring in the first phase promote and enable subsequent slow, but more permanent structural changes (Amedi et al., 2005; Pascual-Leone et al., 2005). This suggests that potentially similar functional outcomes may be mediated by different neural mechanisms whose availability depends on the developmental stage within which they occur.

18.3 Plastic Changes Following Sensory Loss

Regardless of when it occurs, sensory loss drastically affects both cognitive functioning and the anatomy and physiology of the nervous system enabling those functions. Studying changes triggered by sensory impairment provides a unique opportunity for exploring not only the plasticity relevant for pathological states, but also the fundamental principles guiding the formation and functional organization of any nervous system. This is especially true for congenital blindness, deafness, and similar conditions which represent the most documented cases of plasticity, compensatory, and otherwise. Given their early-onset and the fact that they occur in an undeveloped system, these conditions allow large-scale changes that promote full reorganization, which may result in a functional network remarkably different from the one seen in healthy individuals or individuals that sustain brain or peripheral injuries later in life. For example, blindness or deafness resulting from peripheral damage (i.e., dysfunctional retina/cochlea or the sensory tracts) does not injure the brain itself, but withholds parts of the brain from their natural input. However, despite the lack of visual and auditory input, the visual and auditory cortices of the blind and deaf do not simply degenerate but, to some extent, become integrated into other brain networks. Such functional reintegration is enabled through the changed structure, connectivity, and physiology of these areas in comparison to those typically encountered in the majority of the population. A similar reintegration, but to a different extent and partially mediated through different neurophysiological mechanisms, occurs in the case of late sensory loss. Before discussing the differences between these populations in more detail, we present some general findings which are relevant for all sensory impaired individuals. These include recent electrophysiological and neuroimaging studies investigating cognitive and neural processing following loss of sensory (primarily visual) function and the neurophysiological mechanisms underlying these changes.

Just like the sighted, in order to function independently and act efficiently in the world, the blind need to acquire information about their environment and represent it in a way that can constantly be updated and used in different reference frames and for different purposes. Unlike most others, to achieve these goals they rely on fewer sensory modalities and are therefore not privileged to the same qualitative and quantitative richness of information available to the sighted. These individuals need to somehow "compensate for lack of vision," a modality which commonly offers a wide range of information needed for everyday life and draws attention to relevant external events (Millar, 1981). Not having access to this information source, the blind need to identify useable cues from other modalities and/or develop alternative cognitive strategies, allowing them to build a representation of themselves and their environment that can be effectively exploited for (goal-directed) action. This process may, in turn, result in profound changes in higher-order cognitive or sensory functions outside the affected modality. For example, it has been shown that the blind, compared to the sighted, possess superior memory (D'Angiulli and Waraich, 2002; Hull and Mason, 1995; Pozar, 1982; Pring, 1988; Raz et al., 2007; Röder et al., 2001; Smits and Mommers, 1976; Tillman and Bashaw, 1968), as well as tactile, and auditory perceptual abilities: they are, for instance, able to better discriminate between small tactile dots or auditory spatial locations than the sighted and even better identify odors (Collignon et al., 2006; Doucet et al., 2005; Goldreich and Kanics, 2003, 2006; Grant et al., 2000; Hugdahl et al., 2004; Murphy and Cain, 1986; Röder et al., 1999; Smith et al., 2005; Wakefield et al., 2004). This superiority is not always identified (Lewald, 2002; Zwiers et al., 2001), suggesting that optimal development of some aspects of sensory processing in the unaffected modalities may depend on (or at least benefit from) concurrent visual input. Nevertheless, the majority of findings still indicate a compensation for the missing modality through hyper-development of other senses and higher cognitive functions. Once achieved, this advantage could, as indicated by inferior performance of the partially blind (Lessard et al., 1998), even be compromised by the presence of visual information. Comparable to results in the blind, deaf individuals also show improved visual abilities on certain tasks (Bavelier et al., 2006). This clearly runs counter the assumption that sensory loss necessarily leads to general maladjustment and dysfunction in other cognitive domains which cannot develop without the supporting vision. Therefore, this, so-called general-loss hypothesis, can be abandoned to a large extent in favor of the alternative, compensatory view according to which sensory loss leads to the superior development of the remaining senses (Pascual-Leone et al., 2005).

The described changes in cognitive functioning in the blind are necessarily paralleled by changes in many features of neural processing reflecting both the altered computations underlying their unique cognitive functioning and the lack of visual input promoting an atypical organization of the occipital cortex. In the last decades, studies investigating neural processing of blind individuals, as well as more invasive animal experiments, have shown that sensory loss triggers robust modifications of functioning across entire brain networks. Results from electrophysiological studies indicate shorter latencies of event-related potentials (ERP) in auditory and somatosensory tasks in the blind in contrast to the sighted, suggesting more efficient processing in this population (Niemeyer and Starlinger, 1981; Röder et al., 2000). In addition, identified differences in topographies of ERP components in the sighted and the blind suggest a reorganization in the neural implementation of non-visual functions, so as to engage the occipital cortex of the blind (Kujala et al., 1992; Leclerc et al., 2000; Rösler et al., 1993; Uhl et al., 1991). Functional neuroimaging methods characterized by higher spatial resolution corroborate these findings by showing that the occipital cortex is not just neurally active (De Volder et al., 1997) but also functionally engaged in perception in other modalities, namely audition (Gougoux et al., 2005; Kujala et al., 2005) and tactile Braille reading (Büchel et al., 1998; Burton et al., 2002; Gizewski et al., 2003; Sadato et al., 1998, 1996). Even more dramatic are the changes in higher cognitive, verbal, and language functions (Amedi et al., 2004; Burton et al., 2003; Burton et al., 2002; Ofan and Zohary, 2007; Röder et al., 2002) and memory processing (Amedi et al., 2003; Raz et al., 2005). Studies in which processing within the occipital cortex was transiently disrupted using transcranial magnetic stimulation (TMS) confirm the necessity of the targeted occipital engagement in auditory (Collignon et al., 2006) and tactile processing including Braille reading (Cohen et al., 1997; Merabet et al., 2004) as well as linguistic functions (Amedi et al., 2004), suggesting that such processing reflects functionally relevant contributions to these tasks (see Fig. 18.1). Similarly, it has been shown that the auditory cortex of the congenitally deaf is activated by visual stimuli (Finney et al., 2001), particularly varieties of visual movement (Campbell and MacSweeney, 2004). It is important to realize that the involvement of unimodal brain regions in crossmodal perception is not only limited to individuals with sensory impairments but can, under certain circumstances, also be identified in the majority of the population (Amedi et al., 2006; Merabet et al., 2004; Zangaladze et al., 1999). This involvement is much more pronounced in the blind and deaf

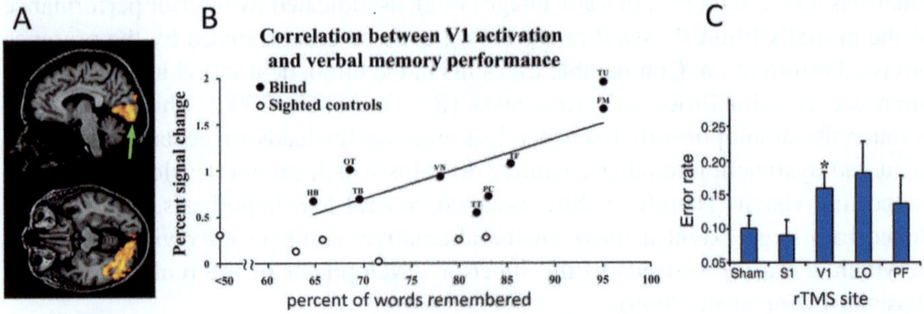

Fig. 18.1 An extreme example of brain plasticity in the primary visual cortex of the blind for verbal memory and language. (**a**) Verbal memory fMRI activation in the early "visual" cortex of the congenitally blind. The results of the congenitally blind group (only) showed robust activation in the visual cortex during a verbal memory task of abstract word retrieval which involved no sensory stimulation. The left lateralized activity was extended from V1 anteriorly to higher order "visual" areas. (**b**) The activation in V1 was correlated with the subjects' verbal memory abilities (*middle panel*). Subjects were tested on the percent of words they remembered 6 months after the scan (or online inside the scanner in an additional study). In general, blind subjects remembered more words and showed greater V1 activation than the sighted controls. Only blind subjects also showed a significant correlation of V1 activity and performance (A and B are modified from Amedi et al., 2003). (**c**) Verb-generation error rates in a blind group show that rTMS over left V1 increased error rates relative to sham and right S1 stimulation, signifying that V1 is functionally relevant to verbal memory task success; error bars, s.e.m. $^{*}P<0.05$ (modified from Amedi et al., 2004)

because sensory areas deprived from their customary sensory input become functionally integrated into other circuits which leads to profound changes within the affected modality and the system as a whole. These include crossmodal, intramodal, multisensory (multimodal), and supramodal changes, namely those pertaining to the involvement of typically visual areas in processing tactile and auditory information in the blind (or typically auditory areas in processing visual information in the deaf), plastic changes occurring within the cortices of unaffected modalities, changes in multisensory regions and global, whole-brain changes involving more than unimodal and multimodal sensory networks, respectively. Although somewhat autonomous, these different types of changes are in reality strongly interdependent and cannot be separated on the level of either cognitive or neural processing.

Multisensory plasticity refers to the reorganization of multisensory areas following sensory loss which arises from the impairment of one modality and compensatory hyper-development of the remaining ones. This altered structure of sensory inputs leads to changes in multisensory areas, the development of which is shaped by the convergence of incoming inputs from unimodal systems (Wallace, 2004). For example, studies investigating the multisensory anterior ectosylvian (AES) cortex in congenitally blind cats indicate an expansion of auditory and somatosensory fields into the area usually housing visual neurons (Rauschecker and Korte, 1993) as well as sharpened spatial filtering characteristics (Korte and Rauschecker, 1993) following blindness. These changes underlie the improved spatial abilities of these animals and may also be crucially important for the development of crossmodal plasticity.

Intramodal plasticity refers to changes occurring within one sensory modality as a consequence of altered, either increased or decreased, use of the respective sensory modality. For example, studies investigating the neural foundations of this phenomenon indicate a high degree of reorganization of sensory maps in different modalities following local peripheral damage, extensive training, or perceptual learning (Buonomano and Johnson, 2009; Kaas, 2000; Recanzone et al., 1992). This reorganization includes a coordinated shrinkage of maps representing the unused, and expansion of those representing the modality/limb experiencing increased use (Rauschecker, 2008), and is determined by the amount of stimulation and structure of the input pattern within which competition between the inputs plays an important role (Buonomano and Johnson, 2009). Another example of intramodal changes may include superior performance of the blind in auditory or tactile tasks. However, these changes can also be a reflection of *crossmodal plasticity* which refers to the reassignment of a particular sensory function to another sensory modality, for instance processing of auditory information to the visual cortex. Numerous invasive studies in animals have shown the vast potential for such reorganization, reflecting the fact that most aspects of structure and function of a given brain area are determined by its inputs, not topographic location. For example, it has been shown that typical auditory areas, after being exposed to visual input through thalamic fibers normally going to primary visual areas, can develop orientation-sensitive cells with the pattern of connectivity resembling the one typically found in the normally developed visual cortex (Sharma et al., 2000) and fulfill the visual functionality of the rewired projections (von Melchner et al., 2000). Similarly, tissue transplanted from the visual cortex into the somatosensory cortex acquires functional properties of its "host" and does not hold on to its genetic predisposition (Schlaggar and O'Leary, 1991). Thus, it can be hypothesized that the crossmodal plasticity observed in the blind is most probably subserved by altered connectivity patterns. It has generally been suggested that all levels of connectivity, including connections within local circuits, long-range cortico-cortical and subcortical connections, are altered in the blind (Bavelier and Neville, 2002). Corroborating this, recent evidence indicates that the visual tracts connecting the visual cortex with the eyes are degenerated in the blind (Noppeney et al., 2005; Pan et al., 2007; Shimony et al., 2006). A similar example of drastic reorganization of afferent nerves going from the eyes to the thalamus and cortex that resulted in changed retinotopic maps in the unaffected hemisphere (such as to represent the ipsilateral in addition to normally represented contralateral visual field), thus enabling almost normal vision in both hemifields, has also recently been reported in a young girl born with only one cerebral hemisphere (Muckli et al., 2009). Furthermore, studies in mammals show that congenital blindness (caused by early enucleation) causes a rewiring of tactile and auditory inputs to the visual cortex of mammals (Chabot et al., 2007; Izraeli et al., 2002; Karlen et al., 2006; Laemle et al., 2006; Piche et al., 2007) which facilitate or enable the involvement of the visual cortex in tactile and auditory tasks. In addition, although it was previously assumed that there are no direct connections between sensory modalities, recent anatomical studies in primates indicate the existence of projections from the auditory to the visual cortex and multisensory feedback connections to primary visual areas (Falchier et al., 2002; Rockland and Ojima, 2003). According to the reverse hierarchy model (Amedi et al., 2003; Büchel, 2003), feedback connectivity may play a crucial role in crossmodal (and supramodal) plasticity. Specifically, connections stemming from temporal, parietal, and frontal lobes may, in the absence of visual input, be responsible for providing non-visual input to the occipital lobe, enabling its engagement in non-visual processing. This is particularly true for areas which are involved in multisensory processing even in the sighted, such as regions within the lateral occipital complex that are naturally active both during tactile and visual object recognition (Amedi et al., 2002; Amedi et al., 2001). Such areas retain some of their original sensory input after the loss of one modality which allows them to preserve their original functions (i.e., tactile shape recognition, including Braille reading). Confirming this notion, it has been shown that the area involved in (visual and auditory) motion processing in the sighted is involved in auditory (Poirieret al., 2006) as well as tactile (Ricciardi et al., 2007) motion processing in the blind. Similar conclusions can be drawn from findings showing the engagement of the ventral visual pathway typically involved in processing information related to the identification of objects and faces (Ungerleider and Mishkin, 1982) in auditorily mediated object recognition, but only if detailed shape information is provided and efficiently extracted (Amedi et al., 2007; Poirieret al., 2006). All of these results are congruent with the more general notion that crossmodal plasticity occurs in situations where the information originally processed within a certain area is similar to the input which is being rerouted into it (Grafman, 2000). This implies that each cortical area operates in a metamodal fashion (Pascual-Leone and Hamilton, 2001) and is specialized in a particular type of computation rather than tied to a specific input modality. However, this type of broad generalization is subject to caution because it is still not clear how such metamodal computations would develop, especially in the case of significantly altered inputs in development.

Supramodal plasticity refers to changes which encompass areas and brain functions that are typically considered non-sensory. Evidence for such plasticity has been revealed in studies showing the involvement of the occipital cortex in memory or language processing in the blind (Amedi et al., 2004; Amedi et al., 2003; Burton et al., 2003; Burton et al., 2002; Ofan and Zohary, 2007; Raz et al., 2005; Röder et al., 2000, 2002). This type of plasticity is comparable to crossmodal plasticity and is enabled by altered connectivity patterns between the visual cortex and other supramodal brain regions. For example, the blind show decreased functional connectivity of the occipital cortex and various other cortical sites, including the supplementary motor area (SMA), pre- and postcentral gyri, superior parietal lobule, the left superior and middle temporal gyri (Yu et al., 2008), which is paralleled with increased functional connectivity with frontal language regions (Liu et al., 2007). These changes in connectivity could account for the altered pattern of inputs reaching the occipital cortex, which may in the end determine morphological and physiological features of this area and enable its functional reassignment to non-sensory tasks.

18.4 Developmental and Adult Plasticity Following Blindness

Age of sensory loss is an important factor which strongly influences the type and extent of subsequent cognitive and neural changes as well as the later potential for rehabilitation. Outcomes of late and early injuries are typically not comparable because the potential for neuroplastic reorganization differs qualitatively and quantitatively in the developing and the adult brain. In addition, the early experience of vision in individuals who become blind later in life significantly shapes their development and possibly limits potential for reorganization of both cognitive and neural functioning following sensory loss. Once they lose their visual function, the late blinded need to relearn and reorganize information processing in order to compensate for the lack of input. However, although visual information is no longer available to them, these individuals can still rely on some visual strategies, e.g., visual imagery (Büchel et al., 1998) which may then be mediated by the visual cortex. The degree to which these individuals use visual strategies and the nature of cortical reorganization following sensory loss is primarily determined by the amount of their prior visual experience, the developmental stage at which sight loss occurred and subsequent experiences. In contrast, in the congenitally blind it is almost impossible to talk about the reorganization of the "visual cortex" because their occipital brain regions have never had experience with visual information and could therefore become integrated into other brain networks in a more flexible way. This flexibility is, of course, limited by the functional needs of the whole system, the idiosyncratic nature and connectivity of the occipital cortex, its potential for coordination with other brain regions as well as the general potential for plasticity in the brain. However, although the mechanisms underlying such integration are still unknown, they are surely less fixed in the congenitally blind than in individuals whose cognition and brain have been shaped by the experience of vision.

Confirmation of the importance of the age at sensory loss for future cognitive and brain reorganization comes from studies comparing performance on non-visual tasks across different populations of the blind. Some of these have suggested that processing advantages and large-scale cortical reorganization are often limited to the congenitally and early blind, with the performance of late blinded resembling more that of the sighted (Fine, 2008). Similarly, Cohen et al. (1999) suggested that the critical period of susceptibility for significant crossmodal plasticity ends at puberty. However, other studies have demonstrated significant reorganization in the occipital cortex of the late blind (Büchel et al., 1998; Burton et al., 2002; Voss et al., 2004). Similarly, it has been shown that cortical maps can be modified to a high degree (Kaas, 1991) even in adulthood. Although these findings suggest that significant potential for plastic changes following sensory loss exists throughout the lifespan, this does not imply that these are also identical in their extent or the underlying neurophysiological mechanisms. On the contrary, they probably reflect the use of different cognitive strategies (Büchel et al., 1998) or changes in different levels/stages of neural processing (Fieger et al., 2006). For example, the brains of different subpopulations of the blind are probably differentially susceptible not just to intramodal, but also to crossmodal and multisensory plasticity. Those who had experience with vision were able to develop fully functioning multisensory systems (Wallace, 2004) which can have a significant role in shaping the cortices of the later impaired modality through feedback connections. In contrast, congenital lack of one sensory system alters the computations underlying multisensory integration and the development of multisensory brain regions, implying that there may be substantial differences across congenitally, early and late blinded individuals. In addition, different functional systems may be differently impaired by sensory loss occurring during certain phases of development. For example, Neville and Bevelier (2000) suggested that systems dedicated to dynamically shifting relations between locations, objects, and events, e.g., the dorsal visual pathway or regions subserving grammatical processing, develop earlier and are differently sensitive to developmental deficits in comparison to, e.g., the areas within the ventral visual stream which are still being modified during early adolescence (Golarai et al., 2007).

Although the early blind may in adulthood show better performance on non-visual tasks, the development of such superior abilities is gradual and often challenging. For example, it has been shown that blind children have difficulties with some tasks, especially those requiring reference to external cues and directions, as they rely mostly on self-reference and movement sequences (Millar, 1981). Recognition of potentially useful information from other modalities can sometimes be more easily accomplished if concurrent visual input is available during learning. In this case, the existence of redundant or overlapping information stemming from more than one modality can guide attention, enhance learning of amodal stimulus features (Lickliter and Bahrick, 2004), and thus facilitate the recognition of helpful auditory or somatosensory features. However, even without the visual cues, one may still learn to recognize these features (Collignon et al., 2009) using different information, for example, those derived from self-motion. If this learning is successful, spatial representations which are generated on the basis of haptic and auditory input in the blind can be considered as equivalent to those of the sighted which are based on visual input (Röder and Rösler, 1998).

In conclusion, although sensory loss triggers plastic changes in all individuals, the age at which it occurs influences the type, extent, and mechanisms supporting these changes as well as the functional outcomes they enable. However, even equivalent functional outcomes do not necessarily imply that they are supported by the same underlying mechanisms. For example, profound changes identified in recently blinded or even sighted, e.g., the engagement of the posterior occipital lobe in tactile discrimination after being blindfolded for only a week (Merabet et al., 2008), may result from "unmasking" of existing connections between the visual and other cortices, which are dormant (inhibited) in normal conditions. Although such unmasking may then promote more stable changes including alterations in connectivity patterns, this might require extended periods of sensory deprivation or even the onset of blindness to be set at the critical or sensitive period in early childhood. This again indicates how different populations of the blind cannot be considered or treated as equivalent. Consequently, the variability in developmental and adult plasticity following blindness necessarily needs to be taken into account when studying the blind as it may also have important consequences for the rehabilitation of these individuals.

18.5 Rehabilitation in the Case of Blindness and Severe Visual Impairment

Sensory loss and blindness in particular decrease the quality of life and impose severe challenges for efficient functioning on millions of individuals. Blindness hinders independent navigation in space in familiar and especially unfamiliar places, reading, and even recognizing people and communicating with them efficiently, using nonverbal communication via hand gesturing or facial expressions such as gaze direction or smiling. An anecdotal indication for this are discussions in forums for the blind, commenting how many years of life would they trade in exchange for regaining vision (though it is important to note that some blind answer that they are happy as they are and do not wish to bother themselves with such questions). Numerous approaches and potential solutions aimed at overcoming these difficulties have been put forward (with various levels of success), offering hope and help to those suffering from sensory (both visual and auditory) impairment. In the blind, these include reading and mobility aids, more advanced sensory substitution devices (SSDs), as well as invasive sensory restoration and neuroprosthetic approaches. In this part of the chapter we present some of these techniques. The main focus is on sensory substitution devices which are gaining increasing popularity thanks to their non-invasiveness, low cost, and high potential for providing systematic rehabilitation solutions for any type of blindness. In addition, we will briefly discuss the potential for medically enabled sensory restoration which, although holding great potential, still needs to overcome many technical and other challenges before it can be truly useful for most of the blind.

18.5.1 Sensory Substitution Devices

Sensory substitution refers to the transformation of the characteristics of one sensory modality into the stimuli of another modality. For example, it is possible to replace vision with touch or audition, audition or vestibular senses with touch. In the case of blindness, SSDs represent a non-invasive rehabilitation approach within which visual information is captured by an external device such as a video camera and communicated to the blind via a human–machine interface in the form of auditory or tactile input. Pioneering work which paved the way to modern SSD was done by Louis Braille (1809–1852) who developed Braille writing, substituting visually read letters by a raised dot code. However, Braille can only work for materials transformed offline from printed visual letters to Braille dots and cannot be used for online reading of regular letters. Recently other reading substitutions have been developed for online reading, e.g., Optacon (print-to-tactual-image device devised for reading embossed letters (Goldish and Taylor, 1974) or various versions of dedicated text-to-speech engines (from the Kurzweil reading machine (Kleiner and Kurzweil, 1977) to current speech software). In addition to these reading aids, a great deal of effort has been invested in developing devices aimed at improving the mobility of the blind. The long cane used to mechanically probe for obstacles represents the simplest, most commonly used device. In recent years its more advanced counterparts, e.g., electronic travel aids designed to be used along with the long cane which extend the distance for environmental preview and thus increase the speed and efficiency of travel provided by the cane, have become available. Such devices, such as the Sonic Pathfinder (Heyes, 1984) or Sonicguide (Kay and Kay, 1983), typically scan the environment acoustically (ultrasonically) or optically (laser light) and transmit spatial information regarding obstacles and objects in the surroundings via vibrotactile or auditory signals.

In contrast to these devices which are typically designed for a limited purpose and successful in substituting for only certain functional aspects of vision, more sophisticated techniques that replace vision through tactile or auditory information have been developed over the last few decades (see Fig. 18.2). The first targeted modality for substituting vision was touch, due to the simplicity and ease of transforming visual into tactile signals which are both characterized by 2-D spatial representations. Pioneering work in this field was done in the 1970s by Paul Bach-y-Rita who devised a tactile display which mapped images from a video camera to a vibrotactile device worn on the subject's back. This device (Bach-y-Rita, 2004; Bach-y-Rita and Kercel, 2003; Bach-Y-Rita et al., 1969), dubbed the tactile vision substitution system (TVSS), provided tactile transformation of black-and-white images at a resolution of 20×20 pixels, and enabled the blind to perform sufficiently well in some visual tasks. However, it was extremely large and non-mobile which motivated the development of smaller mobile tactile devices placed on the tongue and forehead (for a review, see Bach-y-Rita, 2004) that are characterized by better spatial somatosensory resolution. One of these, the Tongue display unit (TDU; Bach-y-Rita et al., 1998), an electrotactile device comprised of a 12×12 matrix of stimulators (measuring approximately 3 cm^2) placed on the subject's tongue, provides blind individuals an initial "visual" acuity (tested by the Snellen E chart) comparable to 20/860, which significantly improves following training. Other studies investigating this device suggest that at least a subgroup of early-onset blind may particularly benefit from its use (Chebat et al., 2007).

Audition emerged as the second candidate to substitute for vision. The development of auditory-based devices was triggered by certain limitations of tactile SSDs, namely their price and the fact that they are inherently limited by the spatial resolution of touch. The first auditory SSD device was The vOICe system (Meijer, 1992),

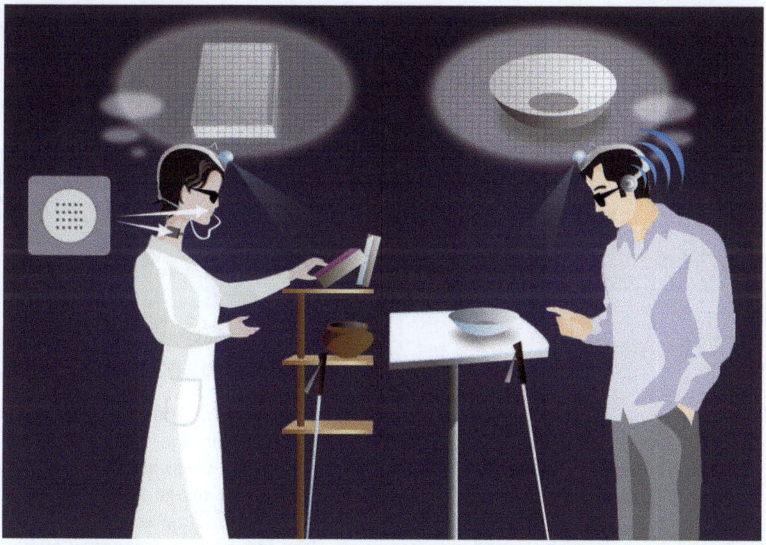

Fig. 18.2 The general concept of sensory substitution (SSD) and a typical visual-to-auditory and visual-to-tactile setup. SSDs typically include a visual capturing device (for example camera glasses), a computational device transforming the visual input into either a tactile or an auditory display using a simple known transformation algorithm, and an output device, transmitting this information to the user. The right side illustrates an example of an auditory SSD (such as The vOICe; Meijer, 1992) transmitting the sensory-transformed information using headphones, and on the left side is an example of a tactile device which may transmit the tactile information via an electrode array targeting the tongue (such as the TDU, Bach-y-Rita et al., 1998) or another skin surface, in this case presented on the neck (copyright of image, by Amir Amedi)

which currently uses a default resolution of 176×64 sampling points. This device is mobile and inexpensive, and it consists of a video camera which provides the visual input, a small computer running the conversion program, and stereo headphones that provide the resulting sound patterns to the user. Given the fact that 87% of the world's visually impaired live in developing countries (WHO, 2009, fact sheet 282), the importance of providing not just high-resolution, but also cheap and accessible solutions for these individuals cannot be underestimated. Visual-to-auditory SSDs fulfill all of these criteria to some extent. On the other hand, these devices possess great challenges both to developers and to the brains of blind individuals trying to use them because the conversion algorithms of these devices are much less intuitive when compared to visual-to-tactile SSDs. For example, in The vOICe, the visual-to-auditory SSD (see Fig. 18.3), the conversion program transforms the visual into auditory information ("soundscapes") based on three simple rules: the vertical axis (i.e., elevation of the object) is represented by frequency, horizontal axis by time and stereo panning, while brightness of the image is encoded by loudness. Although these conversion rules appear relatively simple, it is not trivial to understand even simple shapes without explicit and quite extensive training. Similar but not identical

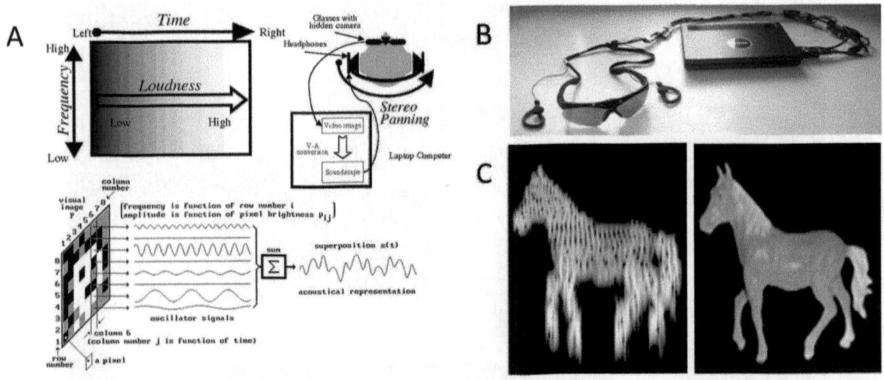

Fig. 18.3 An example of a visual-to-auditory SSD (The vOICe algorithm). (**a**) Schematic summary of the algorithm employed for the visuo-auditory conversion and the components of the system (adapted from Meijer, 1992). The functional basis of the visuo-auditory transformation employed lies in spectrographic sound synthesis from any input image, which is then further perceptually enhanced through stereo panning and other techniques. Time and stereo panning constitute the horizontal axis in the sound representation of an image, tone frequency makes up the vertical axis, and loudness corresponds to pixel brightness. Visual information in the sound representations of complicated gray-scale images is preserved up to a resolution of about 60×60 pixels for a 1 s sound scan and a 5 kHz audio bandwidth. Further technical details can be obtained at http://www.seeingwithsound.com/ where it is also possible to demo the algorithm and get a sense of the generated soundscapes (adapted from Amedi et al., 2007; original algorithm presented in Meijer, 1992). (**b**) An applied setup of The vOICe SSD includes camera glasses, a netbook, and earphones (image courtesy of Peter Meijer). (**c**) Example spectrogram of a one-second sound generated by The vOICe. On the right is the original picture transformed by The vOICe, on the left is the spectrogram of the sound generated from the picture (modified and adapted from Amedi et al., 2007, copyright of Peter Meijer, with permission)

transformations are implemented in two more recently developed auditory SSDs, the Prosthesis Substituting Vision with Audition (PSVA; Capelle et al., 1998), and SmartSight (Cronly-Dillon et al., 1999, 2000). PSVA uses different tones to provide horizontal location directly, whereas SmartSight presents the vertical location information in terms of musical notes. PSVA can break down the "visual sound" into components of vertically and horizontally oriented edges. Additionally, PSVA applies a magnification to the center of the image to simulate the better resolution (magnification factor) of the human fovea.

Although differing to a higher degree, both auditory and tactile SSDs can potentially be very useful for the blind. Recent tests show that blind and/or blindfolded sighted individuals can, especially after training or prolonged use of the device (Poirier et al., 2006), learn to interpret the transmitted information and use it in simple visual discrimination and recognition (Arno et al., 1999, 2001; Sampaio et al., 2001) as well as more complex tasks in which they need to acquire knowledge of spatial locations of objects (Auvray et al., 2007; Proulx et al., 2008) or construct mental images of more complex environments (Cronly-Dillon et al., 2000).

While recent studies demonstrate the behavioral range of SSDs, they also focus on the way blind and sighted brains process such sensory-transformed information. Observing the outcomes of sensory restoration and substitution is not only practically relevant but also offers a unique opportunity for addressing numerous theoretical questions about perception, the nature of qualia and brain plasticity. Functional properties of multisensory regions can easily be studied using SSD as a methodology of choice, as SSDs are naturally processed in a multimodal or transmodal manner. For example, several recent studies used SSDs to test the metamodal processing theory (Pascual-Leone and Hamilton, 2001), which states that each brain region implements a particular type of computation regardless of the modality of its input. These studies showed that object shape information drives the activation of the lateral–occipital complex in the visual cortex, regardless whether it is transmitted in the visual, tactile, or auditory modality (Amedi et al., 2007; Poirier et al., 2006) in sighted as well as blind individuals (see Fig. 18.4). Interestingly, applying TMS to this region can disrupt shape identification using an auditory SSD (Merabet et al., 2008). In the same way, studies conducted using PSVA in the sighted show that auditorily mediated face perception can activate the visual fusiform face area (Plaza et al., 2009), while depth perception activates the occipito-parietal and occipito-temporal regions (Renier et al., 2005). The early visual cortex may to some extent be a multimodal operator for

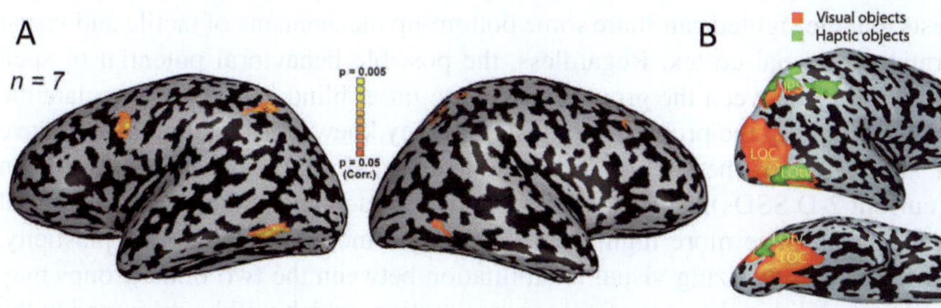

Fig. 18.4 SSDs can be a powerful tool in studying brain plasticity, perception, and multisensory integration. (**a**) A conjunction analysis for shape perception across modalities and experimental conditions in a group of seven expert users of The vOICe SSD (five sighted, one late blind, and one congenitally blind). A conjunction analysis testing for common areas of activation between object recognition using the soundscapes (i.e., using the vOICe SSD to extract the shape information) and touch, but not by typical sounds made by objects (which do not convey shape information) or corresponding sensory controls. The contrast random effect GLM model, corrected for multiple comparisons, showed bilateral LO activation with weaker responses in the right hemisphere, signifying that the LOC region is a multimodal operator for shape (modified and adapted from Amedi et al., 2007). (**b**) Object-related regions in the visual and haptic modalities shown on an inflated right hemisphere (*top*: lateral view; *bottom*: ventral view). Visual object selectivity is relative to scrambled visual images; haptic object selectivity is relative to haptic textures. Visuo-haptic object selectivity in the lateral–occipital complex (LOC) is found within the lateral occipito-temporal sulcus (delineating LOtv), similar to the location of the multisensory object-related area shown in A (modified and adapted from Amedi et al., 2001; and from Lacey et al., 2009)

sensory spatial mapping, as stimulating it in highly trained blind TDU users appears to result in spatially organized tactile sensations on the tongue (Kupers et al., 2006).

Studying the use of SSDs in a longitudinal fashion also provides a good opportunity to monitor in real time how newly acquired information is learned and investigate the accompanying cognitive and neural changes. For example, several studies have looked into differential activation before and after learning how to use a specific SSD. One study showed that after short training, and only in the early-onset blind individuals (but not in sighted), shape discrimination using the TDU SSD generated activation of the occipital cortex (Ptito et al., 2005), suggesting that the occipital lobe of the blind may, even in adulthood, be more prone to plasticity or to crossmodal processing when compared to that of the sighted. Crossmodal activation of the visual cortex of sighted subjects was also demonstrated, following training on the PSVA SSD (Poirier et al., 2007). Although these behavioral and imaging finding benefits have been shown for both early-onset (Arno et al., 2001) and late-onset blind (Cronly-Dillon et al., 1999) and sighted individuals (Poirier et al., 2007), it has recently been suggested that the recruitment of occipital areas during the use of SSDs could be mediated by different mechanisms in different populations. Specifically, while the early blind might exhibit real bottom-up activation of occipital cortex for tactile or auditory perception, in the late blind and sighted this activation might reflect top-down visual imagery mechanisms (Poirier et al., 2007). However, recent evidence of multisensory integration for object recognition, as shown by using a novel crossmodal adaptation paradigm (Tal and Amedi, 2009), suggests that the sighted can share some bottom-up mechanisms of tactile and visual integration in visual cortex. Regardless, the possible behavioral potential of such SSDs may vary between the groups, as the late-onset blind can better associate the crossmodal input to the properties of vision as they knew it (for example, they have better knowledge of the 2-D representation of visual pictures, which is useful in most current 2-D SSDs), while the early blind lack such understanding of the visual world, but may have more highly developed crossmodal networks and plasticity. This difference in utilizing visual rehabilitation between the two blind groups may be even more valid in the case of sensory restoration, which will be discussed in the next section.

Importantly, this differentiation between early- and late-onset blind in SSD use also highlights the importance of introducing such devices as early as possible in development, while the brain is still in its plasticity prime. Similar to the improved outcomes of cochlear implantation in early childhood (Harrison et al., 2005), it may be of particular interest to teach young blind children to utilize such devices. Several early attempts to teach blind infants to use the Sonicguide (Kay and Kay, 1983) showed some promise, as younger subjects showed a more rapid sensitivity to the spatial information provided by the device (Aitken and Bower, 1982, 1983), although with highly variable results (for a discussion see, Warren, 1994). However, to our knowledge, only a few preliminary later attempts (Segond et al., 2007) have been made to adapt the use of SSDs to children.

Overall, although there is still a lot of work to be done in this area, initial experiences with SSDs have shown more than promising results. These devices truly offer new hope for the sensorily impaired in a somewhat non-intuitive, but "brain-friendly" manner, as they use its normal resources and functioning modalities for transmitting previously unavailable information. In order to fully appreciate their value, it might be useful to imagine how exciting it would be (not to mention valuable to many commercial and military applications) to have infrared vision or hear ultrasound frequencies. Interestingly, using one of the future, second-generation SSDs this might just become possible: just like visual information can be transmitted and used by the blind through their functioning auditory or tactile modality, so could infrared or ultrasound frequencies be seen or heard by anyone using functioning vision or audition, making such "super-human" abilities more widely available. Furthermore, observing the outcomes of sensory restoration and substitution is of more than just practical relevance, as it also offers a unique opportunity to address and potentially answer numerous theoretical questions about the fundamental principles of brain organization, neuroplasticity, unimodal sensory processing, and multisensory integration.

18.5.2 Sensory Restoration Approaches

Restoration of sensory input to the visually impaired represents an alternative to sensory substitution devices. Such devices are currently at the forefront of medical, technological, and scientific advances. Conventional sight restoration includes surgical removal of cataracts and treatment of or surgical solutions to vision loss caused by glaucoma. Although highly practical and demonstrating remarkable results (Ostrovsky et al., 2006; Sinha, 2003), these solutions were originally only useful for specific causes and stages of vision loss. Sight restoration in blindness due to other etiologies, such as congenital or late-onset degeneration (for example age-related macular degeneration) of the retina or optic tract, is only nowadays being addressed. The development of such visual prostheses was motivated by early studies in which visual percepts (phosphenes, visual light dots or patterns) were successfully generated by electrical stimulation of the visual cortex (Penfield and Rasmussen, 1950). Although few researchers at the time thought that actual clinical techniques could be developed (Haddock and Berlin, 1950; Newman et al., 1987), in recent years the field of visual prostheses has developed extensively. Today, different approaches are under investigation or are being tested in clinical trials in which visual information is recorded by external (or implanted) devices and transmitted to the sensory tract or secondary processing cells in the retina, ganglion cells, thalamus, or the visual cortex, thereby replacing the healthy receptors of the sensory organs (for several recent reviews of current technologies and remaining challenges see, Dagnelie, 2008; Dowling, 2008; Merabet et al., 2005; Rizzo et al., 2007; Weiland et al., 2005).

There are four main types of approaches in sensory restoration, targeting the retina, optic nerve, lateral geniculate nucleus (LGN), and the visual cortex. The retinal approach is designed to stimulate secondary neurons in the inner retina by an electrode array placed on the inner retinal surface or inserted under the retina (for a description of the different groups and devices developed in recent years see, Dowling, 2008). Such an approach is mainly useful in cases of retinitis pigmentosa and age-related macular degeneration (ARMD) which cause selective degeneration of the photoreceptor layer of the outer retina. In this case the information sent to the visual cortex can still be transmitted over minimally damaged retinal ganglion cells. Optic nerve approaches (Brelen et al., 2005; Veraart et al., 2003) use two forms of stimulation, namely the simultaneous activation of many optic nerve fibers through cuff electrodes and more focused stimulation of small groups of fibers with penetrating microelectrodes. The thalamic prostheses (Pezaris and Reid, 2005, 2009) attempt to stimulate a later station in the visual pathways, i.e., the LGN, but this is currently only under preliminary methodological research in primates. The cortical approach (Fernandez et al., 2002) places electrodes over the central visual field projection in primary visual cortex. Typically this is accomplished using surface (or penetrating) electrodes that may provide relatively good stability of tissue stimulation, but are hard to position in the optimal location based on the known retinotopic mapping of V1. However, this approach can be applied in most cases of blindness, including those (such as glaucoma and diabetic retinopathy, but apart from cortical blindness) which affect the retina and may not benefit from a retinal prosthesis. Devices based on these approaches have so far shown some promising results, as experienced blind users can, to some extent, utilize visual phosphenes generated by some of these devices in order to create meaningful visual percepts and succeed in detecting motion (Weiland and Humayun, 2008) or identifying very simple patterns, shapes, and even letters (Brelen et al., 2005; Dobelle, 2000; Weiland and Humayun, 2008). However, there are still several major issues currently preventing these approaches from becoming true clinical solutions. First of all, their invasive nature makes them prone to risks related to surgical procedures in the brain, such as inflammation, hemorrhage, increased patient morbidity and mortality, and focal seizures induced by direct cortical stimulation. Moreover, retinal prostheses, which currently appear more promising as future solutions for blindness, are not applicable to all populations of the blind, as they require the existence of residual functional retinal ganglion cells. Additionally, these techniques are expensive and have severe technical limitations, such as the relatively low resolution, narrow field of view, and complicated image processing algorithms which compensate for the visual processing taking place in the retina itself. Functionally, these devices typically do not take advantage of eye movements (an exception to this is the system developed by Palanker et al. (2005)) and require large and slow head movements to scan entire visual patterns (Brelen et al., 2005; Chen et al., 2007; Veraart et al., 2003). Therefore, visual prostheses (which are currently not yet available except in preliminary clinical trials) do not provide sight that resembles natural vision and a key milestone in this field has yet to be reached in order to allow generating truly useful and functional vision, at affordable costs. If, however, visual prostheses research succeeds to overcome these difficulties, these approaches could provide a real visual experience and not just the "visual" information provided by SSDs.

As previously mentioned, attempts aimed at restoring audition in the deaf via an artificial receptor, i.e., artificial cochlea (Spelman, 2006), are currently more widely available and show much better rehabilitation outcomes. However, even when surgical restoration procedures in vision or audition become available for commercial use, only the first step in establishing full perceptual abilities in the newly restored modality will be accomplished. Learning how to interpret the new sensory input delivered to the patient's brain will require further behavioral rehabilitation which may allow gradual restoration of the sensory function in question. For example, cochlear implants work because patients learn to associate meanings between sounds and their sources by generating new associations. This is accompanied and enabled by corresponding plasticity in the auditory cortex (Lee et al., 2001). Two case studies of surgical sight restoration after long-term visual deprivation (Fine et al., 2003; Gregory and Wallace, 1963) suggest that in the case of vision as well, restoration of the lost sensory input may not suffice. In both cases the patients showed profound difficulties in recognizing objects, even after a long period of sight and visual training. Thus, visual perception in this case is impaired, even when vision reaches the brain physiologically via a functional retina. In addition to the surgical procedure, specific additional rehabilitation strategies are needed to modulate brain processing and make it capable of extracting relevant and functionally meaningful information from neuroprosthetic inputs. Furthermore, although the behavioral outcomes of cochlear implant patients are encouraging, it is illusory to expect that they can easily generalize to different subpopulations of sensorily impaired. On the contrary, great interindividual variability is to be expected in adapting to the implant (e.g., in cochlear implant patients, speech recognition performance ranges from very poor to near perfect). This is particularly true with regard to the variability in blindness onset, as was discussed for rehabilitation using SSDs. Late-onset blind may particularly benefit from reacquiring visual information, as their visual cortex has developed in a way which would allow it to process such information. Therefore, following sensory loss their visual cortex needs to reorganize to a smaller degree in contrast to the early-onset or congenitally blind who would require greater reorganization and therefore face more difficulties with adapting to visual information. Furthermore, sensory implantation (and, although in a different manner, SSDs) is prone to influence the brain as a system, not just one modality. For example, it was shown that visual information can disrupt the processing of auditory information in newly transplanted cochlear implant patients, most probably due to crossmodal visual processing in the auditory cortex (Champoux et al., 2009). Such interference may occur following visual restoration in all tasks which are functionally dependent on the occipital lobe of the blind (particularly tasks which can be disrupted by occipital TMS, as described in previous sections). At least some of these problems could be minimized, if not solved, through extensive cognitive and behavioral training. In addition, it is possible to speculate that the integration of SSDs and prostheses may also be helpful in this respect, as these provide fairly distinct advantages. Specifically, while prostheses may allow entry of visual signals to the brain, training how to process and interpret this information could be facilitated through some initial training with SSDs which can be very useful in teaching the brain how to interpret the input from the new modality. It is possible to assume that the brains of individuals who have undergone training with SSDs are already sensitive to the structure and able to extract the meaning out of visual signals which could be very beneficial when they first encounter the visual signals delivered through the surgically restored sensory system.

18.6 Concluding Remarks

The main goal of this chapter was to illustrate the effects of sensory loss, including changes in cognitive functioning as well as the radical reorganization of the neural architecture supporting these functions, which varies greatly with regard to the timing of sensory loss in development. Many, but not all, of these changes are beneficial and allow the impaired individual to optimally use the available resources to efficiently function in his/her surroundings. Sensory loss triggers robust remodeling of the whole brain which occurs on all spatial and temporal scales. These different types of changes are supported by different neurophysiological mechanisms which are not equally available at different stages of development. Understanding each of these changes, the factors which influence them, as well as how they relate to each other represents a major challenge for future research. Accomplishing this goal is of great practical importance for hundreds of millions of sensorily impaired (e.g., 314 million of individuals are visually impaired worldwide; WHO report 2009, fact sheet 282), as it will allow further development of rehabilitation techniques. This will, however, be a difficult challenge, as we still need to learn a great deal about how to let the brain adapt following sensory restoration using the same mechanisms which allowed its original adaptation to the lost sensory input. In many ways plasticity can be viewed as a "double-edged sword" which occasionally leads to functionally beneficial, and occasionally to maladaptive outcomes. Mastering such phenomena will not be easy. In fact, it may never be possible if interindividual and intergroup differences characterized by different initial cognitive and neural systems as well as a general potential for change are ignored. Even when the most technological issues have been resolved, introducing visual inputs into the brain via a functional prosthetic retina will not automatically result in normal visual perception. This skill will need to be trained and such learning will be greatly facilitated by the use of inputs from other modalities and multisensory regions which may serve as "teaching signals" for the impaired modality. Observing the outcomes of sensory restoration and substitution is thus not just practically relevant but also offers a unique opportunity to address and potentially answer numerous fundamental theoretical questions about perception, multisensory integration, as well as discover the principles underlying brain reorganization and neuroplasticity.

References

Abraham WC, Bear MF (1996) Metaplasticity: the plasticity of synaptic plasticity. Trends Neurosci 19(4):126–130

Aitken S, Bower TG (1982) Intersensory substitution in the blind. J Exp Child Psychol 33(2): 309–323

Aitken S, Bower TG (1983) Developmental aspects of sensory substitution. Int J Neurosci 19(1–4):13–19

Amedi A, Camprodon J, Merabet L, Bermpohl F, Haligan E, Bass-Pitskel N et al (2006) Highly transient activation of primary visual cortex (V1) for tactile object recognition in sighted following 5 days of blindfolding. Paper presented at 7th annual meeting of the international multisensory research forum, Universtify of Dublin

Amedi A, Floel A, Knecht S, Zohary E, Cohen LG (2004) Transcranial magnetic stimulation of the occipital pole interferes with verbal processing in blind subjects. Nat Neurosci 7(11): 1266–1270

Amedi A, Jacobson G, Hendler T, Malach R, Zohary E (2002) Convergence of visual and tactile shape processing in the human lateral occipital complex. Cereb Cortex 12(11):1202–1212

Amedi A, Malach R, Hendler T, Peled S, Zohary E (2001) Visuo-haptic object-related activation in the ventral visual pathway. Nat Neurosci 4(3):324–330

Amedi A, Merabet LB, Bermpohl F, Pascual-Leone A (2005) The occipital cortex in the blind. Lessons about plasticity and vision. Curr Dir Psychol Sci 14(6):306–311

Amedi A, Raz N, Pianka P, Malach R, Zohary E (2003) Early 'visual' cortex activation correlates with superior verbal memory performance in the blind. Nat Neurosci 6(7):758–766

Amedi A, Stern WM, Camprodon JA, Bermpohl F, Merabet L, Rotman S et al (2007) Shape conveyed by visual-to-auditory sensory substitution activates the lateral occipital complex. Nat Neurosci 10(6):687–689

Arno P, Capelle C, Wanet-Defalque MC, Catalan-Ahumada M, Veraart C (1999) Auditory coding of visual patterns for the blind. Perception 28(8):1013–1029

Arno P, De Volder AG, Vanlierde A, Wanet-Defalque MC, Streel E, Robert A et al (2001) Occipital activation by pattern recognition in the early blind using auditory substitution for vision. Neuroimage 13(4):632–645

Auvray M, Hanneton S, O'Regan JK (2007) Learning to perceive with a visuo-auditory substitution system: localisation and object recognition with 'The vOICe'. Perception 36:416–430

Bach-y-Rita P (2004) Tactile sensory substitution studies. Ann N Y Acad Sci 1013:83–91

Bach-Y-Rita P, Collins CC, Saunders FA, White B, Scadden L (1969) vision substitution by tactile image projection. Nature 221(5184):963–964

Bach-y-Rita P, Kaczmarek KA, Tyler ME, Garcia-Lara J (1998) Form perception with a 49-point electrotactile stimulus array on the tongue: a technical note. J Rehabil Res Dev 35(4):427–430

Bach-y-Rita P, Kercel SW (2003) Sensory substitution and the human-machine interface. Trends Cogn Sci 7(12):541–546

Bavelier D, Dye MW, Hauser PC (2006) Do deaf individuals see better? Trends Cogn Sci 10(11):512–518

Bavelier D, Neville HJ (2002) Cross-modal plasticity: where and how? Nat Rev Neurosci 3(6): 443–452

Brelen ME, Duret F, Gerard B, Delbeke J, Veraart C (2005) Creating a meaningful visual perception in blind volunteers by optic nerve stimulation. J Neural Eng 2(1):S22–S28

Brown JA (2006) Recovery of motor function after stroke. Prog Brain Res 157:223–228

Büchel C (2003) Cortical hierarchy turned on its head. Nat Neurosci 6(7):657–658

Büchel C, Price C, Frackowiak RS, Friston K (1998) Different activation patterns in the visual cortex of late and congenitally blind subjects. Brain 121(Pt 3):409–419

Buonomano DV, Johnson HA (2009) Cortical plasticity and learning: mechanisms and models. In: Squire LR (ed) Encyclopedia of neuroscience. Academic Press, London

Burton H (2003) Visual cortex activity in early and late blind people. J Neurosci 23(10):4005–4011

Burton H, Diamond JB, McDermott KB (2003) Dissociating cortical regions activated by semantic and phonological tasks: a FMRI study in blind and sighted people. J Neurophysiol 90(3):1965–1982. Epub 2003 Jun 1964

Burton H, Snyder AZ, Conturo TE, Akbudak E, Ollinger JM, Raichle ME (2002) Adaptive changes in early and late blind: a fMRI study of Braille reading. J Neurophysiol 87(1):589–607

Burton H, Snyder AZ, Diamond JB, Raichle ME (2002) Adaptive changes in early and late blind: a FMRI study of verb generation to heard nouns. J Neurophysiol 88(6):3359–3371

Campbell R, MacSweeney M (2004) Neuroimaging studies of cross-modal plasticity and language processing in deaf people. In: Calvert G, Spence C, Stein BE (eds) The handbook of multisensory processes). MIT Press, Cambridge, MA, pp. 773–785

Capelle C, Trullemans C, Arno P, Veraart C (1998) A real-time experimental prototype for enhancement of vision rehabilitation using auditory substitution. IEEE Trans Biomed Eng 45(10):1279–1293

Chabot N, Robert S, Tremblay R, Miceli D, Boire D, Bronchti G (2007) Audition differently activates the visual system in neonatally enucleated mice compared with anophthalmic mutants. Eur J Neurosci 26(8):2334–2348

Champoux F, Lepore F, Gagne JP, Theoret H (2009) Visual stimuli can impair auditory processing in cochlear implant users. Neuropsychologia 47(1):17–22

Chebat DR, Rainville C, Kupers R, Ptito M (2007) Tactile-'visual' acuity of the tongue in early blind individuals. Neuroreport 18(18):1901–1904

Chechik G, Meilijson I, Ruppin E (1998) Neuronal regulation: a mechanism for efficient synaptic pruning during brain maturation. Neuroscience Letters, S9-S9

Chen R, Cohen LG, Hallett M (2002) Nervous system reorganization following injury. Neuroscience 111(4):761–773

Chen SC, Hallum LE, Suaning GJ, Lovell NH (2007) A quantitative analysis of head movement behaviour during visual acuity assessment under prosthetic vision simulation. J Neural Eng 4(1):S108

Cohen LG, Celnik P, Pascual-Leone A, Corwell B, Falz L, Dambrosia J et al (1997) Functional relevance of cross-modal plasticity in blind humans. Nature 389(6647):180–183

Cohen LG, Weeks RA, Sadato N, Celnik P, Ishii K, Hallett M (1999) Period of susceptibility for cross-modal plasticity in the blind. Ann Neurol 45(4):451–460

Collignon, O, Lassonde M, Lepore F, Bastien D, Veraart C (2007) Functional cerebral reorganization for auditory spatial processing and auditory substitution of vision in early blind subjects. Cereb Cortex 17(2):457–465

Collignon O, Renier L, Bruyer R, Tranduy D, Veraart C (2006) Improved selective and divided spatial attention in early blind subjects. Brain Res 1075(1):175–182

Collignon O, Voss P, Lassonde M, Lepore F (2009) Cross-modal plasticity for the spatial processing of sounds in visually deprived subjects. Exp Brain Res 192(3):343–358

Cronly-Dillon J, Persaud K, Gregory RP (1999) The perception of visual images encoded in musical form: a study in cross-modality information transfer. Proc Biol Sci 266(1436):2427–2433

Cronly-Dillon J, Persaud KC, Blore R (2000) Blind subjects construct conscious mental images of visual scenes encoded in musical form. Proc Biol Sci 267(1458):2231–2238

D'Angiulli A, Waraich P (2002) Enhanced tactile encoding and memory recognition in congenital blindness. Int J Rehabil Res 25(2):143–145

Dagnelie G (2008) Psychophysical evaluation for visual prosthesis. Annu Rev Biomed Eng 10:339–368

De Volder AG, Bol A, Blin J, Robert A, Arno P, Grandin C et al (1997) Brain energy metabolism in early blind subjects: neural activity in the visual cortex. Brain Res 750(1–2):235–244

Dinse HR, Merzenich MM (2000) Adaptation of inputs in the somatosensory system. In: Fahle M, Poggio T (eds) Perceptual learning. MIT Press, Cambridge, MA, pp 19–42

Dobelle WH (2000) Artificial vision for the blind by connecting a television camera to the visual cortex. ASAIO J 46(1):3–9

Doucet ME, Guillemot JP, Lassonde M, Gagne JP, Leclerc C, Lepore F (2005) Blind subjects process auditory spectral cues more efficiently than sighted individuals. Exp Brain Res 160(2):194–202
Dowling, J (2008) Current and future prospects for optoelectronic retinal prostheses. Nature-Eye 23:1999–2005
Eriksson PS, Perfilieva E, Bjork-Eriksson T, Alborn AM, Nordborg C, Peterson DA et al (1998) Neurogenesis in the adult human hippocampus. Nat Med 4(11):1313–1317
Falchier A, Clavagnier S, Barone P, Kennedy H (2002) Anatomical evidence of multimodal integration in primate striate cortex. J Neurosci 22(13):5749–5759
Fernandez E, Ahnelt P, Rabischong P, Botella C, & Garcia-de Quiros F (2002) Towards a cortical visual neuroprosthesis for the blind. IFMBE Proc 3(2):1690–1691
Fieger A, Röder B, Teder-Salejarvi W, Hillyard SA, Neville HJ (2006) Auditory spatial tuning in late-onset blindness in humans. J Cogn Neurosci 18(2):149–157
Fine, I (2008) The behavioral and neurophysiological effects of sensory deprivation. In: Rieser JJ, Ashmead DH, Ebner FF, Corn AL (eds) Blindness and brain plasticity in navigation and object perception. Taylor & Francis, New York, pp. 127–155
Fine I, Wade AR, Brewer AA, May MG, Goodman DF, Boynton GM et al (2003) Long-term deprivation affects visual perception and cortex. Nat Neurosci 6(9):915–916
Finney EM, Fine I, Dobkins KR (2001) Visual stimuli activate auditory cortex in the deaf. Nat Neurosci 4(12):1171–1173
Gizewski ER, Gasser T, de Greiff A, Boehm A, Forsting M (2003) Cross-modal plasticity for sensory and motor activation patterns in blind subjects. Neuroimage 19(3):968–975
Golarai G, Ghahremani DG, Whitfield-Gabrieli S, Reiss A, Eberhardt JL, Gabrieli JD et al. (2007) Differential development of high-level visual cortex correlates with category-specific recognition memory. Nat Neurosci 10(4):512–522
Goldish LH, Taylor HE (1974) The optacon: a valuable device for blind persons. New Outlook Blind 68(2):49–56
Goldreich D, Kanics IM (2003) Tactile acuity is enhanced in blindness. J Neurosci 23(8): 3439–3445
Goldreich D, Kanics IM (2006) Performance of blind and sighted humans on a tactile grating detection task. Percept Psychophys 68(8):1363–1371
Gougoux F, Zatorre RJ, Lassonde M, Voss P, Lepore F (2005) A functional neuroimaging study of sound localization: visual cortex activity predicts performance in early-blind individuals. PLoS Biol 3(2):e27
Grafman J (2000) Conceptualizing functional neuroplasticity. J Commun Disord 33(4):345–355; quiz 355–346
Grant AC, Thiagarajah MC, Sathian K (2000) Tactile perception in blind Braille readers: a psychophysical study of acuity and hyper acuity using gratings and dot patterns. Percept Psychophys 62(2):301–312
Gregory RL, Wallace JG (1963) Recovery from early blindness: a case study. Reproduced in 2001 from Exp Psychol Soc Mon. 2. Heffers, Cambridge
Haddock JN, Berlin L (1950) Transsynaptic degeneration in the visual system; report of a case. Arch Neurol Psychiat 64(1):66–73
Harrison RV, Gordon KA, Mount RJ (2005) Is there a critical period for cochlear implantation in congenitally deaf children? Analyses of hearing and speech perception performance after implantation. Dev Psychobiol 46(3):252–261
Heyes AD (1984) The sonic pathfinder: a new electronic travel aid. J Vis Impair Blindness 78(5):200–202
Hugdahl K, Ek M, Takio F, Rintee T, Tuomainen J, Haarala C et al. (2004) Blind individuals show enhanced perceptual and attentional sensitivity for identification of speech sounds. Brain Res Cogn Brain Res 19(1):28–32
Hull T, Mason H (1995) Performance of blind-children on digit-span tests. J Vis Impair Blindness 89(2):166–169

Izraeli R, Koay G, Lamish M, Heicklen-Klein AJ, Heffner HE, Heffner RS et al. (2002) Cross-modal neuroplasticity in neonatally enucleated hamsters: structure, electrophysiology and behaviour. Eur J Neurosci 15(4):693–712
Kaas JH (1991) Plasticity of sensory and motor maps in adult mammals. Annu Rev Neurosci 14 137–167
Kaas JH (2000) The reorganization of somatosensory and motor cortex after peripheral nerve or spinal cord injury in primates. Prog Brain Res 128:173–179
Karlen SJ, Kahn DM, Krubitzer L (2006) Early blindness results in abnormal corticocortical and thalamocortical connections. Neuroscience 142(3):843–858
Kay L, Kay N (1983) An ultrasonic spatial sensor's role as a developmental aid for blind children. Trans Ophthalmol Soc N Z 35:38–42
Kleiner A, Kurzweil RC (1977) A description of the Kurzweil reading machine and a status report on its testing and dissemination. Bull Prosthet Res 10(27):72–81
Knudsen EI (2004) Sensitive periods in the development of the brain and behavior. J Cogn Neurosci 16(8):1412–1425
Kolb B (1995) Brain plasticity and behavior. Lawrence Erlbaum Associates, Inc., Mahwah, NJ
Korte M, Rauschecker JP (1993) Auditory spatial tuning of cortical neurons is sharpened in cats with early blindness. J Neurophysiol 70(4):1717–1721
Kujala T, Alho K, Paavilainen P, Summala H, Naatanen R (1992) Neural plasticity in processing of sound location by the early blind: an event-related potential study. Electroencephalogr Clin Neurophysiol 84(5):469–472
Kujala T, Palva MJ, Salonen O, Alku P, Huotilainen M, Jarvinen A et al. (2005) The role of blind humans' visual cortex in auditory change detection. Neurosci Lett 379(2):127–131
Kupers R, Fumal A, de Noordhout AM, Gjedde A, Schoenen J, Ptito M (2006) Transcranial magnetic stimulation of the visual cortex induces somatotopically organized qualia in blind subjects. Proc Natl Acad Sci U S A 103(35):13256–13260
Lacey S, Tal N, Amedi A, Sathian K (2009) A putative model of multisensory object representation. Brain Topogr 21(3–4):269–274
Laemle LK, Strominger NL, Carpenter DO (2006) Cross-modal innervation of primary visual cortex by auditory fibers in congenitally anophthalmic mice. Neurosci Lett 396(2):108–112
Leclerc C, Saint-Amour D, Lavoie ME, Lassonde M, Lepore F (2000) Brain functional reorganization in early blind humans revealed by auditory event-related potentials. Neuroreport 11(3):545–550
Lee DS, Lee JS, Oh SH, Kim SK, Kim JW, Chung JK et al. (2001) Cross-modal plasticity and cochlear implants. Nature 409(6817):149–150
Lessard N, Pare M, Lepore F, Lassonde M (1998) Early-blind human subjects localize sound sources better than sighted subjects. Nature 395(6699):278–280
Lewald J (2002) Vertical sound localization in blind humans. Neuropsychologia 40(12):1868–1872
Lickliter R, Bahrick LE (2004) Perceptual development and the origins of multisensory responsiveness. In: Calvert G, Spence C, Stein BE (eds) The handbook of multisensory processes. MIT Press, Cambridge, MA, pp 643–654
Liu Y, Yu C, Liang M, Li J, Tian L, Zhou Y et al. (2007) Whole brain functional connectivity in the early blind. Brain 130(Pt 8):2085–2096
Malenka RC, Bear MF (2004) LTP and LTD: an embarrassment of riches. Neuron 44(1):5–21
Meijer PB (1992) An experimental system for auditory image representations. IEEE Trans Biomed Eng 39(2):112–121
Merabet L, Thut G, Murray B, Andrews J, Hsiao S, Pascual-Leone A (2004) Feeling by sight or seeing by touch? Neuron 42(1):173–179
Merabet LB, Battelli L, Obretenova S, Maguire S, Meijer P, Pascual-Leone A (2009) Functional recruitment of visual cortex for sound encoded object identification in the blind. Neuroreport 20(2):132–138
Merabet LB, Hamilton R, Schlaug G, Swisher JD, Kiriakopoulos ET, Pitskel NB et al. (2008) Rapid and reversible recruitment of early visual cortex for touch. PLoS ONE 3(8):e3046

Merabet LB, Maguire D, Warde A, Alterescu K, Stickgold R, Pascual-Leone A (2004) Visual hallucinations during prolonged blindfolding in sighted subjects. J Neuroophthalmol 24(2):109–113
Merabet LB, Pitskel NB, Amedi A, Pascual-Leone A (2008) The plastic human brain in blind individuals: the cause of disability and the opportunity for rehabilitation. In: Rieser JJ, Ashmead DH, Ebner FF, Corn AL (eds) Blindness and brain plasticity in navigation and object perception. Taylor & Francis, New York, pp 23–42
Merabet LB, Rizzo JF, Amedi A, Somers DC, Pascual-Leone A (2005) What blindness can tell us about seeing again: merging neuroplasticity and neuroprostheses. Nat Rev Neurosci 6(1):71–77
Michel GF, Tyler AN (2005) Critical period: a history of the transition from questions of when, to what, to how. Dev Psychobiol 46(3):156–162
Millar S (1981) Cross-modal and intersensory perception and the blind. In: Walk RD, Pick HLJ (eds) Intersensory perception and sensory integration. Plenum Press, New York, pp 281–314
Muckli L, Naumer MJ, Singer W (2009) Bilateral visual field maps in a patient with only one hemisphere. Proc Natl Acad Sci U S A 106(31):13034–13039
Murphy C, Cain WS (1986) Odor identification: the blind are better. Physiol Behav 37(1):177–180
Neville HJ, Bavelier D (2000) Specificity of developmental neuroplasticity in humans: evidence from sensory deprivation and altered language experience. In Shaw CA, McEachern JC (eds) Toward a theory of neuroplasticity. New York: Taylor & Francis, pp. 261–275
Newman NM, Stevens RA, Heckenlively JR (1987) Nerve fibre layer loss in diseases of the outer retinal layer. Br J Ophthalmol 71(1):21–26
Niemeyer W, Starlinger I (1981) Do the blind hear better? Investigations on auditory processing in congenital or early acquired blindness. II. Central functions. Audiology 20(6):510–515
Noppeney U, Friston KJ, Ashburner J, Frackowiak R, Price CJ (2005) Early visual deprivation induces structural plasticity in gray and white matter. Curr Biol 15(13):R488–R490
Ofan RH, Zohary E (2007) Visual Cortex Activation in Bilingual Blind Individuals during Use of Native and Second Language. Cereb Cortex 17(6):1249–1259
Ostrovsky Y, Andalman A, Sinha P (2006) Vision following extended congenital blindness. Psychol Sci 17(12):1009–1014
Palanker D, Vankov A, Huie P, Baccus S (2005) Design of a high-resolution optoelectronic retinal prosthesis. J Neural Eng 2(1):S105–S120
Pan WJ, Wu G, Li CX, Lin F, Sun J, Lei H (2007) Progressive atrophy in the optic pathway and visual cortex of early blind Chinese adults: a voxel-based morphometry magnetic resonance imaging study. Neuroimage 37(1):212–220
Pascual-Leone A, Amedi A, Fregni F, Merabet LB (2005) The plastic human brain cortex. Annu Rev Neurosci 28:377–401
Pascual-Leone A, Hamilton R (2001) The metamodal organization of the brain. Prog Brain Res 134:427–445
Penfield W, Rasmussen T (1950) The cerebral cortex of man: a clinical study of localization of function. Macmillan, New York
Pezaris JS, Reid RC (2005) Microstimulation in LGN produces focal visual percepts: proof of concept for a visual prosthesis. J Vis 5(8):367
Pezaris JS, Reid RC (2009) Simulations of electrode placement for a thalamic visual prosthesis. IEEE Trans Biomed Eng 56(1):172–178
Piche M, Chabot N, Bronchti G, Miceli D, Lepore F, Guillemot JP (2007) Auditory responses in the visual cortex of neonatally enucleated rats. Neuroscience 145(3):1144–1156
Pitskel NB, Merabet LB, Ramos-Estebanez C, Kauffman T, Pascual-Leone A (2007) Time-dependent changes in cortical excitability after prolonged visual deprivation. Neuroreport 18(16):1703–1707
Plaza P, Cuevas I, Collignon O, Grandin C, De Volder AG, Renier L (2009) Perceiving faces using auditory substitution of vision activates the fusiform face area. Paper presented at the Belgian Society for Fundamental and Clinical Physiology and Pharmacology, Spring Meeting, 2009
Poirier C, De Volder A, Tranduy D, Scheiber C (2007) Pattern recognition using a device substituting audition for vision in blindfolded sighted subjects. Neuropsychologia 45(5):1108–1121

Poirier C, De Volder AG, Scheiber C (2007) What neuroimaging tells us about sensory substitution. Neurosci Biobehav Rev 31(7):1064–1070

Poirier C, Richard MA, Duy DT, Veraart C (2006) Assessment of sensory substitution prosthesis potentialities in minimalist conditions of learning. Appl Cogn Psychol 20(4): 447–460

Poirier CC, Collignon O, Scheiber C, Renier L, Vanlierde A, Tranduy D et al. (2006) Auditory motion perception activates visual motion areas in early blind subjects. Neuroimage 31(1): 279–285

Poirier CC, De Volder AG, Tranduy D, Scheiber C (2006) Neural changes in the ventral and dorsal visual streams during pattern recognition learning. Neurobiol Learn Mem 85(1):36–43

Pozar L (1982) Effect of long-term sensory deprivation on recall of verbal material. Studia Psychologica 24(3–4):311

Pring L (1988) The 'reverse-generation' effect: a comparison of memory performance between blind and sighted children. Br J Psychol 79 (Pt 3):387–400

Proulx MJ, Stoerig P, Ludowig E, Knoll I (2008) Seeing 'where' through the ears: effects of learning-by-doing and long-term sensory deprivation on localization based on image-to-sound substitution. PLoS ONE 3(3):e1840

Ptito M, Moesgaard SM, Gjedde A, Kupers R (2005) Cross-modal plasticity revealed by electrotactile stimulation of the tongue in the congenitally blind. Brain 128(Pt 3):606–614

Rauschecker JP (2008) Plasticity of cortical maps in visual deprivation. In: Rieser JJ, Ashmead DH, Ebner FF, Corn AL (eds) Blindness and brain plasticity in navigation and object perception. Taylor & Francis, New York, 43–66

Rauschecker JP, Korte M (1993) Auditory compensation for early blindness in cat cerebral cortex. J Neurosci 13(10):4538–4548

Raz N, Amedi A, Zohary E (2005) V1 activation in congenitally blind humans is associated with episodic retrieval. Cereb Cortex 15(9):1459–1468

Raz N, Striem E, Pundak G, Orlov T, Zohary E (2007) Superior serial memory in the blind: a case of cognitive compensatory adjustment. Curr Biol 17(13):1129–1133

Recanzone GH, Merzenich MM, Jenkins WM, Grajski KA, Dinse HR (1992) Topographic reorganization of the hand representation in cortical area 3b of owl monkeys trained in a frequency-discrimination task. J Neurophysiol 67(5):1031–1056

Renier L, Collignon O, Poirier C, Tranduy D, Vanlierde A, Bol A et al. (2005) Cross-modal activation of visual cortex during depth perception using auditory substitution of vision. Neuroimage 26(2):573–580

Ricciardi E, Vanello N, Sani L, Gentili C, Scilingo EP, Landini L et al. (2007) The effect of visual experience on the development of functional architecture in hMT+. Cereb Cortex 17(12): 2933–2939

Rizzo JF 3rd, Snebold L, Keeney M (2007) Development of a visual prosthesis: a review of the field and an overview of the Boston retinal implant project. In: Tombran-Tink J, Barnstable CJ, Rizzo JF 3rd (eds) Visual prosthesis and ophthalmic devices. Humana Press, Totowa, NJ, pp. 71–93

Rockland KS, Ojima H (2003) Multisensory convergence in calcarine visual areas in macaque monkey. Int J Psychophysiol 50(1–2):19–26

Röder B, Rösler F (1998) Visual input does not facilitate the scanning of spatial images. J Ment Imag 22(3 and 4):127–144

Röder B, Rösler F, Neville HJ (2000) Event-related potentials during auditory language processing in congenitally blind and sighted people. Neuropsychologia 38(11):1482–1502

Röder B, Rösler F, Neville HJ (2001) Auditory memory in congenitally blind adults: a behavioral-electrophysiological investigation. Brain Res Cogn Brain Res 11(2):289–303

Röder B, Stock O, Bien S, Neville H, Rösler F (2002) Speech processing activates visual cortex in congenitally blind humans. Eur J Neurosci 16(5):930–936

Röder B, Teder-Salejarvi W, Sterr A, Rösler F, Hillyard SA, Neville HJ (1999) Improved auditory spatial tuning in blind humans. Nature 400(6740):162–166

Rösler F, Röder B, Heil M, Hennighausen E (1993) Topographic differences of slow event-related brain potentials in blind and sighted adult human subjects during haptic mental rotation. Brain Res Cogn Brain Res 1(3):145–159
Sadato N, Pascual-Leone A, Grafman J, Deiber MP, Ibanez V, Hallett M (1998) Neural networks for Braille reading by the blind. Brain 121(Pt 7):1213–1229
Sadato N, Pascual-Leone A, Grafman J, Ibanez V, Deiber MP, Dold G et al. (1996) Activation of the primary visual cortex by Braille reading in blind subjects. Nature 380(6574):526–528
Sampaio E, Maris S, Bach-y-Rita P (2001) Brain plasticity: 'visual' acuity of blind persons via the tongue. Brain Res 908(2):204–207
Schlaggar BL, O'Leary DD (1991) Potential of visual cortex to develop an array of functional units unique to somatosensory cortex. Science 252(5012):1556–1560
Segond H, Weiss D, Sampaio E (2007) A proposed tactile vision-substitution system for infants who are blind tested on sighted infants. J Vis Impair Blindness 101(1):32–43
Sharma J, Angelucci A, Sur M (2000) Induction of visual orientation modules in auditory cortex. Nature 404(6780):841–847
Shaw CA, McEachern JC (2000) Transversing levels of organization: a theory of neuronal stability and plasticity. In: Shaw CA, McEachern JC (eds) Toward a theory of neuroplasticity. Taylor & Francis, New York, pp 427–448
Shimony JS, Burton H, Epstein AA, McLaren DG, Sun SW, Snyder AZ (2006) Diffusion tensor imaging reveals white matter reorganization in early blind humans. Cereb Cortex 16(11): 1653–1661
Sinha (2003) Face classification following long-term visual deprivation. J Vis 3(9):104
Smith M, Franz EA, Joy SM, Whitehead K (2005) Superior performance of blind compared with sighted individuals on bimanual estimations of object size. Psychol Sci 16(1):11–14
Smits B, Mommers MJ C. (1976) differences between blind and sighted children on wisc verbal subtests. New Outlook Blind 70(6):240–246
Spelman FA (2006) Cochlear electrode arrays: past, present and future. Audiol Neurootol 11(2):77–85
Tal N, Amedi A (2009) Multisensory visual-tactile object related network in humans: insights gained using a novel crossmodal adaptation approach. Exp Brain Res 198(2–3):165–182
Tillman MH, Bashaw WL (1968) Multivariate analysis of the WISC scales for blind and sighted children. Psychol Rep 23(2):523–526
Uhl F, Franzen P, Lindinger G, Lang W, Deecke L (1991) On the functionality of the visually deprived occipital cortex in early blind persons. Neurosci Lett 124(2):256–259
Ungerleider LG, Mishkin M (1982) Two cortical visual systems. In: Ingle DJ, M. Goodalec A, Mansfield RJW (eds) Analysis of visual behavior. MIT Press, Boston, pp 549–586
Veraart C, Wanet-Defalque MC, Gerard B, Vanlierde A, Delbeke J (2003) Pattern recognition with the optic nerve visual prosthesis. Artif Organs 27(11):996–1004
von Melchner L, Pallas SL, Sur M (2000) Visual behaviour mediated by retinal projections directed to the auditory pathway. Nature 404(6780):871–876
Voss P, Lassonde M, Gougoux F, Fortin M, Guillemot JP, Lepore F (2004) Early- and late-onset blind individuals show supra-normal auditory abilities in far-space. Curr Biol 14(19): 1734–1738
Wakefield CE, Homewood J, Taylor AJ (2004) Cognitive compensations for blindness in children: an investigation using odour naming. Perception 33(4):429–442
Wallace MT (2004) The development of multisensory integration. In: Calvert G, Spence C, Stein BE (eds) The handbook of multisensory processes. MIT Press, Cambridge, MA, pp 625–642
Warren DH (1994) Blindness and children: an individual differences approach. Cambridge University Press, New York, NY
Weiland JD, Humayun MS (2008) Visual prosthesis. Proc IEEE 96(7):1076–1084
Weiland JD, Liu W, Humayun MS (2005) Retinal prosthesis. Annu Rev Biomed Eng 7: 361–401
WHO (2009) Visual impairment and blindness: World Health Organization, Fact Sheet 282

Wiesel TN, Hubel DH (1963) Single-cell responses in striate cortex of kittens deprived of vision in one eye. J Neurophysiol 26:1003–1017

Wiesel TN, Hubel DH (1965) Comparison of the effects of unilateral and bilateral eye closure on cortical unit responses in kittens. J Neurophysiol 28(6):1029–1040

Yu C, Liu Y, Li J, Zhou Y, Wang K, Tian L et al. (2008) Altered functional connectivity of primary visual cortex in early blindness. Hum Brain Mapp 29(5):533–543

Zangaladze A, Epstein CM, Grafton ST, Sathian K (1999) Involvement of visual cortex in tactile discrimination of orientation. Nature 401(6753):587–590

Zwiers MP, Van Opstal AJ, Cruysberg JR (2001) A spatial hearing deficit in early-blind humans. J Neurosci 21(9):RC142:141–145

Index

M.J. Naumer, J. Kaiser (eds.), *Multisensory Object Perception in the Primate Brain*, DOI 10.1007/978-1-4419-5615-6, © Springer Science+Business Media, LLC 2010

MIX
Papier aus verantwortungsvollen Quellen
Paper from responsible sources
FSC® C105338

If you have any concerns about our products,
you can contact us on
ProductSafety@springernature.com

In case Publisher is established outside the EU,
the EU authorized representative is:
Springer Nature Customer Service Center GmbH
Europaplatz 3, 69115 Heidelberg, Germany

Printed by Libri Plureos GmbH
in Hamburg, Germany